SPRINGER HANDBOOK OF AUDITORY RESEARCH

Series Editors: Richard R. Fay and Arthur N. Popper

Springer
New York
Berlin
Heidelberg
Barcelona
Budapest
Hong Kong
London
Milan
Paris
Santa Clara
Singapore
Tokyo

SPRINGER HANDBOOK OF AUDITORY RESEARCH

Volume 1: The Mammalian Auditory Pathway: Neuroanatomy
Edited by Douglas B. Webster, Arthur N. Popper, and Richard R. Fay

Volume 2: The Mammalian Auditory Pathway: Neurophysiology
Edited by Arthur N. Popper and Richard R. Fay

Volume 3: Human Psychophysics
Edited by William Yost, Arthur N. Popper, and Richard R. Fay

Volume 4: Comparative Hearing: Mammals
Edited by Richard R. Fay and Arthur N. Popper

Volume 5: Hearing by Bats
Edited by Arthur N. Popper and Richard R. Fay

Volume 6: Auditory Computation
Edited by Harold L. Hawkins, Teresa A. McMullen, Arthur N. Popper, and Richard R. Fay

Volume 7: Clinical Aspects of Hearing
Edited by Thomas R. Van De Water, Arthur N. Popper, and Richard R. Fay

---

Forthcoming volumes (partial list)

Development of the Auditory System
Edited by Edwin Rubel, Arthur N. Popper, and Richard R. Fay

The Cochlea
Edited by Peter Dallos, Arthur N. Popper, and Richard R. Fay

Plasticity in the Auditory System
Edited by Edwin Rubel, Arthur N. Popper, and Richard R. Fay

Harold L. Hawkins
Teresa A. McMullen
Arthur N. Popper
Richard R. Fay

Editors

Auditory Computation

With 138 Illustrations

Springer

Harold L. Hawkins
Office of Naval Research
800 N. Quincy Street
Arlington, VA 22217, USA

Teresa A. McMullen
Office of Naval Research
800 N. Quincy Street
Arlington, VA 22217, USA

Arthur N. Popper
Department of Zoology
University of Maryland
College Park, MD 20742-9566, USA

Richard R. Fay
Parmly Hearing Institute
Loyola University of Chicago
Chicago, IL 60626, USA

Series Editors: Richard R. Fay and Arthur N. Popper

Cover illustration: Equivalent electrical circuit model for a cochlear hair cell. This figure appears on p. 125 of the text.

Library of Congress Cataloging in Publication Data
    Auditory computation / editors, Harold L. Hawkins . . . [et al.].
        p.   cm.—(Springer handbook of auditory research : v. 6)
    Includes bibliographical references and index.
    ISBN 0-387-97843-7
        1. Auditory perception—Mathematical models.   I. Hawkins. Harold
L. II. Series.
    QP461.A916   1995
    612.8′5′011—dc20   95-12906

Printed on acid-free paper.

Production managed by Terry Kornak; manufacturing supervised by Jacqui Ashri.
Typeset by Best-set Typesetter, Ltd., Chaiwan, Hong Kong.
Printed and bound by Braun-Brumfield, Ann Arbor, MI.
Printed in the United States of America.

9 8 7 6 5 4 3 2 1

ISBN 0-387-97843-7 Springer-Verlag New York Berlin Heidelberg
ISBN 3-540-97843-7 Springer-Verlag Berlin Heidelberg New York

Series Preface

The *Springer Handbook of Auditory Research* presents a series of comprehensive and synthetic reviews of the fundamental topics in modern auditory research. It is aimed at all individuals with interests in hearing research including advanced graduate students, postdoctoral researchers, and clinical investigators. The volumes will introduce new investigators to important aspects of hearing science and will help established investigators to better understand the fundamental theories and data in fields of hearing that they may not normally follow closely.

Each volume is intended to present a particular topic comprehensively, and each chapter will serve as a synthetic overview and guide to the literature. As such, the chapters present neither exhaustive data reviews nor original research that has not yet appeared in peer-reviewed journals. The series focuses on topics that have developed a solid data and conceptual foundation rather than on those for which a literature is only beginning to develop. New research areas will be covered on a timely basis in the series as they begin to mature.

Each volume in the series consists of five to eight substantial chapters on a particular topic. In some cases, the topics will be ones of traditional interest for which there is a solid body of data and theory, such as auditory neuroanatomy (Vol. 1) and neurophysiology (Vol. 2). Other volumes in the series will deal with topics that have begun to mature more recently, such as development, plasticity, and computational models of neural processing. In many cases, the series editors will be joined by a co-editor having special expertise in the topic of the volume.

Richard R. Fay
Arthur N. Popper

Preface

The purpose of this volume is to provide an overview and understanding of computational analyses of auditory system function. The approach expressed in the chapters of this volume embodies the ideas that complex information processing must be understood at multiple levels of analysis, that it must be addressed from a multidisciplinary perspective that incorporates the constraints imposed by neurobiology, psychophysics, and computational analysis, and that the end product of computational analysis should be the development of formal models.

These ideas and approaches are dealt with in the ten chapters of this volume. In the first chapter, Hawkins and McMullen provide an overview of the volume as well as a brief introduction to the computational approaches that are used to increase understanding of auditory information processing. The next several chapters are organized with respect to levels of auditory analysis and processing. In Chapter 2, Rosowski discusses models of the external and middle ears, while in Chapter 3 Hubbard and Mountain consider computational issues related to inner ear function, primarily cochlear mechanics. In Chapter 4, these same authors (Mountain and Hubbard) discuss computational analysis of the hair cells and eighth nerve fibers that transduce and transmit information to the brain. In Chapter 5 Delgutte treats extraction of acoustic features such as frequency, intensity, and loudness at the eighth nerve. The remaining chapters are concerned with computations involving higher levels of the central auditory system, from the cochlear nucleus to the auditory cortex. In Chapter 6 Lyon and Shamma describe computations underlying the extraction of pitch and timbre, and Mellinger and Mont-Reynaud (Chapter 7) develop the computational basis of auditory scene analysis. In Chapter 8 Colburn summarizes the empirical findings on binaural hearing and describes a number of computational models based on these findings. In Chapter 9 Simmons outlines several computational models of the processes by which spectral and temporal information is combined in the bat auditory system to reconstruct target images. In the final chapter

(10), Lewis describes inferential computations that could underlie the analysis and retention of sequences of acoustic events.

While this volume stands alone, it is important to point out that there are a number of other chapters and volumes in this series that are relevant to the issues discussed in this volume. For example, the psychophysical basis for much of the computational analysis described herein by Lyon and Shamma, Delgutte, and by Colburn is considered in detail in Volume 3 of this series (*Human Psychophysics*, 1993, Yost, Popper and Fay, eds.) and by Long in Volume 4 of the series (*Comparative Hearing: Mammals*, 1994, Fay and Popper, eds.). The structures of the middle and inner ears that form the bases for the chapters by Rosowski and by Hubbard and Mountain are also described in detail in by Rosowski and Echteler, Fay and Popper, respectively, in Volume 4. The discussions of Mountain and Hubbard on computation in the eighth nerve are complemented by papers on eighth nerve anatomy by Ryugo in Volume 1 (*The Mammalian Auditory Pathway: Neuroanatomy*, 1992, Webster, Popper and Fay, eds.) and by Ruggero on physiology of the eighth nerve in Volume 2 (*The Mammalian Auditory Pathway: Neurophysiology*, 1992, Popper and Fay, eds.). The chapter on binaural processing by Colburn is complemented by chapters on the neurophysiological basis of such processing by Irvine in Volume 2, while sound localization is considered in detail in Volume 4 by Brown and by Wightman and Kistler in Volume 3. Finally, the chapter by Simmons et al. on biosonar is paralleled by a chapter on psychophysical aspects of this system in a chapter in Volume 5 (*Hearing by Bats*, 1995, Popper and Fay, eds.).

Harold L. Hawkins
Teresa A. McMullen
Arthur N. Popper
Richard R. Fay

Contents

Contributors

H. Steven Colborn
Biomedical Engineering Department, Boston University, Boston, MA
02215, USA

Steven P. Dear
Department of Psychology, Brown University, Providence, RI 02912,
USA

Bertrand Delgutte
Eaton Peabody Laboratory, Massachusetts Eye & Ear Infirmary, Boston,
MA 02114, USA

Michael J. Ferragamo
Department of Psychology, Brown University, Providence, RI 02912,
USA

Jonathan Fritz
Department of Psychology, Brown University, Providence, RI 02912,
USA

Tim Haresign
Department of Psychology, Brown University, Providence, RI 02912,
USA

Harold L. Hawkins
Office of Naval Research, Arlington, VA 22217, USA

Allyn E. Hubbard
Biomedical Engineering Department, Boston University, Boston, MA
02215, USA

Edwin Lewis
Department of Electrical Engineering and Computer Science, University
of California, Berkeley, Berkeley, CA 94720, USA

Richard Lyon
Apple Computer, One Infinite Loop, Cupertino, CA 95014, USA

Teresa A. McMullen
Office of Naval Research, Arlington, VA 22217, USA

David K. Mellinger
Cornell Laboratory of Ornithology, Ithaca, NY 14850-1999, USA

Bernard M. Mont-Reynaud
Center for Computer Research in Music and Acoustics, Stanford University, Stanford, CA 94305, USA

David C. Mountain
Biomedical Engineering Department, Boston University, Boston, MA 02215, USA

John J. Rosowski
Eaton Peabody Laboratory, Massachusetts Eye & Ear Infirmary, Boston, MA 02114, USA

Prestor A. Saillant
Department of Psychology, Brown University, Providence, RI 02912, USA

Shihab Shamma
Department of Electrical Engineering, University of Maryland, College Park, MD 20742, USA

James A. Simmon
Department of Psychology, Brown University, Providence, RI 02912, USA

1
Auditory Computation: An Overview

HAROLD L. HAWKINS AND TERESA A. MCMULLEN

1. Introduction

Our purpose in this chapter is to provide the reader with a brief introduction to the computational approach for understanding auditory information processing. First we briefly outline the key characteristics, and virtues, of the computational approach. Second, we suggest a general framework within which the approach can be applied to the analysis of auditory function. And finally, we exploit this framework to identify and integrate contributions provided in the chapters comprising the remainder of the volume.

1.1 The Computational Approach: General Formulation

This volume is about the computational analysis of auditory function. The term *computational analysis* has taken on several distinct meanings in the psychological and neuroscientific literature (Boden 1988). The sense of the term reflected in the chapters to follow fits most closely with the approach of Marr (1982) and with subsequent extensions outlined in Churchland and Sejnowski (1988, 1992). This approach embodies three important and closely linked ideas: (1) that complex information processing systems such as the visual or auditory system must be understood at multiple *levels of analysis* before one can be said to understand it fully; (2) that understanding of biological information processing systems should be approached to the extent possible from a *multidisciplinary perspective*, incorporating explicitly the multiple constraints on explanation that are imposed by psychophysical, neurobiological, and formal computational analyses; and (3) that the aim of computational analysis should be the development of *formal models* of sufficient explicitness, internal consistency, and completeness to enable an analytical characterization or computer simulation. Ultimately, perhaps, the aim is the construction of synthetic circuits that emulate the performance characteristics of the biological system under analysis. Let us consider these ideas more closely.

1.2 Levels of Analysis

Churchland and Sejnowski (1988, 1992) distinguished the ideas of *levels of analysis*, as proposed by Marr (1982), and *levels of organization* or spatial scale within the central nervous system. According to Marr, a complete understanding of any complex information processing system requires at the first, most abstract, level of analysis a theoretical treatment of the computational goals or tasks of the system, including determinations of what properties or events in its environment should be extracted, whether the system evidences the capability to extract this putatively desirable information and, if so, on the basis of what input data. At this level the functional competence of the system under study is identified, thereby providing an abstract characterization of the information processing task at hand, together with a specification of the computational constraints involved. An example of an auditory task considered at this level of analysis is to describe or characterize the pattern of acoustic energy arising from a particular source in the listener's environment.

The second, algorithmic, level of analysis is concerned with describing the steps or operations by which the computational goals identified at the first level might be achieved. Central issues at this level concern the nature of the input and output representations of the system in question and the specific operations, or algorithms, by which the input–output transformation is achieved. In theory, the computational problem posed by the task of describing inputs from particular sound sources must be addressed by input segmentation algorithms, including those of acoustic feature extraction and integration.

The third and most concrete level of analysis, implementation, is concerned with how the representations and computational goals specified at the first two levels are realized in system "hardware," that is, in the neural structures and processes forming the system.

These three levels of analysis differ in their system specificity. Analysis at the most abstract level defines the necessary and sufficient computational requirements imposed on any system, biological or nonbiological, faced with an information processing task. However, analysis at the algorithm level is to some extent constrained not only by what computations the system is required to perform, but also by the specific mechanisms available to it to perform these computations. Clearly, physical implementation is highly specific to the system or system class under study. Note, however, that once a biologically inspired algorithm has been postulated and formally characterized, it then may become subject to implementation within systems—for example, electronic circuits—physically different from those in which they were initially derived.

When one attempts to apply Marr's formulation to a particular biological information processing system, one of the initial challenges is to specify the computations carried out by that system. A difficulty arises in attempting to define a consistent set of computations for the system, that is, to specify a set of computations that are carried out at a common level of organization or spatial scale. Churchland and Sejnowski recognized this problem and have suggested a convenient framework for organizing the nervous system into anatomically defined implementation levels. This framework appears in Figure 1.1, along with an estimate of the spatial scale at which each level of implementation is manifested. When applied in the context of such a framework, the Marr analysis leads to a consideration of the goals of computations carried out at a given level of organization, defined in terms of the computational contribution of processes at this level to the higher computational organization of the brain, the formal procedures or algorithms by which those computations are accomplished, and the physical implementation of these algorithms, specified in terms of computations carried out at lower levels of organization. The levels of organization at which modeling effort is focused in this volume extend across virtually the entire implementation range depicted in Figure 1.1. For example, models described in Chapter 4 by Mountain and Hubbard exploit synaptic or molecular-level algorithms to achieve computations at the neuron level, and models described in Chapter 9 by Simmons et al. exploit neuron-level algorithms to achieve computational goals at the network level and network-level algorithms to achieve system-level computations.

1.3 Multidisciplinary Strategy

Many of the theoretical developments described in this volume reflect the constraining influence of multiple disciplines, including psychophysical analysis of the computational limits and informational requirements of specified auditory functions, neurobiological analysis of underlying mechanism, and formal computational modeling of postulated algorithms. This approach has much to be said for it. Because in principle there exists a very large set of possible solutions to a given computational problem, it is not likely that computational modeling in the absence of psychophysical and neurobiological constraints will yield theoretical outcomes accurately capturing the nature of biological computation.

Even if one is not interested in understanding biological function, but rather wishes only to optimize computational solutions to signal processing problems, psychophysical and neurobiological constraints often still have merit: nature has developed solutions over several million years of evolution that may be superior to those derived from extant engineering principles. This argument is instantiated in several of the chapters in this

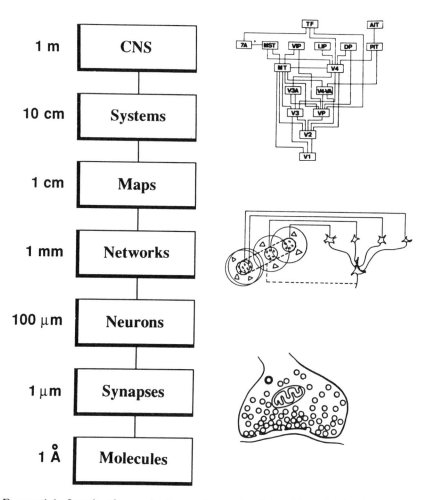

FIGURE 1.1. Levels of organization and associated spatial scales in the nervous system. Three levels of organization at which neural structures might be represented are diagrammed on the *right*. *Top*, system-level representation of a portion of the primate visual system (van Essen and Maunsell 1980). *Middle*, network representation of ganglion and simple cell connections in the retina (Hubel and Wiesel 1962). *Bottom*, representation at the level of a chemical synapse (Kandel and Schwartz 1985). (From Churchland and Sejnowski 1988, with permission.)

volume. Note in particular the active sonar target imaging techniques described in Chapter 9 by Simmons et al. Moreover, as revealed by the history of explanation in perception and cognitive psychology, accounts of behavioral phenomena developed outside the formal constraints of computational modeling can prove very difficult, perhaps impossible, to evaluate because of the inherent vagueness and ambiguity of informal modeling.

Finally, neurobiological investigation in the absence of an appreciation of the functional significance of the phenomena studied, considered initially at higher levels of organization within the central nervous system and ultimately at the behavioral level, arguably reduces neurobiology to a purely reductionist enterprise. Psychophysics provides the computational phenomena that require explanation at the level of neurobiological analysis.

1.4 Formal Computational Modeling

Formal computational modeling of biological information processing systems is the most significant legacy of the multidisciplinary approach outlined in the preceding section. The defining characteristic of modeling of this kind is that the product—the model—is sufficiently specified, internally consistent, and complete to enable formal mathematical characterization or computer simulation. Models lacking these properties will not be computable and therefore will not be subject to computational constraint. Models based on adequately specified but incorrect assumptions may be computable, but are not likely to yield the outcome desired; that is, they may compute outcomes that differ importantly from those of the modeled system.

Churchland and Sejnowski (1992) argued that computer simulations and mathematical models must be considered to be only an interim step in the evolution of understanding of biological computation. Their view is that a more advanced stage of explanation is achieved by the development of synthetic circuits that emulate the performance of the biological system of interest. The merit of synthetic circuits is that they can be constructed to interact with real-world phenomena as input and to generate outputs which are electrical events similar to those of neural systems rather than simply inputting and outputting strings of numbers. These are significant virtues.

Often the specific properties of the real world actually exploited in the biological system's computations are not fully understood by the modeler, who therefore must guess about them when designing and testing a model. If these guesses are incorrect the model may be mechanistically incorrect even when it delivers outcomes generally consistent at first analysis with those oberved in the real system. Moreover, it is often difficult to project from the performance characteristics of computer-simulated models, for example, run time, to the performance of the same models implemented in hardware. Synthetic circuits carrying out computations on real-world phenomena, for example, real acoustic signals embedded in real-world noise, are not subject to these limitations. One of the strongest motivations for hardware implementation of computational models is that the result has the potential for practical application in solving real-world information processing problems. For these reasons, the Churchland and Sejnowski argument has genuine merit. It should be emphasized, however, that computer-simulated modeling is a legitimate enterprise in its own right, a defining goal in the computational approach.

2. The Computational Approach to Understanding Auditory Performance

2.1 Computational Decomposition of Auditory Function

In this section we extend our description of the general concepts of computational modeling outlined in Section 1 and place them in the context of auditory computation. We also discuss in more detail the chapters contained in this volume relative to our conception of the idealized computational approach.

2.1.1 Computational Goals and Algorithms in the Auditory System

The overarching computational tasks or goals of the auditory system are (a) to provide a description or representation of the pattern of acoustic energy arising from separable sources of sound in the external world and (b) to specify the locations of these sources relative to the listener. Sound source descriptions formed in the auditory system are then used in other parts of the central nervous system for sound source interpretation, that is, for identifying and assigning meaning to sources. Sound source interpretation, together with localization information, is used to guide subsequent action, for example, to flee or approach an identified source or to direct other sensors toward it.

In developing a description of a specific sound source, the auditory system often must first segment representations of inputs that are composed of elements arising from many sources. We consider segmentation a system subgoal, which is achieved in the auditory system through operation of the mechanisms underlying feature extraction and feature integration. We have now introduced several key computational concepts: high-level system goals, system subgoals, algorithms, and mechanisms. As many other complex systems, the auditory system may be considered in terms of a hierarchical goal structure, that is, in terms of a set of superordinate goals each of which is realized through the achievement of a set of subgoals, which in turn are realized through the achievement of goals subordinate to them, and so on.

In this scheme, a goal is conceptualized as the desired outcome of a set of computational operations that together can be described as an *algorithm*. Accordingly, algorithms like goals are linked hierarchically: algorithms at one level compute outcomes, or goal states, that serve as inputs to algorithms at the next superordinate level in the system. In the auditory system, computations in fact are carried out or implemented by specific mechanisms, either mechanical or neural processess. Thus, the term algorithm refers to a more or less formal description of a set of hypothesized computational operations and the term mechanism refers to the actual physical processes and structures by which those operations are implemented within a given embodiment of the system, for example, biological or electronic. Illustrations of these concepts—algorithm and mechanism—are given here.

2.1.2 Feature Extraction and Feature Integration

Much of the material in this volume addresses the issue of feature extraction. We define an *auditory feature* as a discriminable attribute of the stimulus to which the computational machinery of the auditory system responds selectively, a *feature extractor* as the filtering process that produces this selective response, and *feature extraction* is as the outcome of this process. These filtering processes, extending from the auditory periphery to the central auditory system, can be differentiated with respect to the features for which they are selective and the operations or algorithms by which they achieve that selectively.

A hierarchy of features can be identified. The formation of higher level features from combinations of lower level, elemental features is accomplished by the process of *feature integration*. The set of neural processes that transforms the representation of two or more features into a single representational entity is a *feature integrator*. For example, azimuthal location is a higher level feature formed by combining more elemental features including interaural time differences and interaural intensity differences (see Chapter 8 by Colburn), and pitch is a higher level feature formed from lower level temporal and spectral features (see Chapter 6 by Lyon and Shamma). Thus, feature integration is a recursive process in which progressively higher level features are formed at successive levels of the system, and it culminates in the formation of separate auditory "objects," that is, sound patterns associated with particular sources.

Features that are bound together to form source representations usually have common properties such as common amplitude modulation, common frequency modulation, common onset and offset occurrence, or common location. Other properties suggesting that features have arisen from a single source are harmonicity of frequency components and the smoothness of the spectral envelope of a set of frequency components. Features that are bound together at lower levels of processing may in fact have arisen from separate sources, and as a consequence listeners somethings misbind features. Misbinding is illustrated in a computational model described by Mellinger and Mont-Reynaud in Chapter 7. Here an algorithm for separating several auditory events, such as the separate tones of a multitoned instrument, forms acoustic events by fusing spectral components that have two features in common, common onset and common frequency modulation (FM). Initially, components representing two separate events are misbound, or incorrectly grouped together, because all components of both events have common onset. Only later are these components separated into separate events as evidence accumulates that they have different FM rates and thus arise from separate sources.

2.1.3 Examples of Features and Feature Extraction

We begin this section with a brief review of the filtering operations underlying feature extraction in the peripheral auditory system before the cochlear nucleus is reached. This review illustrates the remarkable richness and diversity of the filtering operations underlying feature extraction, clarifies the filtering concept as it is represented in this volume, and focuses on it as one of the most throroughly modeled processes in audition. Here we provide only cursory treatment of the less well substantiated filtering operations of the central auditory system, leaving elaboration for the chapters that follow.

At the outer ear, filtering is achieved mechanically in the form of signal shaping by the shadowing, resonance, and absorption characteristics of the head, pinna, and meatus. Spectral shaping by the pinna is directionally dependent and species specific, and is an important cue for sound localization. The specific mechanisms and algorithmic characterization of signal shaping at the outer ear are described in Chapter 2 by Rosowski.

Filtering at the middle ear, which transfers sound energy from the eardrum to the cochlea, is of two forms. One is mechanical, and results in a bandpass transfer function. The low-frequency fall-off results from stiffness of the tympanic membrane and other middle-ear structures, as well as compressibility of air, and the high-frequency fall-off depends on a number of factors including ossicle inertia. A primary function of the middle-ear filtering is to reduce transmission loss by impedance matching. In this way, the middle ear acts as a transformer from the low impedance air to the higher impedance cochlear fluids. A second filtering mechanism in the middle ear, implemented neurally (electrochemically), is the modulation of middle-ear muscle activity to achieve automatic gain control in the low-frequency region during conditions of extreme stimulus intensity. The specific mechanisms of middle-ear filtering are also described by Rosowski in Chapter 2, as are a variety of algorithms (as captured, for example, in lumped-element, continuum, and finite-element models) emulating these mechanisms.

Filtering in the inner ear is mechanical and neural. The input waveform is decomposed into frequency channels by the mechanical filtering properties of the cochlear partition. This process is characterized, by Hubbard and Mountain in Chapter 3 by means of biophysically based models focusing on how the cochlea produces mechanical filtering and signal processing models aimed at mimicking cochlear responses. Transduction of energy from mechanical to electrical occurs in the hair cells. In addition, electromechanical interactions mediated by the outer hair cells are thought to contribute to mechanical amplification of the signal and to the sharpening of the cochlear filters.

Neural filtering in the inner hair cells results in half-wave rectification and lowpass filtering of the (frequency-decomposed) stimulus, which together can be modeled as an envelope detector, allowing the system to process high-frequency signals (see Chapter 4 by Mountain and Hubbard). Other biophysical and micromechanical filtering operations produce a saturating nonlinearity in the hair cell that acts as a nonlinear dynamic range compressor. Neural filtering in the form of auditory nerve adaptation acts as an automatic gain control element to enable preservation of temporal information at high stimulus intensities (Chapter 4, Mountain and Hubbard).

Above the inner ear, filtering is achieved neurally, that is, electrochemically, both at the single neuron and the ensemble levels. Among the low-level features extracted by filtering processes in the central auditory system are intensity and frequency (see Chapter 5 by Delgutte), stimulus onset and stimulus offset, amplitude and frequency modulation, direction of frequency sweep (Chapter 7 by Mellinger and Mont-Reynaud), and interaural time and intensity differences (Chapter 8 by Colburn). Some of the higher level features formed in the central auditory system by the joint processes of feature extraction and feature binding are pitch and timbre (see Chapter 6 by Lyon and Shamma), loudness (Chapter 5 by Delgutte), harmonicity (Chapter 7 by Mellinger and Mont-Reynaud), source location in azimuth and elevation (Chapter 8 by Colburn), and target range fine structure (Chapter 9 by Simmons et al.).

An onset filtering algorithm presented in Chapter 7 by Mellinger can be used to illustrate how one of these types of filtering operation can be modeled. In the Mellinger algorithm, the onset filter is a cross-correlation of the output of a cochlear model, with a time differentiation operator. This time differentiation kernel is motivated by the behavior of onset cells in the cochlear nucleus, which respond at stimulus onset and then are silent. The onset selectivity of these cells is assumed to result from the reception of an excitatory input followed by a delayed inhibitory input. Mellinger points out that an algorithm for offset filtering can be created by reversing the onset filter in time and that an amplitude modulation filter can be derived from the onset and offset filter algorithms simply by extending the kernel of the filters in time. This is possible because the features of the rising and falling phases of amplitude modulation are similar to those of onset and offset events, except for the fact that amplitude modulation exhibits a relatively slower rate of change.

2.1.4 Examples of Feature Integration

Earlier we defined feature integration as the process by which higher level representations are formed by combining multiple lower level features. As noted, this is a recursive process with successively higher level features formed at progressive stages, culminating in the formation of auditory objects. Recursive integration can be illustrated by consideration of the subjective attribute of pitch, a concept treated throughout the volume, but most extensively in Chapters 6 (Lyon and Shamma) and 7 (Mellinger and Mont-Reynaud). Pitch is a high-level feature that is formed by the integration of combinations of lower level features. Moreover, for complex signals with line spectra, pitch is a primary feature by which acoustic objects are segregated. Several algorithms for integrating lower level features to form the perception of pitch are described by Lyon and Shamma in Chapter 6. These algorithms are divided into three classes: those that combine spectral cues (harmonically related components in the input), those that combine temporal cues (complex temporal structures formed by the interaction of harmonic components in the cochlea), and hybrid spectrotemporal algorithms (which exploit the cross-correlation of phase-locked responses from different frequency channels in the cochlea). As noted by Lyon and Shamma in Chapter 6, one variant of the hybrid algorithm for pitch formation has been implemented in hardware, as a very large scale integrated (VLSI) circuit (Lazzaro and Mead 1989) with a cochlear front-end developed by Lyon and Mead (1989).

Pitch appears to play a central role in sound source segregation. Human psychoacoustic studies reveal that the perception of an individual sound source virtually always seems to be accompanied by the perception of a single pitch in the corresponding sound (Hartmann 1988). In fact, there is compelling evidence that pitch and timbre are primary identifiers or tags that the auditory system attaches to subsets of input features that are judged to have arisen from a single source. Thus, it is likely that pitch serves as an organizing percept around which the cues for sound source formation—common frequency and amplitude modulation, common onset and offset time, common source location, similarity in pitch and timbre, harmonicity, smoothness of the spectral envelope of the source components—are integrated. The integratory role of pitch is touched on by Mellinger and Mont-Reynaud in Chapter 7.

2.1.5 Source Localization

The second major computational goal of the auditory system is *localization* of the auditory source. Sound source location is important not only because of its implications for action relative to the sound source; it also serves as an important high-level feature exploited in source formation. Spectral components of common spatial origin are likely to be linked and later associated with a single auditory object.

A difficult computational problem is faced by the auditory system in deriving sound source location. Because source location is not mapped topographically onto the auditory periphery as it is for vision and tactile sensing, the central auditory system must infer it from an array of inherently nonspatial cues present in the inputs at the two ears. To compute source location, algorithms implemented in the central auditory system must extract interaural temporal and level differences over the entire spectrum of the signal to capture the effects of directionally dependent filtering by the head and pinnae. The resulting pattern of signal features must then be interpreted with respect to a presumably learned internal model of the directional transfer function produced by the external filtering (see Chapter 8 by Colburn).

Under optimal listening conditions, with familiar, spectrally extended signals heard in a familiar acoustic environment and the opportunity for head movements, the mammalian auditory system is reasonably good at localizing sound in three-dimensional space. However, as clearly illustrated by Colburn in Chapter 8, most of our understanding of localization derives from analysis of performance with sources varying in azimuth. The capacity of the auditory system to localize radiating sound sources in range and elevation has not been adequately captured in computational modeling for a variety of reasons. Source range judgments are based primarily on changes with distance in sound intensity and to a lesser extent on timbre and reverberation. Interpretation of these cues is highly dependent on familiarity, and presumably requires an internal model both of the sound source and of the acoustic environment. These familiarity effects are too poorly understood to support serious modeling effort at this time. Moreover, while the relationships between range, intensity, and timbre have received some empirical study, they are complex and are not well conceptualized (Gardner 1969).

Although it has been known for many years that localization along the vertical dimension is dependent on analysis of the spectral filtering provided by the head and pinnae, systematic exploration of the spectral shaping yielded by these directional filters at the tympanic membrane has begun only recently (Middlebrooks and Green 1991). It is apparent from the comparatively recent findings that the external filter system produces complex patterns of directionally dependent spectral peaks and valleys and that these are the primary cues exploited for vertical localization. Middlebrooks (1992) has reported a conceptual model of localization based on shaping cues. Briefly put, the key assumption used in modeling elevation judgments is that if the directional filtering of sounds arising from a particular source elevation characteristically yields a peak within a given spectral band, then any narrowband stimulus whose center frequency lies within this frequency region will be perceived as arising from that elevation. This model, which also incorporates consideration of interaural cues for horizontal localization, performs well in describing localization performance with high-frequency narrowband signals. However, it is clearly not a computational model in the sense used here: it is an explicit (and useful) assertion of the spectral shaping hypothesis.

As noted earlier, most of the modeling effort in sound source localization has focused on the horizontal dimension. Several cues to azimuthal source location, both binaural and monaural, are exploited by the auditory system. Binaural cues are interaural intensity differences and interaural temporal differences (both envelope delay and phase). Monaural cues are provided by pinna filtering, head shadowing, and head movements. To our knowledge, published studies on the computational modeling of horizontal localization performance have focused almost exclusively on binaural cues.

Two classes of binaural models for localization are described Colburn in Chapter 8. One class of models is based on psychoacoustic data. These models in turn are separated into "black-box" models, which do not incorporate physiological data, and "pink-box" models, which do. The underlying algorithm in most of these models, both black-box and pink-box, is an interaural cross-correlation applied to individual frequency bands. The neural implementation of most of these algorithms is an interaural coincidence network in which interaural time differences are encoded spatially across an array of binaural neurons. Interaural correlation models of this general class have been extended successfully to include the effects of interaural intensity differences as well as interaural time differences.

The second class of models discussed in Chapter 8 is focused on binaural processing in the brainstem where neurons coding interaural time differences and neurons coding interaural intensity differences have been identified. These models focus at the implementation level of analysis. In contrast, black-box models are focused at the algorithm level in that they are tied to neural mechanism only abstractly. Pink-box models include algorithms that are clearly and directly linked to the neural level of implementation.

2.2 Current State of the Art in Auditory Computation

The approach described in Section 1 should be considered the goal of computational analysis, an ideal that is often difficult to achieve given our current level of understanding of the mechanistic basis of sensory computation. In fact, few of the analyses reported in this volume fully capture the ideal. All these analyses address critical issues in the formal computational modeling of selected auditory function, but few describe effort to implement these models in electronic circuitry (several that do are those by Hubbard and Mountain in Chapter 3, Lyon and Shamma in Chapter 6, and Lewis in Chapter 10). Several report modeling efforts incorporating the multiple constraints of psychophysics, neurobiology, and computation. Appropriately, most analyses reflect a level of understanding that extends from definition of the computational goals of the system considered at a given level or organization to implementation of the operational algorithms underlying achievement of these goals. In all, the idealized approach is probably most closely approximated by work reported in chapters 6 (Lyon and Shamma), 7 (Mellinger and Mont-Renaud), and 9 (Simmons et al.).

Another significant facet of the current state of the art in the computational analysis of sensory function is reflected in this volume. Although a good deal is known about the neural mechanisms of peripheral and early central auditory system function, understanding of later central auditory processes (at or above the olivary complex) is extremely limited. For this reason, the most neurobiologically well informed computational modeling reported in this volume, and in the literature, focuses on preolivary mechanisms. A noteworthy exception is the computational work on bat echolocation and target imaging described in Chapter 9 by Simmons et al., a unique state of affairs attributable in large measure to the elegant and detailed neurophysiological and neuroanatomical work of Suga and his coworkers (e.g., Suga 1984) on the central mechanisms of bat echolocation and Simmons' systematic psychophysical explorations of bat target imaging performance. Because most of the computational analyses reported in the literature on audition focus at or below early central processes, there exists little or no computational literature on the important phenomena of central auditory system plasticity, echoic and auditory memory, sound source identification, auditory attention, and the effects of uncertainty, source familiarity, and context on auditory performance (see Hawkins and Presson 1986). It is to be expected that as understanding of the neural underpinnings of these phenomena evolves over the years ahead they will receive increasing attention in the literature on auditory computation.

References

Boden MA (1988) Computer Models of Mind: Computational Approaches in Theoretical Psychology. Cambridge: Cambridge University Press.

Churchland PS, Sejnowski TJ (1988) Perspectives in cognitive neuroscience. Science 242:741–745.

Churchland PS, Sejnowski TJ (1992) The Computational Brain. Cambridge: MIT Press.

Gardner MB (1969) Distance estimation of 0° or apparent 0° oriented speech in anechoic space. J Acoust Soc Am 45:47–53.

Hartmann WM (1988) Pitch perception and the segregation and integration of auditory entities. In: Edelman GM, Gall WE, Cowan WM (eds) Auditory Function: Neurobiological Bases of Hearing. New York: Wiley, pp. 623–645.

Hawkins HL, Presson JC (1986) Auditory information processing. In: Boff KR, Kaufman L, Thomas JP (eds) Handbook of Perception and Human Performance. New York: Wiley, pp. 1–64.

Hubel DH, Wiesel TN (1962) Receptive fields, binocular interaction and functional architecture in the cat's visual cortex. J Physiol (Lond) 160:106–154.

Kandel ER, Schwartz J, eds (1985) Principles of Neural Science, 2nd Ed. New York: Elsevier.

Lazzaro J, Mead C (1989) Silicon modeling of pitch perception. Proc Natl Acad Sci USA 86:9587–9601.

Lyon RF, Mead C (1989) Cochlear hydrophonics demystified. Tech. Rep. Caltech-CS-TR-88-4, California Institute of Technology, Pasadena, CA.

Marr D (1982) Vision. New York: Freeman.

Middlebrooks JC (1992) Narrow-band sound localization related to external ear acoustics. J Acoust Soc Am 92:2607–2624.

Middlebrooks JC, Green DM (1991) Sound localization by human listeners. Annu Rev Psychol 42:135–159.

Suga N (1984) The extent to which biosonar information is represented in the bat auditory cortex. In: Edelman GM, Gall WE, Cowan WM (eds) Dynamic Aspects of Neocortical Function. New York: Wiley, pp. 315–373.

van Essen D, Maunsell JHR (1980) Two-dimensional maps of the cerebral cortex. J Comp Neurol 191:255–281.

2
Models of External- and Middle-Ear Function

JOHN J. ROSOWSKI

1. Introduction

The primary function of the external and middle ear is to gather sound energy and conduct it to the inner ear. How this goal is achieved depends almost entirely on the *passive* acoustical and mechanical properties of the ear's most peripheral structures (Fletcher 1992; Rosowski 1994). (There are *active* components within the middle ear, i.e., the middle-ear muscles, but these structures principally work by modulating the passive properties of the middle ear [Møller 1983; Pang and Peake 1985, 1986].) Comprehension of the function of each of the ear's peripheral components necessitates a physical description of the relevant acoustical and mechanical properties of the components as well as quantitative schemata for how the components interact. Such schemata serve two purposes: (1) they crystallize our understanding of how the structures work and provide testable hypotheses for further refinements, and (2) they supply approximations of external- and middle-ear function which can act as prefilters in studies of the inner ear and central auditory nervous system.

This chapter discusses and develops various mathematical models of external- and middle-ear function. Although several animal phyla have well-developed middle- and external-ear-like structures (Henson 1974; Fletcher and Thwaites 1979; Michelsen 1992), we focus on the ears of terrestrial mammals. Because various reviews of middle- and external-ear structure already exist (Henson 1974; Shaw 1974a; Rosowski 1994), none are presented here although a few specific structural features relevant to the model discussions are described. A brief discussion of the history of mathematical description of the external and middle ear is presented, but the main emphasis is presenting a unified scheme for future analyses. Several topics of recent interest are discussed in detail, but for the most part readers interested in an in-depth analysis of specific issues are directed to relevant papers.

The mathematical descriptions developed here are essentially estimates of the equations of motion that define the relationship between external-

and middle-ear pressures and forces and the velocity of the tympanic membrane and ossicles as well as the volume velocity of the air in the external ear and middle-ear cavity. In defining and discussing the relevant equations, we generally use equivalent electrical circuits. Such circuits provide a common shorthand for the equations of motion (Zwislocki 1975; Beranek 1986; Fletcher 1992).

2. Overview of the Auditory System

The auditory periphery of most terrestrial vertebrates consists of three closely connected components (Fig. 2.1):

1. The external ear includes the visible pinna flange or flap, the funnel-like concha, and the tubelike external auditory canal or meatus. Scattering

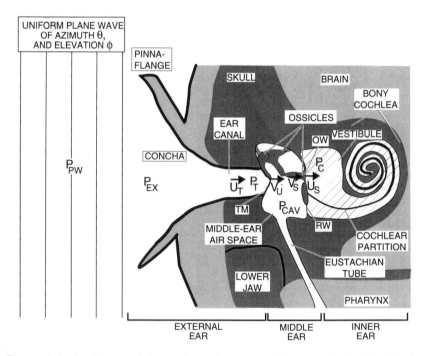

FIGURE 2.1. Auditory periphery of a typical terrestrial mammal. Included in the schematic are the external ear (pinna flange, concha, and ear canal), middle ear (tympanic membrane, ossicles, middle-ear air space, and Eustachian tube), and the fluid-filled inner ear (vestibule, cochlear windows, and cochlear partition). *TM, OW,* and *RW* refer to the tympanic membrane, oval window, and round window, respectively. The middle-ear muscles (not illustrated) attach to the malleus (the tensor tympani muscle) and to the stapes (the stapedius muscle). The various mechanical and acoustic variables are defined in the legend of Figure 2.2.

and diffraction of sound by the head, body, and torso also contribute to external-ear function (Shaw 1974a; Kuhn 1979; Blauert 1983).

2. The middle ear incorporates the tympanic membrane at the medial end of the external auditory canal, the middle-ear air spaces, the ossicles (in mammals, these include the malleus [hammer], incus [anvil], and stapes [stirrup]) and the Eustachian tube, which aerates the middle-ear air spaces and equalizes middle-ear static pressure to that of the atmosphere. The ossicles are supported by ligaments including the anterior and superior mallear ligament, the posterior incudal ligament, and the annular ligament, which holds the stapes footplate in the oval window.

3. The inner ear is fluid filled, contains the sensory structures of the cochlear partition, and is connected to the middle ear by the two (oval and round) cochlear windows.

Although the basic structures of the periphery are present in nearly all terrestrial vertebrates, the structures themselves vary greatly in size and shape among the various vertebrate species (Fleischer 1973, 1978; Henson 1974; Khanna and Tonndorf 1978; Rosowski and Graybeal 1991; Rosowski 1994).

Environmental sound reaches the inner ear through a variety of mechanisms, but the primary path for sound conduction to the cochlea is through the coupled motion of the tympanic membrane, ossicles, and stapes footplate (Wever and Lawrence 1950, 1954; von Békésy 1960; Peake, Rosowski, and Lynch 1992; Shera and Zweig 1992a). One can view the "ossicular coupling" of sound to the inner ear as a cascade of interdependent acoustical and mechanical processes with outputs that act as inputs to subsequent stages (Fig. 2.2). The input to the ear can be defined in terms of the sound pressure P_{PW} and direction (noted by the effective azimuth θ and elevation ϕ) of an equivalent uniform plane wave. Diffraction of the plane wave by the head, body, and pinna results in a sound pressure P_{EX} and volume velocity U_{EX} at the concha entrance that are dependent on direction (Wiener 1947a; Shaw 1974a,b; Kuhn 1977, 1987). Waves of sound pressure and volume velocity travel through the concha and ear canal and interact with the mechanics of the tympanic membrane and middle ear to produce a pressure P_T and volume velocity U_T acting on the membrane.

The directional dependence observed at the entrance to the ear canal remains essentially unaltered throughout later signal transmission stages (Wiener and Ross 1946; Wiener, Pfeiffer, and Backus 1966; Hudde and Schröter 1980; Rabbitt and Holmes 1988; Middlebrooks, Makous, and Green 1989). The tympanic membrane converts the pressure and volume velocity acting on its lateral surface into an effective force F_U ON and mechanical velocity V_U of the umbo (the umbo is the tip of the malleus handle, which is embedded in the tympanic membrane near the membrane's center; see Fig. 2.1). The umbo velocity is converted by trans-

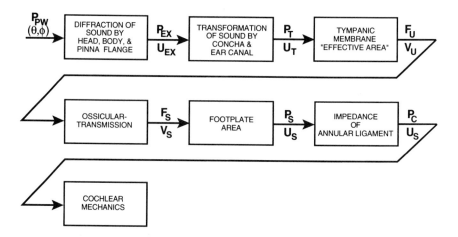

FIGURE 2.2. A "cascade" representation of the processes of the external and middle ear. Each block has associated input and output variables, and the processes within each box relate the inputs and outputs in a manner that is dependent on the load applied by "downstream" processes. The input to the ear is a plane wave of sound pressure P_{PW} and propagation direction of azimuth θ and elevation ϕ. The variables associated with the external ear include the volume velocity U_{EX} and average pressure P_{EX} at the entrance to the external ear, and the sound pressure P_T and volume velocity U_T at the tympanic membrane. The middle-ear variables include the sound pressure within the middle-ear air spaces P_{CAV}, the effective force F_U, and velocity V_U at the *umbo* (the most peripheral place in the ossicular chain located at the tip of the malleus handle in the center of the TM), the effective force F_S and velocity V_S at the stapes head, and the volume velocity U_S and effective pressure P_S of the stapes footplate. The inputs to the inner ear are U_S and the sound pressure P_C within the fluid-filled vestibule just medial to the footplate. All the output variables are similarly dependent on source azimuth θ and elevation ϕ where these angles are defined as by Shaw (1974a), such that θ and $\phi = 0$ describes the propagation direction from a sound source placed in front of the listener at the intersection of the midsagittal plane with the horizontal plane which contains the interaural axis. These variables represent complex amplitudes (defined by **bold** symbols) that are implicitly dependent on frequency. Each complex quantity can be described in terms of a magnitude $|X|$ and an angle $\angle X$. The time wavefom of the pressure and motion variables can be reconstructed as $x(t) = |X| \cos(\omega t + \angle X)$, where $\omega = 2\pi f$ is the radian frequency of the tonal stimulus.

lations and rotations of the ossicular chain into a piston-like motion of the stapes of velocity V_S and a force on the stapes footplate F_S (Guinan and Peake 1967; Gyo, Aritomo, and Goode 1987). The stapes velocity and force are integrated over the area of the footplate to produce a volume velocity U_S and pressure P_S, which work against the acoustic impedance of the annular ligament and cochlea to produce sound pressure within the cochlear vestibule P_C. (Acoustic impedance is a comples, time-invariant quantity defined by the ratio of a sound pressure to a volume velocity, $Z = P/U$, and has units of acoustic ohms such that 1 acoustic ohm = $1 \, \text{Pa sec m}^{-3}$.) The sound pressure and volume velocity within the vestibule produce the inner-ear fluid motions associated with the transduction of sound to hair cell and neural responses.

3. Physical Considerations

Before discussing some of the issues involved in modeling the ear, one needs to appreciate the forces and factors that contribute to the propagation of sound in acoustical and mechanical systems. One also needs to appreciate the wide range of wavelengths and sound amplitudes involved in hearing. The human auditory area covers three decades of frequency (20 Hz–20 kHz) and six orders of magnitude in sound pressure 2×10^{-5} to 20 Pa), and the average displacement of air particles in an audible plane wave stimulus can vary from subatomic dimensions of less than $10^{-10} \, \text{m}$ to motions that are nearly of macroscopic magnitude $10^{-4} \, \text{m}$ (Zwislocki 1965). Given a sound propagation velocity of 343 m/sec in air, the wavelength of audible sounds varies from 17 m to 1.7 mm. This range includes wavelengths λ that are much larger than the dimensions of many ear structures through wavelengths that are much smaller than many ear dimensions (Shaw 1974a; Rosowski 1994). Because it is simpler to deal with wave–structure interactions when wavelength is at least 10 fold the dimensions of the structure (e.g., Beranek 1986; Fletcher 1992) it is often useful to produce simple models that work well at low frequencies and then expand the models to include high-frequency effects.

3.1 The Properties of Sound in Air and Other Fluids

A complete discussion of sound is not within the scope of this chapter, but we do need to appreciate some of the physical properties that influence how sound propagates through air and other fluids (Table 2.1). (There are many texts available for those interested in more detailed discussions of acoustics. Beginning readers may find Fletcher [1992] useful, while more advanced readers may prefer Kinsler et al. [1982] or Beranek [1986].) Because air has a low viscosity, the two physical features that play the largest role in determining how sound propagates within it are its static density ρ_0 and bulk modulus of compressibility B. The compressibility of a fluid depends on whether the temperature of the fluid changes as it is rarefied or compressed. In the case of isothermal pressure changes, any change in the heat content of the fluid is immediately corrected by heat transfer to and from the fluid's surround. In such conditions the compressibility of the fluid is determined by the isothermal bulk modulus B_I. In cases in which no heat is allowed to flow between the

TABLE 2.1. The acoustical properties of air and water at 1 atmosphere (atm) and 20°C (Kinsler et al. 1982).

Property	Air	Fresh water
ρ_0, Density ($kg\,m^{-3}$)	1.21	998
B_I, Isothermal bulk modulus (Pa)	1.013×10^5	2.18×10^9
γ, Ratio of specific heats	1.402	1.004
B_A, Adiabatic bulk modulus (Pa)	1.42×10^5	2.19×10^9
c, Propagation velocity ($m\,sec^{-1}$)	343	1480
$\rho_0 c$, Characteristic impedance ($Pa\,sec\,m^{-1}$)	415	1.48×10^6
σ, Coefficient of shear viscosity ($Pa\,sec$)	1.8×10^{-5}	0.001

fluid and the surround, the temperature of the fluid increases and decreases with each compression and rarefaction, and the compressibility of the fluid is determined by the adiabatic bulk modulus B_A. For sounds of all but the lowest audible frequencies, the changes in temperature generated by sound pressure changes alternate so rapidly with each compression and rarefaction of the media that there is no time for heat to flow between the fluid and the surround, and we can consider the pressure changes to be adiabatic. The propagation velocity of sound c in a low-viscosity fluid can be calculated from its adiabatic bulk modulus and density:

$$c = \sqrt{\frac{B_A}{\rho_0}}, \tag{1}$$

where $B_A = \gamma B_I$, and γ is the ratio of specific heats. In the case of an ideal gas such as air, $B_I = P_0$, the static pressure, and $\gamma \approx 1.4$. Note that the density of the air, isothermal bulk modulus, and the propagation velocity are dependent on the static temperature and pressure such that climatic alterations in temperature and pressure as well as location-dependent variations in altitude have small but noticeable effects on c (at 0°C and 1 atm, $c = 331\,m\,s^{-1}$, while at 20°C and 1 atm, $c = 343\,m\,s^{-1}$). In water or other liquids the adiabatic bulk modulus depends not only on pressure but also on the molecular properties that hold the liquid molecules together and $B_A^{water} \gg B_A^{air}$ (Table 2.1). Furthermore, the isothermal bulk modulus, ratio of specific heats, and density of water all vary with temperature and pressure (Kinsler et al. 1982).

Because the assumptions of low viscosity and adiabatic compression hold for most situations in hearing, we can generally think of the air in the external-ear tube and middle-ear air spaces as a lossless media. However, viscous losses can be significant in very small tubes (Beranek 1986), and measurable loss of heat energy can occur in tubes and cavities at low frequencies (Daniels 1947; Egolf 1977; Beranek 1986; Zuercher, Carlson, and Killion 1988; Ravicz, Rosowski, and Voigt 1992). Other physical characteristics of air important to sound are its characteristic impedance $z_0 = \rho_0 c$ with units of pressure/velocity (Pa sec m^{-1}, or rayls, after Lord Rayleigh). The characteristic *acoustic* impedance of an air-filled tube is the characteristic impedance normalized by the cross-scetional area of the tube, $Z_0 = \rho_0 c/A$ with units of acoustic ohms ($1\,\Omega = 1\,\text{Pa sec m}^{-3}$).

3.2 Lumped Acoustic and Mechanical Elements and Their Electric Analogs

Lumped acoustic and mechanical elements provide a shorthand notation for describing the equations of motions in a fluid or a mechanical system (Table 2.2). The force needed to compress a spring is proportional to the spring's displacement, $f(t) = \int v(t)dt/C^M$, the force needed to compress a shock absorber or dash pot (a mechanical resistor) is proportional to the velocity of compression, $f(t) = v(t)R^M$, while the force needed to displace a mass is proportional to the acceleration $f(t) = L^M dv(t)/dt$, where C^M is

TABLE 2.2. Mechanical, acoustical, and electrical analogies.

	Mechanical	Acoustical	Electrical
Generalized force and motion	Force: $f(t)$, $\mathbf{F}(\omega)$ Velocity: $v(t)$, $\mathbf{V}(\omega)$	Pressure: $p(t)$, $\mathbf{P}(\omega)$ Volume Velocity: $u(t)$, $\mathbf{U}(\omega)$	Voltage: $e(t)$, $\mathbf{E}(\omega)$ Current: $i(t)$, $\mathbf{I}(\omega)$
Impedance analogies	Mechanical impedance: $\mathbf{Z}^M(\omega) = \mathbf{F}(\omega)/\mathbf{V}(\omega)$	Acoustical impedance: $\mathbf{Z}^A(\omega) = \mathbf{P}(\omega)/\mathbf{U}(\omega)$	Electrical impedance: $\mathbf{Z}^E(\omega) = \mathbf{E}(\omega)/\mathbf{I}(\omega)$
Ideal element values	Compliance of a spring: $\mathbf{F}(\omega) = \mathbf{V}(\omega)/j\omega C^M$	An acoustic compliance: $\mathbf{P}(\omega) = \mathbf{U}(\omega)/j\omega C^A$	A capacitor: $\mathbf{E}(\omega) = \mathbf{I}(\omega)/j\omega C^E$
	Resistance of a dash pot: $\mathbf{F}(\omega) = R^M\mathbf{V}(\omega)$	An acoustic resistor: $\mathbf{P}(\omega) = R^A\mathbf{U}(\omega)$	An electric resistor: $\mathbf{E}(\omega) = R^E\mathbf{I}(\omega)$
	Inertance of a mass; $\mathbf{F}(\omega) = j\omega L^M\mathbf{V}(\omega)$	An acoustic inertance: $\mathbf{P}(\omega) = j\omega L^A\,\mathbf{U}(\omega)$	An electric inductance: $\mathbf{E}(\omega) = j\omega L^E\,\mathbf{I}(\omega)$
	Transformer ratio of an ideal lever: $T^M = \mathbf{F}_1/\mathbf{F}_2 = \mathbf{V}_2/\mathbf{V}_1$	Transformer ratio of an ideal pneumatic piston system: $T^A = \mathbf{P}_1/\mathbf{P}_2 = \mathbf{U}_2/\mathbf{U}_1$	Transformer ratio of an ideal electric transformer: $T^E = \mathbf{E}_1/\mathbf{E}_2 = \mathbf{I}_2/\mathbf{I}_1$

Top rows, time-dependent variables of generalized force and motion. Lowercase italicized variables are used to define the time waveforms. Uppercase bold variables define the frequency-dependent complex amplitudes $\mathbf{F}(\omega) = |\mathbf{F}(\omega)|e^{j\angle \mathbf{F}(\omega)}$ associated with the time waveforms. For sinusoidal time dependencies, the time waveforms and the complex amplitudes are simply related:

$$f(t) = |\mathbf{F}(\omega)|\cos(\omega t + \angle \mathbf{F}(\omega))$$

where ω is the radian frequency $2\pi f$. More complex time waveforms can be constructed from the sum of sinusoids of different frequencies. Middle row, time-invariant complex amplitudes of the impedance in each system. Botton rows, relationships between the analogous variables for ideal mechanical, acoustical, and electrical elements.

the mechanical compliance of the spring, R^M is the mechanical resistance of the dash pot, and L^M is the mechanical inertance or mass. With sinusoidal stimuli these relationships can be written in terms of complex amplitudes (Table 2.2). (Remember that integration of a complex amplitude is analogous to division by $j\omega$ [where j is the imaginary number $\sqrt{-1}$ and $\omega = 2\pi f$ is radian frequency], and differentiation is analogous to multiplication by $j\omega$.) Similar relationships between pressure and volume displacement, volume velocity, and volume acceleration can be used to define acoustic compliances, resistances, and inertances. If we consider pressure or force to be analogous to voltage and consider volume velocity or velocity analogous to current, there are also obvious analogies between the compliances and an electric capacitor, between mechanical, acoustic, and electrical resistances, and between the inertances and inductance. There are also simple analogies between ideal transformers in the three systems.

Although mechanical and electrical elements—springs, dash pots, masses, capacitors, resistors, and inductors—are familiar to many, acoustic elements require some introduction (Fig. 2.3). In looking at Figure 2.3, the reader should be aware that in comparison to electric elements, it is much more difficult to construct ideal acoustic elements, especially inductances or resistors. Nevertheless, there are some common structures that approximate acoustic elements. Acoustic compliances are of two general types (Beranek 1986): (i) the "series" compliances associated with elastically supported membranes or pistons (e.g., the tympanic membrane) that influence the flow of volume velocity within open tubes and air spaces, and (ii) the "grounded" compliances associated with the compression of air in enclosed spaces. In the series type of acoustic compliance, volume velocity flows across the membrane and displacements of the membrane cause equal displacements in the fluid on either side, just as the same current flows through electric elements that are in series. In the "grounded" type of compliance, volume velocity is expended in compressing the air within the closed space and flows nowhere else; it is as if one side of the compliance were attached to ground. Acoustic resistors can be constructed from screens, slits, or long narrow tubes in which the viscosity of air dominates the forces that impede the flow of volume velocity, while acoustic inertances can be constructed from short, wide, open tubes where much of the sound pressure is used to accelerate the mass of air within the tube. It is difficult to build screens or slits that do not have a significant inertance (especially at high frequencies) or tubes that do not have a significant resistive component (especially at low frequencies). Beranek (1986) and Fletcher (1992) give more detailed discussions of acoustic elements.

Although others have used mechanical velocity and specific acoustic impedance (**P/V**) to define sound flow through the ear (Allen 1986), here I consistently use volume velocities and acoustic impedance (**P/U**) as acoustic variables. This choice of variables permits some useful simplifications. (1) As is discussed later, the units of acoustic power and energy are identical to those for their mechanical and electrical analogs. (2) Because the volume velocity of the tympanic membrane is a spatial average of the motion of the entire membrane, variations in motion across the drum are already accounted for and knowledge of the "effective area" of the tympanic membrane (Wever and Lawrence 1954) is not necessary to specify the input to the ear.

3.3 Distributed Parameters

At frequencies at which the wavelength of sound is less than 10 fold the dimensions of the ear structures (this happens above 2 kHz in the human middle ear and at lower frequencies in the human ear canal), the pressures and forces acting on the structures are no longer uniform and the structures can no longer be adequately described with lumped parameters. One approach is to subdivide the structure into smaller lumped components. A more general approach is to describe the structures in terms of distributed parameters; for example, the ear canal can be modeled as an acoustic transmission line or tube (Møller 1965; Rabinowitz 1981; Chan and Geisler 1990), the concha can be approximated by a horn (Coles and Guppy 1986; Fletcher and Thwaites 1988; Rosowski, Carney, and Peake 1988), and the tympanic membrane can be modeled as a membrane or shell (Esser 1947; Funnell and Laszlo 1976; Rabbit and Holmes 1986). Systems of distributed parameters allow the forces and motions to vary in well-defined manners along the lengths and cross sections of tubes, surfaces of the membrane, or lengths of nonrigid ossicles. Descriptions of distributed systems can take many guises including finite-element models or continuum equations. Geometrically simple structures such as tubes or horns can be described in terms of well worked out, two-port parameters (Fig. 2.3B) (Malecki 1969; Egolf 1980; Benade 1988; Fletcher 1992).

3.4 Impedance, Power, and Energy

Table 2.2 also defines analogous impedances. The unit of mechanical impedance is force/velocity -1 mechanical ohm ($= 1\,\text{N sec m}^{-1}$), $-$ the unit of acoustic impedance is pressure/volume $\times$ velocity $- 1$ acoustic ohm ($= 1\,\text{Pa sec m}^{-3}$), $-$ and the unit of electrical impedance is volts/amps or the electrical ohm. The instantaneous or *complex power* Π of all three analogies is defined by the product of the root mean square (rms) values of generalized force and the complex conjugate (*) of the velocity. Because we define complex amplitudes in terms of their peak value

$$\Pi = \frac{FV^*}{2} = \frac{PU^*}{2} = \frac{EI^*}{2}. \tag{2}$$

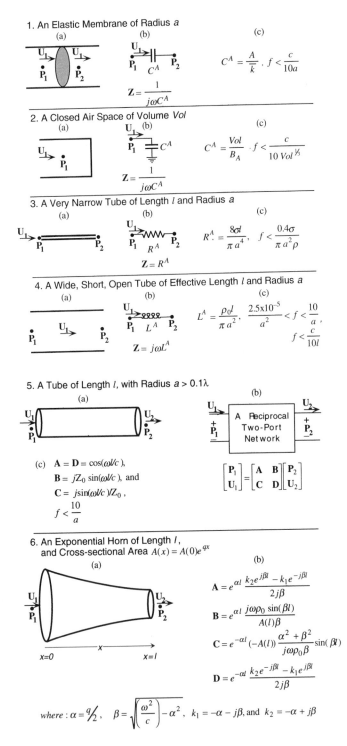

1. An Elastic Membrane of Radius a

(a) (b) (c)

$U_1 \rightarrow$ $U_1 \rightarrow$ P_1 P_2
P_1 P_2

C^A

$$Z = \frac{1}{j\omega C^A}$$

$$C^A = \frac{A}{k} \cdot f < \frac{c}{10a}$$

2. A Closed Air Space of Volume Vol

(a) (b) (c)

$U_1 \rightarrow$ P_1 $U_1 \rightarrow$ P_1 C^A

$$Z = \frac{1}{j\omega C^A}$$

$$C^A = \frac{Vol}{B_A} \cdot f < \frac{c}{10\, Vol^{1/3}}$$

3. A Very Narrow Tube of Length l and Radius a

(a) (b) (c)

$U_1 \rightarrow$ P_1 P_2

$U_1 \rightarrow$ P_1 R^A P_2

$$Z = R^A$$

$$R^A = \frac{8\sigma l}{\pi a^4}, \quad f < \frac{0.4\sigma}{\pi a^2 \rho}$$

4. A Wide, Short, Open Tube of Effective Length l and Radius a

(a) (b) (c)

P_1 $U_1 \rightarrow$ P_2

$U_1 \rightarrow$ P_1 L^A P_2

$$Z = j\omega L^A$$

$$L^A = \frac{\rho_0 l}{\pi a^2}, \quad \frac{2.5 \times 10^{-5}}{a^2} < f < \frac{10}{a},$$

$$f < \frac{c}{10l}$$

5. A Tube of Length l, with Radius $a > 0.1\lambda$

(a)

$U_1 \rightarrow$ P_1 $U_2 \rightarrow$ P_2

(b)

$U_1 \rightarrow$ $+$ P_1 $-$ | A Reciprocal Two-Port Network | $U_2 \rightarrow$ $+$ P_2 $-$

(c) $A = D = \cos(\omega l/c)$,
$B = jZ_0 \sin(\omega l/c)$, and
$C = j\sin(\omega l/c)/Z_0$,
$$f < \frac{10}{a}$$

$$\begin{bmatrix} P_1 \\ U_1 \end{bmatrix} = \begin{bmatrix} A & B \\ C & D \end{bmatrix} \begin{bmatrix} P_2 \\ U_2 \end{bmatrix}$$

**6. An Exponential Horn of Length l,
and Cross-sectional Area $A(x) = A(0)e^{qx}$**

(a)

$U_1 \rightarrow$ P_1 $U_2 \rightarrow$ P_2

$x=0$ x $x=l$

(b)

$$A = e^{\alpha l} \frac{k_2 e^{j\beta l} - k_1 e^{-j\beta l}}{2j\beta}$$

$$B = e^{\alpha l} \frac{j\omega\rho_0 \sin(\beta l)}{A(l)\beta}$$

$$C = e^{-\alpha l}(-A(l)) \frac{\alpha^2 + \beta^2}{j\omega\rho_0 \beta} \sin(\beta l)$$

$$D = e^{-\alpha l} \frac{k_2 e^{-j\beta l} - k_1 e^{j\beta l}}{2j\beta}$$

where : $\alpha = q/2$, $\beta = \sqrt{\left(\frac{\omega^2}{c}\right) - \alpha^2}$, $k_1 = -\alpha - j\beta$, and $k_2 = -\alpha + j\beta$

(The conjugate has equal magnitude but negative angle.) In all three analogies, the unit of power is the watt, where $1\,W = 1\,N\,m\,sec^{-1} = 1\,Pa\,m^3\,sec^{-1} = 1\,VA$. The complex power includes not only power that is absorbed but also power that is stored within the circuit elements. (Compliances and inertances store power during half the cycle of a sinusoidal stimulus and deliver it back to the circuit during the other half-cycle.) The *average* power is a measure of only the absorbed power and is defined as the real part of the complex power:

$$\bar{\Pi} = \frac{\text{Re}\{\mathbf{PU*}\}}{2} = \frac{|\mathbf{U}|^2\text{Re}\{\mathbf{Z}\}}{2}. \tag{3}$$

FIGURE 2.3A,B. (A) Representations of four lumped acoustic elements. For each element, (a) is a schematic of the element, (b) shows an electrical analog and a mathematical representation of the analog, and (c) defines the element value in terms of some dimensional, acoustical, or mechanical parameters and also notes the frequency limits of the approximation. 1. An elastic membrane of radius a acts as an acoustic compliance defined by the average stiffness $\bar{k}$ and area $A = \pi a^2$ of the membrane. The lumped analysis breaks down at frequencies at which the wavelengths are less than $10\,a$. 2. A closed air space acts as an acoustic compliance connected to ground, where the compliance is the ratio of the volume of the enclosed air space Vol and the adiabatic compressibility of air B_A. The lumped analysis breaks down at frequencies at which the linear dimensions of the air space become larger than 0.1λ. 3. A very narrow tube is one example of a lumped acoustic resistor so long as the frequency of the sound is less than some radius-dependent factor. At higher frequencies, the tube acts as a mixed resistance and inertance. The value of the resistance depends on the tube's length, radius, and the shear viscosity σ of air (Beranek 1986). 4. A wide, open, short tube is an example of a lumped acoustic inertance. The value of the inertance is dependent on length, density, and radius (Beranek 1986). Below the low-frequency limit, a resistive component becomes significant. The lumped analysis breaks down at frequencies at which the tube dimensions are $>0.1\lambda$. (B) Representations of two distributed acoustic elements. 5. A tube with length $>0.1\lambda$ can be modeled as an acoustic transmission line of distributed elements (a). (b) An electrical analog of such a system is a reciprocal two-port network (Fletcher 1992). The solution of the two-port equations is easily expressed in terms of a transmission matrix (c) (Desoer and Kuh 1969; Egolf 1977; Benade 1988; Shera and Zweig 1991). If we assume lossless media and adiabatic conditions, the coefficients of the transmission matrix are simple functions of the tube length l, the propagation velocity of sound c, the radial frequency of the sound ω, and the characteristic acoustic impedance of the tube $Z_0 = \rho_0 c/(\pi a^2)$. At frequencies above the upper frequency limit, nonplanar modes of propagation complicate the analysis. At low frequencies where $l < 0.1\lambda$, the tube can be approximated by lumped elements such as those just described. 6. Malecki's (1969) two-port model of an exponential horn. Similar models of other horn types are described by others (Benade 1988; Fletcher and Thwaites 1988).

The unit of energy (power integrated over time) in all three analogies is the joule (J), computed from force times distance or from power times time ($1\,\mathrm{J} = 1\,\mathrm{N\,m} = 1\,\mathrm{W\,sec}$).

3.5 Two-Ports

Two-ports, like those described in Figure 2.3B, are common tools for the analysis of electric, acoustic, and mechanical circuits (Desoer and Kuh 1969; Malecki 1969; Fletcher and Thwaites 1979; Egolf 1980; Benade 1988; Shera and Zweig 1991, 1992a,b,c; Fletcher 1992). At the basis of the two-port is the idea that the relationships between force and voltage (or pressure and volume velocity, or between voltage and current) at the inputs and outputs of a system box can be used to define how the box works. The main advantage of two-ports is that they allow a complete description of the relationships between paired input and output variables in terms of four frequency-dependent parameters. The small number of parameters makes them an effective tool in quantifying specific functional questions. Shera and Zweig (1991, 1992a,b,c) used two-port analyses of the tympanic membrane and middle ear to investigate questions concerning the transformation of sound pressure and volume velocity at the tympanic membrane into ossicular force and motion, the influence of the cochlear load, and how sound generated within the cochlea is transferred to the external ear. Figure 2.3B includes a two-port representation of a uniform tube and an exponential horn. More information about two-ports and the equations that define them can be found in general circuit texts (e.g., Desoer and Kuh 1969).

3.6 Linearity

Although it is possible to describe a system in which the outputs are not proportional to the inputs, the constraint of proportionality (more generally, linearity) greatly simplifies the task; if a system is linear we need only determine the system response for one level of stimulation. Sounds within the normal audible range can be described in terms of linear acoustic equations (Kinsler et al. 1982), and many measurements indicate that the external and middle ear act as linear transducers over the normal audible range (except when under active control of the middle-ear muscle system [Silman 1984]). At high intensities (>120 dB sound pressure level [SPL]), however the middle-ear apparatus appears to saturate, and the input to the inner ear grows less than proportionally with the input to the ear (Guinan and Peake 1967; Nedzelnitsky 1980; Buunen and Vlaming 1981). One probable site of nonlinear response is the annular ligament of the stapes, which appears to stiffen in response to larger than normal stimulus levels (Lynch, Nedzelnitsky, and Peake 1982; Price and Kalb 1991). In the case of very intense sound pressures

(>156 dB SPL) the variations in sound pressure begin to approach atmospheric pressure (196 dB SPL), and one must also consider nonlinear acoustic effects.

3.7 Reciprocity

Acoustical, mechanical, and electrical systems that are (1) linear, (2) made up of passive elements (no power sources), (3) time invariant (the response to a constant stimulus does not vary with time), and (4) contain no gyrators or moving-coil electroacoustic transducers exhibit reciprocity (Desoer and Kuh 1969; Kinsler et al. 1982; Shera and Zweig 1991). (Passive mechanomagnetic and electromagnetic elements such as moving-coil transducers and gyrators exhibit antireciprocity [Kinsler and Frey 1962]). Reciprocal systems are constrained in that the transfer relationships between input and output variables are independent of the port location of the source and load. Because of this constraint, only three parameters are needed to define the function of a reciprocal two-port. Middle and external ears can be considered reciprocal in that they are generally linear (as we have discussed) and contain no sound sources or magnetic elements, and their responses generally show no adaptation or effect of previous stimuli. There are exceptions to this reciprocal behavior: (1) the middle ear is not linear at very high sound levels; (2) middle-ear muscle contractions are a source of acoustic and mechanical power; and (3) there is a clear hysteresis in the effect of large static pressures on the ear (Margolis and Smith 1977; Wada and Kobayashi 1990).

4. A Brief Review of Quantitative Descriptions of the External and Middle Ear

4.1 The Transformer Description of External- and Middle-Ear Function

The common view of external- and middle-ear function dates back at least to the time of von Helmholtz, who in 1877 in the fourth edition of his *Sensation of Tones* (p. 134 of the English translation reprinted in 1954) suggested that the external and middle ear *transform* sound energy between the low acoustic impedance of air and the high impedance of cochlear fluid. Several different transformer mechanisms have been discussed in the literature.

4.1.1 The External-Ear Horn

The shapes of many mammalian external ears approximate a tapered tube or horn (Coles and Guppy 1986; Fletcher and Thwaites 1988; Rosowski,

Carney, and Peake 1988; Rosowski 1994) with the larger end of the
horn open to the outside (Fig. 2.4). This ear horn acts to transform
the acoustic impedances at its two ends. At frequencies higher than a
dimension-specific cutoff frequency (the cutoff frequency depends on the
horn's length and cross-sectional areas of the mouth and throat; the cutoff
of the human ear is near $1-2\,\mathrm{kHz}$), the ratio of the acoustic impedances
at the two ends of the horn is approximately equal to the inverse ratio of
the cross-sectional areas at the two ends of the horn (Malecki 1969;
Benade 1988; Fletcher and Thwaites 1988). For the case of sound entering
the pinna opening and traveling toward the middle ear:

$$\frac{\mathbf{Z_T}}{\mathbf{Z_{EX}}} \approx \frac{A_P}{A_{EC}}, \tag{4}$$

where A_P is the cross-sectional area of the pinna opening, $\mathbf{Z_{EX}}$ is the
acoustic impedance looking into the pinna opening, A_{EC} is the cross-
sectional area at the termination of the ear canal near the tympanic
membrane, and $\mathbf{Z_T}$ is the acoustic impedance of the middle ear. A similar

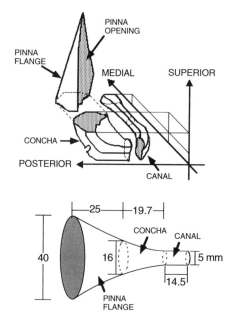

FIGURE 2.4. A comparison of the air-filled external-ear tube of the cat and a
simple horn and uniform tube model that has been demonstrated to reproduce
some of the ear's function (Rosowski, Carney, and Peake 1988). The cross-
sectional area of the concha and pinna flange sections of the horn vary exponentially with pinna length. All dimensions are in millimeters.

transformation of the sound pressures occurs at frequencies above cutoff such that the ratio of the pressures at the tympanic membrane and pinna opening is related to the square root of the inverse area ratio:

$$\frac{P_T}{P_{EX}} \approx \sqrt{\frac{A_P}{A_{EC}}}. \tag{5}$$

Equations 4 and 5 only approximate the behavior of the external ear at frequencies between about 2 and 10 times cutoff. At higher frequencies, where the wavelengths approximate the smaller dimensions of the horn, nonuniform modes within the horn greatly complicate the pressure gain and impedance transformation (Fletcher 1992). In the low-frequency limit, where the wavelengths of sound are much larger than the dimensions of the external ear, there is no horn effect and $P_T/P_{EX} \approx 1$. At low to moderate frequencies, the transformer action of the external ear is frequency dependent and smaller in magnitude than is predicted by Eqs. 4 and 5.

4.1.2 The Area Ratio

von Helmholtz (1877) pointed out that the difference in area between the larger tympanic membrane and smaller stapes footplate allows the middle ear to act as a pneumatic lever. In pneumatic levers, volume velocities and pressures acting on a larger input surface area are converted to smaller volume velocities and larger pressures at a smaller output surface area where the transformer ratio is the ratio of the areas of the two surfaces. Transformer models of the middle ear (Fig. 2.5) approximate the tympanic membrane and stapes footplate as coupled pistons and use the areas of these structures (A_{TM} and A_{FP}, respectively) to predict the ratio of pressure and volume velocities at the input and output of the middle ear:

$$\frac{P_C}{P_T} = \frac{U_T}{U_S} \propto \frac{A_{TM}}{A_{FP}}. \tag{6}$$

The large magnitude of the apparent pneumatic lever (the area ratio is generally between 10:1 and 40:1 in terrestrial vertebrate ears [Rosowski and Graybeal 1991; Rosowski 1992]) suggests the area ratio is the primary mechanism of middle-ear function (Wever and Lawrence 1954; Zwislocki 1965, 1975; Tonndorf and Khanna 1976). Measurements of the motion of the tympanic membrane (Khanna and Tonndorf 1972; Tonndorf and Khanna 1972) at frequencies less than 1 kHz in cat and man are consistent with piston-like motion in that all parts of the membrane appear to move in phase. At higher frequencies, tympanic membrane motion is inconsistent with the pneumatic lever in that sections of the membrane move with different phases.

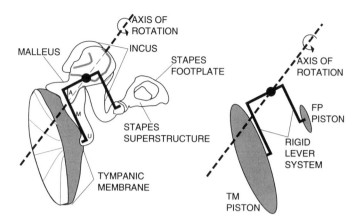

FIGURE 2.5. Hypothesized area and lever ratio mechanisms of the middle-ear transformer. *Left*, a schematic representation of the human tympanic membrane and ossicles viewed from in front and slightly lateral. The *thick black lines* represent the hypothesized rigid-body rotation of the ossicles about a fixed axis. A two-dimensional projection of the axis of rotation is included; in reality the axis goes into and comes out of the plane of the paper. *Right*, a schematic representation of the coupled piston-lever model.

4.1.3 The Ossicular Lever

The second middle-ear transformer discussed by Helmholtz was a mechanical lever mechanism. He assumed the ossicles rotated around a fixed axis such that the velocity and force at the umbo and the stapes (Fig. 2.5) were related by the ratio of the lengths of the malleus and incus lever arms l_M and l_I:

$$\frac{\mathbf{F_S}}{\mathbf{F_U}} = \frac{\mathbf{V_U}}{\mathbf{V_S}} \propto \frac{l_M}{l_I} \, . \tag{7}$$

Measurements of the motion of the malleus and stapes at low sound frequencies in cat (Guinan and Peake 1967), man (Gyo, Aritomo, and Goode 1987), and guinea pig (Manley and Johnstone 1974) predict lever ratios that are consistent with the middle-ear anatomy in these species, where l_M/l_I approximates 2.5 in cat and guinea pig and 1.2 in humans (Wever and Lawrence 1954). Clearly the ossicular lever ratio is small compared to the area ratios. At higher frequencies, there is clear evidence that the ratio of umbo and stapes motion is frequency dependent (Guinan and Peake 1967; Manley and Johnstone 1974), and there is growing evidence that the motion of the malleus is more complicated than simple rotation at frequencies above 800–1000 Hz (Gyo, Aritomo, and Goode 1987; Decraemer, Khanna, and Funnell 1989, 1990, 1991; Donahue 1989; Donahue, Rosowski, and Peake 1991; Decraemer and Khanna 1994).

4.1.4 The Catenary Lever

Observations of the distinct inward curvature of the mammalian tympanic membrane led Helmholtz to propose catenary action as a third type of middle-ear transformer mechanism. The catenary hypothesis suggests that small deformations of a taut curved membrane are associated with relatively large forces. Such trading of increased force for decreased motion is similar to the transformation proposed for the ossicular lever mechanism (Tonndorf and Khanna 1970; Zwislocki 1975), and it is difficult to distinguish the two mechanisms. Wever and Lawrence (1954) compared the displacement of the umbo with the volume displacement of the tympanic membrane in cats and concluded that catenary action did not occur in the middle ear; an analysis of the motion of the tympanic membrane (Tonndorf and Khanna 1970, 1972) suggested that a catenary mechanism could contribute a small transformer ratio of about a factor of two.

4.1.5 The Ideal Transformer Hypothesis

The association of simple transformers with specific anatomical features suggests that one can predict middle-ear function just by knowing a few anatomical ratios, the ratio of the tympanic membrane and footplate areas, and the ratio of the malleus and incus lever arms. More specifically, Eqs. 6 and 7 can be combined to yield:

$$\frac{\mathbf{P_C}}{\mathbf{P_T}} = \frac{A_{TM} l_M}{A_{FP} \; l_I} = \frac{\mathbf{U_T}}{\mathbf{U_S}} \tag{8}$$

Equation 8 defines what Dallos (1973) called the ideal transformer model, which approximates middle-ear function with a simple, frequency-*in*dependent transformer ratio. But, as Dallos pointed out, the ideal transformer model is too simple; it neither accounts for the forces needed to overcome the mass, stiffness, and damping within the middle-ear components nor does it account for nonpiston-like motion of the tympanic membrane or nonrotational behavior of the ossicles. The errors within the simplification are detailed in Figure 2.6, which compares the ratio of $\mathbf{P_C}/\mathbf{P_T}$ predicted by Eq. 8 with actual measurements in cats. Unlike the ideal ratio, the magnitude of the measured ratio is frequency dependent, with the largest magnitudes in the middle-frequency range. The measured magnitude is at least 8 dB less than the predicted transformer ratio at any frequency, and the difference between measured and predicted is larger than 20 dB in the low- and high-frequency limit of the measurements. Because anatomical areas and lengths are real numbers, Eq. 8 also predicts that $\mathbf{P_C}$ and $\mathbf{P_T}$ are in phase, but the lower panel of Figure 2.6 shows that the measured angle of the pressure ratio varies by more than a period.

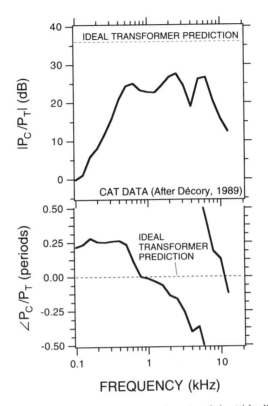

FIGURE 2.6. Comparison of the magnitude and angle of the "ideal" and measured middle-ear pressure transfer function (the ratio of sound pressure within the entrance of the inner ear P_C and the sound pressure at the input to the middle ear P_T). The *thick lines* are the results of measurements in cats by Décory (1989) with the calculated effect of the intact middle-ear cavity included (after Rosowski 1994). The *dashed lines* represent the magnitude and angle of the transfer ratio predicted by the "ideal-transformer" model. (Although the axis of the angle plots only covers a full period, the angle measurements vary by more than a period and are illustrated by two lines that intersect at ±0.5 period.)

The difference in magnitude and angle between the ideal theory and real measurements result from failure to account for the mechanics and acoustics of the middle-ear structures. Indeed, although the ideal transformer model suggests that the middle ear has little effect on the frequency response of the auditory system, theories of external- and middle-ear function that account for the mechanical and acoustical properties often lead to the opposite conclusion, that is, that the frequency-dependent acoustical and mechanical constraints of the external and middle ear shape the auditory threshold function (von Waetzmann and Keibs 1936; Khanna and Tonndorf 1969; Dallos 1973; Zwislocki 1975;

Shaw and Stinson 1983; Rosowski et al. 1986; Rosowski 1991a,b). The influence of the nonzero impedances of the middle-ear components also negates the concept of a single pressure–velocity transfer ratio that is implied by Eq. 8. Analyses of more realistic models of a piston-and-lever system point out that the magnitude of the transfer ratio for sound pressure generally is less than that of the inverse transfer ratio for volume velocity, that is, $|P_C/P_T| < |U_T/U_S|$ (Shera and Zweig 1991).

4.2 Model Frameworks That Account for the NonIdeal Acoustical and Mechanical Properties of the External and Middle Ear

Schemes of varying complexity have been used to account for the non-ideal middle- and external-ear transformation of sound power from the environment to the inner ear.

4.2.1 Block Models of the Ear

Simple block models such as the schematics of Figures 2.2 and 2.7 can be quite useful in analyzing external- and middle-ear function in that the blocks define subprocesses with specific inputs and outputs. The more general cascade description of Figure 2.2 traces the transformation of specific acoustic and mechanical variables but says little about the processes within the blocks, while the model of Figure 2.7 is a hybrid acoustic-electrical analog in which the block structure reflects some assumptions about the internal processes of each box. In the circuit of Figure 2.7, pressure is analogous to voltage and volume velocity is analogous to current. A special feature of Figures 2.2 and 2.7 is that all the inputs and outputs are defined by *measurable* quantities.

The leftmost block in Figure 2.7 defines external-ear function in terms of the transformation of plane wave sound pressure P_{PW} from various directions to sound pressure P_T and volume velocity U_T at the tympanic membrane. The middle block defines middle-ear function with a two-port network (with paired inputs at the left-hand port and paired outputs on the right) that describes the transformation of P_T and U_T to pressure P_C and volume velocity U_S at the input to the inner ear. The rightmost block of Figure 2.7 describes the cochlear load on the middle ear in terms of a one-port. Such a description assumes that the cochlear acoustic input impedance $Z_C = P_C/U_S$ totally defines the entire contribution of the inner ear to middle-ear mechanics. This assumption appears reasonable for normal middle-ear function in humans and common laboratory mammals (Wever and Lawrence 1950, 1954; Zwislocki 1965, 1975) but need not be the case (Shera and Zweig 1992c).

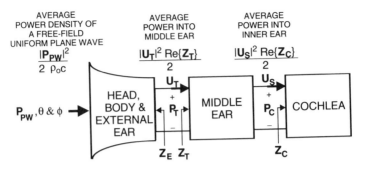

FIGURE 2.7. The flow of average sound power from the stimulus to the cochlea. The *left block* accounts for the action of the head, body, and external ear, and has a uniform plane wave of sound pressure P_{PW} and direction θ and ϕ as its input, with outputs of volume velocity and pressure at the tympanic membrane U_T and P_T. The middle-ear block (*middle*) is a two-port network defined in terms of its inputs, outputs (the volume velocity of the stapes U_S and the sound pressure in the cochlear vestibule P_C), and its load. The cochlea (*right*) is a one-port defined completely by its input impedance Z_C. The radiation impedance looking out the external ear from the tympanic ring Z_E and the middle-ear input impedance Z_T are also illustrated. The equations at the *top* define the average power or power density at the inputs to the three stages of the system. $\text{Re}\{Z\}$ is the real part of the impedance, where $\text{Re}\{Z\} = |Z|\cos(\angle Z)$; ρ_0 is the density of air at normal atmospheric pressure; c is the speed of sound; and $\rho_0 c$ is the characteristic impedance of air. (After Rosowski 1991a.)

Also included in Figure 2.7 are definitions for various acoustic impedances and average powers. The impedances are the ratio of sound pressure to volume velocity looking into the cochlea Z_C as well as into Z_T and out from Z_E the tympanic membrane. The quantification of average power (or power density) at the entrance to each of the model blocks allows computations of power transfer between the different blocks and can be used to quantify power collection by the external and middle ear (Shaw and Stinson 1983; Rosowski et al. 1986; Rosowski 1991a,b). (The power in the plane wave stimulus is defined in terms of average power density, power per unit of area, because by definition a plane wave is of infinite extent and the average power in such a wave is infinite.) The practical significance of analyzing power flow in this manner is twofold: all the variables used to define the inputs and outputs of the different ear parts have been measured in a number of animals, and these measurements allow one to define and compare the power that enters each of the blocks. Calculations of power flow through the ear have been made for several mammalian species including humans, cats, chinchillas, and guinea pigs (Shaw and Stinson 1983; Rosowski et al. 1986; Rosowski 1991a,b). The results of these calculations generally support the hypothesis that the filtering of acoustic power by the external and middle ear plays a large role in determining the shape of the audiogram.

4.2.2 Lumped and Distributed Models of External-Ear Structure and Function

The transformation of sound power by the head, body, and external ear has been investigated by Wiener (1947b), Shaw and his coworkers (Teranishi and Shaw 1968; Shaw 1974a,b; Shaw and Stinson 1983), and Kuhn (1977, 1979, 1987) using measurements of ear canal sound pressure in physical model systems of increasing complexity. The results of these analyses point out that diffraction and scattering of sound by the head, pinna flange, and concha introduce direction-dependent variations in the sound pressure at the entrance to the ear canal P_{EC} that can be used by listeners to determine the location of sound sources in space (Hudde and Schröter 1980; Blauert 1983; Middlebrooks, Makous, and Green 1989; Wightman and Kistler 1989; Musicant, Chan, and Hind 1990; Rice et al. 1992). Power collection by the external ear can be reasonably approximated, at least for sound frequencies less than about 5–10 kHz, by simple concatenations of uniform tubes and horns each of which can be associated with some structural features of the ear (Siebert 1970, 1973; Fukudome 1980; Fletcher and Thwaites 1988; Rosowski, Carney, and Peake 1988; Keefe et al. 1994). However, at higher frequencies and smaller wavelengths of sound, the propagation of sound in the ear canal is no longer well approximated by simple plane waves and the smaller structures in the ear become more important, necessitating more complicated structural models of the ear canal, concha, and pinna flange (Shaw 1974a,b, 1982; Stinson 1985, 1986; Stinson and Khanna 1989; Rabbitt 1988; Rabbitt and Holmes 1988).

4.2.3 Lumped-Element Models of Middle-Ear Function

Just as attempts to model the external ear generally use simple tubes and horns with well-defined acoustic characteristics, the most basic attempts to relate middle-ear structures and function define those structures in terms of lumped electrical, mechanical, or acoustic elements (Onchi 1961; Zwislocki 1962, 1963, 1975; Peake and Guinan 1967; Dallos 1973; Shaw and Stinson 1983; Rosowski et al. 1985; Wada and Kobayashi 1990). The use of lumped elements implies that the forces on and velccities of these sturctures do not vary within the structure. For example, modeling the middle-ear air spaces by a single acoustic compliance implies that there are no spatial variations in sound pressure within the air spaces. Similarly, modeling the malleus by a simple mechanical inertance implies that the malleus is rigid and that its mode of motion is independent of frequency. While these approximations may be appropriate at frequencies where the dimensions of the structures are small compared to sound wavelength, they fail to capture the complexities of middle-ear structure, which can contribute to middle-ear behavior at high frequencies. Such more complicated behaviors can be modeled by increasing the number of lumped elements used to represent a specific feature.

4.2.4 Continuum and Finite-Element Models of Middle-Ear Function

Finite-element and continuum models allow distributions of velocity and force within specific structures. For example, while lumped-element models of the tympanic membrane often assume that it is made up of one or two rigid plates (Zwislocki 1962; Shaw and Stinson 1983), continuum and finite-element models allow more realistic breakdowns of nonrigid structural elements. Continuum models include the tympanic membrane models of Esser (1947) and Rabbitt and Holmes (1986), which use estimates of the radial and longitudinal stiffness of the membrane and the fibers within it to describe tympanic membrane function. The related finite-element models (Funnel and Laszlo 1977; Funnell 1983; Funnell, Decraemer, and Khanna 1987, 1992; Wada, Metoki, and Kobayashi 1992) model the tympanic membrane and ossicles by a series of small plates and blocks of defined mass, stiffness, and damping constrained by the rigid tympanic ring and the ossicular ligaments. These models allow a much more detailed description of middle-ear and external-ear function, but are computationally more difficult, tend to have more free parameters, and can be more difficult to interpret.

5. A Model of External- and Middle-Ear Structure and Function

Although the model schemes of Figures 2.2 and 2.7 describe frameworks for quantifying external- and middle-ear function, they are not complete descriptions of the acoustics and mechanics of the ear. The two-port description of Figure 2.7 does not localize function in specific structures, and the cascade model of Figure 2.2 does not quantify the effects of the mechanisms it invokes. Therefore, before we can determine the effects of the middle-ear cavity, tympanic membrane stiffness, ossicular mass, and other structural features on power flow, we need to define a set of rules that relate these features to auditory function. Another way of looking at this task is to replace each of the boxes in Figures 2.2 and 2.7 with a set of equations or network of elements that specifies the role of individual structures on the function of the box. Figure 2.8 is an example of such an improved network description of the human external and middle ear. (The network of Figure 2.8 is not meant to be a complete description of the external- and middle-ear acoustics and mechanics; it is an approximation that attempts to locate form and function better than simple block models.) The network topology comes from models proposed by many others including Onchi (1961), Zwislocki (1962), Peake and Guinan (1967), Siebert (1973), Kuhn (1977), Lynch (1981), Shaw and Stinson (1983), Matthews (1983), Kringlebotn (1988), Fukudome and Yamada (1989), and Wada and Kobayashi (1990). The anatomical variables used to define the external-ear components are from Shaw (1974a). The element values used to define middle-ear function are essentially those of Kringlebotn (1988). The placement of the middle-ear cavity block in the lower branch of the circuit follows Peake, Rosowski, and Lynch (1992) and Shera and Zweig (1992a). The rest of this section describes the network of Figure 2.8 and some of its predictions and limitations.

5.1 The External Ear Model

5.1.1 The Equivalent Pressure Source: Diffraction and Scattering by the Head and Body

The leftmost block of Figure 2.8a contains a Thévenin equivalent sound source where the equivalent source pressure is the product of the sound pressure of the uniform plane wave stimulus and a directional term G_S^{OC} that accounts for the diffraction and scattering of sound by the head, body, and pinna (Siebert 1970). (Although the processes of scattering and diffraction differ, it is difficult to isolate their effects and I shall discuss them together.) Kuhn (1979) has demonstrated that the torso and shoulder influence the magnitude and phase of low-frequency sound entering the ear, and it has long been known that the head as well as the pinna flange scatter and diffract sounds of audible frequency. Teranishi and Shaw (1968) explained how the finer structures of the pinna flange and concha significantly affect sounds in the top half of the audible frequency range.

Any analysis of scattering and diffraction of sound entering the ear canal is greatly complicated by the irregular and asymmetric geometry of the head, body, pinna flange, and concha. The model analysis I performed uses a common simplification to approximate diffraction and scattering. I assume that the head is a rigid sphere (e.g., Ballentine 1928; von Schwarz 1943; Wiener 1947b; Kuhn 1987). The source directional term G_S^{OC} is then equal to the open-circuit pressure gain that relates the sound pressure on the surface of the rigid sphere P_{SS} with the sound pressure of a plane wave stimulus P_{PW}. this gross simplification leads to a direct solution for the directional term (Kuhn 1987):

$$G_S^{OC}(\omega,\Theta) = \frac{P_{SS}(\omega,\Theta)}{P_{PW}(\omega)} = \frac{1}{(ka)^2} \sum_{m=0}^{\infty} \frac{j^{m+1}L_m(\cos\Theta)(2m+1)}{J_m'(ka) - jN_m'(ka)}, \quad (9)$$

where Θ is the angle describing the position on the sphere relative to the direction of propagation of the plane wave (as the sphere is symmetrical,

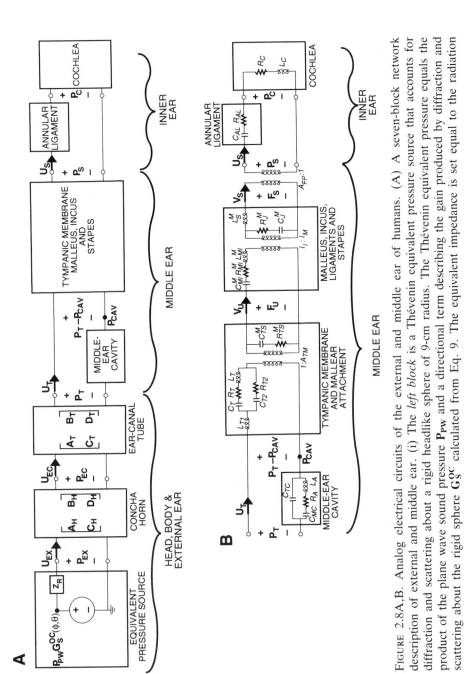

FIGURE 2.8A,B. Analog electrical circuits of the external and middle ear of humans. (A) A seven-block network description of external and middle ear. (i) The *left block* is a Thévenin equivalent pressure source that accounts for diffraction and scattering about a rigid headlike sphere of 9-cm radius. The Thévenin equivalent pressure equals the product of the plane wave sound pressure P_{PW} and a directional term describing the gain produced by diffraction and scattering about the rigid sphere G_S^{OC} calculated from Eq. 9. The equivalent impedance is set equal to the radiation

impedance looking out from a circular opening of area equal to the area of the concha opening. Similar descriptions of diffraction have been proposed by Siebert (1970, 1973) and Fukudome and Yamada (1989). (ii and iii) The external ear concha and meatus are described by two-port descriptions of a tube and horn. The concha horn is approximated by an exponential horn (part 6 of Fig. 2.3) with a length of 0.01 m, a wide end of 4.4×10^{-4} m^2, and a narrow end of 4.4×10^{-5} m^2 (Shaw 1974a). The meatus is approximated by a tube 0.023 m in length and 4.4×10^{-5} m^2 in cross section (part 5 of Fig. 2.3). (iv–vii) The structure of the four *rightmost model blocks* is defined in (B). The middle-ear cavity is included as a one-port placed in series with the rest of the middle ear. The reasons for placing the cavity impedance in the lower branch of the circuit are described in the text (e.g., Lynch 1981; Peake, Rosowski, and Lynch 1992; Shera and Zweig 1992a). The tympanic membrane and ossicles are described by a two-port with inputs of the volume velocity of the $\mathbf{TM}$ $\mathbf{U_T}$ and the difference in sound pressure between the ear canal and the middle-ear cavity $\mathbf{P_T} - \mathbf{P_{CAV}}$. The outputs of the TM ossicular two-port are stapes volume velocity $\mathbf{U_S}$ and the effective pressure acting on the stapes $\mathbf{P_S}$. The impedance of the annular ligament is included as a one-port, which acts to divide the pressure of the stapes between the ligament and the cochlea. The cochlea is modeled as a one-port in which the volume velocity of the stapes equals the volume velocity of the round window and in which the drive to the cochlea is specified by $\mathbf{U_S}$ and $\mathbf{P_C}$. (B) The middle and inner ear blocks are defined in terms of 24 elements including three transformers. The elements and their values are essentially those of Kringlebotn (1988). The tympanic transformer. 1: $A_{TM} = 1/(60 \times 10^{-6})$ m^{-2}, has dimensions of inverse area and converts the acoustic variables at the input of the tympanic membrane-malleus block into mechanical forces and velocities of the ossicles. The footplate transformer, A_{FP}: $1 = 3.2 \times 10^{-6}$ m^2, has dimensions of area and converts the ossicular forces and velocities into a pressure and volume velocity at the stapes. The ossicular transformer, $l_I : l_M = 1/1.3$, is dimensionless and accounts for the mechanical transformation of ossicular force and velocity that results from rotation of the ossicles. Elements with no superscript have units of acoustic compliance, inertance, or resistance. Elements with a superscript M are mechanical elements. The middle-ear cavity elements are the compliance of the mastoid air spaces $C_{MC} = 3.9 \times 10^{-11}$ m^{5}N^{-1}, the compliance of the tympanic air space $C_{TC} = 4 \times 10^{-12}$ m^{5}N^{-1} and the resistance and inertance of the *aditus* that connects the two air spaces $R_A = 6 \times 10^6$ Pa s m^{-3} and $L_A = 100$ kg m^{-4}. The elements that describe the tympanic membrane and mallear attachment include a series branch with $L_{T1} = 750$ kg m^{-4}, $L_T = 6.6 \times 10^3$ kg m^{-4}, $C_T = 3 \times 10^{-12}$ m^{5}N^{-1}, $C_{T2} = 1.3 \times 10^{-11}$ m^{5}N^{-1}, $R_T = 2 \times 10^6$ Pa s m^{-3}, and a parallel branch with $C_{TS}^M = 1.1 \times 10^{-3}$ m N^{-1}, $R_{TS}^M = 4.3 \times 10^{-2}$ N s m^{-1}. The ossicles and their supporting ligaments are modeled by four series and two parallel elements. The malleus and incus are modeled by an RLC branch with an infinite mechanical compliance $C_{MI}^M = \infty$, a resistance $R_{MI}^M = 7.2 \times 10^{-3}$ N s m^{-1}, and a mass of $L_{MI}^M = 7.9 \times 10^{-6}$ kg. The stapes mass is $L_S^M = 3.0 \times 10^{-6}$ kg. Compression of the ossicular joints, which results in reduced motion of the stapes, is modeled by the parallel ossicular branch with $R_J^M = 3.6$ N s m^{-1} and $C_J^M = 4.9 \times 10^{-4}$ m N^{-1}. The annular ligament and cochlear are to the right of the mechanoacoustic transformation of the footplate and are in acoustic units with $C_{AL} = 9.4 \times 10^{-15}$ m^{5}N^{-1}, $R_{AL} = 0$, $R_C = 2.0 \times 10^{10}$ Pa s m^{-3}, and $L_C = 2.4 \times 10^6$ kg m^{-4}.

only one angle is needed to describe the source direction; see inset in Fig. 2.8A)

$\omega = 2\pi f$, the radian frequency of the stimulus
$k = \omega/c$, the wave number
a is the radius of the sphere
L_m is the Legendre polynomial of order m
J'_m is the derivative of the mth order spherical Bessel function
N'_m is the derivative of the mth order spherical Neumann function.

(This is an open-circuit gain because the impedance of the load on the sound source, the impedance at the surface of the rigid sphere, is infinite.) Examples of the open-circuit gain computed for a human headlike sphere of 0.09 m in radius are illustrated in Figure 2.9. At low frequencies, $|G_S^{OC}|$ is close to 0 dB regardless of the direction of the sound source. At higher frequencies, gains with magnitudes as large as 6 dB occur when the sound is directed toward the same side of the sphere ($80° > \Theta > -80°$), while sounds that are directed toward the opposite side of the sphere ($90° > \Theta < 270°$) tend to produce gains of negative dB value (sometimes called sound shadowing). Note, however, that sounds directed exactly at the opposite side of the sphere ($\Theta = 180°$) produce a region of positive gain (Wiener 1947b; Kuhn 1977). The positive and negative gains are analogous to the bright and dark bands produced by the diffraction and scattering of light.

The open-circuit pressure gains predicted by Eq. 9 are complex and vary in both magnitude and angle. Because the reference point for the plane wave pressure P_{PW} is the position of the center of the sphere (Kuhn 1977), sounds from directions of $-90° < \Theta < 90°$ result in a phase lead at the measurement point (the sound reaches the measurement point before it would reach the point at the center of the sphere) and the angle of G_S^{OC} is positive, while sounds from the other hemisphere result in a phase lag and negative angles. The contribution of scattering and diffraction on the phase of the pressure at different locations on the sphere can be quite significant even at low frequencies. Kuhn (1977, 1987) discusses how scattering and diffraction have a profound effect on interaural phase and time differences used to help localize sounds in space.

Although Eq. 9 accurately predicts diffraction and scattering about a sphere, it does not account for the additional diffraction and scattering produced by the torso, shoulders, pinna, concha, and concha opening (Wiener 1947a; Batteau 1967; Teranishi and Shaw 1968; Shaw 1974a; Calford and Pettigrew 1984), nor does it account for the asymmetry of the head and ear structures that affects the sensitivity of the ear to sound sources placed above or below the horizon (Searle et al. 1975; Butler and Belendiuk 1977; Middlebrooks, Makous, and Green 1989; Wightman and Kistler 1989; Musicant, Chan, and Hind 1990; Rice et al. 1992). the

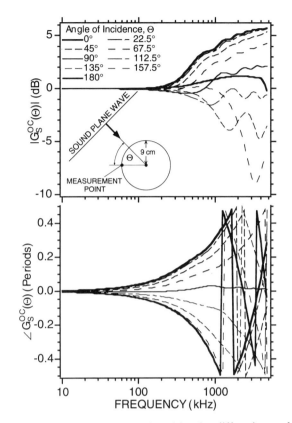

FIGURE 2.9. Gain in sound pressure produced by the diffraction and scattering of sound by a rigid sphere of 0.09-m radius computed from Eq. 9. The inset describes the angle Θ, which relates the position on the sphere and the direction of propagation of a plane wave stimulus. (Because the sphere is radially symmetric, only one angle is needed to describe the direction of the plane wave.) Accurately computing Bessel and Neumann functions with large arguments ($ka > 10$) is not trivial; the accuracy of the methods we used decreased at frequencies above 5 kHz.

effect of the torso and shoulders is small and appears to be restricted to the lower frequencies (Kuhn 1979). However, the pinna, concha, and asymmetries are prominent at frequencies higher than a few kilohertz where the wavelength of sound is less than 10fold the dimensions of these structures (Teranishi and Shaw 1968; Shaw 1974a; Calford and Pettigrew 1984). An alternative to Eq. 9 is to use measurements of the directional gain made in real or model ears to describe $\mathbf{G_S^{OC}}$ (Wiener 1947a; Teranishi and Shaw 1968; Siebert 1970, 1973; Shaw 1974b, 1982; Kuhn 1977).

5.1.2 The Equivalent Pressure Source: The Radiation Impedance

The Thévenin equivalent pressure $P_{PW}G_S^{OC}$ of the model source predicts the "open-circuit" pressure, *not* the sound pressure at the opening of the external ear P_{EX}. Equation 9 assumes the sphere is rigid and does not compensate for the noninfinite impedance of the ear opening. The model of Figure 2.8 compensates for this noninfinite impedance, relating the source pressure and the pressure at the opening of the external ear P_{EX} via a voltage divider formula:

$$P_{EX} = P_{PW}G_S^{OC}\frac{Z_{EX}}{Z_{EX} + Z_R}, \tag{10}$$

where $Z_{EX} = P_{EX}/U_{EX}$, the acoustic impedance looking into the ear opening, P_{EX} and U_{EX} are the sound pressure and volume velocity at the entrance to the ear, and Z_R is the Thévenin impedance of the equivalent source. The Thévenin impedance of the source is the impedance looking out from the load when the source is off and is equivalent to the radiation impedance looking out from the ear opening into the environment (Siebert 1970; Shaw 1988; Fukudome and Yamada 1989). Descriptions of the radiation impedance for simple structures can be found in Kinsler et al. (1982) and Beranek (1986). The radiation impedance is somewhat dependent on the surround; for example, placing the ear opening in an infinite baffle allows sound to radiate into only half the environment and could increase the radiation impedance by 3 dB. In estimating Z_R, I have assumed the head approximates a rigid baffle (Siebert 1970); Fukudome and Yamada (1989) have demonstrated that such an assumption gives a reasonable fit to measurements of sound pressure at the opening of model ears. Both the open-circuit gain and the computed pressure gain at the entrance of the ear P_{EX}/P_{PW}, for a sound source pointed directly at the opening of the ear (azimuth of 90° and elevation of 0°), are plotted in Figure 2.10. Note that $P_{EX}/P_{PW} \approx G_S^{OC}$, over much of the lower frequency range (where $Z_R \ll Z_{EX}$), while at frequencies above 8–9 kHz (where $Z_R \approx Z_{EX}$), $P_{EX}/P_{PW} \neq G_X^{OC}$.

5.1.3 Models of the Concha and External Canal

If we ignore the directional effects introduced by the fine structures of the pinna and concha, much of the power collection function of the external ear can be approximated by distributed acoustic elements such as simple horns and uniform tubes (Siebert 1970; Fletcher and Thwaites 1988; Rosowski, Carney, and Peake 1988). Acoustics texts that describe sound flow through tubes and horns include Malecki (1969), Kinsler et al. (1982), and Fletcher (1992). Research papers dealing with the same topic include Benade (1968, 1988), Egolf (1977), and Keefe (1984). Computations of tube and horn functions are easily made using the "transfer matrix" description of the two-ports used in Figure 2.3B. One

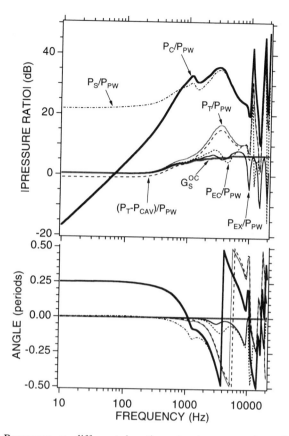

FIGURE 2.10. Pressures at different locations in the external- and middle-ear model of Figure 2.8 normalized by the pressure in the plane wave stimulus. The direction of propagation of the plane wave stimulus is $\Theta = 0°$ (equivalent to $\theta = 90°$ and $\phi = 0°$). $\mathbf{G_S^{OC}}$ is the normalized pressure at the surface of a rigid sphere defined in Eq. 9. (We have assumed $|\mathbf{G_S^{OC}}| = 6\,dB$ at frequencies above $5\,kHz$.) $\mathbf{P_{EX}}$ is the pressure at the entrance of the concha (Eq. 10); $\mathbf{P_{EC}}$ is the pressure at the entrance to the ear canal; $\mathbf{P_T}$ is the pressure on the lateral surface of the tympanic membrane; $\mathbf{P_T} - \mathbf{P_{CAV}}$ is the pressure difference across the tympanic membrane; $\mathbf{P_S}$ is the effective sound pressure of the stapes acting on the annular ligament; and $\mathbf{P_C}$ is the pressure in the vestibule of the cochlea.

issue of recent interest has been the effect of viscous and thermal losses in sound transmission through tubes and horns (Egolf 1977; Keefe 1984; Zuercher, Carlson, and Killion 1988). Because such losses are dependent on interactions at the tube walls, they are most significant in narrow tubes and horns with large surface area-to-volume ratios (Beranek 1986). These losses are probably only significant at frequencies below 1 kHz and in animals with small ear canals (Rosowski 1994).

In the model of Figure 2.8, I assume that the pinna has little effect on power gathering and model the concha and external canal as a horn and tube of dimensions similar to those of the human external ear (Shaw 1974). The horn and tube transform P_{EX} and U_{EX} into P_T and U_T. The dependence of this transformation on the middle-ear input impedance is made explicit in the transfer matrix computations defined in Figure 2.3B. The results of such horn and tube transformations can be observed in Figure 2.10. The pressure transformation produced by the model horn and the head P_{EC}/P_{PW} shows an 8·dB peak in magnitude at 2.8 kHz and a small valley at 4.5 kHz. The calculated peak in $|P_{EC}/P_{PW}|$ is of smaller magnitude and at a lower frequency than that suggested by measurements in human ears (Shaw 1974a); the match could be improved by adjustments in the horn dimensions. However, the computed pressure gain of the horn and tube P_T/P_{PW} is much like measurements in human ear canals (Wiener and Ross 1946; Shaw 1974a,b) with a magnitude peak of about 17 dB near 3 kHz. At frequencies above 10 kHz, the computed P_T/P_{PW} and P_{EC}/P_{PW} show sharp variations in magnitude and angle that result from large reflections produced at the model tympanic membrane in this frequency range. These reflections occur in range where the wavelengths of the sound begin to approximate the radius of the tube and the plane wave propagation assumption of the tube and horn models is no longer valid. More accurate model treatments of the ear canal at frequencies above 8–10 kHz require tube models that account for variations in ear canal area with length (Stinson 1986) and nonplanar modes within the canal and concha (Rabbitt and Holmes 1988; Rabbitt 1988).

5.1.4 Comparisons of Measured and Predicted External-Ear Gain

The total external-ear gain predicted by the model of Figure 2.8 (P_T/P_{PW} in Fig. 2.10) is similar to measurements made by Shaw and others (Shaw 1974a,b). Low-frequency sounds with large wavelengths produce pressures at the tympanic membrane that are similar to the pressure in the plane wave such that $P_T/P_{PW} \approx 1$ (0 dB). In the middle frequencies, diffraction about the head and the horn and tube properties of the concha and canal result in a broad peak in pressure magnitude at the tympanic membrane of about 17 dB. (Shaw's measurement of P_T/P_{PW} in human ears suggested an even broader peak. Modifications in the concha model might produce a better fit.) At higher frequencies, plane wave models of the ear fail and measurements become complicated by variations in the sound pressure that occur within single cross sections of the canal (Khanna and Stinson 1985; Stinson 1985, 1986; Rabbitt and Holmes 1988). One of the biggest assumptions we have made is to lump all the directional response into a source term G_S^{OC}. Such lumping is consistent with the notion that the directional response is produced by structures peripheral to the canal (Wiener and Ross 1946; Wiener, Pfeiffer, and Backus 1966; Shaw

1974a,b; Blauert 1983; Rabbitt 1988) but ignores the role of the concha in producing elevational and azimuthal cues for localization (Teranishi and Shaw 1968; Shaw 1974a, 1982; Searle et al. 1975; Middlebrooks, Makous, and Green 1989; Musicant, Chan, and Hind 1990).

5.2 The Middle-Ear Input Impedance and the Series Model of Middle-Ear Cavity Function

The outputs of the external ear are the pressure outside the tympanic membrane P_T and the membrane volume velocity U_T. The ratio of these two variables defines the middle-ear input impedance $Z_T = P_T/U_T$, where this impedance represents the combination of the impedances of the tympanic membrane, ossicles, cochlea, and the middle-ear air space or cavity. The air spaces contribute to this impedance because motion of the tympanic membrane changes the volume of the closed air space behind it (the Eustachian tube is normally closed), thereby producing a sound pressure P_{CAV} within the space that pushes on the medial surface of the tympanic membrane. In terms of our circuit, the acoustic impedance of the middle-ear cavity Z_{CAV} is "in series" with the input impedance of the rest of the middle ear Z_T^{OC} (Onchi 1961; Zwislocki 1962; Møller 1965; Guinan and Peake 1967; Lynch 1981; Peake, Rosowski, and Lynch 1992) such that

$$Z_T = Z_T^{OC} + Z_{CAV}. \tag{11}$$

(Z_T^{OC} is the input impedance of the tympanic membrane, ossicles, and cochlea measured with "open cavities.")

The circuit topology I have chosen (placing the cavity impedance in the lower branch between the rest of the middle ear and ground) allows P_{CAV} to be defined explicitly and also correctly illustrates the pressure difference across the tympanic membrane $P_T - P_{CAV}$ (Lynch 1981; Ravicz, Rosowski, and Voigt 1992; Peake, Rosowski, and Lynch 1992; Shera and Zweig 1992a,c). This pressure difference, which is somewhat smaller in magnitude than P_T (Fig. 2.10), acts as the drive to the rest of the middle ear. A simple way of thinking about the reduction in middle-ear drive by the cavity is that some of the pressure in the ear canal is used to compress the air in the middle-ear spaces. Equation 11 points out that the effect of the middle-ear air spaces on middle-ear function depends on the relative magnitude of Z_T^{OC} and Z_{CAV}. In humans, $Z_T^{OC} > Z_{CAV}$, and the effect of the cavities is small. In other mammals, for example, chinchilla, guinea pig, and hamster, $Z_{CAV} > Z_T^{OC}$ and the effect of the cavities is large (Dallos 1970; Zwillenberg, Konkle, and Saunders 1981; Ravica, Rosowski, and Voigt 1992; Rosowski 1994).

The four acoustic elements I used to represent the human middle-ear air spaces (Fig. 2.8B) are those of Kringlebotn (1988) and include two parallel capacitors, one for the tympanic air space located directly behind the tympanic membrane C_{TC}, and one for the mastoid air volume C_{MC}, which is separated from the other space by the *aditus ad antrum* (McElveen et al. 1982). The mastoid air space is placed in series with an acoustic inertance that approximates the inertance of the connecting aditus L_A and a resistance R_A associated with viscous losses in the aditus and mastoid spaces. Similar models have been proposed for other animals with separated middle-ear air spaces (Onchi 1961; Zwislocki 1963; Møller 1965; Peake, Rosowski, and Lynch 1992). The reasons for including the resistance in the cavity model are not completely straightforward. The theoretical bases for the resistance are (1) the tubelike aditus between the tympanic and mastoid cavity introduces a resistive as well as mass component (Beranek 1986; Peake, Rosowski, and Lynch 1992), and (2) the large surface area of the mastoid air spaces (in humans the mastoid spaces are subdivided into numerous small spaces) multiplies the small viscous losses that occur at the surface of the walls (Zwislocki 1962). However, the effect of R_A in the model of Figure 2.8 is small. Indeed, impedance measurements in cats and gerbils indicate that except for a slight damping of sharp middle-ear cavity resonances (Lynch 1981; Peake, Rosowski, and Lynch 1992) the resistance of the cavity is insignificant except at very low frequencies (Ravicz, Rosowski, and Voigt 1992; Lynch, Peake, and Rosowski 1994), and that malleus motion measured in human temporal bones is not greatly affected by opening the middle-ear cavities (McElveen et al. 1982; Gyo, Aritomo, and Goode 1987).

5.3 The Tympanic Membrane Model

5.3.1 The Model Network

Refering back to Figure 2.8B, the pressure difference across the tympanic membrane $P_T - P_{CAV}$ acts as the drive on the rest of the model. The first block the pressure difference works against represents the action of the tympanic membrane in converting the pressure P_T and volume velocity U_T at the tympanic membrane into force F_U and velocity V_U at the umbo. Nine elements are used to model this conversion (Fig. 2.8B). A series branch with two compliances, two resistances, and two inertances accounts for the damping, mass, and stiffness associated with "coupled" motions of the tympanic membrane and malleus. An ideal transformer $1:A_{TM}$ with units of inverse area converts the acoustic variables of sound pressure P, volume velocity U, and acoustic impedance $Z = P/U$ to mechanical variables of force F, velocity V, and mechanical impedance $Z^M = F/V$. The two mechanical elements of the parallel branch are a compliance C_{TS}^M and a resistance R_{TS}^M that account for relative ("uncoupled") motions between the tympanic membrane and malleus (Zwislocki 1962). Velocity moving through the parallel branch is shunted to ground and does not contribute to the motion of the malleus V_U. Although we can associate each of the nine elements with some approximate feature of the tympanic membrane—the capacitors are related to the tympanic membrane compliance (1/stiffness), the resistors are associated with losses and damping, the inductors are associated with acoustic and mechanical masses and the transformer associated with the area of the tympanic membrane—the precise relationships between the elements and TM structure are far from understood. Several mathematical formulations attempt to relate functional attributes such as stiffness, damping, and mass with the membrane's three-dimensional shape (Funnel and Laszlo 1977; Funnel 1983; Funnel, Decraemer, and Khanna 1987) or specific fiber populations within the membrane (Rabbit and Holmes 1986); however, our knowledge of the actual physical parameters and structures need to be improved before these analyses can be completely tested.

It is important to note that the velocity of the umbo V_U used in the model (Fig. 2.8) is the translational velocity that would be measured using current experimental techniques (Buunen and Vlaming 1981; Ruggero et al. 1990; Cooper and Rhode 1992). The mechanisms associated with this motion can be simple ossicular translation in the mediolateral direction or small rotations of the malleus about a fixed axis (Gyo, Aritomo, and Goode 1987; Decraemer, Khanna, and Funnell 1989; Donahue, Rosowski, and Peake 1991). Other motions such as translations of the umbo in the dorsoventral or anteroposterior directions (Hüttenbrink 1988; Donahue 1989; Decraemer, Khanna, and Funnell 1991; Decraemer and Khanna 1994) are not included.

5.3.2 The Topology and the Transformer

The topology of Kringlebotn's (1988) tympanic membrane-malleus network (see Fig. 2.8B) with the series branch preceding the parallel is similar to that used by Peake and Guinan (1967), and differs from that of Zwislocki (1962, 1975), who places the parallel branch first. As Shera and Zweig (1991) have pointed out, such topological differences interact with conceptions of tympanic membrane function. For example, Zwislocki's circuits suggest that part of the tympanic membrane is completely decoupled from the rest and that volume velocity in the decoupled branch is simply defined by the sound pressure across the tympanic membrane and the branch impedance. In the model of Figure 2.8, however, the flow of velocity through the parallel branch depends on the stimulus pressure, the branch impedance, *and* the load of the rest of the ear. Because our conceptions of the relationship between membrane structure and function are relatively primitive, Shera and Zweig (1991) suggested a more general two-port description of the tympanic membrane.

The use of a transformer with a turns ratio equal to one over the area of the tympanic membrane is reminiscent of the ideal transformer model. However, the model of Figure 2.8 also contains elements that account for the nonzero impedance of the membrane and its connections. The additional elements have a large effect on the transformation from pressure to force and volume velocity to umbo velocity. Model calculations (Fig. 2.11) demonstrate that two estimates of the tympanic membrane transformer's "effective area" F_U/P_T and U_T/V_U are (a) frequency dependent, (b) not equal to the transformer ratio A_{TM}, and (c) not equal to each other (Shera and Zwig 1991).

5.3.3 The Pars Flaccida

The tympanic membrane of many mammals including humans is composed of two different membrane segments, including the *pars tensa*, which

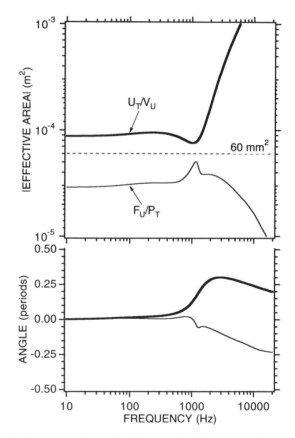

FIGURE 2.11. Estimates of the "effective area" of the tympanic membrane in the model of Figure 2.8. Model estimates of the transformation between umbo force and pressure at the tympanic membrane F_U/P_T and tympanic volume velocity and umbo velocity U_T/V_U are compared. Both these ratios have units of area. The area of the model's tympanic membrane is a real constant of $6 \times 10^{-5} \text{m}^2 = 60 \text{mm}^2$.

appears stiff and surrounds the manubrium of the malleus, and the *pars flaccida*, which is floppy and generally either superior or posterior to the ossicular attachment (Henson 1974; Kohllöffel 1984). In humans the flaccida is about one-thirtieth the area of the tensa, but the relative area of the flaccida varies greatly within mammal species (Kohllöffel 1984). In some goats the flaccida is nearly 50% of the total membrane area, while it is nonexistent in guinea pig and chinchilla. The flaccida is usually considered to be uncoupled from the ossicles and the tensa, and most discussions of middle-ear function ignore it. However, sound acting on the flaccida can influence the sound pressure within the middle-ear cavity and thereby affect the pressure difference that acts on the tympanic membrane and ossicles (Kohllöffel 1984). I have not included any effect of the pars flaccida in the model of Figure 2.8.

5.4 The Ossicular Model

5.4.1 The Ossicular Network Model

The next stage of the middle-ear model of Figure 2.8B is the conversion of motion of the umbo to a piston-like volume velocity of the stapes characterized by the transfer function U_S/V_U. This function is performed by the malleus, incus, and stapes block. The circuit element representation of this block includes (a) a series resistance, compliance and mass that models the mass L_{MI}^M, compliance C_{MI}^M, and damping R_{MI}^M within the malleus, incus, and their supporting ligaments; (b) a dimensionless transformer $l_I : l_M$ that accounts for rotation of the malleus and incus complex, (c) a shunt branch with a capacitor C_J^M and resistor R_J^M that accounts for the loss of stapes velocity from compression of the ossicular joints; (d) the mass of the stapes L_S^M; and (e) a second transformer $A_{FP} : 1$. The second transformer is the stapes footplate area that converts the force F_S and piston-like velocity of the stapes V_S into pressure P_S and volume velocity U_S. Compression of the incudostapedial joint has been inferred in man (Zwislocki 1962, 1975) and demonstrated in guinea pig (Zwislocki 1963; Manley and Johnstone 1974) and chinchilla (Ruggero et al. 1990). Others have demonstrated such compressions in the incudomallear joint in cat (Guinan and Peake 1967) and man (Gyo, Aritomo, and Goode 1987; Hüttenbrink 1988). Again the relationship between single structures within the middle ear and single model elements is far from perfect.

5.4.2 The Element Values

Most of the element values have been chosen by fitting the model to various middle-ear data (Kringlebotn 1988). One exception is my introduction of a stapes mass of 3 mg equal to measurements of this mass

(Wever and Lawrence 1954). This mass value is consistent with the view that the stapes translates in a piston-like manner. In contrast, the 7.9-mg value for L_{MI}^{M}, the mechanical inertance of the incus and malleus, is much smaller than would be predicted by translation of the 60-mg mass of these ossicles (Wever and Lawrence 1954); however, if the ossicles rotated and the incus and malleus mass were ideally distributed about the axis of rotation the ossicular inertance could be very small. In fact, both the ossicular mass elements are so small that they have little effect on the model predictions. The model fit also suggests that the compliances of the ligaments that support the malleus and incus are so large that they have no effect on the circuit. This conclusion is highly dependent on the model topology. Clearly there are animal species in which the stiffness of the ossicles plays a significant role in middle-ear mechanics (Aitkin and Johnstone 1972; Rosowski 1994).

5.4.3 The Transformers

The ossicular model includes two ideal transformers, but again the transformation performed by the ossicular subcircuit depends greatly on the other elements. For example, data showing that the ossicular lever is frequency dependent (Manley and Johnstone 1974; Gyo, Aritomo, and Goode 1987) can be modeled by adjusting the relative value of the joint impedance and the impedance looking into the stapes. The action of the model's footplate transformer is predicated on a piston-like action of the stapes. Other modes of stapes motion (von Békésy 1960; Gyo, Aritomo, and Goode 1987) could also be compensated for by adjusting the joint impedance.

5.5 The Cochlear Load on the Middle Ear

The rightmost blocks of Figure 2.8A and B are one-ports that account for the impedance of the annular ligament and cochlea. These two blocks were originally combined into a single RLC by Kringlebotn (1988). The new topology separates out the nonzero impedance of the ligament (Zwislocki 1962, 1975; Lynch 1981) and better defines the input to the cochlea. The effect of the annular ligament impedance can be assessed by comparing the model estimates of the pressure acting on the ligament (P_S) and the pressure at the cochlear entrance P_C (see Fig. 2.10). The series combination of the ligament and cochlea acts as a voltage divider. At low frequencies, where the stiffness of the ligament dominates the total impedance, much of the pressure is used to move the ligament and little reaches the cochlea $P_S \gg P_C$. At high frequencies, the impedance of the cochlea dominates the total impedance and little of the pressure is spent moving the ligament $P_S \approx P_C$.

The model of Figure 2.8 assumes that the cochlea is a one-port such that all the volume velocity that enters the cochlea via the oval window leaves by way of the round window. This assumption is very common in the literature, but has only been loosely tested (Wever and Lawrence 1950; Tonndorf and Tabor 1962). Indeed, von Békésy (1960) came to a different conclusion when he attempted to explain residual hearing after removal of the middle ear. The theoretical analysis of Shera and Zweig (1992c) suggested that only small amounts of volume velocity, if any, leave the cochlea through paths other than the cochlear windows.

The cochlear and annular ligament element values used in the model are Kringlebotn's (1988) variations on the annular ligament compliance and cochlear resistance and inertance of Zwislocki (1975). Because of some indications that the annular ligament of humans has a significant resistive component (Merchant, Ravicz, and Rosowski 1992), we have included an annular ligament resistance but have set its value to zero. The Kringlebotn and Zwislocki models of the annular ligament and cochlear impedances are consistent with measurements of this impedance in cat (Lynch, Nedzelnitsky, and Peake 1982) and human (Nakamura, Aritomo, and Goode 1992; Merchant, Ravicz, and Rosowski 1992) and with other cochlear models (Puria and Allen 1991).

5.6 The Effect of Middle-Ear Muscle Contractions

The model of Figure 2.8 does not include the action of the middle-ear muscles; however, a few modifications to the model can imitate muscle function. Rabinowitz (1977) and Pang and Peake (1985, 1986) have suggested that stapedius muscle contractions work by altering the impedance of the annular ligament. The contracted muscle pulls on the stapes and stretches the ligament, thereby increasing the ligament's impedance. Increases in the stiffness of the model's annular ligament produce alterations in middle-ear input impedance and middle-ear transmission that are quite similar to those produced by stapedius contraction (Rabinowitz 1977; Pang and Peake 1985). Contraction of the *tensor tympani* could likewise be modeled by an increase in the impedance of the tympanic membrane.

5.7 Summary

The block and circuit element models of Figure 2.8 represent attempts to associate external- and middle-ear functions with various structural features. It should be clear, however, that these associations are far from complete. The simple two-port representation of the external-ear components ignores details in the external-ear structure that may be important in the gathering of moderate and high-frequency sound. The elements associated with the tympanic membrane and malleus account adequately for their average function but are not associated with particular fiber populations within the membrane or connecting ligaments. Most of the other model blocks suffer from similar failings. More detailed associations between structure and function are being investigated (Funnel and Laszlo 1977; Rabbit and Holmes 1986; Puria and Allen 1991; Funnel, Decraemer, and Khanna 1987, 1992), but much work needs to be done. Indeed, the major deficiency facing the study of the middle and external ear is a lack of detailed functional and structural measurements, because computational and physical models can only be as good as the data on which they are based.

In spite of these difficulties, the predicted function of the circuit model is similar to existing measurements of function in humans and animals. The external-ear transformation of the stimulus produces a peak in $|P_T/P_{PW}|$ which is similar in magnitude and frequency to that measured by Shaw and others (Shaw 1974b), but the predicted peak is less broad than the measured. The middle-ear input impedance P_T/U_T and stapes motion U_S/P_T predicted by the model match measurements quite well (Kringlebotn 1988), and the predicted middle-ear pressure transfer ratio P_C/P_T is similar to that measured in animals (Nedzelnitsky 1980; Décory 1989).

6 Other Mechanisms for Sensing Sound

The model of Figure 2.8 implies that stapes motion is the only input to the inner ear, yet it is well known that interruption of the incudostapedial joint produces only a 50–60 dB hearing loss (Peake, Rosowski, and Lynch 1992). Clearly, there must be other mechanisms for sound to enter the inner ear. These "other" mechanisms are thought to include bone conduction (von Békésy 1960; Tonndorf 1972) and direct acoustic stimulation of the cochlear windows (Wullstein 1956; Peake, Rosowski, and Lynch 1992; Shera and Zweig 1992c). The greatly decreased hearing sensitivity after interruption of the ossicular chain (Wever and Lawrence 1954) and model analyses (Peake, Rosowski, and Lynch 1992; Shera and Zweig 1992a,c) suggest that these other paths are not significant in normal ears, but can become important in middle-ear pathology.

7 Future Work

I have summarized the present thinking about how the external and middle ear work together to gather sound power and conduct it to the inner ear. I have also presented a rudimentary model that exemplifies much of our present knowledge about external- and middle-ear processes. Much work still needs to be done. More measurements of ear function and better descriptions of external- and middle-ear anatomy are required to refine existing ideas, and more sophisticated analyses are also required.

Suggestions for future work include (1) investigations of the role of tympanic membrane fiber structure in its function, (2) understanding how the compartmentalization of the middle-ear air spaces affects function at high frequencies, (3) determination of the function of the ossicular joints, (4) investigations of nonossicular sound conducting mechanisms, (5) determination of the causes of middle-ear nonlinearity and hysteresis, (6) measurements of the forces acting on the ossicles, and (7) determination of the backward transmission function that relates sounds generated within the cochlea with the resultant sound pressure in the ear canal.

Acknowledgments. I thank my colleagues at the Eaton-Peabody Laboratory for their assistance and support in preparing this chapter, especially Bill Peake, Mike Ravicz, and Christopher Long. This work was supported by National Institutes of Health (NIH) grants RO1-DC00194 and PO1-DC00119.

List of Symbols

γ	ratio of specific heats	l_M	length of the malleus lever arm
λ	wavelength	$p(t)$	time waveform of a pressure
ϕ	elevational angle	$v(t)$	time waveform of a velocity
θ	azimuthal angle	$u(t)$	time waveform of a volume velocity
ρ_0	static density of a fluid	z_0	characteristic impedance of a fluid
σ	coefficient of shear viscosity	A	an area
ω	radian frequency	B	bulk modulus of compressibility
Θ	a spherical angle	C or C^A	an acoustic compliance
Ω	an acoustical, electrical, or mechanical ohm	C^M	a mechanical compliance
m	meters	L or L^A	an acoustic inertance
s,sec	seconds	L^M	a mechanical mass
A	amps of current	P_0	static pressure
J	joules of energy	R or R^A	an acoustic resistance
N	newtons of force	R^M	a mechanical resistance
Pa	pascals of pressure, $1\,\mathrm{N\,m}^{-2}$	Z_0	characteristic acoustic impedance of a tube
SPL	sound pressure re $2 \times 10^{-5}\,\mathrm{Pa}$	**A, B, C, D**	complex two-port parameters

V	volts of voltage	E	complex amplitude of a voltage
W	watts of power	F	complex amplitude of a force
a	radius	G	complex amplitude of a gain function
c	propagation velocity of sound	I	complex amplitude of an electrical current
$f(t)$	time waveform of a force	P	complex amplitude of a pressure
j	the imaginary number $\sqrt{-1}$	U	complex amplitude of a volume velocity
k	wave number, ω/c	V	complex amplitude of a point velocity
$\bar{k}$	average mechanical stiffness	Z	complex impedance
l	length	Π	complex power
l_I	length of the incus lever arm	$\bar{\Pi}$	average power

References

Aitkin LM, Johnstone BM (1972) Middle-ear function in a monotreme: the echidna (*Tachyglossus aculeatus*). J Exp Biol 180:245–250.

Allen J (1986) Measurements of eardrum acoustic impedance. In: Allen JB, Hall JL, Hubbard A, Neely ST, Tubis A (eds) Peripheral Auditory Mechanism. New York: Springer-Verlag, pp. 44–51.

Ballentine S (1928) Effect of diffraction around the microphone in sound measurements. Phys Rev 32:988–992.

Batteau DW (1967) The role of the pinna in human localization. Proc R Soc Lond B 168:158–180.

Benade AH (1968) On the propagation of sound waves in a cylindrical conduit. J Acoust Soc Am 44:616–623.

Benade AH (1988) Equivalent circuits for conical waveguides. J Acoust Soc Am 83:1764–1769.

Beranek LL (1986) Acoustics. New York: McGraw-Hill.

Blauert J (1983) Spatial Hearing. Cambridge: MIT Press.

Butler RA, Belendiuk K (1977) Spectral cues utilized in the localization of sound in the median sagittal plane. J Acoust Soc Am 61:1264–1269.

Buunen TJF, Vlaming MSMG (1981) Laser-Doppler velocity meter applied to tympanic membrane vibrations in cat. J Acoust Soc Am 69:744–750.

Calford MB, Pettigrew JD (1984) Frequency dependence of directional amplification at the cat's pinna. Hear Res 14:13–19.

Chan JCK, Geisler CD (1990) Estimation of eardrum acoustic pressure and of ear canal length from remote points in the canal. J Acoust Soc Am 87:1237–1247.

Coles RG, Guppy A (1986) Biophysical aspects of directional hearing in the Tammar wallaby, *Macropus eugenii*. J Exp Biol 121:371–394.

Cooper NP, Rhode WS (1992) Basilar membrane mechanics in the hook region of cat and guinea-pig cochleae: sharp tuning in the absence of baseline position shifts. Hear Res 63:163–190.

Dallos P (1970) Low-frequency auditory characteristics. J Acoust Soc Am 48: 489–499.

Dallos P (1973) The Auditory Periphery. New York: Academic Press.

Daniels FB (1974) Acoustical impedance of enclosures. J Acoust Soc Am 19: 569–572.

Décory L (1989) Origine des différences interspecifiques de susceptibilité an bruit. Thése de Doctorat de l'Université de Bordeaux, France.

Decraemer WF, Khanna SM (1994) Modelling the malleus vibration as a rigid body motion with one rotational and one translational degree of freedom. Hear Res 72:1–18.

Decraemer WF, Khanna SM, Funnell WRJ (1989) Interferometric measurement of the amplitude and phase of tympanic membrane vibrations in cat. Hear Res 38:1–18.

Decraemer WF, Khanna SM, Funnell WRJ (1990) Heterodyne interferometer measurements of the frequency response on the manubrium tip in cat. Hear Res 47:205–218.

Decraemer WF, Khanna SM, Funnell WRJ (1991) Malleus vibration mode changes with frequency. Hear Res 54:305–318.

Desoer CA, Kuh ES (1969) Basic Circuit Theory. New York: McGraw-Hill.

Donahue KM (1989) Human middle-ear motion: models and measurements. M.S. thesis, Department of Electrical Engineering and Computer Science, Massachusetts Institute of Technology, Cambridge, MA.

Donahue KM, Rosowski JJ, Peake WT (1991) Can the motion of the human malleus be described as pure rotation? In: Abstracts of the 14th Midwinter Meeting of the Association for Research in Otolaryngology, p. 52.

Egolf DP (1977) Mathematical modeling of a probe-tube microphone. J Acoust Soc Am 61:200–205.

Egolf DP (1980) Techniques for modeling the hearing aid receiver and associated tubing. In: Studebaker GA, Hochberg I (eds) Acoustical Factors Affecting Hearing Aid Performance. Baltimore: University Park Press, pp. 297–319.

Esser MHM (1947) The mechanics of the middle ear: II. The drum. Bull Math Biophys 9:75–91.

Fleischer G (1973) Studien am Skelett des Gehörorgans der Säugetiere, einschliesslich des Menschen. Säugetierkd Mitt (München) 21:131–239.

Fleischer G (1978) Evolutionary principles of the mammalian middle ear. Adv Anat Embryol Cell Biol 55:3–69.

Fletcher NH (1992) Acoustic Systems in Biology. Oxford: Oxford University Press.

Fletcher NH, Thwaites S (1979) Physical models for the analysis of acoustical systems in biology. Q Rev Biophys 12:25–65.

Fletcher NH, Thwaites S (1988) Obliquely truncated horns: idealized models for vertebrate pinnae. Acustica 65:195–204.

Fukudome K (1980) Equalization for the dummy-head-headphone system capable of reproducing true directional information. J Acoust Soc Jpn (E) 1:59–67.

Fukudome K, Yamada M (1989) Influence of the shape and size of a dummyhead upon Thévenin acoustic impedance and Thévenin pressure. J Acoust Soc Jpn (E) 10:11–22.

Funnell WR (1983) On the undamped natural frequencies and mode shapes of a finite-element model of the cat eardrum. J Acoust Soc Am 73:1657–1661.

Funnell WR, Laszlo CA (1977) Modeling of the cat eardrum as a thin shell using the finite-element method. J Acoust Soc Am 63:1461–1467.

Funnell WR, Decraemer WF, Khanna SM (1987) On the damped frequency response of a finite-element model of the cat eardrum. J Acoust Soc Am 81:1851–1859.

Funnell WR, Decraemer WF, Khanna SM (1992) On the degree of rigidity of the manubrium in a finite-element model of the cat eardrum. J Acoust Soc Am 91:2082–2090.

Guinan JJ Jr, Peake WT (1967) Middle-ear characteristics of anesthetized cats. J Acoust Soc Am 41:1237–1261.

Gyo K, Aritomo H, Goode RL (1987) Measurement of the ossicular vibration ratio in human temporal bones by use of a video measuring system. Acta Otolaryngol 103:87–95.

Henson OW Jr (1974) Comparative anatomy of the middle ear. In: Kiedel WD, Neff WD (eds) Handbook of Sensory Physiology: The Auditory System, Vol. 1. New York: Springer-Verlag, pp. 39–110.

Hudde H, Schröter J (1980) The equalization of artificial heads without exact replication of eardrum impedance. Acoustical 44:301–307.

Hüttenbrink KB (1988) The mechanics of the middle-ear at static air pressures. Acta Otolaryngol Suppl 451:1–35.

Keefe DH, Bulen JC, Campbell SL, Burns EM (1994) Pressure transfer function and absorption cross section from the diffuse field to the human infant ear canal. J Acoust Soc Am 95:355–371.

Khanna SM, Stinson MR (1985) Specification of the acoustical input to the ear at high frequencies. J Acoust Soc Am 77:577–589.

Khanna SM, Tonndorf J (1969) Middle ear power transfer. Arch Klin Exp Ohren-Nasen Kehlkopfheilkd 193:78–88.

Khanna SM, Tonndorf J (1972) Tympanic membrane vibrations in cats studied by time-average holography. J Acoust Soc Am 51:1904–1920.

Khanna SM, Tonndorf J (1978) Physical and physiological principles controlling auditory sensitivity in primates. In: Noback R (ed) Neurobiology of Primates. New York: Plenum Press, pp. 23–52.

Kinsler LE, Frey AR (1962) Fundamentals of Acoustics. New York: Wiley.

Kinsler LE, Frey AR, Coppens AB, Sanders JV (1982) Fundamentals of Acoustics. New York: Wiley.

Kohllöffel LUE (1984) Notes on the comparative mechanics of hearing. III. On Shrapnell's membrane. Hear Res 13:83–88.

Kringlebotn M (1988) Network model for the human middle ear. Scand Audiol 17:75–85.

Kuhn GF (1977) Model for the interaural time differences in the azimuthal plane. J Acoust Soc Am 62:157–167.

Kuhn GF (1979) The pressure transformation from a diffuse sound field to the external ear and to the body and head surface. J Acoust Soc Am 65:991–1000.

Kuhn GF (1987) Physical acoustics and measurements pertaining to directional hearing. In: Yost WA, Gourevitch G (eds) Directional Hearing. New York: Springer-Verlag, pp. 3–25.

Lynch TJ III (1981) Signal processing by the cat middle ear: admittance and transmission, measurements and models. Sc.D. Thesis, Massachusetts Institute of Technology, Cambridge, MA.

Lynch TJ III, Nedzelnitsky V, Peake WT (1982) Input impedance of the cochlea in cat. J Acoust Soc Am 72:108–130.

Lynch TJ III, Peake WT, Rosowski JJ (1994) Measurements of the acoustic input impedance of cat ears: 10 Hz–22 kHz. J Acoust Soc Am 96:2184–2209.

Malecki I (1969) Physical Foundations of Technical Acoustics. Oxford: Pergamon Press.

Manley GA, Johnstone BM (1974) Middle-ear function in the guinea pig. J Acoust Soc Am 56:571–576.

Margolis RH, Smith P (1977) Tympanometric asymmetry. J Speech Hear Res 20:437–466.

Matthews JW (1983) Modeling reverse middle ear transmission of acoustic distortion signals. In: deBoer E, Viergever MA (eds) Mechanics of Hearing. Delft: Delft University Press, pp. 11–18.

McElveen JT, Goode RL, Miller C, Falk SA (1982) Effect of mastoid cavity modification on middle ear sound transmission. Ann Otol Rhinol Laryngol 91:526–532.

Merchant SN, Ravicz ME, Rosowski JJ (1992) The acoustic input impedance of the stapes and cochlea in human temporal bones. In: Abstracts of the 15th Midwinter Meeting of the Association for Research in Otolaryngology, p. 98.

Michelsen A (1992) Hearing and sound communication in small animals: evolutionary adaptations to the laws of physics. In: Webster DB, Popper AN, Fay RR (eds) The Evolutionary Biology of Hearing. New York: Springer-Verlag, pp. 61–77.

Middlebrooks JC, Makous JC, Green DM (1989) Directional sensitivity of sound-pressure levels in the human ear canal. J Acoust Soc Am 86:89–108.

Møller AR (1965) Experimental study of the acoustic impedance of the middle ear and its transmission properties. Acta Otolaryngol 60:129–149.

Møller AR (1983) Auditory Physiology. New York: Academic Press.

Musicant AD, Chan JCK, Hind JE (1990) Direction-dependent spectral properties of cat external ear: new data and cross-species somparisons. J Acoust Soc Am 87:757–781.

Nakamura K, Aritomo H, Goode RL (1992) Measurements of human cochlear impedance. In: Yanagihara N, Suzuki JI (eds) Transplants and Implants in Otology II. New York: Kugler, pp. 227–230.

Nedzelnitsky V (1980) Sound pressures in the basal turn of the cat cochlea. J Acoust Soc Am 68:1676–1689.

Onchi Y (1961) Mechanism of the middle ear. J Acoust Soc Am 33:794–805.

Pang XD, Peake WT (1985) A model for changes in middle-ear transmission by stapedius-muscle contraction. J Acoust Soc Am 78:S13.

Pang XD, Peake WT (1986) How do contractions of the stapedius muscle alter the acoustic properties of the middle ear? In: Allen JB, Hall JL, Hubbard A, Neely ST, Tubis A (eds) Peripheral Auditory Mechanisms. New York: Springer-Verlag, pp. 36–43.

Peake WT, Guinan JJ Jr (1967) Circuit model for the cat's middle ear. MIT Q Prog Rep 84:320–326.

Peake WT, Rosowski JJ, Lynch TJ III (1992) Middle-ear transmission: acoustic vs. ossicular coupling in cat and human. Hear Res 57:245–268.

Price GR, Kalb JT (1991) Insight into hazard from intense impulses from a mathematical model of the ear. J Acoust Soc Am 90:219–227.

Puria S, Allen JB (1991) A parametric study of cochlear input impedance. J Acoust Soc Am 89:287–309.

Rabbitt RD (1988) High-frequency plane waves in the ear canal: application of a simple asymptotic theory. J Acoust Soc Am 84:2070–2080.

Rabbitt RD, Holmes MH (1986) A fibrous dynamic continuum model of the tympanic membrane. J Acoust Soc Am 80:1716–1728.

Rabbitt RD, Holmes MH (1988) Three-demensional acoustic waves in the ear canal and their interaction with the tympanic membrane. J Acoust Soc Am 83:1064–1080.

Rabinowitz WM (1977) Acoustic-Reflex Effects on the Input Admittance and Transfer Characteristics of the Human Middle Ear. Ph.D. thesis, Massachusetts Institute of Technology, Cambridge MA.

Rabinowitz WM (1981) Measurement of the acoustic input admittance of the human ear. J Acoust Soc Am 70:1025–1035.

Rhode WS (1978) Some observations on cochlear mechanics. J Acoust Soc Am 64:158–176.

Ravicz ME, Rosowski JJ, Voigt HF (1992) Sound-power collection by the auditory periphery of the Mongolian gerbil *Meriones unguiculatus*. I. Middle-ear input impedance. J Acoust Soc Am 92:157–177.

Rice JJ, May BJ, Spirou GA, Young ED (1992) Pinna-based spectral cues for sound localization in cat. Hear Res 58:132–152.

Rosowski JJ (1991a) The effects of external- and middle-ear filtering on auditory threshold and noise-induced hearing loss. J Acoust Soc Am 90:124–135.

Rosowski JJ (1991b) Erratum: "The effects of external- and middle-ear filtering on auditory threshold and noise-induced hearing loss." J Acoust Soc Am 90:3373.

Rosowski JJ (1992) Hearing in transitional mammals: predictions from the middle-ear anatomy and hearing capabilities of extant mammals. In: Webster DB, Popper AN, Fay RR (eds) The Evolutionary Biology of Hearing. New York: Springer-Verlag, pp. 625–631.

Rosowski JJ (1994) The external and middle ear. In: Popper AN, Fay RR (eds) Springer Handbook of Auditory Research, Vol. IV, Comparative Mammalian Hearing. New York: Springer-Verlag, pp. 172–247.

Rosowski JJ, Graybeal A (1991) What did *Morganucodon* hear? Zool J Linn Soc 101:131–168.

Rosowski JJ, Garney LH, Peake WT (1988) The radiation impedance of the external ear of cat: measurements and applications. J Acoust Soc Am 84:1695–1708.

Rosowski JJ, Carney LH, Lynch TJ III, Peake WT (1986) The effectiveness of the external and middle ears in coupling acoustic power into the cochlea. In: Allen JB, Hall JL, Hubbard A, Neely ST, Tubis A (eds) Peripheral Auditory Mechanisms. New York: Springer-Verlag, pp. 3–12.

Rosowski JJ, Peake WT, Lynch TJ III, Weiss TF, Leong R (1985) A model for signal transmission in an ear having hair cells with free-standing stereocilia. II. Macromechanical stage. Hear Res 20:139–155.

Ruggero MA, Rich NC, Robles L, Shivapuja BG (1990) Middle ear response in the chinchilla and its relationship to mechanics at the base of the cochlea. J Acoust Soc Am 87:1612–1629.

Searle CL, Braida LD, Cuddy DR, Davis MF (1975) Binaural pinna disparity: another localization cue. J Acoust Soc Am 57:448–455.

Shaw EAG (1974a) The external ear. In: Keidel WD, Neff WD (eds) Handbook of Sensory Physiology, Vol. 1, Auditory System. New York: Springer-Verlag, pp. 455–490.

Shaw EAG (1974b) Transformation of sound pressure level from the free field to the eardrum in the horizontal plane. J Acoust Soc Am 56:1848–1860.

Shaw EAG (1982) External ear response and sound localization. In: Gatehouse R (ed) Localization of Sound: Theory and Application. Groton: Amphora Press, pp. 30–41.

Shaw EAG (1988) Diffuse field response, receiver impedance and the acoustical reciprocity principle. J Acoust Soc Am 84:2284–2287.

Shaw EAG, Stinson MR (1983) The human external and middle ear: models and concepts. In: deBoer E, Viergever MA (eds) Mechanics of Hearing. Delft: Delft University Press, pp. 3–10.

Shera C, Zweig G (1991) Phenomenological characterization of eardrum transduction. J Acoust Soc Am 90:235–262.

Shera C, Zweig G (1992a) Middle-ear phenomenology: the view from the three windows. J Acoust Soc Am 92:1356–1369.

Shera C, Zweig G (1992b) Analyzing reverse middle-ear transmission: non-invasive *Gedankenexperiments*. J Acoust Soc Am 92:1371–1381.

Shera C, Zweig G (1992c) An empirical bound on the compressibility of the cochlea. J Acoust Soc Am 92:1382–1388.

Siebert WA (1970) Simple model of the impedance matching properties of the external ear. Quarterly Progress Report of the Research Laboratory of Electronics, Massachusetts Institute of Technology, Cambridge, MA, pp. 236–242.

Siebert WM (1973) Hearing and the ear. In: Brown JHU (ed) Engineering Principles in Physiology, Vol. 1. New York: Academic Press, pp. 139–184.

Silman S (1984) The Acoustic Reflex: Basic Principles and Clinical Applications. New York: Academic Press.

Stinson MR (1985) The spatial distribution of sound pressure within scaled replicas of the human ear. J Acoust Soc Am 78:1596–1602.

Stinson MR (1986) Spatial distribution of sound pressure in the ear canal. In: Allen JB, Hall JL, Hubbard A, Neely ST, Tubis A (eds) Peripheral Auditory Mechanisms. New York: Springer-Verlag, pp. 13–20.

Stinson MR, Khanna SM (1989) Specification of the geometry of the human ear canal for the prediction of sound-pressure level distribution. J Acoust Soc Am 85:2492–2503.

Teranishi R, Shaw EAG (1968) External ear acoustic models with simple geometry. J Acoust Soc Am 44:257–263.

Tonndorf J (1972) Bone conduction. In: Tobias JV (ed) Foundations of Auditory Theory, Vol. II. New York: Academic Press, pp. 197–237.

Tonndorf J, Khanna SM (1970) the role of the tympanic membrane in middle ear transmission. Ann Otol 79:734–753.

Tonndorf J, Khanna SM (1972) Tympanic-membrane vibrations in human cadaver ears studied by time-averaged holography. J Acoust Soc Am 52:1221–1233.

Tonndorf J, Khanna, SM (1976) Mechanics of the auditory system. In: Hinchcliffe R, Harrison D (eds) Scientific Foundations of Otolaryngology. London: Heineman, pp. 237–252.

Tonndorf J, Tabor JR (1962) Closure of the cochlear windows. Ann Otol Rhinol Laryngol 71:5–29.

von Békésy G (1960) Experiments in Hearing. New York: McGraw-Hill.

von Helmholtz HL (1877) The Sensation of Tones. (Translated by Ellis AJ, 1954). New York: Dover.

von Schwarz L (1943) Theorie der Beugung einer ebenen Schallwelle an der Kugel. Akust Z 8:91–117.

von Waetzmann E, Keibs L (1936) Theoretischer und experimenteller Vergleigh von Hörschwellenmessungen. Akust Z 1:1–12.

Wada H, Kobayashi T (1990) Dynamical behavior of the middle ear: theoretical study corresponding to measurement results obtained by a newly developed measuring apparatus. J Acoust Soc Am 87:237–245.

Wada H, Metoki T, Kobayashi T (1992) Analysis of dynamic behavior of human middle ear using a finite-element method. J Acoust Soc Am 92:3157–3168.

Wever EG, Lawrence M (1950) The acoustic pathways to the cochlea. J Acoust Soc Am 22:460–467.

Wever EG, Lawrence M (1954) Physiological Acoustics. Princeton: Princeton University Press.

Wiener FM (1947a) On the diffraction of a progressive sound wave by the human head. J Acoust Soc Am 19:143–146.

Wiener FM (1947b) Sound diffraction by rigid spheres and circular cylinders. J Acoust Soc Am 19:444–451.

Wiener FM, Ross DA (1946) The pressure distribution in the auditory canal in a progressive sound field. J Acoust Soc Am 18:401–408.

Wiener FM, Pfeiffer RR, Backus ASN (1966) On the sound pressure transformation by the head and auditory meatus of the cat. Acta Otolaryngol 61: 255–269.

Wightman FL, Kistler DJ (1989) Headphone simulation of free-field listening. I: Stimulus synthesis. J Acoust Soc Am 85:858–867.

Wullstein H (1956) The restoration of the function of the middle ear, in chronic otitis media. Ann Otol Rhinol Laryngol 65:1020–1041.

Zuercher JC, Carlson EV, Killion MC (1988) Small acoustic tubes: new approximations including isothermal and viscous effects. J Acoust Soc Am 83:1653–1660.

Zwillenberg D, Konkle DF, Saunders JC (1981) Measures of middle ear admittance during experimentally induced changes in middle-ear volume in hamster. Otolaryngol Head Neck Surg 89:856–860.

Zwislocki J (1962) Analysis of the middle-ear function. Part I: Input impedance. J Acoust Soc Am 34:1514–1523.

Zwislocki J (1963) Analysis of the middle ear function. Part II. Guinea-pig ear. J Acoust Soc Am 35:1034–1040.

Zwislocki J (1965) Analysis of some auditory characteristics. In: Luce RD, Bush RR, Galanter E (eds) Handbook of Mathematical Psychology, Vol. III. New York: Wiley, pp. 3–97.

Zwislocki J (1975) The role of the external and middle ear in sound transmission. In: Tower DB (ed) The Nervous System, Vol. 3, Human Communication and Its Disorders. New York: Raven Press, pp. 45–55.

3
Analysis and Synthesis of Cochlear Mechanical Function Using Models

ALLYN E. HUBBARD AND DAVID C. MOUNTAIN

1. Introduction

This chapter begins with a brief review of cochlear anatomy and physiology (see also Echteler, Fay, and Popper 1994), and then progresses through the modeling of the cochlea. We relate anatomy and physiology to models, thus broadening the base of potentially interested readers. Most state-of-the-art cochlear models include an embodiment of outer hair cells that have been shown to change length in response to transmembrane voltage. This electromotility is hypothesized to underlie a process of mechanical amplification that increases the ability of mammals to detect faint sounds by a hundredfold. Chapter 4 (Mountain and Hubbard) of this volume further considers the biophysics of hair cell motility, while in this chapter, the ability of hair cells to provide a feedback force in response to the motion of their own stereocilia is considered to be the engine that underlies the most sensitive aspects of mammalian hearing. This work is not exhaustive, but instead attempts to cover in some detail those models that are either of significant historical interest or represent the current state of the art in cochlear modeling. We attempt to interpret the current significance of both data and models for the reader.

2. Review of Cochlear Anatomy and Physiology

The mammalian cochlea consists of a fluid-filled duct that is coiled like a snail shell and is embedded in bone. Uncoiled, the cochlea is portrayed as having a longitudinal dimension, a vertical dimension, and a radial dimension that translates to the width of the duct when the cochlea is uncoiled. The cochlea is divided into three compartments, or scalae, by two longitudinal membranes. Figure 3.1 illustrates some of the key features of the anatomy. The upper membrane, Reissner's membrane, divides the scala vestibuli above from the scala media below while the lower membrane, the basilar membrane, separates the scala media from

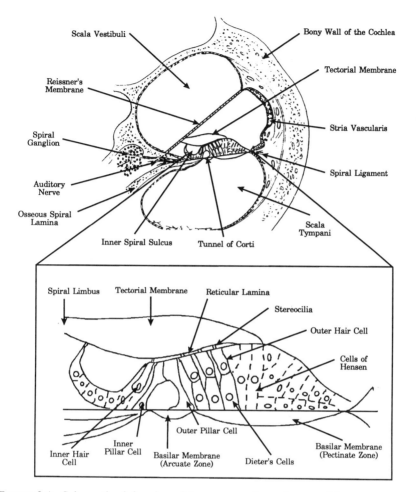

FIGURE 3.1. Schematized drawing of the cross section of one cochlear turn. *upper panel*, fluid-filled scalae and the organ of Corti; *lower panel*, magnified view of the organ of Corti.

the scala tympani. Reissner's membrane is very compliant, and has little impact on cochlear hydromechanics. Its main function is to serve as an ionic barrier between the scala vestibuli and the scala media. The fluid in the scala media has an electrical potential (endocochlear potential) that is 90 mV more positive than that of the scala vestibuli and the scala tympani. These ionic and potential gradients are maintained by specialized cells located in the stria vascularis, which makes up much of the outer wall of the scala media.

The basilar membrane and the organ of Corti, which sits upon it, comprise the cochlear partition. Today it is known that the cochlear partition is the key element that interacts with the pressures in the cochlear fluids to provide the mechanical filtering exhibited by the cochlea. The basilar membrane is attached to a bony shelf called the spiral lamina on its inner side and to a specialized tissue called the spiral ligament on its outer edge. The spiral ligament is attached to the bone by anchoring cells that contain a full complement of contractile proteins and may serve to create or maintain tension in the spiral ligament–basilar membrane complex (Henson et al. 1985). The width of the cochlear duct and the spiral lamina decreases from base to apex while the width of the basilar membrane increases from base to apex.

The cells of the organ of Corti are usually divided into two groups: the sensory cells (hair cells and nerve endings), and the supporting cells. One group of supporting cells, the pillar cells, form an arch that divides the organ of Corti along its radial (width) dimension into two regions, one containing a single row of inner hair cells (IHCs) and the other containing three or more rows of outer hair cells (OHCs). The pillar cells are quite rigid and appear to play an important structural role within the organ of Corti. The IHCs provide the input to approximately 95% of the auditory nerve fibers that have their cell bodies located in the spiral ganglion. The IHCs get their name from their location on the inner side of the cochlear spiral. The OHCs are located toward the outer side of the cochlear spiral and are believed to play a motor function in the mechanical performance of the cochlea. They receive significant efferent innervation from fibers that originate in the medial superior olivary nucleus of the brainstem, providing neural control of cochlear mechanics (Mountain 1980). The OHCs are always located at a significant angle to the basilar membrane and are approximately parallel to the outer pillar cells.

The basilar membrane is divided along its radial dimension into two regions, the arcuate and pectinate zones. The arcuate zone extends from the spiral lamina to the foot of the outer pillar cell, and the pectinate zone extends from the outer pillar cell to the spiral ligament. The force needed to displace a small (25-μm-diameter) region of the basilar membrane a constant distance has been measured as a function of radial position (Olson and Mountain 1991, 1993, 1994). The region of the basilar membrane directly under the foot of the outer pillar cells is stiffer than the regions to either side. The pectinate zone was found to be stiffer than the arcuate zone. These results suggest that the cellular elements of the organ of Corti provide much of the resistance to displacement of the basilar membrane because the more compliant region, the arcuate zone, adjoins the fluid-filled arch (or tunnel) of Corti whereas the stiffer pectinate zone is directly beneath the Dieter's and Henson's cells.

The apical membranes of the pillar cells, hair cells, and some other supporting cells are tightly joined together to form a platelike structure on the top of the organ of Corti called the reticular lamina. These tight junctions form an ionic barrier between the scala media and the organ of Corti. In contrast, the basilar membrane appears to be reasonably permeable to ions with the result that the ionic composition of the fluid within the organ of Corti is very much like that of perilymph, the fluid that fills the scala vestibuli and the scala tympani. One unusual group of cells that make up part of the reticular lamina are the Dieter's cells. Their cell bodies are located beneath the OHCs and form cups within which the OHCs are seated. Each Dieter's cell has a long thin process that projects to the reticular lamina. Each of these processes ends in a platelike structure, and these phalangeal processes are interdigitated between the OHCs (not shown in Fig. 3.1).

The limbus is a structure that protrudes from the upper surface of the spiral lamina and provides the major attachment point for the tectorial membrane. The tectorial membrane is an extracellular matrix that runs the length of the cochlea and extends from the limbus to slightly beyond the outermost row of OHCs. The tallest hairs of the OHCs appear to be embedded in the tectorial membrane (Engström and Engström 1978). Whether the IHC stereocilia are in contact with the tectorial membrane remains unresolved (Lim 1980). Another unresolved issue is whether the outer edge of the tectorial membrane is connected to the reticular lamina (Kronester-Frei 1979).

The stapes (see Rosowski, Chapter 2, this volume) provides the sound pressure input to the scala vestibuli from the middle ear. The stapes is located not at the basal end of the scala vestibuli but rather in the wall of the scala 1–2 mm from the base (Plassmann, Peetz, and Schmidt 1987). At the apical end of the cochlea, the scala vestibuli is connected to the scala tympani via the helicotrema, which equalizes the static pressures between the two scalae. Pressure relief for the scala tympani is provided by the round window, which is a compliant membrane that provides an opening to the middle ear cavity. The round window forms the wall of the basalmost 5%–10% of the scala tympani.

Next we identify key experimental data that models should be able to replicate. From one perspective, the data are very complex, and we present a cursory if not trivialized view. On the other hand, the data are sparse because it is extremely difficult to access the cochlea in good physiological condition and obtain measurements. Only a few locations can be accessed in various species. The bulk of what we know about the motion of the cochlea *in vivo* comes from two basal locations along the longitudinal dimension of the basilar membrane in the cat (Khanna and Leonard 1982; Rhode and Cooper 1993), one basal location in the squirrel monkey (Rhode 1971), one location in the guinea pig (Johnstone and Boyle 1967; Johnstone, Taylor, and Boyle 1970; Sellick, Patuzzi, and Johnstone 1982; Sellick, Yates, and Patuzzi 1983; Nuttall and Dolan 1993), one location in the chinchilla (Robles, Ruggero, and Rich 1986; Ruggero et al. 1990; Ruggero, Rich, and Recio 1992), and one location in the gerbil (Xue, Mountain, and Hubbard 1993). This motion is resolved into one dimension, nominally the up-down dimension, because of the nature of the measurement techniques. If the motion of the particular piece of basilar membrane under measurement moved in another dimension, it would not be detected. Moreover, the exact location of the probes on the basilar membrane is usually not reported. The fact that measurements are made on the basilar membrane rather than on the reticular lamina, where there is critical interaction between the sensory stereocilia and the adjacent moving structures, underscores the rather limited state of our knowledge.

The first report on the dynamic behavior of the basilar membrane is credited to von Békésy (summarized in von Békésy 1960, although the findings date to the 1920s). von Békésy stroboscopically observed a wave that traveled down the basilar membrane. Traveling waves were not predicted by models of the day, and the discovery was not anticipated. As the wave traveled down the cochlea, its forward velocity decreased, which meant that the distance between spatial peaks, the spatial wavelength, got shorter. As the wave slowed, its amplitude peaked, and this place of maximum response has come to be called the characteristic place. The characteristic place has an equivalent counterpart in the frequency domain: given fixed place, that place has a maximum response at a frequency called the characteristic frequency (CF). More basal regions of the cochlea have higher CFs, while the characteristic place of lower frequencies shifts progressively toward the apex.

When the auditory nerve was found to possess sharply tuned behavior (Galambos and Davis 1943) that was not accounted for by the tuning found in the basilar membrane, investigators presumed that an unknown filtering process existed between the mechanical cochlea and the electrical behavior found in the nerve. Thus there developed a search for a mechanism of neural or mechanical origin known as the second filter. The first filter, of course, was the basilar membrane tuning as described by von Békésy.

The second filter search was essentially abandoned as it became evident over a period of decades that the motion of the basilar membrane accounts for much of the behavior demonstrable in the auditory nerve. Using the Mössbauer technique, Johnstone and Boyle (1967) initially determined that the ratio of basilar membrane velocity to stapes velocity was "consistent" with the findings of von Békésy. (The reader should keep in mind that von Békésy worked at low frequencies, e.g., 50–300 Hz in the cochlear apex, while the majority of later measurements were made near the base at much higher frequencies.) The Mössbauer technique is a velocity-sensitive measure made possible by the motion of a tiny radioactive source whose gamma radiation is Doppler-shifted by the motion of the source. Johnstone, Taylor, and Boyle (1970) later (see Fig. 3.2) determined the basilar membrane was more sharply tuned than had

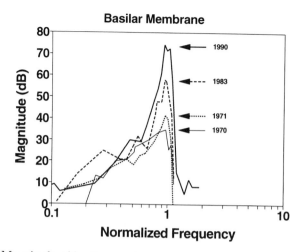

FIGURE 3.2. Magnitude of basilar membrane responses versus stapes from several experimental studies over many years. Data plotted with *thin solid lines* are from Johnstone, Taylor, and Boyle (1970); with *dots*, from Rhode (1971); and with dashes, from Sellick, Patuzzi, and Johnstone (1983). The top curve (*heavy solid line*) is an average of five sets of data while other curves are from individual animals from Ruggero et al. (1990).

been claimed by von Békésy. The region of sharp tuning came to be called the "tip." Rhode (1971), in the squirrel monkey (Fig. 3.2), also showed that the pectinate zone was much more sharply tuned than claimed by von Békésy. Moreover, the velocity response of the basilar membrane demonstrated nonlinear growth with increasing sound level. Previously the motion was thought to grow linearly as a function of increasing sound level, and the new findings were viewed with skepticism for some time, especially since the Mössbauer technique is nonlinear and requires a linearizing correction when used for determining velocity. Also using the Mössbauer technique, Sellick and his colleagues (Sellick, Patuzzi, and Johnstone 1982; Sellick, Yates, and Patuzzi 1983) reported velocity responses from the guinea pig that showed ratios of basilar membrane velocity to stapes velocity upward of 55 dB (see Fig. 3.2). Khanna and Leonard (1982), using a laser interferometer (which measures small displacements by making use of interference patterns), also found sharp tuning in the cat basilar membrane. Robles, Ruggero, and Rich (1986) reported data acquired using the Mössbauer technique from the chinchilla that showed basilar membrane/stapes ratios as high as 75 dB (summarized in Ruggero et al. 1990; see Fig. 3.2).

In summary, the measured sizes of the peaks in the frequency–response curves were growing in such a way as to begin to account for auditory nerve tuning. The peaks becoming larger probably resulted from both the physiological condition of the animal studied and the use of lower sound levels. Good data obtained at low sound levels show higher sensitivity and sharper tuning in the tips than data obtained at high sound levels. Species differences may also account for some variations in the data.

Phase angle data are given in Figure 3.3. There is no systematic change that correlates with the systematic changes in the magnitude data. We believe from examining all available phase data to date (not just Fig. 3.3 data) that "good" phase data are quite flat or show gradual lags (especially in the primate) versus frequency below about 0.9 CF. Then there is a rather abrupt breakpoint at which steeper slopes occur. The phase continues to lag until hitting a maximum lag (plateau) whose value varies considerably. Rhode's (1971) phase (Fig. 3.3) demonstrates a gradual drop in phase at lower frequencies but breaks sharply around CF and finally falls to around 1300°. Sellick, Yates, and Patuzzi's (1983) phase shows slight leads through frequencies below the CF and drops to around 900° lag at the phase plateau. The Ruggero et al. (1990) data show a nearly flat phase angle up until the CF that then drops to approximately 900° and alternatively, one cycle more at about 1260°. Rhode (1978, data not shown here) once found a phase plateau of about 2000°.

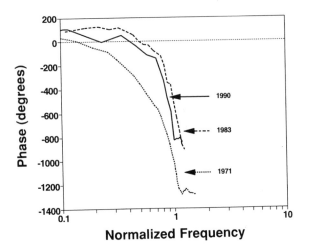

FIGURE 3.3. Phase angle of basilar membrane responses versus stapes from studies shown in Figure 3.2. The Rhode (1971) phase curve (*dotted line*) does not correspond to the magnitude response shown in Figure 3.3; rather it was from a different animal in the same study; the Sellick, Patuzzi, and Johnstone (1983) phase curves (*dashed line*) correspond to the magnitude data shown in Figure 2.2; the Ruggero et al. (1990) phase responses (*solid line*) are from two of the animals used in the average magnitude response shown in Figure 3.2. The phase responses from the remaining animals fall between the two curves shown. No phase responses were reported in the study of Johnstone, Taylor, and Boyle (1970).

The phase characteristics beyond CF are difficult to nail down because the magnitude of the velocity response becomes small and therefore the signal-to-noise ratio becomes poor. Additionally, measurements must be made at very small frequency increments to correctly interpret phase angles that are measured to be between 0° and 360°. A typical rule is to interpret less than 180° of phase drop between adjacent frequency sample points as a lag; more than 180° lag should then be considered a lead. For example, 270° lag would be interpreted as a 90° lead. However, being strongly influenced by the traveling-wave theory of the cochlea, which only produces lags, auditory investigators might plot the observed 270° shift as a lag under the presumption that the frequency increment used in the experiment was simply too great.

Figure 3.4 shows multiple velocity-versus-frequency curves obtained at varying sound levels (Ruggero, Rich, and Recio 1992). The data are scaled relative to input sound pressure such that segments of curves that lie over each other must be linearly increasing with sound pressure. The relative size of the tip, and the "tip-to-tail" ratio, decrease with increasing sound level, while the tail itself remains at a nearly constant relative response level. The frequency at which the maximum response occurs shifts to lower frequencies with increasing sound level. The high sound level condition is taken to be the response of the passive mechanics of the

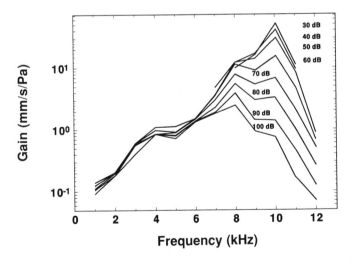

FIGURE 3.4. Basilar membrane velocity scaled by input sound pressure for a number of sound levels. Vertical overlap of the curves indicates linearity of the response. At lower frequencies, the curves align within approximately a factor of 2, and are characterized as being essentially linear. By contrast, in the region of the characteristic frequency (CF), the response is remarkably nonlinear. (Data from Ruggero, Rich, and Recio 1992.)

cochlear partition, while the low sound level condition that produces large tips demonstrates an active process thought to be produced by the OHCs. Clearly the tip region of the response demonstrates marked nonlinearity while the tail is nominally linear.

Figure 3.5A shows basilar membrane velocity (Ruggero and Rich 1991b) versus sound pressure level (SPL) when the cochlea is in "normal" physiological condition and when the endocochlear potential is reduced using the loop diuretic furosemide. The two input–output curves under normal conditions are a recapitulation of the obvious aspects of Figure 3.4, namely that the tip is linear and the tail is nonlinear. Before furosemide administration, the CF curve shows approximately 40 μm/sec response at 20 decibels (dB) SPL, marked nonlinearity between 40 and 80 dB SPL, and a linear region above 90 dB SPL. In the time period 11–19 min after furosemide is injected intravenously, the CF curve shifts to the right such that 40 μm/sec is achieved at approximately 46 dB SPL; the curve remains linear and nearly aligns with the "before" curve at high sound levels. By comparison, velocity responses at 1000 Hz increase linearly with sound level and are less affected by furosemide. Lowering endocochlear potential has the apparent effect of lowering the "gain" of the hypothesized cochlear amplifier, as seen in the shift to higher sound levels of the "after" curve at CF.

The same kind of shift (Fig. 3.5B) is seen in the response to one (probe) tone in the presence of a second tone called the suppressor. The reduced response to the probe is called two-tone suppression. Figure 3.6 shows that the growth of suppression, that is, the separation between the responses along the abscissa in Figure 3.5, differs for suppressor tones above versus below the CF of the probe tone. Suppressors at frequencies above CF have a lower threshold of suppression than suppressors below CF. However, the growth of suppression for suppressor frequencies above CF is of the order of 0.36 dB per dB (Ruggero, Robles, and Rich 1992). Suppressors having frequencies below CF have a higher threshold of suppression, and suppression increases at an average of 0.96 dB per dB.

Another phenomenon produced by the nonlinear interactions between two tones are distortion-product tones at frequencies other than those presented. In particular, for two tones, f_1 and f_2, the new tones are at frequencies $nf_1 \pm mf_2$, for all integer n and m. Two-tone distortion is the subject of a vast body of neural and psychoacoustic literature; far less is presently known about the mechanical response of the basilar membrane as a result of direct measurement. The issue is whether the distortion products exist at the level of the basilar membrane or whether they arise from other processes, for example, neural. Two early studies (Wilson and Johnstone 1973; Rhode 1977) failed to confirm the distortion product, $2f_1 - f_2$, in the response of the basilar membrane, partially because of nonlinearity inherent to the Mössbauer technique. We now know (Nuttall,

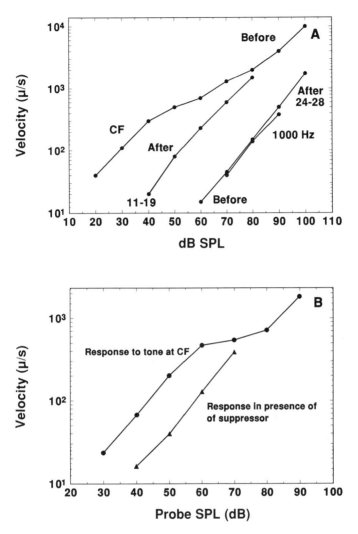

FIGURE 3.5A,B. (A) Effects of furosemide on cochlear sensitivity as reflected in basilar membrane velocity. Furosemide shifts the input–output curve at CF to the right (*After*) in the time period 11–19 min after furosemide injection; sensitivity (responsiveness) at lower sound levels in reduced and the curve is more linear than before furosemide delivery. Furosemide has little effect at 1000 Hz. (Data from Ruggero and Rich 1991b.) (B) Suppressive effects on the response to a tone at CF (8 kHz) by a 10-kHz tone at 80 dB (sound pressure level) (SPL). The effect of the second tone is to shift the response to the CF tone to the right on the abscissa. The amount of shift is nominally the size in decibels (dBs) of the suppressive effect. (Data from Ruggero, Rich, and Recio 1992.)

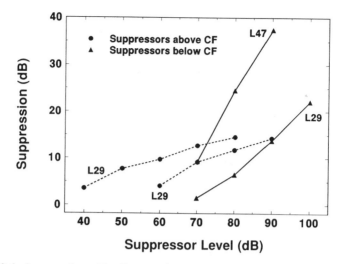

FIGURE 3.6. Suppression of basilar membrane velocity response (see Fig. 3.5B) as a function of suppressor tone level (dB) for suppressors above (*dots, dotted line*) and below (*triangles, solid line*) the CF. Suppressor tones below CF are associated with faster growth of suppression and greater maximum suppression. Suppression was computed as the horizontal shift in input–output curves at an average velocity of 126 μm/sec. (Data from Ruggero, Rich, and Recio 1992.)

Dolan, and Avinash 1990; Robles, Ruggero, and Rich 1991; Rhode and Cooper 1993) that various distortion products including $2f_1 - f_2$ exist at the level of the basilar membrane. They are generated in the cochlea and travel from their site of generation to their own characteristic place. These measurements have been made using various laser interferometry techniques that determine velocity or displacement of a reflecting bead placed on the basilar membrane. The laser velocimeter (Ruggero and Rich 1991a) has also been used to acquire data on the motion of the basilar membrane under impulsive sound excitation (Ruggero, Rich, and Recio 1992), which, when transformed into the frequency domain, resembles but does not duplicate the responses shown in Figure 3.4.

The International Team for Ear Research (ITER) (1989) developed a laser interferometer coupled to a confocal microscope that allowed reconstruction of the motion of selected spatial regions of explanted cochleae. They reported that different portions of the organ of Corti were moving with different motions. The picture of the cochlea as developed by this group continues to emerge (Khanna, Flock, and Ulfendahl 1989; Khanna, Ulfendahl, and Flock 1989, 1993; Brundin et al. 1991; Ulfendahl, Khanna, and Flock 1991; Teich et al. 1993; Ulfendahl and Khanna 1993; Ulfendahl, Khanna, and Lofstrand 1993). One goal of the investigators is to obtain data on the motion of various parts of the organ of Corti, for example, the reticular lamina, the tectorial membrane, and the basilar membrane. In contrast with almost all other measurements on the basilar membrane, which are near the cochlear base, the ITER group is working in the apex.

Recent work involving electrical stimulation of the cochlea sheds light on the forces and mechanical properties as well as the motion of individual portions of the organ of Corti. Xue, Mountain, and Hubbard (1993) measured the electrically evoked motion of the basilar membrane *in vivo*. By comparing these data with the stiffness measurements of Olson and Mountain (1991), they arrived at an estimate of the force per OHC of the order of 0.2 nN/mV receptor potential. They also found that the arcuate and pectinate zones move out of phase over a range of frequencies. Mammano and Ashmore (1993) found that when hair cells are stimulated electrically, the reticular lamina and basilar membrane move out of phase. *In vitro*, the reticular lamina moves much more than the basilar membrane. The experimental finding implies that the reticular lamina is much less stiff than the basilar membrane. Therefore under normal sound excitation when the hair cells are thought to be actively pushing within the organ of Corti and providing amplification of the motion caused by sound, the motion of the reticular lamina is probably very different from the motion of the basilar membrane.

The state of cochlear data is continually evolving. At this time, we believe that reasonable baseline data by which mechanical models should be judged are those just described. Unfortunately, to date no model has been subjected to the battery of validation tests we recommend. We think that in the near future target data should include the motion of various parts of the organ of Corti. In addition, models must incorporate realistic OHC force production as well as all known mechanical parameters.

3. Goals and Rationale

The goal of cochlear modeling is to advance the field of auditory science. Modeling is a potential shortcut to understanding how the cochlea works before it is completely characterized by data: the latter could require significant time. On the whole, cochlear modeling has been driven by experimental data. As more and better data became available, models have changed considerably (see Hawkins and McMullen, Chapter 1, this volume).

Computational modeling of the cochlea serves a number of specific purposes. The validation of existing hypotheses is of prime importance. Everyone builds conceptual models as a means of translating the reality (e.g., the ear works well) into a structure that the mind perceives as an explanation (e.g., it works like an amplifier). The computational model solidifies the conceptual model such that responses can be computed and compared against existing data. On the basis of comparison, the model validates or fails to validate the hypothesis. If a model fits an existing body of data well, then the model can be used to predict what can happen under an unexplored set of stimulus/measurement conditions. The predictions lead to new experiments, the results of which further validate or disprove the model. This cycling of challenges advances the field in a quantitative systematic way.

Models can also be used to interpret data. If the model fits the data, then the mechanism of the model might be analogous to the mechanism that produced the experimental data. We can look inside the model to understand how it works. The alternative is that the model has nothing to do with the underlying biophysics. A modeler should not presuppose that a solution is unique or necessarily related to the actual mechanism.

Finally, models should challenge the current state of understanding of the field. From the modeling side, the challenge is for experimentalists to obtain information regarding motion within the cochlea and physical parameters. From the experimental side, the challenge comes in the form of data that are exceedingly difficult to reproduce. Unfortunately, models can often reproduce *certain* aspects of experimental data, and therefore many different models can claim to be correct. What modelers need from experimentalists are data that rule out individual models if not general classes of models. Today, there are too many potentially viable models and relatively few that are absolutely unacceptable.

4. Modeling the Cochlea

We consider the cochlea from several broad perspectives. There are models that seek to explain how the cochlea works, and these are biophysically based to a greater or lesser degree. Of the biophysical models, there are two types: the macromechanical and the micromechanical. The principal effort in cochlear modeling has been directed toward biophysical models. Other models have been created with the goal of efficiently mimicking cochlear-like responses for use in applications such as speech processing or cochlear prosthetics.

This chapter is concerned principally with biophysical models. Macromechanical models are biophysical models in which the fluid of the cochlea plays a major role, while the characteristics of the cochlear partition are abstracted to a few mechanical parameters. This perspective dominated early work and persisted well into the 1980s. We shall consider the linear one-, two-, and three-dimensional hydromechanical situations. We also consider passive and active nonlinear macromechanical models. Passive means that the model only dissipates energy that comes in via the stapes, while active means that the model can be energy producing and thereby dissipates more energy than enters via the stapes. The biological basis for the active models is that the cochlear OHCs carry out both reverse (i.e., electrical-to-mechanical) transduction (Hubbard and Mountain 1983; Brownell et al. 1985) and forward transduction (see Davis 1983; Kim 1986a,b; Dallos 1992 for discussions of the "active cochlea"; Mountain and Hubbard, Chapter 4, this volume). In the micromechanical models, fluid is still significant, but the mechanical details of the organ of Corti play the major role. Today all investigators recognize that the organ of Corti is capable of deforming in various ways owing to its diverse subcomponents. These submotions are considered to be *degrees of freedom*, for simplicity usually excited by a one-dimensional hydrodynamic environment. In multimode models, additional wave propagation modes are possible. Multimode models may be either micromechanical or macromechanical.

4.1 Macromechanical Models

The cochlea is obviously a three-dimensional structure with considerable detail found in the partition separating the major channels, the scala tympani, and the scala vestibuli. But rather surprisingly, much time has been devoted to studying principally the fluid or simple boundary interactions in one, two, and three dimensions. Here, following the style attributable to the classic literature, dimensionality refers to the representation in terms of the cochlear encasement, and there is a clear ordering of importance. The first dimension is the length down the cochlea. The second dimension is the fluid height above and below the cochlear partition: without this metric, the pressure in the scalae could not vary at different locations above the cochlear partition, and there could be no vertical variation in the flow of the fluid. In particular, the representation of the cochlea in a region near the best frequency, that is, where it responds maximally to sound stimulation, requires on the basis of hydrodynamic principles (Lighthill 1981, 1991) that the fluid flows vary with distance above the cochlear partition. The third dimension is radial. This is the dimension across the width of the cochlear partition, but is considered to be "radial" because the cochlea itself is wrapped in a spiral. Variations in pressure and flow in the radial dimension have not been thought to be of considerable importance, although this aspect is of potential interest for a number of reasons involving micromechanics. Dimensionality is often confused with degrees of freedom of motion, which we reserve to refer to the motion of a particular element of the cochlear partition resolved along a particular coordinate axis. For example, the cochlear partition as a whole might be moving principally in the vertical dimension while the tectorial membrane moves radially. However, the hydrodynamic forces operating on the structures could be calculated using a one-, two-, or three-dimensional representation of the fluid. Propagation modes are different: modes of propagation are associated

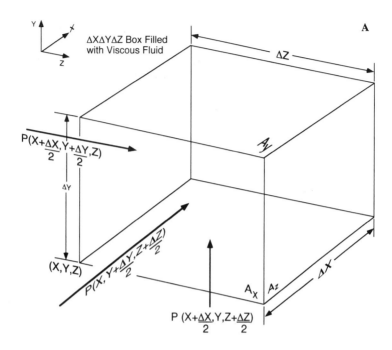

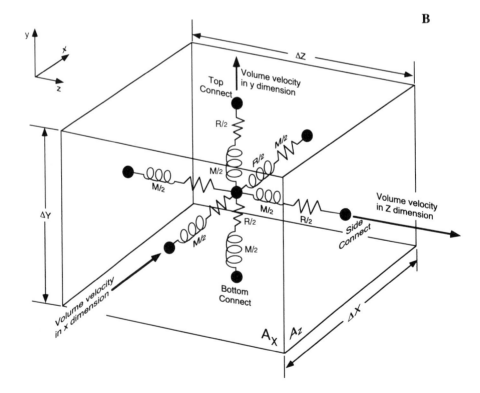

with velocities of a particular kind of wave propagation, for example, the speed of a planar sound wave in water.

We have chosen to present hydromechanical cochlear models chiefly in the framework of analogous electrical circuits of the impedance type. This particular analogy results simultaneously in a circuit topology and a picture that can be easily understood. The circuit diagrams specify the equations of the system in difference form. The widespread availability of circuit simulation software makes it easier to simulate the circuits than to solve the differential equations of the system. Moreover, such simulation tools offer the potential to include nonlinear or "active" elements that may confound mathematical solution.

Consider the three-dimensional volume depicted in Figure 3.7A to be filled with viscous, compressible fluid having a certain density. The density allows one to calculate the mass when the volume is known. Consider a pressure on the front face denoted by the arrow, and a different pressure on the back face at location $(x + \Delta x)$. This pressure difference will act on the fluid, tending to move it in the x direction if it were constrained from moving in other directions. In the simplest case in which the fluid has only the property of mass, pressure times area is a force, and force is equal to mass times acceleration ($F = ma$). What actually moves is the volume of fluid, and so an appropriate variable in this acoustic system is volume velocity, that is, how much volume moves how fast. Similarly, the mass would be called an acoustic mass, the

---

FIGURE 3.7A,B. (A) Quantal-sized box containing viscous fluid has a pressure acting on each face. The pressures are indicated as *directed arrows* (not intended to represent vectors because pressure is a scalar quantity) acting at different locations. (B) In the acoustic impedance analogy, pressure at a node corresponds to voltage. At each *large black dot* is a voltage that is analogous to the pressures depicted in A. In the electrical analog are depicted electrical currents (*arrows*) that are analogous to fluid flows. Hence the amount of fluid accumulating in the box is zero, as is the sum of currents at the center node. Inductance elements represent fluid mass and resistance elements represent acoustic resistance, that is, fluid viscosity. The elements are divided in half so that a node can be placed in the center of the box, and the fluid properties in any dimension are represented by M and R. The formulation shown using identical Ms and Rs presupposes that the box is a cube, although in simulations the boxes could be sized differently in various dimensions. In three dimensions, boxes could be placed side by side and on top of each other to form a network. In two dimensions, boxes could be stacked in the x- and y dimensions, leaving the terminals in the z dimension open circuited so that no current can flow. This would be the equivalent of placing a rigid constraint on the faces in A that are orthogonal to the z axis. Similarly, the circuits could be placed in a line down the x axis, thus representing a one-dimensional fluid; this would be the equivalent of adding rigid faces orthogonal both the z- and y axes in A.

damping an acoustic resistance, and the compressibility an acoustic compliance. The latter is the reciprocal of stiffness. We could attach a mechanical element at the boundary perpendicular to the y dimension if it were a solid plate, for example, as in the case of a mechanical piston interfacing a fluid-filled chamber. When force is applied via the piston, a pressure is created. Scaling the mechanical force over the area results in a pressure, which is an acoustic unit. Hence, mechanical elements can readily be represented in the acoustic analogy by way of scaling over an appropriate area.

Just as one can work in acoustic rather than mechanical units, we can also transform variables into the electrical domain in a way so that the original governing differential equations are unchanged. Hence the different representations are called analogous systems. A natural analogy is that pressure is the analog of voltage and volume velocity is the analog of electric current. This "acoustic impedance" analog carries with it the benefit that the picture is intuitive: that is, volume flows through and pressure acts upon surfaces. Thus, the box of Figure 3.7A can be redrawn as the electrical circuit of Figure 3.7B. Acoustic mass becomes an electrical inductance. Acoustic damping is an electrical resistor. Notice that we have centered a node in the box and drawn the elements as symmetric with respect to each axis, but when many such sections are concatenated, it makes no difference whether the node is centered or at the edges of an incremental section. An acoustic compliance representing the compressibility of the fluid would be (not shown) a capacitor referenced to ground (ambient) potential.

4.1.1 One-Dimensional Model

The one-dimensional perspective (Peterson and Bogert 1950; Zwislocki 1950) is the simplest to understand and to implement as a computational model. The one-dimensional model can be simply derived using the acoustic analogy. Figure 3.8 shows an acoustic impedance analog of the cochlea superimposed over the cochlear scalae. Notice that a segment of length Δx is depicted as a single collection of elements at location x. The incremental aspects of the vertical and width dimensions have been suppressed, which is equivalent to saying that a particular variable as a function of x is constant from top to bottom and side to side. The acoustic impedance of a Δx segment of fluid in the channels is represented as an inertial term:

$$L_S(x) = \rho \Delta x / A \tag{1}$$

where A is the area of the scala and ρ is the fluid density. A spring, $C_{CP}(x)$, a mass, $L_{CP}(x)$, and a damper, $R_{CP}(x)$, referenced over the surface area of the incremental portion, represent the acoustic impedance of the cochlear partition x. The capacitors to ground extending from each

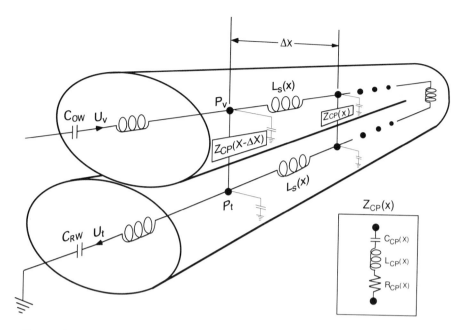

FIGURE 3.8. A schematic of the bony shell of the cochlea encapsulates a circuit representation of the acoustic properties of the system. P_t and P_v are voltage signals that are analogous to pressure in the scala tympani and the scala vestibuli, respectively. Currents (U) represent volume velocities into and out of the cochlea through the capacitors that represent the compliances of the oval (OW) and round window (RW). The fluid flows and pressures are represented in the x dimension only. To reduce detail, the cochlear partition impedance is represented as $Z_{CP}(x)$. The *inset* shows that the properties of the cochlear partition can be represented as acoustic compliance, viscosity, and mass, which are equivalent to capacitance, resistance, and inductance, respectively, in the electrical analog. The small capacitances to ground represent compressibility of the cochlear fluid. The inductance at the far end of the circuit represents the fluid acoustic mass in the region of the helicotrema, a small hole that connects the scala vestibuli to the scala tympani.

node in the cochlear scalae represent compressibility of the cochlear fluid. These are rarely included in models today, as the fluid is nominally incompressible. However, the effects of fluid compressibility may be important above approximately 7 kHz (Geisler and Hubbard 1972), and many experimental data have been obtained at such frequencies. As models are improved so that they better approximate data, second-order effects that were previously omitted might have to be included to account for small discrepancies.

In a one-dimensional model, parameters can be easily chosen using ways that make sense anatomically, physiologically, and mechanically.

Notice that there are three acoustic masses, one in the upper channel, another in the lower, and one in the cross-branch, $Z_{CP}(x)$. The acoustic masses associated with the channel can be calculated directly from the known area of the scalae as in Eq. 1. The acoustic mass $L_{CP}(x)$ in the middle branch represents the mass of the cochlear partition. It can be estimated in the same way as the scala mass (Eq. 1) except that the volume in question moves vertically rather than down the scala, and its surface size is approximately the surface area of the cochlear partition in an incremental section. Then the tuning function, $F_{CF}(x)$ (Greenwood 1961, 1990) allows one to compute the stiffness of the cochlear partition by solving the equation for resonance for the variable $C_{CP}(x)$. Accordingly:

$$C_{CP}(x) = 4\pi^2/(L^2_{CP}(x) \cdot F_{CF}(x)) \tag{2}$$

where $C_{CP}(x)$ is the acoustic compliance and $F_{CP}(x)$ is given by the Greenwood map.

The range of CFs on the basilar membrane spans roughly two to three orders of magnitude, and as a consequence the stiffness parameter varies over four to six orders of magnitude. This is difficult to reconcile because the cochlear partition does not appear to change so drastically from base to apex. To give the reader a feeling for this variation, low-density polyethylene is about four orders of magnitude more compliant than tungsten (Askeland 1989).

It is common for models to be formulated in terms of the pressure difference between upper and lower scalae at the same x location. This can be accomplished provided one wishes to compute only the current through the cross-branch, which is the analog of cochlear partition volume velocity. Obviously, it is this variable that is of high interest. The pressure difference construct is attributed to Peterson and Bogert (1950), who split the pressure waves into a $P^+(x)$ wave that was the average of $P_v(x)$ and $P_t(x)$ and traveled at the velocity of a plane wave in the fluid, and a pressure difference wave, $P_-(x)$, that was the difference between $P_v(x)$ and $P_t(x)$. The P^+ wave could be solved analytically, while the P_- wave required numerical techniques to calculate. Peterson and Bogert noted the analog circuit formulation as an alternative solution for the P_- wave.

Incremental springs and masses, or alternatively, capacitors and inductors, connected in a chain-ladder arrangement form transmission line structures that propagate waves: this basic construction holds by analogy for electromagnetic waves, hydrodynamic waves, and mechanical waves. In Figure 3.8, one can recognize the scala mass and compliance (compressibility) to ground as constituting one chain ladder, and from these elements arises one propagation mode. Because the compressibility is very small, the P_v (also P_t and P^+) wave is a *fast propagation mode*, traveling essentially at the speed of sound in water. The P_- wave that depends on the mass of the scala and the compliance of the cochlear partition is a *slow propagation mode* because of the relatively high compliance of the cochlear partition. Hubbard and Geisler (1972) combined the P_- wave from an analog computer model with an analytical solution for the P^+ wave having new boundary conditions (Geisler and Hubbard 1972), and used (Geisler and Hubbard 1975) the formulation in conjunction with a middle-ear model to calculate P_v and P_t and approximately replicate Nedzelnitski's (Nedzelnitski 1974, 1980) pressure measurements that were referenced to eardrum pressure. Since that time, almost all model formulations have been according to the pressure difference scheme, discarding the P^+ solution that is necessary to recover P_t and P_v.

Results from the one-dimensional model were somewhat satisfactory in terms of the early data. As better data emerged, the passive model's damping was reduced to push up its peak response. A result from the passive model using a Q of 300 is shown in Figure 3.9. The quality factor, Q, generally denotes how underdamped a system is. A Q factor of unity or less means that the system does not ring. Data from Ruggero et al. (1990) are also plotted. The experimental data are not actually "sharply" tuned. In fact, it took some time for the field to appreciate that for the most part the newer data showed high cochlear sensitivity (high peaks) that were rather broadly tuned. High, relatively broad peaks were the

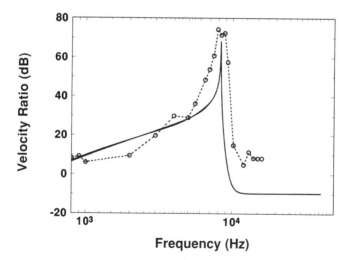

Frequency (Hz)

FIGURE 3.9. Magnitude results from a passive one-dimensional model (*solid line*) with a Q (quality factor) of 300 are compared with data (*dashed line*) from Ruggero et al. (1990). The model was computed using 2000 sections to eliminate most of the fine "ripples" that are an artifact of computing this difficult case in which an abrupt change in velocity occurs over a narrow region on the simulated cochlear partition. The experimental curve was obtained by averaging similar data from five different animals.

Achilles' heel of the classical model (de Boer 1983a). To produce high peaks, the model peaks became unrealistically narrow. The phase angle of the model's response shows reasonable agreement with the experimental data but does not show enough total lag (Fig. 3.10). Moreover, the rapid phase drops observed in the data imply short spatial wavelengths on the basilar membrane. Theoretically (Lighthill 1981, 1991), such short wavelengths necessitate a two-dimensional formulation so that the model corresponds to the physical system intended.

4.1.2 Two-Dimensional Model

The two-dimensional hydrodynamic model takes into account both the distance down the cochlea and the height of the fluid above and below the cochlear partition. A circuit formulation of the two-dimensional case is shown in Figure 3.11. Here we partition the fluid above and below the cochlear partition into additional subdivisions, resulting in a topology that allows different fluid flows in the longitudinal dimension as a function of distance from the cochlear partition. Notice that at the boundary nodes which represent the cochlear bone, the current paths are open circuited, indicating that no flow occurs at the bony periphery. This is the equivalent of a boundary condition in the differential equations governing the system.

Let us examine the circuit. The stapes volume velocity input is represented by the directed arrows at y coordinates located above the cochlear partition. Assuming the stapes moves as a piston, these nodes all receive

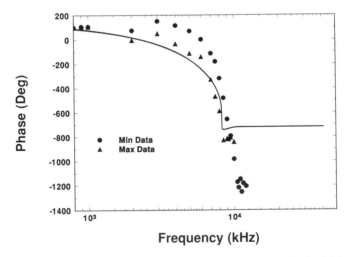

FIGURE 3.10. Phase angles from a passive one-dimensional model (*solid line*) with a Q of 300 are compared with data from Ruggero et al. (1990). The two sets of experimental data represent the maximum (*triangles*) and minimum (*circles*) phase curves found in data from five animals described in Figure 3.9.

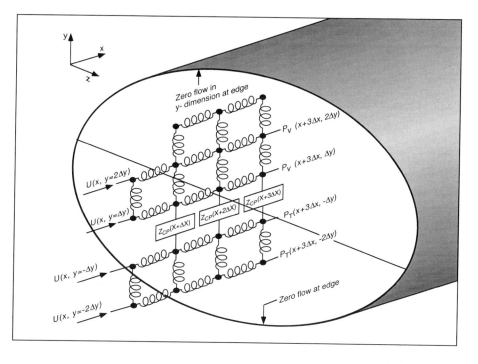

FIGURE 3.11. A two-dimensional fluid model in electrical analog (acoustic impedance analogy) form sketched within a representation of the bony walls of the cochlea. The fluid flows and pressures are represented in the x and y dimensions. To reduce detail, the cochlear partition impedance is represented as $Z_{CP}(x)$. It is the same impedance as depicted in Figure 3.8. The *arrows* pointing into the circuit on the left face are currents. The zero coordinate in the y dimension is considered to be at the cochlear partition. The representation allows differing fluid flows in both the x and the y directions. In general, there are current flows in all branches, through all elements. When a wire is left open circuit, no current flows; to do this in the circuit thus amounts to a boundary condition of zero flow. In the case that the wave propagating down the cochlea is planar, which occurs for low frequencies in basal regions, the two-dimensional model reduces to a one-dimensional model.

equal volume velocities. Thus signals that begin to propagate down the scalae are planar in nature; that is, at a particular location slightly down the cochlea from the stapes, the pressures and flows are equal at every height coordinate. At frequencies below resonance, little volume velocity is present in the cross-branch that represents the cochlear partition. Hence, there will be little vertical flow of fluid.

Near the resonant location of the cochlear partition impedance, however, there is relatively large flow in the cochlear partition branch because the cross-branch has a low impedance. If the impedance were zero, all the volume velocity would be short circuited across the cochlear partition branch. In a similar but less definitive fashion, the fluid in the scala vestibuli moves toward the cochlear partition in the case of a low but nonzero impedance that occurs at resonance. However, all flow is not directed through the cochlear partition branch, and instead there is a concentration of flow closer to the cochlear partition as compared with that farther above it. This is the equivalent of saying that the pressure wave in the scala vestibuli is no longer planar, that is, it takes more than the x coordinate to describe the pressure or flow.

Consider the two-dimensional model compared to the one-dimensional model: as the forward propagating wave makes use of less than the full scalae cross section, the one-dimensional equivalent scala mass term (see Fig. 3.8) increases because the area term (see Eq. 1) decreases. The ratio of scala mass to the cochlear partition mass is a critical factor in determining the steepness of the high-frequency fall-off in simulated cochlear partition response. A way to visualize this effect is to realize that the cochlear partition impedance becomes mass dominated beyond the resonant frequency. Because the scalae impedance is always mass dominated, it becomes a contest whether the volume velocity travels down the cochlea or goes through the cochlear partition branch. Translated into the frequency domain, the high-frequency slope becomes steeper as the scala impedance dominates. Therefore, it follows that the two-dimensional model cuts off more sharply than the one-dimensional model.

The first serious mathematical formulation of the (von Békésy) cochlea was two dimensional. Ranke (1931, 1950) resorted to highly simplifying assumptions and was able to find an analytical solution for the cochlear partition in a fluid of infinite depth. Siebert (1974) found closed-form solutions for the short-wave formulation that compared favorably with magnitude data but poorly with the phase data of Rhode (1971). Lesser and Berkley (1972) found a two-dimensional solution for a cochlear partition in a rectangular constraint, but did not show comparisons with basilar membrane tuning data. Allen (1977) showed results from the Lesser and Berkley (1972) formulation of the cochlea that matched fairly well the magnitude data available at the time. Neely (Neely 1981a; Neely and Kim 1983, 1986) checked two-dimensional against one-dimensional hydrodynamic formulations, and concluded that the two-dimensional effort slightly changed the results of greatest interest, which at that time were micromechanical not hydrodynamic. However, from a hydrodynamic point of view (Lighthill 1981, 1991), the two-dimensional formulation is of fundamental importance in the region of the cochlear partition where the traveling wave peaks.

4.1.3 Three-Dimensional Model

The three-dimensional formulation is depicted in Figure 3.12. Only mass elements are used to represent the fluid, for the sake of simplicity. An

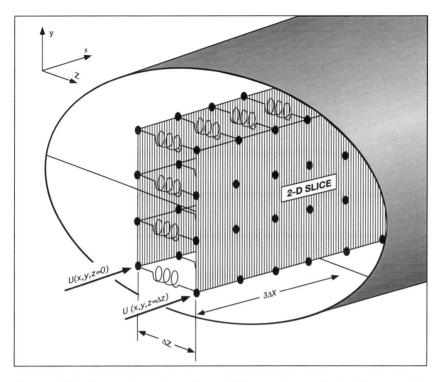

FIGURE 3.12. Planar circuits from Figure 3.11 are placed side by side across the width of the cochlea. The circuits from Figure 3.11 are simplified such that only nodes (*block circles*) are preserved. Nodes in the circuit are interconnected in the z dimension using acoustic masses on a one-to-one correspondence with nodes in the x- and y dimensions. Only two *arrows* representing volume velocity are shown to reduce clutter. The three quantized units depicted in the x dimension are for orientation purposes only.

opportunity afforded by this model is that the pressure and fluid flows can vary across the width (the radial dimension) of the cochlear partition. This potential variation is consistent with the notion that the arcuate and pectinate regions of the basilar membrane have different properties. In light of the obvious possibilities afforded by the three-dimensional formulation, it may surprise the reader that early work on the three-dimensional model ignored three-dimensional variation in the physical properties of the cochlear partition. Probably the reason was that the analytical difficulty was so extreme that to deal with even simplifications of the total picture represented a major accomplishment.

Much of the three-dimensional work was enveloped in difficult mathematics (Steele 1974, 1976; Chadwick and Cole 1979; Steele and Taber 1979a,b, 1981; Holmes 1980) that required simplifying assumptions which denied much of the physiological reality of the cochlea. An exception was the work of Taber and Steele (1981), who augmented their three-dimensional formulation to include abstractions of anatomically relevant structures, many of which are at the forefront of interest today. In particular, their model included three-dimensional fluid and four modalities of partition flexibility that included the motion of the bony shelf and arches of Corti, as well as the pectinate zone of the basilar membrane. Lien (Lien 1973; Lien and Cox 1974) made simplifying assumptions and collapsed the three-dimensional formulation to a one-dimensional formulation that included a frequency-dependent impedance in the cochlear partition branch (cf. Fig. 3.8) whose magnitude increased at 3 dB per octave. Potentially, this gave credibility to the one-dimensional approach, which from a practical standpoint was about as much as the computers of the day could handle.

4.1.5 Enhanced One-Dimensional Macromodels

We next consider modifications to the one-dimensional macromechanical model that can be viewed (Fig. 3.13) as adding to the cochlear partition impedance additional series elements that may be passive or active. In the first case we consider passive nonlinearities that were in vogue for the

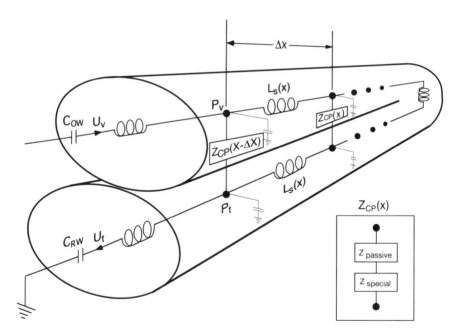

FIGURE 3.13. A generic depiction of the one-dimensional formulation indicating basilar membrane impedances, $Z_{passive}$ and $Z_{special,}$ in series. All other features are as in Figure 3.8.

better part of a decade. Passive nonlinear models were superseded by active models that were often extended to include nonlinearity.

Nonlinear passive macromodels

While it was generally thought that the higher dimensional formulations must be more accurate than the one-dimensional formulation, they did not do a better job matching data. Moreover, almost all the higher dimensional models were linear despite the fact that the emerging data showed marked cochlear nonlinearity. What ensued was a case of modelers being challenged by the data but lacking the analytical tools to handle nonlinearity. Computer simulations were the logical alternative.

The one-dimensional hydrodynamic formulation was particularly advantageous for modelers of the time (Hubbard and Geisler 1972; Kim, Molnar, and Pfeiffer 1973; Hall 1974, 1977a,b), principally because of the primitive state of computing. First of all, the one-dimensional model simply had fewer components (compare Figs. 3.8, 3.11, and 3.12). Fewer components meant reduced memory requirements and shorter computing times. Because the exploration of model nonlinearity required computing in the time domain, simulations were required for several input levels for many frequencies. (For a recent discussion of time-domain solutions in the nonlinear, two-dimensional case, see Diependaal and Viergever 1989.) To compute model responses for a set of frequencies and levels necessary to fully characterize the model for one fixed-parameter assignment took days. Storage of data was another severe constraint that was balanced against long recompute times.

Relatively few nonlinear passive models were tried throughout the 1970s. Hubbard and Geisler (1972) developed and computed the first nonlinear cochlear model based on a variation of the one-dimensional formulation. Because Rhode's (1971) results showed the strongest nonlinear effects in the tip region where damping was a crucial parameter, Hubbard and Geisler (1972) used a nonlinear damping. Kim, Molnar, and Pfeiffer (1973) digitally solved a system of coupled, nonlinearly damped, second-order differential equations, which were cascaded by making the input variable corresponding to the vertical deflection of the cochlear partition the input to the next section. This system had no hydrodynamic counterpart and was not a wave-propagation formulation. However, this model was interesting because of extensive comparisons with experimental data demonstrating cochlear nonlinearity, especially the paired-click experiments of Goblick and Pfeiffer (1969) that were the puzzles of the time. Hall (1974, 1977a,b) studied distortion products and two-tone suppression using a damping nonlinearity. Hubbard (1976), using a similar nonlinear formulation, was able to mimic the nonlinear basilar membrane response to click presented at multiple intensity levels (Robles 1973; Robles, Rhode, and Geisler 1976).

The model used by both Hall and Hubbard was the one-dimensional hydrodynamic formulation with nonlinear damping that increased with the level of the cochlear partition velocity. Hall rigorously tested his model using a variety of stimuli. The model had several features: it did a reasonable job of mimicking the effects of suppressor tones, which were higher in frequency than the test tone (Hall 1974). Also, the model mimicked the exact alignment of zero crossings in the basilar membrane response to click excitation (Hubbard 1976; Robles 1973; Robles, Rhode, and Geisler 1976; see also Ruggero, Rich, and Recio 1992). However, the model also had major shortcomings: first, one had to increase the Q of the tuned sections so that the nonlinearity could have the desirable effect of subsequently reducing the Q in a way that agreed with the data. This caused the model to have tips that were too sharp at low sound levels (see Fig. 3.9) but were reasonable at higher sound levels. Second, the model could not account for suppressors below CF, because to produce sufficient nonlinearity, the model response to the suppressor had to be very large. This was in conflict with (neural) experimental evidence (Sachs and Kiang 1968; Abbas and Sachs 1976; Sachs and Hubbard 1980). Hall (1977b) invoked a kind of second filter that extracted the huge response to the suppressor tone. This was acceptable for the era, as a lore had built up around the existence of a "second filter" that would take the poorly tuned (as many still thought at this time) basilar membrane response and modify it to account for the highly tuned auditory nerve responses. Matthews (Matthews 1980; Matthews and Molnar 1986) also employed a similar nonlinear damping formulation to study distortion product emissions.

Further study indicated nonlinear damping alone could not account for cochlear data. Furst and Goldstein (1982) tested a nonlinear damping model versus a nonlinear damping and stiffness model. The latter was necessary to explain both amplitude and phase of the cubic distortion product $2f_1 - f_2$. This is important to note not only for its relevance to the particular data modeled, but in consideration of the more general problem that modeling *both* magnitude and phase was a supreme challenge even in the single-tone case (Viergever and Diependaal 1983). Hubbard (1986) concluded that both nonlinear compliance and nonlinear damping must be used to obtain reasonable comparisons with electrically evoked otoacoustic emission data (Hubbard and Mountain 1983). A similar model was used to simulate dc displacement of the cochlear partition (LePage and Hubbard 1986; LePage 1987). Allen and Fahey (1993) also argued that compliance must be the nonlinear element so that models correctly reproduce distortion product emission data.

Active one-dimensional macromodels

Unquestionably the principal data that motivated the development of active models were the findings of evoked otoacoustic emissions (Kemp

1978) and spontaneous otoacoustic emissions (Wilson 1980), which could only be interpreted as meaning that the cochlea was sound energy producing. No passive model can replicate such a result. Motivation for the development of active micromechanical models also included other experimental data and theoretical work. Basilar membrane data from additional experimental groups showed highly sensitive tips (Khanna and Leonard 1982; Sellick, Patuzzi, and Johnstone 1982; Sellick, Yates, and Patuzzi 1983). These were precisely the kind of data that passive, higher dimensional hydromechanical models and passive, micromechanical models of the time could not reproduce well. The discovery of *in vitro* (Brownell et al. 1985) and likely *in vivo* (Hubbard and Mountain 1983) electromotility attributed to cochlear OHCs was also a factor because these results pointed to the active component from which an active mechanism could be constructed. In addition, theoretical work (de Boer 1983b) on apparent cochlear partition impedance indicated that the cochlea must be energy producing. At present, the speculation is that the cochlea is linear and active at low sound levels; the activity saturates causing nonlinear effects at higher sound levels, and at highest sound levels the cochlea is passive, possibly linear (see Figs. 3.4 and 3.5).

Contrasted with early models in which the passive damping was an increasing function of cochlear partition velocity, negative damping (undamping) was used by early active models (see Davis 1983 for a summary of the state of the active modeling field around 1983) to produce forces with the right phase angle so as to cancel damping. Thus undamping and active forces were synonymous. Undamping was envisioned to exist at sound levels near threshold for auditory nerve fibers. When a cochlea was "inactive," its response reverted to the passive, high-damping condition (see Fig. 3.4 or Fig. 3.5 at the highest sound level). Although one might have expected the concept of the active cochlea undamping to have arisen about 1980, Gold (1948) had much earlier proposed an active feedback model in which feedback force was proportional to velocity and canceled damping forces around the best frequency. Zwicker (1979) described a nonlinear feedback model that fed back force proportional to velocity. That model was later realized as an electronic circuit (Zwicker 1986) that used discrete elements, electronic amplifiers, and diode nonlinearities. As seen from the perspective of the loading on the cochlear partition, the feedback circuit (likened to the OHCs) nonlinearity appeared to be a saturating acoustic damping. The model behaved as a high-Q circuit for low levels of input and a low-Q circuit for high input levels. Mountain, Hubbard, and McMullen (1983) proposed a negative-feedback model. Negative feedback means that the effect of the feedback is subtractive in the case of low-frequency excitation. The actual phase angle of the feedback then varies as a function of frequency. The negative-feedback model increased cochlear partition stiffness (relative to passive stiffness) at frequencies below CF, but included a phase shift in the feedback loop such that the circuit reduced damping around CF.

Furst (Furst and Lapid 1988; Furst 1990; Furst, Reshef, and Attias 1992; Cohen and Furst 1993) also used an active macromechanical model that employed nonlinear damping and nonlinear stiffnesses. The models have primarily been tested against otoacoustic emission data and account well for a number of difficult-to-understand effects. For example, the latest model (Cohen and Furst 1993) accurately mimics the level dependence of some otoacoustic emissions as well as interference interactions between distortion products and spontaneous emissions.

Kolston et al. (Kolston 1988; Kolston and Viergever 1989; Kolston et al. 1989), concerned over the large amounts of power used by active models, pursued passive, micromechanical models (see later discussion of Kolston's models in the micromechanical section of this chapter) to demonstrate that "active" models were not required to reproduce the correct shape of the cochlear tuning function. Kolston et al. (1990) also considered what kind of forces an active cochlear amplifier must produce to achieve the sensitivity (correct absolute levels of outputs) that was lacking in the passive models. The work developed from the standpoint that a particular form of active impedance in series with the nominal spring–mass–damper impedance (see Fig. 3.13) could produce the desired cochlear partition response. The Kolston et al. (1990) active impedance formulation used negative reactive elements although an earlier formulation (Kolston and Smoorenburg 1990) used dependent sources and positive elements. The negative reactive elements partially canceled the cochlear partition reactance. The total resistance in the circuit was such that the circuit manifested low damping, hence very high sensitivity, at the tuned frequency. The unloading (high admittance) of the reactive passive component of impedance in frequency regions just flanking the tuned peak then made the whole peak more broad. This desirable response occurred with one-thirtieth of the power required in the Neely and Kim model (1986). Magnitudes of the cochlear partition responses produced by the model compared favorably with experimental basilar membrane data, but the phase responses were very far off the mark. The model could simulate "damage" by the method of eliminating the unusual circuit that represented the cochlear amplifier. When the model changed from passive (damaged) to active, there was not a corresponding change in the tuned frequency.

Zweig (1991) calculated the cochlear partition impedance needed to fit data provided by Rhode that Zweig had extrapolated to very low sound levels. Zweig found that a negatively damped harmonic oscillator representing the active cochlear partition impedance could be stabilized by a force proportional to velocity at a previous time. The fast feedback force that created the oscillator was an instantaneous negative damping that had similarities with previous negative (damping) feedback models.

A slow feedback force (delayed one and three-fourths periods at the characteristic frequency) stabilized the model against uncontrolled oscillation.

4.2 Micromechanical Models

In the introduction to this chapter we pointed out that the anatomy of the organ of Corti is quite complex. In spite of this complexity, much modeling work has ignored or greatly simplified the cellular architecture of the organ of Corti. In the macromechanical models that have been described, the entire organ was reduced to a single mass and its viscoelastic properties were represented by a single spring plus a damping element. In this section we review some of the more recent attempts to include in models at least a portion of the anatomical detail. The typical approach is to develop a schematic model of the organ of Corti made up of rigid bodies interconnected by hinges and viscoelastic elements. The motion of the rigid elements in some cases may be composed of both rotational and radially varying components. In the active models, OHC motility is usually represented by a dependent force or velocity source. Further simplifying assumptions are usually made, such as transforming the rotational motions into equivalent colinear displacements, thereby eliminating unnecessary components.

In Figure 3.14 we illustrate how one goes about the development of a micromechanical model. Figure 3.14A depicts a typical drawing of the organ of Corti. The geometry is then simplified, as shown in Figure 3.14B, by choosing the structures that will be represented by rigid bodies. In this case we have chosen to represent the tectorial membrane, the reticular lamina, the basilar membrane, and the inner and outer pillar cells. The springs in Figure 3.14B represent viscoelastic properties (for simplicity, both dampers [visco] and elastic elements have been represented by a symbol that would normally be a spring according to formal mobility definitions). In some cases, for example, the tectorial membrane and the reticular lamina, the viscoelastic elements may oppose both longitudinal and radial forces. K_p and K_a are the viscoelasticity of the pectinate and arcuate regions. K_{ip} allows rotational elasticity of the inner pillar around the inner hinge point. Any solid surface can have the property of mass, although the enclosure (bone) is also rigid relative to the frame of reference. The hinge shown in the tectorial membrane can be infinitely stiff, thereby implying a one-piece tectorial membrane rather than a segmented structure.

Although the generic picture is of a two-dimensional structure having apparent height and width, the models are still one dimensional and the pressure wave propagates into the plane of the paper via scala fluid in the same way that is depicted in Figure 3.8, which employs an oblique viewing angle. In the micromechanical models, what amounts to an

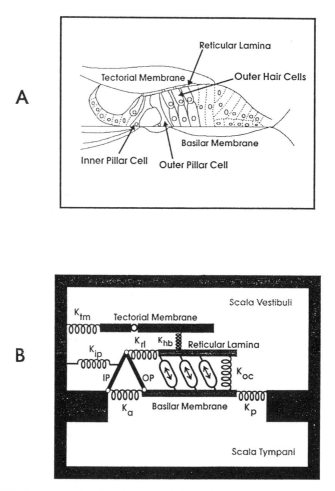

FIGURE 3.14A,B. (A) Simplified anatomical depiction of the organ of Corti. (B) Mechanical drawing of a generic model for cochlear micromechanics seen in cross section. The picture shown is the equivalent of a snapshot of an apparatus assembled from physical springs, hinges, and masses. Here, coils represent viscoelastic elements, not inductors as in previous figures. The cross section could be thought of as being similar to $Z_{CP}(x)$ in Figure 3.8. The *scala vestibuli* and the *scala tympani* are also depicted, and one assumes that the cochlear hydromechanics plays a fundamental role. $L_S(x)$, the fluid acoustic mass (not shown in this figure), comprises the pressure wave transmission in the x dimension, which is into the plane of the page.

augmented impedance formulation, similar to $Z_{CP}(x)$ in Figure 3.8, allows various portions of the contributing structures to move in various ways. Hence there are multiple degrees of freedom, not higher dimensionality. The models compute motion of various structures at a location denoted by a single spatial variable, x, the longitudinal distance along the cochlear scalae.

The models to be discussed next are similar to that shown in Figure 3.14B. Most investigators ignored the inner and outer pillar cells. Others included or excluded various pieces of the generic model. For example, Neely (1993) ignored the properties of the reticular lamina, and had the OHC active forces acting through the hair bundles (K_{hb}) on the tectorial membrane. Geisler (1993), for example, assumed that the tectorial membrane was connected to a rigid hinge, thereby eliminating K_{tm}, and also that the reticular lamina was connected via a rigid hinge, which eliminated K_{rl}.

Another type of simplification, found in all but a few micromechanical models, was to neglect the fluid flows within the organ of Corti. Because the fluid within the organ of Corti is essentially incompressible, a change in organ of Corti cross-sectional area in one cochlear region must be compensated by area changes in other cochlear regions; that is, if one squeezes a tube in one spot it bulges out elsewhere. A similar situation could hold for the inner sulcus region as well. Longitudinal fluid flow within the organ of Corti is potentially significant. It is hydrodynamic, not micromechanic, by nature. Therefore, we have chosen to review models that include such flows in a separate section called multimode models (Section 4.3).

4.2.1 Passive Micromechanical Models

Early work on micromechanics focused on ways that the shearing motion of hair cell cilia could be sharply tuned. The principal interest was not in mimicking basilar membrane mechanical responses; rather, it was on reproducing neural responses. This makes it difficult in many cases to assess model performance in terms of the target basilar membrane data described in Figures 3.1–3.6. The neural magnitude data were quite well known, and the modelers made cases for the validity of their models on the basis of such data. However, the substantial number of workers involved with the details of modeling mechanical-to-electrical transduction, the synapse, and auditory nerve firing patterns argues that these stages of auditory processing are nontrivial from a modeling standpoint (see Chapter 4, Mountain and Hubbard, of this volume).

Zwislocki and Kletsky (1979) devised a model in which the shear between the reticular lamina and the tectorial membrane was enhanced by a longitudinally stiff tectorial membrane interacting with the shortening spatial wavelengths on the basilar membrane near resonance. This model was followed (Zwislocki 1980) by one using a radially resonant tectorial membrane system. The tectorial membrane provided mass, while its elastic connection to the limbus provided the stiffness component. The plausibility of such a system was demonstrated using a mechanical model

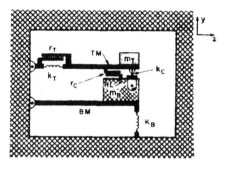

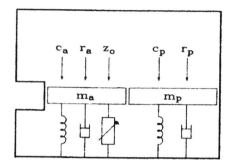

(A)

(B)

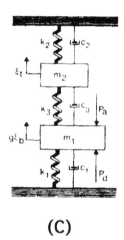

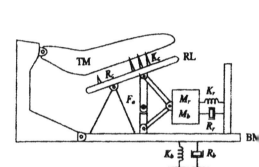

(C)

(D)

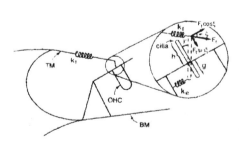

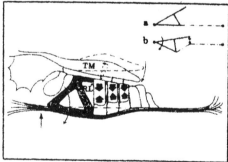

(E)

(F)

composed of physical elements. Basically, the reticular lamina moved up and down while the tectorial membrane moved from side to side. The frequency selectivity of the shear provided to OHC stereocilia was considerably enhanced relative to the shear caused by the vertical motion alone.

Allen (1980) elaborated and simulated (Fig. 3.15A) a conceptually similar micromechanical model. Allen developed a "transduction filter" based on micromechanics and included this transduction function in a hydromechanical model (Allen and Sondhi 1979). To obtain the filter characteristics, the basilar membrane and the tectorial membrane were represented as resonant structures coupled by viscoelastic stereocilia. The tectorial membrane essentially vibrated in a radial dimension while the basilar membrane vibrated up and down. This particular work was one of the most detailed depictions of cochlear microstructure to the present

---

FIGURE 3.15A–F. Micromechanical models from various studies (reprinted with permission). In all cases, the pressure wave propagates into the plane of the paper, and only the micromechanical aspects of the system are depicted. (A) Allen's (1980) model (passive) is characterized by a radially resonant tectorial membrane system coupled to a vertically resonant basilar membrane system. The hydrodynamic drive to the system (not depicted) is assumed to act on the basilar membrane. (B) The Kolston et al. (1989) formulation is that of the arcuate and pectinate zones, each responding to the pressure difference across the cochlear partition. Each system is a spring–mass–damper combination; however, the arcuate zone impedance contains a hair cell component that gives the arcuate zone model response a realistic shape. (C) Neely and Kim's (1986) model depicts a vertically resonant tectorial membrane system coupled to a vertically resonant basilar membrane. Interconnecting the systems are a spring and damper associated with hair cell stereocilia. Pressure difference P_d acts on the basilar membrane system, and the hair cell force is modeled as a pressure acting on the basilar membrane just as a pressure difference. (D) Neely and Stover's (1993) model shows mechanical representation of many aspects of cochlear anatomy. The tectorial membrane (TM) and reticular lamina (RL) are free to rock around pivoting anchor points. The scissoring hinge allows for resolution of forces in both the horizontal and vertical dimensions. (E) Geisler's (1993) model shows details of the feedback structure implemented by the outer hair cells (OHCs). A stiff tectorial membrane attaches through the cilia, which exerts rotational stiffness that opposes its motion. The reticular laminar rotates about a hinge point located at the intersection of the pillar cells. The force created by OHCs acts along the axis of the OHC. (F) The pseudoanatomical sketch of Mammano and Nobili's (1993) model depicts some of the assumptions that go into the equations of motion. When the OHCs contract, the tunnel rotates downward around a pivot point inward from the inner pillar cell. The tectorial membrane characteristics are mass dominated while the stereocilia stiffness is large in relation to the tectorial membrane stiffness. These two factors constitute a resonant system. The basilar membrane is also resonant.

date. The model showed reasonable comparisons with neural tuning curve data, that is, the lowest sound pressure levels required at the threshold of neural excitation.

Kolston (1988) hypothesized that the arcuate and pectinate regions of the basilar membrane responded in parallel to the pressure difference provided by the one-dimensional hydrodynamic environment (a sketch is shown in Fig. 3.15B). Hair cell stiffness, assumed to originate from stereocilia, dominated the impedance of the arcuate region in cochlear regions basal to the characteristic place for a given frequency. To make the model work for all cochlear regions and for all acoustic frequencies, the stiffness was allowed to be a function of both frequency and displacement such that the resulting impedance was a frequency-dependent reactance and a frequency-dependent resistance. The effect of the second (parallel) branch was to reduce the responsiveness of the basilar membrane in the tail region but not in the tip such that the remaining peak was broad, not sharp, and the tip-to-tail ratio was in approximate agreement with data. However, the overall response of the model was much too low (~40 dB) to be in *absolute* agreement with data. Nonetheless, this model was the first to get right the shape of the magnitude of the cochlear partition response-versus-frequency profile.

In a later work (Kolston et al. 1989), the hair cell was realized using a circuit element that sensed motion of basal segments, filtered the response, and passed on a force. The model appeared to the more "real" than the earlier version, because the hair cell was elaborated as a physical structure that had its base in the pectinate region and its apex in the arcuate region of a more basal location along the length of the cochlea. Kolston et al. (1989) compared model results with data of Robles, Ruggero, and Rich (1986) and Sellick, Yates, and Patuzzi (1983) using different parameter sets. While the selectivity was quite good, meaning that the shape of the tuning matched data quite well, the sensitivity, meaning the absolute level of the response, remained far too small. The effect that was produced by the mechanism was still (Kolston 1988) basically a subtractive one that achieved desirable shapes of the cochlear partition response versus frequency.

4.2.2 Active Micromechanical Models

Neely and Kim developed (Neely 1981b; Neely and Kim 1983) a model in which, at any point in the longitudinal dimension, the mechanical impedance consisted of two spring–mass–damper systems that were coupled via a negative damping element. The first subsystem, which corresponded to the basilar membrane, was driven by pressure changes in the fluid. The second system was associated with the stereocilia coupled to the tectorial membrane and was tuned higher in frequency than the basilar membrane system. This model was similar to the passive tectorial membrane/basilar membrane model of Zwislocki (1980) and to the formulation of Allen (1980), although the tectorial membrane was not a radially compliant element and, of course, the active element was present. The active generator was a negative resistance combined with a spring, and this apparatus was associated with OHC stereociliary forces. The Neely and Kim model (1983) showed quite excellent agreement with neural tuning curves, that is, the sound pressure required to produce the onset of neural responses versus frequency.

Their next model (Neely and Kim 1986; see Fig. 3.15C) was an improvement in that the active elements were better detailed, but the model was different both in concept as well as in implementation. In concept, the force (Neely and Kim 1986) originated as a pressure difference acting on the basilar membrane. This force was associated with the notion that the hair cell length change translated into hair cell pressure, P_a, which in turn was transmitted isometrically to the surrounding fluid. The active element was implemented as a pressure source the value of which was a filtered version of the shear between the tectorial membrane and the reticular lamina. In the diagram of the model shown in Figure 3.15C, P_d is the pressure difference, which can be computed using either a one- or two-dimensional formulation of the cochlear fluid, and the mechanical load. Thus, pressure difference acts on the mass of the basilar membrane that was coupled, via the spring and damper (passive) that represented the stereocilia, to the tectorial membrane. The model really contains three filtering subsystems: the tectorial membrane, the basilar membrane (reticular lamina), and the hair cell function.

This model has been criticized as having a pressure source acting on the basilar membrane that reacts against nothing. However, Neely and Kim (1986) suggested that OHC pressure resulted in a trans (basilar) -membrane pressure. Therefore, the force (depicted as a pressure in Fig. 3.15C) reacts against the hydrodynamics, and the model created what could be viewed as *increased* pressure difference. If one could measure an enhanced pressure difference, such a result would be strong evidence in support of this particular model formulation. The apparent power gain in the model was of the order of 30,000. This power gain was calculated as the ratio of the power entering the system at a particular frequency relative to the power dissipated at the particular CF location.

Neely (1993; Neely and Stover 1993) published another improved model (see Fig. 3.15D) that mimicked IHC excitation as well as basilar membrane motion. This model did an excellent job of reproducing the neural tuning curves of Liberman (1978) and the basilar membrane velocity magnitudes of Sellick, Patuzzi, and Johnstone (1982). Basilar membrane phase data were not shown, but model results were compared with neural group delay. The model functionally portrayed forward (mechanical-to-electrical) transduction and also reverse (electrical-to-mechanical) transduction. The model exhibited negative (Mountain, Hubbard, and

McMullen 1983) feedback and a region of negative damping as did the earlier models by Neely; however, these were now apparently outcomes, not a priori assumptions. The model functioned as a cochlear amplifier and a second filter. The second-filter aspect of the model produced a notch called a spectral zero in the output at a frequency about a half-octave below CF. Because of this, the model was consistent with a new hypothesis that a second frequency map (Allen and Fahey 1993) exists in the cochlea. The model parameters were algorithmically optimized. Neely noted that some of the resulting parameters may have been unphysio-logical and hair cell power output may have been unrealistically high. Such a consideration represents a maturation of the field toward greater sensitivity to physiological issues and challenges experimentalists to confirm or dispute parameter values used by modelers. A nonlinear version of the Neely (1993) model has been used to study distortion products (Neely and Stover 1993).

Geisler also developed several feedback models. One model (Geisler 1986) used a force acting between the tectorial membrane and the reticular lamina that was proportional to ciliary deflection. The work demonstrated that the tectorial membrane system need not be tuned (as in the case of Neely and Kim 1983) for a model to produce data were physiologically reasonable. Later models included a delayed feedback (Geisler and Shan 1990; Geisler 1991) that was similar in form to that worked out by Zweig (1991). A range of feedback force delays were explored. A schematic of the most recent version of the model (Geisler 1993) is presented in Figure 3.15E. The feedback force acted between the tectorial membrane and the basilar membrane. The principal mechanical property of the tectorial membrane was that of a relatively stiff spring. There was no damping between the tectorial membrane and the basilar membrane. The feedback function amounted to an all-pass circuit with delay, as contrasted with a tuned feedback function. The earlier frequency-domain representation (Geisler and Shan 1990; Geisler 1991) nominally showed signs of instability above CF. The improved version of the model (Geisler 1993) included an additional filter function that preserved the essence of the delay and stabilized the model. The model produced magnitude results that were reasonably good fits to experimental data, but did not show a shift in the tuned frequency as the gain parameter was increased. Geisler's (1993) model generated power over roughly 2.2 mm of basilar membrane, and at the peak of the response the power absorbed was about 15 times that entering the cochlea. A nonlinear version of the model (Geisler, Bendre, and Liotopoulos 1993) has been used to study two-tone suppression and distortion-product generation.

Figure 3.15F shows Mammano and Nobili's (1993) model. Although it appears similar in form to other contemporary models, differing a priori assumptions were made regarding the way that the cochlear partition and tectorial membrane could move (see insert of Fig. 3.15F). In particular, when the OHCs shortened, the basilar membrane moved toward the scala tympani by pivoting downward around the inner pillar, which necessitated that the pillars be very rigid. While there is evidence that the pillars may be relatively rigid (Olson and Mountain 1991, 1993, 1994), other data (Xue, Mountain, and Hubbard 1993) indicate the basilar membrane moves toward the scala vestibuli when OHCs shorten. The model's tectorial membrane was a tuned system; however, the tectorial membrane affected the cochlear partition primarily by virtue of its mass. Model results compared well with basilar membrane velocity data (Sellick, Yates, and Patuzzi 1983).

4.3 Multimode Models

We call a particular formulation a multimode model if the organ of Corti or associated structures are allowed the ability to propagate waves. One formulation has been called a "sandwich" (de Boer 1990a), as the fluid-filled tunnel of Corti and the OHCs are sandwiched between the upper and lower surfaces of the cochlear partition. A schematic is shown in Figure 3.16A. Pressure in the scala vestibuli and the scala tympani (P_1, P_2) squeeze the cross section of the organ of Corti, which supports pressure P_3. OHCs located within the organ of Corti push with equal and opposite forces on the reticular lamina (upper surface) and the basilar membrane (lower surface), thereby affecting pressures in each of the three channels. Pressures added by the hair cells and membranes are shown as P_{HC} and P_{OC}.

The formulation clearly contrasts with the micromechanical models previously discussed because in the present case the hair cells are not hypothesized to act on the tectorial membrane. This is possibly a meritorious feature of the model, as it seems possible that the stereocilia of the OHCs are not stiff enough to move the organ of Corti (Howard and Hudspeth 1988). Under the assumption that the reticular lamina presented zero stiffness, de Boer (1990a) showed using the sandwich model that the basilar membrane could remain essentially stationary while the reticular lamina moved considerably. This potentially explained the report that the reticular lamina in explanted cochleae moved considerably while the basilar membrane moved little (ITER 1989).

Using a similar model, basically a cylinder within a cylinder, de Boer (1990b) considered alternatives to the problem that insufficient reaction forces via the OHC stereocilia left the basilar membrane unstimulated. Two types of shape deformations were explored. One, which kept the net cross section of the organ of Corti constant as the organ deformed, had no interaction with cochlear fluids for long waves and was an unlikely candidate for the cochlear amplifier. The other, which did allow variation in the cross section of the organ, was capable of exciting both the basilar

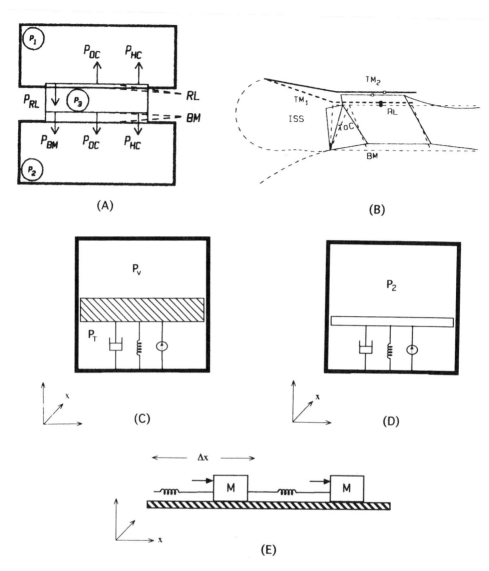

FIGURE 3.16A–E. Multimode models (reprinted with permission). (A) The basic ingredients of the sandwich model (de Boer 1990a). (B) The deflection of the sulcus model (de Boer 1993) away from its equilibrium position is shown in *solid lines*. The *open dots* indicate the final relative positions of the *solid dots* that in the resting position were vertically aligned with one another. (C) The traveling-wave amplifier model (Hubbard 1993). Parts shown in C together with (D), or alternatively C together with (E), constitute a physical representation of the traveling-wave amplifier model. In C, a spring, damper, and a mass comprise the basilar membrane (cf. Fig. 3.8), and the basilar membrane is acted on by a controlled pressure source as well as P_t and P_v, which are pressure waves that propagate into the plane of the page. In D, a spring and damper characterize a surface having negligible mass that is acted on by a controlled pressure source and P_2, which is a pressure wave propagating into the plane of the page. In (E), controlled force sources (*arrows*) act on masses M that are longitudinally connected (in the x dimension) by springs. The masses are subject to friction forces.

membrane and the scala fluids but seemingly spent much of its energy, for no good purpose, simply squeezing fluid through the organ of Corti.

Seeking something that the OHC forces could react against, de Boer (1993) explored a situation in which a hinged tectorial membrane was attached to the stereocilia of the OHC, providing a connection to the tubelike inner spiral sulcus (see Fig. 3.16B). The outer portion of the reticular lamina (RL) was allowed to flex such that the cross-sectional area of the organ of Corti could remain constant under excitation conditions. The RL was also allowed to rotate around the more central connection to the modiolus. The tectorial membrane (TM) was doubly hinged so that it could execute a rotational movement around a point located over the tunnel of Corti and around a point above the internal spiral sulcus (ISS). The TM was assumed to be composed of two stiff elements, TM_1 and TM_2.

de Boer's (1993) "sulcus connection" model was both micromechanical as well as capable of a second propagation mode (although this capability was secondary to the micromechanics). The assumed micromechanical situation circumvented the problem that the tectorial membrane as measured by Zwislocki and Cefaratti (1989) was so compliant that it probably could not provide adequate reactive force against even the hair cell stereocilia stiffnesses. In de Boer's sulcus connection model, the tectorial membrane could freely rotate upward because of a compliant hinge and therefore it did not disagree, by assumption, with Zwislocki and Cefaratti's (1989) measurement. However, the tectorial membrane was still assumed to be stiff. The tectorial membrane was pulled out in the radial dimension by the stereocilia while the opposing force provided by the inner spiral sulcus pulled the tectorial membrane back toward the modiolus. The feedback function in the model was that used by Geisler and Shan (1990). The model produced active tips that peaked at almost exactly the same frequency as the passive model. Side-by-side comparisons with data were not presented.

Hubbard (1993) proposed a traveling-wave amplifier model that hypothesized the existence of two separate wave-propagation modes in the cochlea (see Fig. 3.16) that were coupled via the OHCs. The model was realized using a circuit simulation of two tightly coupled transmission lines, each of which represented a different mode of energy propagation in the cochlea. The coupling was by means of active circuit elements that represented the OHCs. This model afforded a new mechanism for cochlear amplification that differed from other models because the feedback coupled two separate *transmission systems* rather than simply coupling *local elements* that were not interconnected themselves in a way that constituted a second transmission system.

One transmission line in the new model was the same as the classical one-dimensional model having a tuned shunt branch (cf. $Z_{CP}(x)$, Fig. 3.8). In Figure 3.16C, P_v, P_t, and an active force generator act on the spring–mass–damper characterization of the organ of Corti. The second transmission line (Fig. 3.16D or Fig. 3.16E) did not have tuned shunt branches, thereby further distinguishing it from models that employed coupled, tuned segments. The second transmission line was not defined as corresponding to a particular cochlear structure.

Two physical interpretations of the model were possible, because the second transmission structure could be viewed as either an acoustic analog (pressure analogous to voltage, volume velocity analogous to current) or as the dual mechanical analog (velocity analogous to voltage, force analogous to current). Therefore one physical interpretation of the second mode was hydromechanical and the other was mechanical. The acoustic analogy (Fig. 3.16D) implied the second transmission line represented a tube with pressure wave P_2 propagating down the length of the cochlea (cf. Fig. 3.8), that is, into the plane of the page, as does the pressure wave of all models shown in Figures 3.15 and 3.16 (except Fig. 3.16E). Pressure in this tube was created by motion of the cochlear partition (Fig. 3.16C) via the controlled pressure source shown (Fig. 3.16D), and, reciprocally, force that was proportional to the pressure P_2 acted on the cochlear partition (Fig. 3.16C). The mechanical analogy (Fig. 3.16E) implied a mechanical transmission line that propagated mechanical (not pressure) waves down the cochlea. Note the change in coordinate systems such that Figure 3.16E is not a cross section as in all previous cases shown in Figures 3.16 and 3.15, but instead depicts an assembly of masses that are longitudinally interconnected with springs and are subject to friction drag. It was suggested (Hubbard 1993; Hubbard, Gonzales, and Mountain 1993) that many cochlear structures were candidates for the second transmission line.

The model produced distributed amplification in spatial regions where constructive addition of waves occurred. This happened when the forward-traveling wave on the classical transmission line slowed down near resonance (cf. Fig. 3.10 in which rapid phase drops imply a slow forward-propagation velocity) to approximately match the wave-propagation velocity of the second transmission line. The amplification mechanism of the coupled lines operated over different regions of the cochlea, depending on the frequency of excitation.

The traveling-wave amplifier model produced data that matched well the experimental results of Ruggero et al. (1990) in both the magnitude (Fig. 3.17) and the phase (Fig. 3.18) responses. This model is currently one of the best with regard to reproducing *both* the magnitude and the phase responses of the basilar membrane. A nonlinear version of the model (Hubbard, Gonzales, and Mountain 1993) showed favorable comparisons to click-evoked basilar membrane data. The active-to-passive CF shift that was of the order of one-half octave in the base (Gonzales 1994) was consistent with physiological data. A decrease in tail sensitivity accompanied the increase in tip sensitivity when the model was passive as

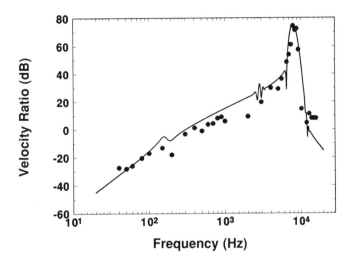

FIGURE 3.17. Magnitude responses from the traveling-wave amplifier model (*solid line*) are compared with experimental data averaged from five animals (Ruggero et al. 1990). Variations in the curve that occur around 200 Hz result from reflections at the helicotrema. Similarly, the variation seen at approximately 3000 Hz stems from a purposely introduced impedance anomaly at the 3000-Hz place in the simulated cochlea that produced a simulated otoacoustic emission of that frequency (described in Hubbard 1993).

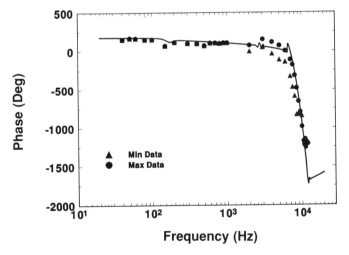

FIGURE 3.18. Phase responses from the traveling-wave amplifier model compared with experimental data. *Circles* indicate the most positive phases obtained from one of the five animals used in Figure 3.16; *triangles* are the most negative phases obtained from one of the five animals. Phase variation in the model results clearly accompany salient aspects of the model's response curve shown in Figure 3.17.

compared to active. This decreased tail sensitivity resulted from the active stiffness increase at frequencies below CF. However, the tail decreased only a few decibels while the tip increased 30–50 dB. This behavior is consistent with experimental data from the auditory nerve and probably with the behavior of the basilar membrane.

In the traveling-wave amplifier model, power generation was very local to the tip (~400 μm for high frequencies). Some experimental evidence (Cody 1992) suggests that the cochlear amplifier does work over a cochlear region of the order of 300–600 μm in length. Other investigators (Allen and Fahey 1992) have potentially ruled out the existence of a cochlear amplifier that works over a spatial region greater than 2 mm. Thus it is a possible constraint on cochlear amplifier models that amplification take place over a region less than 2 mm or even less than 600 μm long.

4.4 Active Admittance

An essential feature of all one-dimensional active formulations is that they modify basilar membrane admittance relative to the passive case. Because admittance is the ratio of velocity to pressure, a model might somehow amplify pressure and keep impedance constant, thereby producing a peaked velocity response that compares favorably with experimental data. At the other extreme a model might produce an increased admittance so that the desired velocity response is achieved without the pressure also peaking. The extent to which the admittance function does not show the same characteristics as the velocity response is the extent to which the pressure difference peaks.

Figure 3.19 shows a reconstruction of point admittances from a number of models. In every panel a passive admittance is plotted so that the reader can observe shifts in the tuned frequency as well as the relative sizes of the active-versus-passive admittances. The curves are normalized to the passive compliance. Figure 3.19A shows the simplest type of admittance transformation (passive to active) obtained using negative damping feedback of the type used by Zwicker (1979, 1986). The effect simply creates an admittance peak the same way that an increase in Q (or equivalently, a drop in the damping factor) would affect a passive model. There is no shift in the tuning of the admittance. Figure 3.19B (Mountain, Hubbard, and McMullen 1983) shows admittance from a negative feedback model in which the phase angle of the feedback changes with frequency such that below CF the feedback stiffens the basilar membrane, thus decreasing admittance. Therefore, this model is also called an active stiffness model, which contrasts with simple negative damping. There is a shift in the frequency of the maximum of the active model admittance compared to the passive model because the increased stiffness pushes up the resonant frequency. The phase angle of the feedback is shifted at frequencies around CF, such that it cancels damping there, pushing up the magnitude of the peak admittance. Notice that the admittance peak is still narrow in the frequency dimension. The Kolston et al. (1990) formulation was specifically designed to broaden the width of cochlear partition velocity response in the region of passive CF, and one can clearly see why this happens by examining the admittance function shown in Figure 3.14C. Negative reactive elements (Kolston et al. 1990) increase admittance on either side of the passive CF and therefore pull up the cochlear partition velocity response in a way such that the frequency response is broadened. The "dip" in the admittance around CF can be minimally seen in the cochlear partition response (see Kolston et al. 1990; Fig. 3.5).

Figure 3.19D, based on Zweig's (1991) formulation, shows an admittance peak that is broad, not narrow. The magnitude of the active admittance both increases and decreases relative to the passive admittance because of phase changes in the feedback attributed to the fixed, one and three-fourths cycle delay at CF. There is no shift in the active CF because (by assumption) there was no attempt to demonstrate a shift. Figure 3.19E is a reconstruction of the Geisler (1991) formulation, computed using a feedback delay of a half period at the tuned frequency. The fixed delay in the fed-back force causes oscillations in the nature of the feedback as a function of frequency as seen in Figure 3.19D. The admittance indicates a stiffer basilar membrane in the region from ~0.3CF up to ~0.8CF. The peak admittance shifts slightly to lower frequencies, although the velocity response (Geisler 1991) of the basilar membrane does not appear to change in CF as the model goes from active to passive. Finally, Hubbard's (1993) admittance (Fig. 3.19F) shows active stiffness at low frequencies and a large, broad admittance increase in the region of shifted CF. The model achieves its cochlear partition velocity tuning characteristic mainly as a result of an admittance increase and not a pressure increase. The maximum cochlear partition velocity response occurs at a frequency lower than the frequencies of the small, sharp peaks in the admittance. The peaks result from numerical inaccuracy in the simulation enhanced by the division operation (the calculation of admittance) at frequencies for which both the model velocity and the pressure are small. In all other cases (Fig. 3.19A–E), the impedance was calculated by means of evaluating published admittance functions, rather than by simulating the full models to obtain pressures and velocities for subsequent scaling into admittance.

4.5 van der Pol Oscillator Models

van der Pol limit-cycle oscillators have been used to generate otoacoustic emissions and to mimic cochlear nonlinearlity (Bialek and Wit 1984; Jones et al. 1986; Talmadge et al. 1990; van den Raadt and Duifhuis 1990; Long, Tubis, and Jones 1991). The "oscillators" are actually oscil-

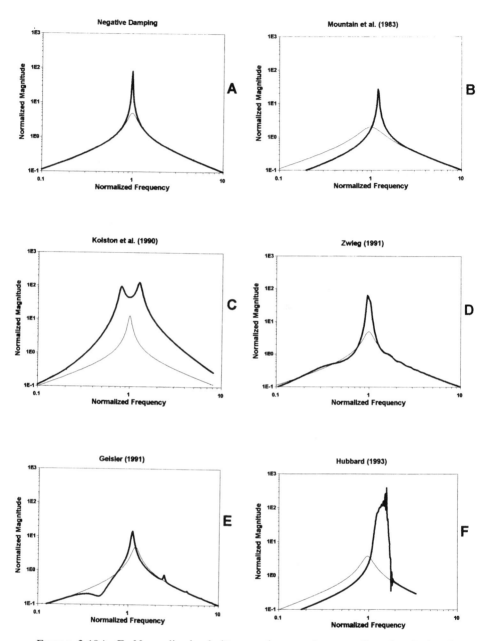

FIGURE 3.19A–F. Normalized admittances (re: passive compliance) calculated using a number of one-dimensional cochlear models in the active (*heavy line*) and passive (*light line*) cases. (A) Generic negative damping produces the equivalent of a high-Q spring–mass–damper, second-order system. The effect is only seen at the tuned frequency because the damping only has an effect there. The admittance curve is precisely the same as the passive curve using a small value of damping.

latory solutions to the nonlinear differential equation of van der Pol (see, for example, Minorsky 1962). The approach has at its foundation the noted similarities between the properties of van der Pol oscillators and the active nonlinear behavior displayed by the cochlea (Johannesma 1980). In particular, oscillations can build up but instead of growing without bound, they stabilize or self limit. The style of modeling appears to have great appeal for auditory scientists principally interested in otoacoustic emissions. Some scientists interested principally in micromechanics have seemingly found it more difficult to identify with the approach. Talmadge and Tubis (1993) made progress toward bridging this apparent gap by investigating the relationship between spontaneous and evoked otoacoustic emissions using a one-dimensional model and Zweig's feedback function in conjunction with a van der Pol nonlinearity that stabilized the otherwise unstable circuit.

5. Alternative Cochlear Models

5.1 Signal Processing Models

Signal processing models of the cochlea use filtering constructs from engineering signal processing to produce cochlear-like output signals to be used in various applications (Lyon 1982; Lyon and Lauritzen 1985; Lyon and Dyer 1986; Meddis 1986; Ghitza 1986; Seneff 1988; Shamma 1988; Meddis, Hewitt, and Shackleton 1990; Hewitt and Meddis 1991; Kates 1991). For example, investigators interested in modeling responses higher in the auditory system than the cochlea often seek cochlear-like input to their models. Also, there is interest in cochlear-like preprocessors for cochlear prosthetic implants or for speech recognition systems. Most

---

(B) Negative feedback from the model of Mountain, Hubbard, and McMullen (1983) produces enhanced stiffness, a shift in tuned frequency, and a sharp peak in admittance. (C) The admittance calculated using the model of Kolston et al. (1990) shows two admittance maxima on either side of the original tuned admittance, thus resulting in a high, broad peak in the response of the cochlear partition. In the working model, the pressure waveform apparently causes the velocity response to be flatter than the exaggerated trough seen in the admittance. (D) Admittance calculated using a delayed-feedback model (Zweig 1991) shows a high, relatively broad peak centered at CF and small, cyclic effects at frequencies away from CF. (E) Admittance calculated using Geisler's (1991) model shows enhanced stiffness below CF, a slightly shifted tuned frequency relative to the passive condition, and additional peaks at frequencies above the CF. (F) Admittance of the traveling-wave amplifier (Hubbard 1993) model. At frequencies below CF, cochlear partition stiffness is enhanced. Active CF occurs at a higher frequency than passive. CF.

of the signal processing models are not aimed at explaining the mechanism of the cochlea.

There are two common formulations of signal processing models of the cochlea. One formulation is a serially arranged cascade of filters such that each filter has a lower tuned frequency than the one preceding. Each filter in sequence introduces a phase shift that approximates the traveling-wave delay in the cochlea. The cascade approach yields responses to speech sounds that are similar to but not as physiologically realistic as responses from a one-dimensional cochlear model (Kates 1993). Alternatively, the input signal can be fed simultaneously to a parallel bank of filters, each tuned to a different frequency. Because low-frequency filters have slower rise times than high-frequency filters, this somewhat approximates cochlear delay. However, additional delays are needed to make the outputs more cochlear like.

An easy-to-implement, cochlear-like filter function is the gamma-tone filter. This function was first introduced by Johannesma (Johannesma 1972; Aertsen and Johannesma 1980) to mimic the shape of the impulse response function of the auditory nerve as estimated by the linear reverse correlation function (de Boer and Kuyper 1968). The gamma function can be implemented (Holdsworth et al. 1988) as a cascade of recursive filters, and as such it is possible to achieve real-time signal processing using modest computer resources. Moreover, the characteristics of the gamma-tone filter can be simply (and dynamically) manipulated. Carney (1993), for example, dynamically modified the filter parameters via a feedback mechanism to simulate the compressive nonlinearity found in cochlear mechanics.

An alternative to strategies used in earlier signal processing models is the wave digital filter approach (Strube 1982, 1985; Friedman 1990; Giguere and Woodland 1994), which involves translation of analog networks into time-domain computational structures (Fettweis 1986). This results in a computationally efficient algorithm and access to internal variables of the original analog formulation. For example, a wave digital formulation of the one-dimensional model can preserve all variables of interest, for example, pressures and volume velocities. The filter bank and filter cascade strategies simply handle inputs and outputs, while internal variables are irrelevant. Moreover, there is no pretense that the output variables strictly represent physical variables, even if the output might have similar frequency-domain or time-domain characteristics.

Goldstein (1990a,b, 1991) has attempted to shed light on cochlear function and cochlear nonlinearity using bandpass and nonlinear filtering techniques. This is an alternative strategy of using an abstract model to guide the biophysical models. Goldstein's signal processing work has apparently helped lead the way in modeling work of Furst and colleagues (see Section 4.1.5.2). Also, Goldstein's recent model that demonstrated "additive amplification" created by bidirectionally coupled, cascaded filters (Goldstein 1992a,b, 1993) has similarities with the traveling-wave amplifier (Hubbard 1992, 1993).

5.2 Silicon Implementations

A number of workers have implemented silicon preprocessors that produced responses to electrical stimuli that were similar to responses from the cochlea under sound stimulation. Lyon and Mead (1988) designed a chip that used a cascade of filters implemented by means of subthreshold complementary metal oxide semiconductor (CMOS) circuits (Mead 1989). The chip mimicked group delay and magnitudes of cochlear responses in the time domain. Watts et al. (1992) presented an improved version of the earlier chip. A collection of modules implemented using subthreshold CMOS circuits (Liu, Andreou, and Goldstein 1992) model the middle ear, basilar membrane, and hair cell synapse in a system for nonlinear processing of speech signals. Lazarro et al. (1993) considered the possibility of silicon chips as auditory computer peripherals. Indeed it would seem that the minute detail so laboriously considered by the micromechanical modelers is not an impediment to the development of the silicon realizations.

6. Summary and Areas of Future Research

A number of competing hypotheses remain outstanding while a few disagreements have been resolved. Macromechanical models demonstrated that fluid mechanics alone could not account for salient features found in the experimental data. However, as micromechanical models mature, the two- and three-dimensional pressure formulations will prove useful by adding to the total authenticity of the simulations. In the case of the more recent active micromechanical models, modelers have looked at the same anatomy yet have come away with diverse assumptions that went into their models. More experimental work is needed to establish which assumptions are correct. Modelers also used differing parameter values, and as physical parameter determinations are made by experimentalists, comparative study of the parameter values used by various models will likely rule out some formulations. Most modelers claim their formulations replicate certain experimental data. No modelers have addressed the battery of tests recommended here. Some modelers have been more interested in exploring new concepts rather than with comparisons to data. This strategy also challenges experimentalists to refute the a priori assumptions used. Models using assumptions that are shown to be wrong should be dropped. In the micromechanical arena, for example, experimental data on which way the tectorial membrane moves under sound excitation would validate or exclude various models.

The multimode formulations are truly different from earlier macro- and micromechanical models, and the experimental finding of a new mode of wave propagation in the cochlea would validate these models. It is unlikely that micromechanical models will ever be unimportant, and micromechanical, multi-propagation-mode models should be the focus of modeling in the near future, as new data on the motion of individual components of the organ of Corti emerge. Clearly there is an overwhelming need that experimental techniques such as those used by the ITER group be extended to the basal turn of the explanted cochlea and to the *in vivo* cochlea in various locations. In addition to measurements of velocity and displacement, accurate measurement of intracochlear pressure can be an important means to validate or exclude models: Some models show large pressure peaks to produce an amplified basilar membrane velocity response while others do not. Simultaneous pressure and velocity measurements (see Fig. 3.19) will allow computation of active and passive cochlear partition admittance, and according to Figure 3.19, this will differentiate models quite clearly.

List of Symbols and Terms

A	area of cross section of cochlear scala
CF	characteristic frequency
$C_{CP}(x)$	acoustic compliance of the cochlear partition at location x
Δx	delta x, a small variation in the x dimension
CMOS	complementary metal oxide semiconductor
f_1	frequency
f_2	frequency
$F_{CF}(x)$	characteristic frequency of a particular location, x
ISS	internal spiral sulcus
kHz	kilohertz
K_a	viscoelasticity of the arcuate region
K_{hb}	viscoelasticity of the hair bundle
K_{ip}	rotational viscoelasticity of the inner pillar
K_p	viscoelasticity of the pectinate region
K_{rl}	viscoelasticity of the reticular lamina
K_{tm}	viscoelasticity of the tectorial membrane
$L_{CP}(x)$	acoustic mass of the cochlear partition at location x
$L_S(x)$	acoustic mass of the cochlear scala at location x
P_a	active pressure
P_d	pressure difference
$P_v(x)$	pressure in scala vestibuli
$P_t(x)$	pressure in scala tympani
P_{HC}	pressure exerted by hair cell in sandwich model
P_{OC}	pressure exerted by organ of Corti in sandwich model

$P_-(x)$	pressure difference
$P^+(x)$	average pressure
$P_2(x)$	a second pressure in the traveling-wave model
P_1	pressure in scala vestibuli in the sandwich model
P_2	pressure in scala tympani in the sandwich model
P_3	pressure in organ of Corti of the sandwich model
ρ	density
Q	quality factor of a resonant system
RL	reticular lamina
$R_{CP}(x)$	acoustic damping of the cochlear partition at location x
TM	tectorial membrane
TM_1	portion of tectorial membrane
TM_2	portion of tectorial membrane
x	spatial dimension down the cochlea
y	spatial dimension above the cochlear partition
z	spatial dimension across the width of the cochlea
$Z_{CP}(x)$	acoustic impedance of the cochlear partition

References

Abbas PJ, Sachs MB (1976) Two-tone suppression in auditory-nerve fibers: extension of a stimulus-response relationship. J Acoust Soc Am 59:112–122.

Aertsen AMJH, Johannesma PIM (1980) Spectro-temporal receptive fields of auditory neurons in the grassfrog. I. Characterization of tonal and natural stimuli. Biol Cybern 38:223–234.

Allen JB (1977) Two-dimensional cochlear fluid model: new results. J Acoust Soc Am 61:110–119.

Allen JB (1980) Cochlear micromechanics—a physical model of transduction. J Acoust Soc Am 68:1660–1679.

Allen JB, Fahey PF (1992) Using acoustic distortion products to measure the cochlear amplifier gain on the basilar membrane. J Acoust Soc Am 92(1): 178–188.

Allen JB, Fahey PF (1993) Distortion product emission as a probe of cochlear mechanical resonance. In: Duifhuis H, Horst JW, van Dijk P, van Netten SM (eds) Biophysics of Hair Cell Sensory Systems. Singapore: World Scientific, pp. 296–303.

Allen JB, Fahey PF (1993) A second cochlear-frequency map that correlates distortion product and neural tuning measurements. J Acoust Soc Am 94(2): 809–816.

Allen JB, Sondhi MM (1979) Cochlear macromechanics: time domain solutions. J Acoust Soc Am 66:123–132.

Askeland DR (1989) The Science and Engineering of Materials, 2nd Ed. Boston: PWS-KENT.

Bialek WS, Wit HP (1984) Quantum limits to oscillator stability: theory and experiments on acoustic emissions from the human ear. Phys Lett 104A:173–178.

Brownell W, Bader C, Bertrand D, de Ribaupierre Y (1985) Evoked mechanical responses of isolated cochlear outer hair cells. Science 227:194–196.

Brundin L, Flock Å, Khanna SM, Ulfendahl M (1991) Frequency-specific position shift in the guinea pig organ of Corti. Neurosci Lett 128:77–80.

Carney LH (1993) A model for the responses of low-frequeny auditory-nerve fibers in cat. J Acoust Soc Am 93(1):401–417.

Chadwick RS, Cole JD (1979) Modes and waves in the cochlea. Mech Res Commun 6:177–184.

Cody AR (1992) Acoustic lesions in the mammalian cochlea: implications for the spatial distribution of the "active process". Hear Res 62:166–172.

Cohen A, Furst M (1993) Cochlear model for rate suppression based on cochlear amplifiers dynamics. In: Duifhuis H, Horst JW, van Dijk P, van Netten SM (eds) Biophysics of Hair Cell Sensory Systems. Singapore: World Scientific, pp. 323–329.

Dallos P (1992) The active cochlea. J Neurosci 12:4575–4585.

Davis H (1983) An active process in cochlear mechanics. Hear Res 9:79–90.

de Boer E (1983a) No sharpening? A challenge for cochlear mechanics. J Acoust Soc Am 73:567–573.

de Boer E (1983b) On active and passive cochlear models—Towards a generalized analysis. J Acoust Soc Am 73:574–576.

de Boer E (1990a) Wave propagation modes and boundary conditions for the Ulfendahl-Flock-Khanna preparation. In: Dallos P, Geisler CD, Matthews JW, Ruggero MA, Steele CR (eds) The Mechanics and Biophysics of Hearing. New York: Springer-Verlag, pp. 333–339.

de Boer E (1990b) Can shape deformations of the organ of Corti influence the travelling wave in the cochlea? Hear Res 44:83–92.

de Boer E (1993) The sulcus connection. On a mode of participation of outer hair cells in cochlear mechanics. J Acoust Soc Am 93:2845–2859.

de Boer E, Kuyper P (1968) Triggered correlation. IEEE Trans Biomed Eng BME 15-3:169–179.

Diependaal RJ, Viergever MA (1989) Nonlinear and active two-dimensional cochlear models: time-domain solution. J Acoust Soc Am 85:803–812.

Echteler SM, Fay RR, Popper AN (1994) Structure of the mammalian cochlea. In: Fay RR, Popper AN (eds) Comparative Hearing in Mammals. New York: Springer-Verlag, pp. 134–171.

Engström H, Engström B (1978) Structure of hairs on cochlear sensory cells. Hear Res 1:49–66.

Fettweis A (1986) Wave digital filters: theory and practice. Proc IEEE 74:270–327.

Friedman DH (1990) Implementation of a nonlinear wave-digital-filter cochlear model. In: Proceedings of the IEEE International Conference on Acoustics and Speech Signal Process, Albuquerque, NM, pp. 397–400.

Furst M (1990) Nonlinear transmission line model can predict the statistical properties of spontaneous otoacoustic emission. IN: Dallos P, Geisler CD, Matthews JW, Ruggero MA, Steele CR (eds) The Mechanics and Biophysics of Hearing. New York: Springer-Verlag, pp. 380–386.

Furst M, Lapid M (1988) A cochlear model for acoustic emissions. J Acoust Soc Am 72:717–726.

Furst M, Goldstein JL (1982) A cochlear nonlinear transmission-line model compatible with combination tone psychophysics. J Acoust Soc Am 72:717–726.

Furst M, Reshef (Haran) I, Attias J (1992) Manifestation of intense noise stimulation on spontaneous otoacoustic emission and threshold microstructure: experiment and model. J Acoust Soc Am 91:1003–1014.

Galambos R, Davis H (1943) The response of single auditory-nerved fibers to acoustic stimulation. J Neurophysiol 6:39–57.

Geisler CD (1986) A model of the effect of OHC motility on cochlear vibrations. Hear Res 18:125–131.

Geisler CD (1991) A cochlear model using feedback from motile outer hair cells. Hear Res 54:105–117.

Geisler CD (1993) A realizable cochlear model using feedback from motile outer hair cells. Hear Res 68:253–262.

Geisler CD, Hubbard AE (1972) New boundary conditions and results for the Peterson-Bogert model of the cochlea. J Acoust Soc Am 52:1629–1634.

Geisler CD, Hubbard AE (1975) The compatibility of various measurements on the ear as related by a simple model. Acustica 33:220–222.

Geisler CD, Shan X (1990) A model for cochlear vibrations based on feedback from motile outer hair cells. In: Dallos P, Geisler CD, Matthews JW, Ruggero MA, Steele CR (eds) The Mechanics and Biophysics of Hearing. New York: Springer-Verlag, pp. 85–95.

Geisler CD, Bendre A, Liotopoulos FK (1993) Time-domain modeling of a nonlinear, active model of the cochlea. In: Duifhuis H, Horst JW, van Dijk P, van Netten SM (eds) Biophysics of Hair Cell Sensory Systems. Singapore: World Scientific, pp. 330–337.

Ghitza O (1986) Auditory nerve representation as a front-end for speech recognition in a noisy environment. Comput Speech Lang 1:109–130.

Giguere C, Woodland PC (1994) A computational model of the auditory periphery for speech and hearing research. I. Ascending path. J Acoust Soc Am 95: 331–342.

Goblick TJ, Pfeiffer RR (1969) Time-domain measurements of cochlear non-linearities using combination click stimuli. J Acoust Soc Am 46:924–938.

Gold T (1948) Hearing II. The physical basis of the action of the cochlea. Proc R Soc Lond Ser B 135:492–498.

Goldstein JL (1990a) Mathematical analysis of a nonlinear model for hybrid filtering in the cochlea. In: Dallos P, Geisler CD, Matthews JW, Ruggero M, Steele CR (eds) Mechanics and Biophysics of Hearing. Berlin: Springer, pp. 387–394.

Goldstein JL (1990b) Modeling rapid waveform compression on the basilar membrane as multiple-bandpass-nonlinearity filtering. Hear Res 49:39–60.

Goldstein JL (1991) Modeling the nonlinear cochlear mechanical basis of psycho-physical tuning. J Acoust Soc Am 90(A):2267–2268.

Goldstein JL (1992a) Identification of the cochlear mechanical response effective in hearing. In: Abstracts, Association for Research in Otolaryngology 15th Midwinter Meeting, St. Petersburg, FL, Feb. 2–6, 1992, p. 273.

Goldstein JL (1992b) Changing roles in the cochlea: bandpass filtering by the organ of Corti and additive amplification on the basilar membrane. J Acoust Soc Am 92:s2407.

Goldstein JL (1993) Exploring new principles of cochlear operation: bandpass filtering by the organ of Corti and additive amplification on the basilar membrane. In: Duifhuis H, Horst JW, van Dijk P, van Netten SM (eds) Biophysics of Hair Cell Sensory Systems. Singapore: World Scientific, pp. 315–322.

Gonzales DA (1994) A traveling-wave amplifier model of the cochlea. Thesis, Boston University, Boston, MA.

Greenwood DD (1961) Critical bandwidth and the frequency coordinates of the basilar membrane. J Acoust Soc Am 33:1344–1356.

Greenwood DD (1990) A cochlear frequency-position function for several species— 29 years later. J Acoust Soc Am 87:2592–2605.

Hall JL (1974) Two tone distortion products in a nonlinear model of the basilar membrane. J Acoust Soc Am 56:1818–1828.

Hall JL (1977a) Two-tone suppression in a nonlinear model of the basilar membrane. J Acoust Soc Am 61:802–810.

Hall JL (1977b) Spatial differentiation as an auditory "second filter": assessment on a nonlinear model of the basilar membrane. J Acoust Soc Am 61:520–524.

Henson MM, Burridge K, Fizpatrick D, Jenkins DB, Pillsbury HC, Henson OW (1985) Immunocytochemical localization of contractile and contraction associated proteins in the spiral ligament of the cochlea. Hear Res 20:207–214.

Hewitt MJ, Meddis R (1991) An evaluation of eight computer models of mammalian inner hair-cell function. J Acoust Soc Am 90:904–917.

Holdsworth J, Nimmo-Smith I, Patterson RD, Rice P (1988) Implementing a gammatone filter bank. SVOS Final Report: The Auditory Filterbank. Annex C Report 2341 MRC APU, Cambridge, MA.

Holmes M (1980) An analysis of a low-frequency model of the cochlea. J Acoust Soc Am 68:482–488.

Howard J, Hudspeth AF (1988) Compliance of the hair bundle associated with gating of mechanoelectric transduction channels in the bullfrog's saccular hair cell. Neuron 1:189–199.

Hubbard AE (1976) A digital simulation of cochlear hydrodynamics. Model Simul 7:1278–1281.

Hubbard AE (1986) Cochlear emissions in a one-dimensional model of cochlear hydromechanics. Hear Res 21:74–81.

Hubbard AE (1992) A traveling-wave amplifier model of the cochlea. In: Abstracts, 15th Association for Research in Otolaryngology Midwinter Meeting, St. Petersburg, FL, Feb. 2–6, 1992, p. 156

Hubbard AE (1993) A traveling-wave amplifier model of the cochlea. Science 259:68–71.

Hubbard AE, Geisler CD (1972) A hybrid-computer model of the cochlear partition. J Acoust Soc Am 51:1895–1903.

Hubbard AE, Mountain DC (1983) Alternating current delivered into the scala media alters sound pressure at the eardrum. Science 22:510–512.

Hubbard AE, Gonzales D, Mountain DC (1993) A nonlinear traveling-wave amplifier model of the cochlea. In: Duifhuis H, Horst JW, van Dijk P, van Netten SM (eds) Biophysics of Hair Cell Sensory Systems. Singapore: World Scientific, pp. 370–376.

International Team for Ear Research (ITER) (1989) Cellular vibration and motility in the organ of Corti. Acta Otolaryngol Suppl 467.

Johannesma PIM (1972) The pre-response stimulus ensemble of neurons in the cochlear nucleus. In: Proceedings of the Symposium on Hearing Theory. Eindhoven: IPO, pp. 58–69.

Johannesma PIM (1980) Narrow band filters and active resonators. Comments on papers by D.T. Kemp & R.A. Chum, and H.P. Wit & R.J. Ritsma. In: van den Brink G, Bilson FA (eds) Psychophysical, Physiological and Behavioural Studies in Hearing. Delft: Delft University Press, pp. 62–63.

Johnstone BM, Boyle AJ (1967) Basilar membrane vibration examined with the Mössbauer technique. Science 158:389–390.

Johnstone BM, Taylor KJ, Boyle AJ (1970) Mechanics of the guinea pig cochlea. J Acoust Soc Am 47:504–509.

Jones K, Tubis A, Long GR, Burns EM, Strickland EA (1986) Interactions among multiple spontaneous otoacoustic emissions. In: Allen JB, Hall JL, Hubbard A, Neely ST, Tubis A (eds) Peripheral Auditory Mechanisms. Berlin: Springer, pp. 266–273.

Kates JM (1991) A time-domain digital cochlear model. IEEE Trans Sig Proc 39:2573–2592.

Kates JM (1993) Accurate tuning curves in a cochlear model. IEEE Trans Speech Aud Proc 1(4):453–462.

Kemp DT (1978) Stimulated acoustic emissions from within the human auditory system. J Acoust Soc Am 64:1386–1391.

Khanna SM, Leonard DGB (1982) Basilar membrane tuning in the cat cochlea. Science 215:305–306.

Khanna SM, Flock Å, Ulfendahl M (1989) Comparison of tuning of outer hair cells and the basilar membrane in the isolated cochlea. Acta Otolaryngol Suppl 467:151–156.

Khanna SM, Ulfendahl M, Flock Å (1989) Waveforms and spectra of cellular vibrations in the organ of Corti. Acta Otolaryngol Suppl 467:189–193.

Khanna SM, Ulfendahl M, Flock (1993) Level dependence of cellular responses in the guinea-pig cochlea. In: Duifhuis H, Horst JW, van Dijk P, van Netten SM (eds) Biophysics of Hair Cell Sensory Systems. Singapore: World Scientific, pp. 266–271.

Kim DO (1986a) A review of nonlinear and active cochlear models. In: Allen JB, Hall JL, Hubbard A, Neely ST, Tubis A (eds) Lecture Notes in Biomathematics, vol. 64, Peripheral Auditory Mechanisms. New York: Springer-Verlag, pp. 239–247.

Kim DO (1986b) Active and nonlinear biomechanics and the role of the outer-hair-cell subsystem in the mammalian auditory system. Hear Res 22:105–114.

Kim DO, Molnar CE, Pfeiffer RR (1973) A system of nonliner differential equations modeling basilar-membrane motion. J Acoust Soc Am 54:1517–1529.

Kolston PJ (1988) Sharp mechanical tuning in a micromechanical cochlear model. J Acoust Soc Am 83:1481–1486.

Kolston PJ, Smoorenburg GF (1990) Does the cochlear amplifier produce reactive or resistive forces? In: Dallos P, Geisler CD, Matthews JW, Ruggero MA, Steele CR (eds) The Mechanics and Biophysics of Hearing. New York: Springer-Verlag, pp. 96–105.

Kolston PJ, Viergever MA (1989) Realistic basilar membrane tuning does not require active processes. In: Wilson JP, Kemp DT (eds) Cochlear Mechanisms—Structure, Function and Models. London: Plenum Press, pp. 415–424.

Kolston PJ, Viergever MA, de Boer E, Diependaal RJ (1989) Realistic mechanical tuning in a micromechanical cochlear model. J Acoust Soc Am 86:133–140.

Kolston PJ, Viergever MA, de Boer E, Smoorenburg GF (1990) What type of force does the cochlear amplifier produce? J Acoust Soc Am 88:1794–1801.

Kronester-Frei A (1979) Localization of the marginal zone of the tectorial membrane in situ, unfixed and in vivo-like ionic milieu. Arch Otorhinolaryngol 224:3–9.

Lazzaro J, Wawrzynek J, Mahowald M, Sivilotti M, Gillespie D (1993) Silicon auditory processors as computer peripherals. IEEE Trans Neural Networks 4:523–528.

LePage EL (1987) Frequency-dependent self-induced bias of the basilar membrane and its potential for controlling sensitivity and tuning in the mammalian cochlea. J Acoust Soc Am 82:139–154.

LePage EL, Hubbard AE (1986) Basilar membrane motion in guinea pig cochlea exhibits frequency-dependent DC offset. In: Allen JB, Hall JL, Hubbard A, Neely ST, Tubis A (eds) Peripheral Auditory Mechanisms. Berlin: Springer, pp. 274–281.

Lesser MB, Berkley DA (1972) Fluid mechanics of the cochlea. J Fluid Mech 51:497–512.

Liberman MC (1978) Auditory-nerve response from cats raised in a low-noise chamber. J Acoust Soc Am 63:442–455.

Lien M (1973) A mathematical model of the mechanics of the cochlea. D.Sc. dissertation, Washington University, St. Louis, MO.

Lien MD, Cox JR (1974) A mathematical model of the mechanics of the cochlea. J Acoust Soc Am 55(A):432.

Lighthill MJ (1981) Energy flow in the cochlea. J Fluid Mech 106:149–213.

Lighthill MJ (1991) Biomechanics of hearing sensitivity. J Vibr Acoust 113:1–13.

Lim DJ (1980) Cochlea anatomy related to cochlear micromechanics. A review. J Acoust Soc Am 67:1686–1685.

Liu W, Andreou AG, Goldstein MH (1992) Voiced-speech representation by an analog silicon model of the auditory periphery. IEEE Trans Neural Networks 3:477–487.

Long GR, Tubis A, Jones KL (1991) Modeling synchronization and suppression of spontaneous otoacoustic emissions using Van der Pol oscillators: effects of aspirin administration. J Acoust Soc Am 89:1201–1212.

Lyon RF (1982) A computational model of filtering, detection, and compression in the cochlea. In: Proceedings of the IEEE International Conference on Acoustics, Speech, and Signal Processing, Paris, France, May 1982.

Lyon RF, Dyer L (1986) Experiments with a computational model of the cochlea. In: Proceedings of the IEEE International Conference on Acoustics, Speech, and Signal Processing, Tokyo, Japan, April 1986.

Lyon RF, Lauritzen N (1985) Processing speech with the multisignal serial signal processor. In: Proceedings of the IEEE International Conference on Acoustics, Speech, and Signal Processing, Tampa, FL, March 1985.

Lyon RF, Mead C (1988) An analog electronic cochlea. IEEE Trans Acoust Speech Sig Proc 36:1119–1134.

Mammano F, Ashmore JF (1993) Reverse transduction measured in the isolated cochlea by laser Michelson interferometry. Nature 365:838–841.

Mammano F, Nobili R (1993) Biophysics of the cochlea: linear approximation. J Acoust Soc Am 93:3320–3332.

Matthews JW (1980) Mechanical Modeling of Nonlinear Phenomena Observed in the Peripheral Auditory System. Ph.D. thesis, Washington University, St. Louis (MO).

Matthews JW, Molnar CE (1986) Modeling of intracochlear and ear canal distortion product $2f_2$-f_1. In: Allen JB, Hall JL, Hubbard A, Neely ST, Tubis A (eds) Peripheral Auditory Mechanisms. New York: Springer-Verlag, pp. 258–265.

Mead C (1989) Analog VLSI and Neural Systems. Reading: Addison-Wesley.

Meddis R (1986) Simulation of mechanical to neural transduction in the auditory receptor. J Acoust Soc Am 79:702–711.

Meddis R, Hewitt MJ, Shackleton TM (1990) Non-linearity in a computational model of the response of the basilar membrane. In: Dallos P, Geisler CD, Matthews JW, Ruggero MA, Steele CR (eds) The Mechanics and Biophysics of Hearing. New York: Springer-Verlag, pp. 403–410.

Minorsky N (1962) Nonlinear Oscillations. Princeton: Van Nostrand.

Mountain DC (1980) Changes in endolymphatic potential and crossed olivocochlear bundle stimulation alter cochlear mechanics. Science 210:71–72.

Mountain DC, Hubbard AE, McMullen TA (1983) Electromechanical processes in the cochlea. In: de Boer E, Viergever MA (eds) Mechanics of Hearing. Delft: Delft University Press, pp. 119–126.

Nedzelnitsky V (1974) Measurements of sound pressure in the cochlea of anesthetized cats. In: Zwicker E, Terhardt E (eds) Facts and Models in Hearing. New York: Springer-Verlag, pp. 45–53.

Nedzelnitsky V (1980) Sound pressures in the basal turn of the cat cochlea. J Acoust Soc Am 68:1676–1689.

Neely ST (1981a) Finite difference solution of a two-dimensional mathematical model of the cochlea. J Acoust Soc Am 69:1386–1393.

Neely ST (1981b) Fourth-order partition dynamics for a two-dimensional model of the cochlea. Ph.D. thesis, Washington University, St. Louis, MO.

Neely ST (1993) A model of cochlear mechanics with OHC motility. J Acoust Soc Am 94:137–146.

Neely ST, Kim DO (1983) An active cochlear model showing sharp tuning and high sensitivity. Hear Res 9:123–130.

Neely ST, Kim DO (1986) A model for active elements in cochlear biomechanics. J Acoust Soc Am 79:1472–1480.

Neely ST, Stover LJ (1993) Otoacoustic emissions from a non-linear, active model of cochlear mechanics. In: Duifhuis H, Horst JW, van Dijk P, van Netten SM (eds) Biophysics of Hair Cell Sensory Systems. Singapore: World Scientific, pp. 64–71.

Nuttall AL, Dolan DF (1993) Two tone suppression of inner hair cell and basilar membrane responses in the guinea pig. J Acoust Soc Am 93:390–400.

Nuttall AL, Dolan DF, Avinash G (1990) Measurements of basilar membrane tuning and distortion with laser Doppler velocimetry. In: Dallos P, Geisler CD, Matthews JW, Ruggero MA, Steele CR (eds) The Mechanics and Biophysics of Hearing. New York: Springer-Verlag, pp. 288–295.

Olson E, Mountain DC (1991) In vivo measurement of basilar membrane stiffness. J Acoust Soc Am 89:1262–1275.

Olson E, Mountain DC (1993) Probing the cochlear partition's micromechanical properties with measurements of radial and longitudinal stiffness variations. In: Duifhuis H, Horst JW, van Dijk P, van Netten SM (eds) Biophysics of Hair Cell Sensory Systems. Singapore: World Scientific, pp. 280–287.

Olson E, Mountain DC (1994) Mapping the cochlear partition's stiffness to its cellular architecture. J Acoust Soc Am 95:395–400.

Peterson BP, Bogert LC (1950) A dynamical theory of the cochlea. J Acoust Soc Am 22:3–381.

Plassmann W, Peetz W, Schmidt M (1987) The cochlea in gerbelline rodents. Brain Behav Evol 30:82–101.

Ranke OF (1931) Die Gleichrichter-Resonanztheorie. Munich: Lehmann.

Ranke OF (1950) Theory of operation of the cochlea: a contribution to the hydrodynamics of the cochlea. J Acoust Soc Am 22:772–777.

Rhode WS (1971) Observations of the vibration of the basilar membrane in squirrel monkeys using the Mössbauer technique. J Acoust Soc Am 64:158–176.

Rhode WS (1977) Some observations on two-tone interaction measured with the Mössbauer effect. In: Evans EF, Wilson JP (eds) Psychophysics and Physiology of Hearing. London: Academic Press, pp. 27–38.

Rhode WS (1978) Some observations on cochlear mechanics. J Acoust Soc Am 64:158–176.

Rhode WS, Cooper NP (1993) Two-tone suppression and distortion production on the basilar membrane in the hook region of cat and guinea-pig cochleae. Hear Res 66:31–45.

Robles LR (1973) Measurements on the transient response of the basilar membrane using the Mössbauer effect. Ph.D. Thesis, University of Wisconsin, Madison, WI.

Robles L, Rhode WS, Geisler CD (1976) Transient response of the basilar membrane measured in squirrel monkeys using the Mössbauer effect. J Acoust Soc Am 59:926–939.

Robles L, Ruggero MA, Rich NC (1986) Basilar membrane mechanics at the base of the chinchilla cochlea. Input-output functions, tuning curves, and response phases. J Acoust Soc Am 80:1364–1374.

Robles L, Ruggero MA, Rich NC (1991) Two-tone distortion in the basilar membrane of the cochlea. Nature 349:413–414.

Ruggero MA, Rich NC (1991a) Application of a commercially-manufactured Doppler-shift laser velocimeter to the measurement of basilar-membrane vibration. Hear Res 51:215–230.

Ruggero MA, Rich NC (1991b) Furosemide alters organ of Corti mechanics: evidence for feeback of outer hair cells upon the basilar membrane. J Neurosci 11:1057–1067.

Ruggero MA, Rich NC, Recio A (1992) Basilar membrane responses to clicks. In: Cazals Y, Demany L, Horner K (eds) Auditory Physiology and Perception. Oxford: Pergamon Press, pp. 85–92.

Ruggero MA, Robles L, Rich NC (1992) Two-tone distortion in the basilar membrane of the cochlea: mechanical basis of auditory-nerve rate suppression. J Neurophysiol (Bethesda) 68:1087–1099.

Ruggero MA, Rich NC, Robles L, Shivapuja B (1990) Middle ear responses in the chinchilla and its relationship to the mechanics at the base of the cochlea. J Acoust Soc Am 87:1612–1629.

Sachs MB, Kiang NYS (1968) Two-tone inhibition in auditory nerve fibres. J Acoust Soc Am 43:1120–1128.

Sachs MB, Hubbard AE (1980) Temporal aspects of two-tone suppression: low frequency suppressors effect on high characteristic frequency fibers. Hear Res 4:309–324.

Sellick PM, Patuzzi R, Johnstone BM (1982) Measurement of basilar membrane motion in the guinea pig using the Mössbauer technique. J Acoust Soc Am 72:131–141.

Sellick PM, Yates GK, Patuzzi R (1983) The influence of Mössbauer source size and position on phase and amplitude measurements of the guinea pig basilar membrane. Hear Res 10:101–108.

Seneff S (1988) A joint synchrony/mean-rate model of auditory speech processing. J Phonetics 16(1):55–76.

Shamma S (1988) The acoustic features of speech sounds in a model of auditory processing: vowels and voiceless fricatives. J Phonetics 10:77–91.

Siebert WM (1974) Ranke revisited—a simple short-wave cochlear model. J Acoust Soc Am 56:594–600.

Steele CR (1974) Behavior of the basilar membrane with pure-tone excitation. J Acoust Soc Am 55:148–162.

Steele CR (1976) Cochlear mechanics. In: Keidel WD, Neff WD (eds) Handbook of Sensory Physiology, Vol. 3, Auditory System: Clinical and Special Topics. Berlin: Springer, pp. 443–478.

Steele CR, Taber LA (1979a) Comparison of WKB calculations and finite difference calculations for a two-dimensional cochlear model. J Acoust Soc Am 65:1001–1006.

Steele CR, Taber LA (1979b) Comparison of WKB calculations and experimental results for a three-dimensional cochlear model. J Acoust Soc Am 65:1007–1018.

Steele CR, Taber LA (1981) Three-dimensional model calculations for guinea pig cochlea. J Acoust Soc Am 69:1107–1111.

Strube HW (1982) Time-varying wave digital filters for modeling analog systems. IEEE Trans Acoust Speech Sig Proc 30:864–868.

Strube HW (1985) A computationally efficient basilar membrane model. Acustica 58:207–214.

Taber LA, Steele CR (1981) Cochlear model including three-dimensional fluid and four modes of partition flexibility. J Acoust Soc Am 70:426–436.

Talmadge CL, Tubis A (1993) On modeling the connection between spontaneous and evoked otoacoustic emissions. In: Duifhuis H, Horst JW, van Dijk P, van Netten SM (eds) Biophysics of Hair Cell Sensory Systems. Singapore: World Scientific, pp. 25–32.

Talmadge CL, Long GR, Murphy WJ, Tubis A (1990) Quantitative evaluation of limit-cycle oscillator models of spontaneous otoacoustic emissions. In: Dallos P, Geisler CD, Matthews JW, Ruggero MA, Steele CR (eds) The Mechanics and Biophysics of Hearing. New York: Springer-Verlag, pp. 235–242.

Teich MC, Heneghan C, Khanna SM, Flock Å, Brundin L, Ulfendahl M (1993) Analysis of dynamical motion of sensory cells in the organ of Corti using the spectrogram. In: Duifhuis H, Horst JW, van Dijk P, van Netten SM (eds) Biophysics of Hair Cell Sensory System. Singapore: World Scientific, pp. 272–279.

Ulfendahl M, Khanna SM (1993) Tuning characteristics of the hearing organ in an in vitro preparation of the gerbil temporal bone. Pflugers Archiv 424:95–104.

Ulfendahl M, Khanna SM, Flock Å (1991) Effects of opening and resealing the cochlea on the mechanical response in the isolated temporal bone preparation. Hear Res 57:31–37.

Ulfendahl M, Khanna SM, Lofstrand P (1993) Changes in the mechanical tuning characteristics of the hearing organ following acoustic overstimulation. Eur J Neurosci 5:713–723.

Viergever MA, Diependaal RJ (1983) Simultaneous amplitude and phase match of cochlear model calculations and basilar membrane vibration data. In: de Boer E, Viergever MA (eds) Mechanics of Hearing. The Hague: Martinus Nijhoff/Delft: Delft University Press, pp. 53–61.

van den Raadt MPMG, Duifhuis H (1990) A generalized Van der Pol-oscillator cochlea model. In: Dallos P, Geisler CD, Matthews JW, Ruggero MA, Steele CR (eds) The Mechanics and Biophysics of Hearing. New York: Springer-Verlag, pp. 227–234.

von Békésy G (1960) Experiments in Hearing. New York: McGraw-Hill.

Watts L, Kerns D, Lyon R, Mead C (1992) Improved implementation of the silicon cochlea. IEEE J Solid State Circ 27:692–700.

Wilson JP (1980) Evidence for a cochlear origin for acoustic re-emissions, threshold fine-structure and tinnitus. Hear Res 2:233–252.

Wilson JP, Johnstone JR (1973) Basilar membrane correlates of the combination tone $2f_1 - f_2$. Nature 241:206–207.

Xue S, Mountain D, Hubbard A (1993) Direct measurement of electrically-evoked basilar membrane motion. In: Duifhuis H, Horst JW, van Dijk P, van Netten SM (eds) Biophysics of Hair Cell Sensory Systems. Singapore: World Scientific, pp. 361–369.

Zweig G (1991) Finding the impedance of the organ of Corti. J Acoust Soc Am 89:1221–1254.

Zwicker E (1979) A model describing nonlinearities in hearing by active processes with saturation at 40 dB. Biol Cybern 35:243–250.

Zwicker E (1986) A hardware cochlear nonlinear preprocessing model with active feedback. J Acoust Soc Am 80:146–153.

Zwislocki JJ (1950) Theory of the acoustical action of the cochlea. J Acoust Soc Am 22:778–784.

Zwislocki JJ (1980) Five decades of research on cochlear mechanics. J Acoust Soc Am 67:1679–1685.

Zwislocki JJ, Cefaratti LK (1989) Tectorial membrane. II: Stiffness measurements in vivo. Hear Res 42:211–227.

Zwislocki JJ, Kletsky EJ (1979) Tectorial membrane: a possible effect on frequency analysis in the cochlea. Science 203:639–641.

4
Computational Analysis of Hair Cell and Auditory Nerve Processes

DAVID C. MOUNTAIN AND ALLYN E. HUBBARD

1. Introduction

The mammalian cochlea must transduce an acoustic signal that has a dynamic range of 100 dB and a bandwidth that ranges up to 150 kHz in some species. The high information rate of the acoustic environment must be encoded in the activity of auditory nerve fibers that have average firing rates of only 200 pulses/sec or less. This encoding process must preserve the temporal information in the acoustic signal because interaural time delays in the microsecond range are used for source localization and because many natural acoustic sources produce signals with similar spectral characteristics, but different temporal characteristics.

The processing and encoding of information in the cochlea takes place in several stages. The first level of processing is implemented mechanically and involves amplification and filtering of the acoustic waveform along the cochlear partition. This process appears to be the result of electrically induced contractions of the outer hair cells (OHC). The second stage involves the transduction and rectification of the outputs of the mechanical filters by the inner hair cells (IHC). The final stage involves the encoding of the IHC receptor potential as firing patterns in auditory nerve (AN) fibers by the IHC synapse.

Accurate models of how these processes work are necessary for the development of theories for signal detection and classification in the central nervous system. Accurate models can also be used as the basis for efficient signal processing algorithms in human-engineered signal detection and classification systems. To this end, many groups have been developing models that range from detailed, biophysically based models of single cells or ionic channels within cells to comprehensive signal processing-based models of the cochlea. These models are important not only because they allow more accurate prediction of auditory nerve firing patterns but also because they provide insights into the algorithms used by the physiological system. An understanding of auditory physiology at the algorithmic level also allows easy transfer of the physiological findings to signal processing applications.

This chapter begins with a general overview of cochlear anatomy and physiology and then discusses some of the goals of modeling work related to the hair cell membrane and synaptic properties. The chapter then progresses through a review of hair cell membrane properties followed by biophysical models of these properties. Synaptic physiology is reviewed next, followed by biophysically based models of the IHC synapse. The biophysically based models are then compared to signal processing based models. This review is not intended to be exhaustive, but instead attempts to cover in some detail those models that either are of significant historical interest or represent the current state of the art in auditory modeling.

2. Anatomical and Physiological Overview

As was discussed in Chapter 3 of this volume, two membranes (Reissner's and basilar) divide the cochlear duct into three longitudial compartments: the scala vestibuli, the scala media, and the scala tympani. The basilar membrane separates the scala media from the scala vestibuli, and, together with the organ of Corti, forms the cochlear partition. Reissner's membrane separates the scala vestibuli and the scala media and appears to serve as an ionic barrier between them. The fluid in the scala vestibuli and scala tympani, called the perilymph, is similar in ionic composition to the cerebrospinal fluid. It is high in Na^+ and low in K^+. The ionic composition of the endolymph in the scala media, however, is more similar in concentration to intracellular fluids. It is high in K^+ and low in Na^+ as well as being quite low in Ca^{2+}. The fluid in the scala media has an electrical potential that is 90 mV more positive than that of scala vestibuli and scala tympani. These ionic and potential gradients are maintained by specialized cells located in the stria vascularis, which makes up much of the outer wall of the scala media.

The apical membranes of the hair cells, and some of the supporting cells in the organ of Corti, are tightly joined together to form a platelike structure on top of the organ of Corti called the reticular lamina. These tight junctions form an ionic barrier between the scala media and the organ of Corti. In contrảst, the basilar membrane appears to be reasonably permeable to ions with the result that the ionic composition of the fluid within the organ of Corti is very much like that of perilymph. As a result of these anatomical features, the electrochemical gradient across the apical membranes of the hair cells differs considerably from the gradient across the basolateral membranes.

The inner and outer hair cells have quite different innervation patterns (Ryugo 1992). Each IHC is innervated by 10–30 afferent fibers, which are unbranched except possibly in the cochlear apex (Liberman, Dodds, and Pierce 1990). These fiber originate from type I neurons (large cell bodies) in the spiral ganglion of the cochlea and project radially, only synapsing with the IHCs. The OHCs are innervated by afferent fibers from the type II neurons (small cell bodies) of the spiral ganglion, which project radially to the OHC region and then turn basally to follow the cochlear spiral; finally, each fiber innervates several OHCs. Each OHC is innervated by 5–15 afferent fibers.

The organ of Corti also receives an extensive efferent innervation (Warr 1992). These fibers have their origin in the superior olivary complex and are usually divided into two groups, the lateral olivocochlear (LOC) and medial olivocochlear (MOC) systems. The principal postsynaptic targets of the LOC fibers are the peripheral processes of the radial fibers that innervate the IHCs. Very little is known about the physiology of the LOC system, so it is not considered in this chapter. The MOC system has as its principal target the OHCs. The MOC fibers branch to innervate a number of OHCs spanning 1–2 mm. As a result, even though the MOC fibers innervating one cochlea may only number in the hundreds, a single OHC may receive a number of efferent endings. The MOC system has received considerable attention since it was demonstrated that stimulation of the MOC could modify cochlear mechanics (Mountain 1980), and the MOC is now postulated to be a key element in the regulation of cochlear sensitivity (Kim 1986).

The mechanical properties of the cellular elements of the organ of Corti have become an important area of study. The OHCs have become the focus of most of the work since the discovery of voltage-dependent length changes (Brownell et al. 1985). This novel form of motility is believed to be the basis for the high sensitivity of the mammalian cochlea as well as other phenomena such as otoacoustic emissions. These "sounds from the ear" can be observed to arise spontaneously or in response to acoustic or electrical stimulation (Mountain and Hubbard 1989; Probst, Lonsbury-Martin, and Martin 1991). The supporting cells are also coming under scrutiny because these, along with the OHCs, may form the major contribution to the stiffness of the cochlear partition (Olson and Mountain 1991, 1994).

3. Modeling Goals

Models of cochlear hair cells and the auditory nerve are usually developed with one of three goals in mind. The most common goal is to accurately represent what is known about the biophysics of specific cellular properties. Models of this type include representations of membrane conductances, synaptic transmission (both afferent and efferent), and OHC voltage-dependent length changes. A second goal is to use models of hair cell receptor potentials to infer the mechanical input to the cell so as to gain greater insight into cochlear micromechanics, because direct mechanical measurements are difficult with existing technology. Signal processing models are aimed at the third goal, which is to capture the salient features of the information processing performed by the hair cells and auditory nerve. These efforts are often intended to produce a computationally efficient preprocessor, either for use with automatic recognition systems such as speech processors or as a way to predict the input to higher level auditory structures for use in interpreting physiological data from those structures. Signal processing models of the auditory periphery are also being used with increasing frequency as preprocessors for psychophysical models of perceptual phenomenon.

4. Hair Cell Membrane Physiology

4.1 Background

The inner and outer hair cells are located in a complex electrochemical environment. Their apical membranes face the endolymph of scala media with its intracellular-like ionic concentrations, while the basolateral membranes face a fluid with ionic concentrations similar to the perilymph. In addition, because the electrical potentials of the endolymph and perilymph are quite different, the voltage gradients across the apical and basolateral membranes differ. The IHCs have a resting potential of approximately $-35\,\mathrm{mV}$ (with respect to an extracochlear electrode) while the OHCs have a resting potential closer to $-70\,\mathrm{mV}$ (Dallos, Santos-Sacchi, and Flock 1982; Russell, Cody, and Richardson 1986). As a result of the similarity between the ionic composition of the cell interior and the endolymph, the driving force for ionic current flow through the apical membrane results almost entirely from the electrical gradient. In contrast, the driving force for ionic current flow through the basolateral membrane is a combination of the electrical and chemical gradients.

Ionic current flow in hair cells is caused by a combination of three types of ionic channels: tension-gated channels, voltage-gated channels, and ligand-gated channels. Ionic current flow through the hair cell apex appears to pass almost entirely through tension-gated channels, which open and close in response to hair-bundle displacement. Ionic current flow through the basolateral membrane is probably caused by all three channel types. The properties of the OHCs are further complicated by the fact that they can undergo voltage-dependent and Ca^{2+}-induced length changes.

A generic hair cell membrane model is illustrated in Figure 4.1. The elements g_A and C_A represent the conductance and capacitance of the apical membrane; E_{SM} represents the potential in scala media; and the elements g_1 through g_N represent conductances in the basolateral

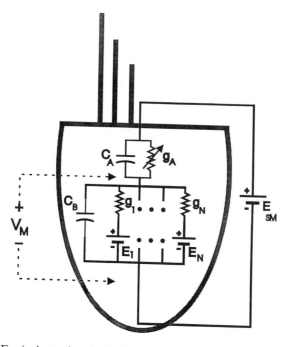

FIGURE 4.1. Equivalent electrical circuit model for a cochlear hair cell. The individual circuit elements are described in the text.

membrane. These can be, for example, stretch-activated conductances, voltage- or ion-dependent conductances, or conductances associated with efferent synapses. E_1 and E_N represent the electrochemical gradients of the ion or ions associated with each conductance. The capacitance of the basolateral membrane is represented by C_B. Most authors assume that the potential in scala media is a constant, which allows C_A and C_B to be combined into a single capacitance, which will be called C_T.

4.2 Tension-Gated Conductances

Davis (1958) first proposed that hair cell mechanoelectric transduction is the result of a conductance change in the apical membrane caused by movement of the sensory hairs. Cochlear hair cells are now known to have more than one type of mechanically sensitive conductance in their membranes. The best known is the tension-gated conductance of the apical membrane (see Howard, Roberts, and Hudspeth 1988 for a review). In addition there are stretch-activated conductances in the OHC basolateral membrane (Ding, Salvi, and Sachs 1991; Iwasa et al. 1991). The tension-gated channels of the apical membrane appear to be located in or near the tips of the stereocilia (Hudspeth 1982; Assad, Shepherd, and Corey 1991; Jaramillo and Hudspeth 1991). The sensory hairs, or stereocilia, are arranged in parallel rows with the heights of the rows changing in a staircase fashion. The stereocilia are interconnected through tip links, which are thin filaments running from the top of a stereocilium to the side of the stereocilium in the next, taller row (Pickles, Comis, and Osborne 1984). When the hair bundle is deflected in the direction of the tallest row of stereocilia, then the tip links will be in tension, and it is believed that this tension is applied to the transduction channels and causes them to open (Pickles and Corey 1992).

The tension-gated channels are most commonly modeled using (see, for example, Sachs and Lecar 1991) energy barrier (Boltzman) models. The free energy of a channel is assumed to be the result of the combination of its conformation and of external forces acting on the channel. The proteins that make up channels are known to fold up into quite complicated configurations. A protein with a given amino acid sequence has many possible conformations that differ through rotations of internal bonds, but most are energetically unfavorable. Most proteins are believed to have only one preferred conformation, but the proteins that make up tension- or voltage-gated channels appear to have several conformations which are energetically favorable.

A schematized energy diagram for a tension-gated channel with two preferred conformations is illustrated in Figure 4.2. The ordinate is the internal energy of the channel and the abscissa is an arbitrary conformational coordinate that represents mechanical distortion of the channel. In the upper panel, the stereocilia are assumed to be in their resting position (X_0) and the energy diagram exhibits two energy wells. The left well corresponds to the channel-closed conformation and the right well to the channel-open conformation. The left well has a minimum energy (G_A) that is less than the minimum energy of the right well (G_B), so it is more likely that the channel will be in the left well or closed state. Thermal vibration of the molecule will cause it to oscillate within the left well, and if the energy level becomes high enough then the channel can cross the energy barrier (G_0) between the wells and move to the right well, causing the channel to be open. Subsequent oscillations will ultimately cause the channel to move randomly back and forth across the barrier, and the channel will randomly open and close. If an external force is applied and the stereocilia are displaced to some new position (X_1), then the energy diagram will change (Fig. 4.2, lower panel). In this example, the external force has resulted in the open state becoming more energetically favorable than the closed state, and so the channel will spend a greater percentage of time in the open state than the closed state.

In computational models of tension-gated channels, each individual channel is assumed to have either two (Howard and Hudspeth 1988; Sachs and Lecar 1991) or three (Corey and Hudspeth 1983) conforma-

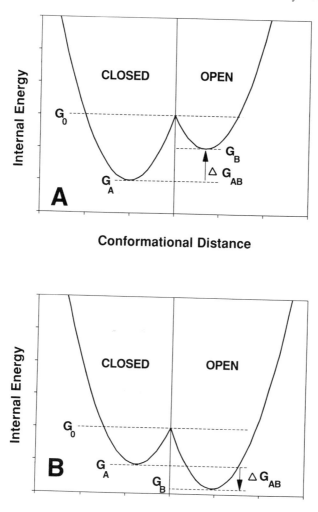

FIGURE 4.2A,B. Energy diagram for a tension-gated channel with two preferred conformations. The ordinate is the internal energy of the channel, and the abscissa is an arbitrary conformational coordinate that represents mechanical distortion of the channel. (A) The stereocilia are assumed to be in their resting position (X_0). (B) The stereocilia are assumed to be displaced to a position (X_1) that increases the channel open probability.

tional states. The channel switches from one state to another with a rate constant that depends on the relative energies of the current state and the height of the energy barrier between the two states. Schematically this is represented by

$$A \underset{K_2}{\overset{K_1}{\rightleftarrows}} B \underset{K_4}{\overset{K_3}{\rightleftarrows}} C \,, \tag{1}$$

where

A = the fraction of channels in state A (a closed state)
B = the fraction of channels in state B (a closed state)
C = the fraction of channels in state C (the open state)
K_1 = the rate constant for transitions from A to B
K_2 = the rate constant for transitions from B to A
K_3 = the rate constant for transitions from B to C
K_4 = the rate constant for transitions from C to B.

In the case of a two-state model, the intermediate state would be eliminated and only two rate constants would be needed. To minimize the number of free parameters in the model, a direct path between state A and state C is usually not included. The rates of change of B and C with respect to time are determined by the following equations:

$$\frac{dB}{dt} = K_1A - K_2B - K_3B + K_4C \tag{2}$$

$$\frac{dC}{dt} = K_3B - K_4C \tag{3}$$

The total number of channels in all states is constant, therefore:

$$A + B + C = 1 \tag{4}$$

Combining Eqs. 2 and 4, one gets the following:

$$\frac{dB}{dt} = -(K_1 + K_2 + K_3)B + (K_4 - K_1)C + K_1 \tag{5}$$

The response of the transduction channels to changes in the displacement of the hair bundle is extremely fast (Corey and Hudspeth 1979), so the kinetics of the channel are not usually included in most computational models. The steady-state solution for the fraction of channels in the open state can be found by setting the derivatives of B and C to zero (Eqs. 3 and 5) and solving for C:

$$C = \frac{1}{1 + \dfrac{K_3}{K_4}\left(1 + \dfrac{K_2}{K_1}\right)} \tag{6}$$

For a two-state model, Eq. 6 simplifies to the following:

$$C = \frac{1}{1 + \dfrac{K_2}{K_1}} \tag{7}$$

The ratios of the rate constants are exponential functions of the energy difference between the states. For example, in Figure 4.2 the energy difference is ΔG_{AB} so the probability (P_{open}) that the channel is open (in state C) is given by

$$P_{open} = \frac{1}{1 + e^{\frac{\Delta G_{AB}}{RT}}} \tag{8}$$

and for a three-state model:

$$P_{open} = \frac{1}{1 + e^{\frac{\Delta G_{AB}}{RT}}\left(1 + e^{\frac{\Delta G_{BC}}{RT}}\right)} \tag{9}$$

where R is the gas constant and T is absolute temperature (K).

The energy differences are most commonly assumed to be a linear function of cilia displacement (Howard and Hudspeth 1988), so for a two-state model the open probability as a function of displacement becomes:

$$P_{open} = \frac{1}{1 + e^{\frac{-x - x_0}{S_0}}} \tag{10}$$

and for the three-state model:

$$P_{open} = \frac{1}{1 + e^{-\frac{x - x_0}{S_0}}\left(1 + e^{-\frac{x - x_1}{S_1}}\right)} \tag{11}$$

where S_0 and S_1 are sensitivity constants and x_1 and x_2 are offset constants. The two-state model has proved to fit data from frog saccular hair cells well (Howard and Hudspeth 1988), but the three-state model provides a much better fit to the data from cochlear hair cells. This difference is because the cochlear hair cell transfer characteristic exhibits asymmetrical saturation, especially for the inner hair cells, while the two-state model exhibits symmetrical saturation. Figure 4.3 illustrates the fit of a three-state model to data from both inner and outer hair cells.

4.3 Voltage-Dependent Conductances

The membranes of hair cells appear to have a variety of conductances in addition to the mechanically sensitive ones described in Section 4.2. The IHCs have a relatively rapid voltage-dependent K^+ conductance (Russell, Cody, and Richardson 1986; Ashmore 1989; Kros and Crawford 1990). In addition the IHCs exhibit a rapidly inactivating current with a positive reversal potential, implying that it is carried by either Na^+ or Ca^{2+}. If

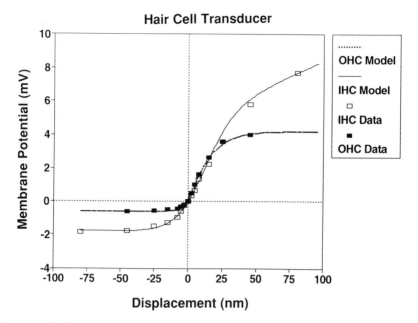

FIGURE 4.3. Comparison of the three-state channel model results to experimental data for inner and outer hair cells. Model parameters for the outer hair cell (OHC): displacement-sensitivity constant (S_0) = 12 nm; displacement-offset constant (x_0) = 5 mm; S_1 = 5 nm; x_1 = 5 nm. Model parameters for the inner hair cell (IHC): S_0 = 85 nm; x_0 = 17 nm; S, = 5 nm; x_1 = 17 nm. To obtain voltage values for comparison purposes, the OHC open probability was multiplied by 4.8 mV and an offset of 0.6 mV subtracted from the product. For the IHC, the open probability was multiplied by 14 mV and an offset of 1.8 mV subtracted from the product. (Data from Russell, Cody, and Richardson 1986.)

the latter is the case, then this current could play a role in synaptic transmission (Crawford and Kros 1989; Rusch et al. 1991). The OHCs have a slower K^+ conductance, which appears to be a Ca^{2+}-activated K^+ conductance (Ashmore and Meech 1986; Gitter, Frömter, and Zenner 1992).

There have been few attempts to develop computational models of voltage-dependent couductances in mammalian hair cells, partly because of the lack of detailed information about the properties of these conductances. Zagaeski (1991) developed a simple model of the voltage-dependent K^+ conductance of the IHCs. He modeled the basolateral IHC membrane with a parallel combination of a linear leakage conductance (g_1) and a voltage-dependent conductance (g_v). The voltage-dependent conductance was modeled using a two-state model (see Eq. 8). This conductance (in nS) as a function of membrane potential (V_m in mV) is given by:

$$gv = \frac{82}{1 + e^{\frac{-(V_m + 29)}{6.5}}} \tag{12}$$

The K^+ reversal potential in the Zagaeski model is $-79\,mV$, g_1 is $13.3\,nS$, and the reversal potential for the leakage current is $-17.3\,mV$.

Little published information is available on the Ca^{2+} conductance of the IHCs. There appears to be both a transient and a noninactivating component in the Ca^{2+} current (Crawford and Kros 1989; Rusch et al. 1991). Mountain, Wu, and Zagaeski (1991) proposed a three-state model (see Eq. 1) for the IHC Ca^{2+} conductance. Although the IHC Ca^{2+} current may be the result of a combination of two different conductances, a single conductance was chosen for simplicity. In this model, the A state is the closed state, the B state is the open state, and the C state is an inactive state. Depolarization of the IHC causes a transient increase in Ca^{2+} conductance as channels move from the A state to the B state. The conductance then decreases as channels close by moving to the C state. K_1 and K_2 are assumed to be voltage dependent, and K_3 and K_4 were made voltage independent to minimize the free parameters. K_1 and K_2 are assumed to be exponential functions of membrane potential, as is the usual case in Boltzmann models. This model was tested extensively as part of an IHC synaptic model by D'Angelo (1992). D'Angelo computed K_1 and K_2 as follows:

$$K_1 = 1000\,e^{\frac{(V_m + 42.1)}{1.25}}$$

$$K_2 = 1000\,e^{\frac{(-42.1 - V_m)}{1.25}} \tag{13}$$

K_3 was $375\,s^{-1}$ and K_4 was $125\,s^{-1}$.

No attempt has been made to model in detail the voltage-dependent Ca^{2+} conductance of the OHCs. Because the OHCs appear to have the combination of a voltage-dependent Ca^{2+} conductance and a Ca^{2+}-activated K^+ conductance, a model proposed by Hudspeth and Lewis (1988a,b) for a similar combination in hair cells of the frog sacculus may be relevant. They used a Hodgkin–Huxley (1952) type of model for the Ca^{2+} current (I_{Ca}):

$$I_{Ca} = g_{Ca}m^3(V_m - E_{Ca}) \tag{14}$$

The gating variable m is found by solving a first-order differential equation:

$$\frac{dm}{dt} = \beta_m(1 - m) - \alpha_m m \tag{15}$$

where the coefficients are voltage dependent.

4.4 Ligand-Gated Conductances

The OHCs would appear to have at least two classes of ligand-gated conductances: the Ca^{2+}-activated K^+ conductance mentioned previously as well as a conductance or conductances associated with the efferent synapses on the OHCs. In contrast, the IHCs appear not to have any ligand-gated conductances. Again one must look to the Hudspeth and Lewis (1988a,b) model for the saccular hair cells for guidance as to how one might model the Ca^{2+}-activated K^+ conductance of the OHC. In their model, two intracellular Ca^{2+} ions must bind to a channel to open it, and a third Ca^{2+} ion can then bind to prolong the opening:

$$C_0 \underset{K_{-1}}{\overset{K_1[Ca^{2+}]}{\rightleftharpoons}} C_1 \underset{K_{-2}}{\overset{K_2[Ca^{2+}]}{\rightleftharpoons}} C_2 \underset{\alpha_C}{\overset{\beta_C}{\rightleftharpoons}} O_2 \underset{K_{-3}}{\overset{K_3[Ca^{2+}]}{\rightleftharpoons}} O_3 \qquad (16)$$

where the subscripts of the Cs and Os refer to the number of Ca^{2+} ions bound and the C_s are the closed states and the Os are the open states. The transition rates in the forward direction (except for C_2 to O_2) depend on the product of a rate constant and the internal Ca^{2+} concentration.

A much simpler approach has been taken by Roddy et al. (1994), who used a piecewise linear approach to model the combination of the OHC Ca^{2+} and Ca^{2+}-activated K^+ conductances as a single instantaneous conductance. These fit the data current–voltage curve of Russell, Cody, and Richardson (1986) with two straight-line segments and a breakpoint at the resting potential. This model combined with a model of the OHC voltage–length relationship (Santos-Sacchi 1991) successfully reproduced the effect of electrical biasing on electrically evoked emissions.

4.5 Electromechanical Transduction

Hair cells can exhibit several forms of motility. The form that has attracted the most interest in the mammalian cochlea is the voltage-dependent length change of the OHCs. First reported by Brownell et al. (1985), it is now under active study in a number of laboratories. The OHC voltage-dependent length changes have been shown to depend on cell membrane potential and not on whole cell current (Santos-Sacchi and Dilger 1988). Ashmore (1989) demonstrated that the length change was associated with charge movement within the cell membrane, which manifests itself as a nonlinear membrane capacitance. This nonlinear capacitance contributes up to 40% of the cell's total capacitance (Santos-Sacchi 1991). It is now generally believed that the voltage-induced length changes are the result of charged transmembrane proteins within the OHC lateral wall undergoing conformational changes in response to changes in the membrane potential (Dallos, Evans, and Hallworth 1991).

Santos-Sacchi (1991) demonstrated that both the charge movement and the length change could be predicted by a two-state Boltzmann distribution. Dallos, Hallworth, and Evans (1993) extended this idea by proposing a model for the voltage-sensitive proteins (motor molecules) that is similar to those used for voltage-dependent channels (see Section 4.3). In their model, the motor molecules exist in one of two conformational states (short and long). The probability of a molecule being in a given state is dependent on the membrane potential of the OHC. A good fit to the available experimental data from isolated OHCs (Fig. 4.4) can be obtained by assuming the probability (P_{short}) of a molecule being in the short state is given as follows (Dallos, Hallworth, and Evans 1993):

$$P_{short} = \frac{1}{1 + e^{-(V_m - V_h)/S_v}} \tag{17}$$

where S_v and V_h are constants. The overall length of an OHC can then be computed:

$$L_{OHC} = L_{max}(1 - \Delta L\, P_{short}) \tag{18}$$

where L_{max} is the maximum OHC length and ΔL is the difference between the maximum and minimum lengths. The sensitivity of the OHC length to changes in membrane potential is greatest when the membrane potential

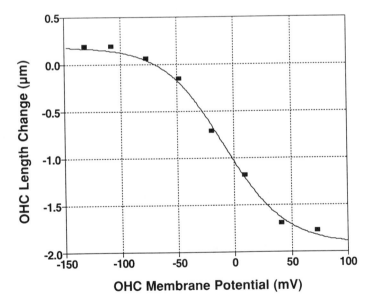

FIGURE 4.4. Comparison of the two-state Boltzmann model for the OHC voltage-dependent length change to experimental data from Santos-Sacchi (1991). Model parameters: voltage-sensitivity constant (S_v) = 26.3 mV; voltage-offset constant (V_h) = −8 mV; difference between maximum and minimum OHC length (ΔL) = 2.1 μm.

is in the range of -8 to $-40\,\mathrm{mV}$ and yet the normal resting potential of the OHCs is $-70\,\mathrm{mV}$. The OHC resting potential actually corresponds to a region of the voltage–length curve where changes in resting potential can have a significant impact on the sensitivity of the voltage-dependent length change. It may be that the cochlear efferents alter cochlear sensitivity by hyperpolarizing the OHCs and as a result reducing the sensitivity of voltage-dependent length change (Roddy et al. 1994).

The modeling approach illustrated in Eqs. 17 and 18 does not fully capture the interaction between the mechanical system and the electrical systems of which the OHCs are a part. An alternative approach has been proposed by Mountain and Hubbard (1994). The fundamental premise is that the OHCs act as piezoelectric transducers. The piezoelectric model emphasizes the bidirectional coupling between the mechanical and electrical system. Weiss (1982) pointed out that in bidirectional coupling the mechanical system affects the electrical impedance of the cell membrane and vice versa. Standard models of piezoelectric devices (Mason 1950) consist of an electrical circuit representing the passive electrical characteristics of the device coupled through a transformer to a second circuit, which is the electrical analog of the passive mechanical properties of the device and its load. In the mechanical circuit voltages correspond to force and currents to velocity.

The piezoelectric model is illustrated in Figure 4.5. The mechanically sensitive channels in the apical membrane produce a receptor current (I_A). The receptor current entering the cell must be balanced by the currents leaving the cell. Some of this current represents ionic current, some of this current leaves as a displacement current in the cell membrane capacitance (C_T), and the rest exits via the transformer. This transformer current represents the charge movement associated with cell motility. The output of the transformer produces a force (F_{OHC}), which acts on the mechanical system resulting in the length change velocity (I_V). The model as shown assumes that the conformational changes in the motor molecules result in changes in cell membrane surface area that translate to a length change because the cell volume is constant. In the case of perfect coupling between the molecular conformation changes and cell length, the OHC impedance (Z_{OHC}) represents the effective mechanical impedance of the motor molecules. The coupling between cell-surface area and cell length can be diminished if the cell membrane buckles or can be compressed. This loss of efficiency can be modeled by adding an additional impedance in parallel with the organ of Corti impedance (Z_{OC}). This membrane buckling impedance would shunt some of I_V, thereby decreasing the motion of the organ of Corti.

The fraction of the receptor current that is transformed into mechanical motion depends on the impedance of the mechanical circuit and the transformer ratio. If the mechanical system is a resonant one, then it is possible that its impedance at resonance, when viewed through the

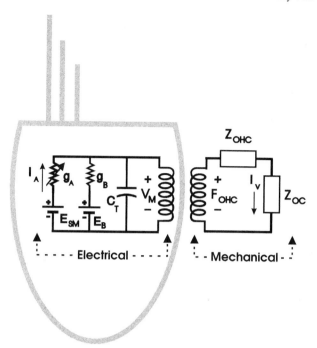

FIGURE 4.5. Equivalent circuit for the OHC piezoelectric model adapted from Mountain and Hubbard (1994). Components within the cell outline correspond to the electrical properties of the OHC; components to the right of the cell outline correspond to the mechanical properties of the OHC and the organ of Corti. The electrical circuit represents a simplification of the Figure 4.1 in which the apical and basal capacitances have been combined into total membrane capacitance (C_T) and the basolateral conductance branches have been combined into (g_B) and (E_B).

transformer, will be quite low and that most of the receptor current will be coupled into mechanical motion. In the extreme case, virtually all the receptor current could be coupled into mechanical velocity; the mechanical system then would act as a short circuit and no receptor potential would be measurable in the OHC. Using the data of Santos-Sacchi (1991), Mountain and Hubbard (1994) estimated the transformer ratio to be 0.125 dyne/V and the cell stiffness to be 63 dyne/cm.

Two different experimental approaches have been used to directly measure the transformer ratio for the OHC electromotility. Iwasa and Chadwick (1992) estimated the membrane compliance by observing changes in cell shape induced by intracellular pressure changes and then used the voltage–length relationship for the OHC to arrive at a value of 0.05 dyne/V. Xue, Mountain, and Hubbard (in press) measured electrically evoked basilar membrane motion in the gerbil cochlea to estimate a displacement/voltage ratio and then used the basilar membrane stiffness measurements of Olson and Mountain (1991) to arrive at a transformer ratio of 0.02 dyne/V. This latter estimate is a lower bound because the total OHC force will be divided between the displacement of the basilar membrane and the compression of the organ of Corti.

The relationship of the OHC forces in the cellular models to the active elements used in the hydromechanical models can be found by converting the OHC force to an equivalent pressure (see Hubbard and Mountain, Chapter 3). There are usually three rows of OHCs with the cells spaced 10 μm apart within a row. The width of the basilar membrane varies with species and cochlear location, but a typical value in the high-frequency region would be 200 μm. If the force estimate of Iwasa and Chadwick (1992) is used, then three hair cells acting on an area of 2000 μm^2 would produce a pressure/voltage ratio of 7.5×10^3 dyne cm^{-2} V^{-1}.

An equivalent transfer stiffness (the force produced by the OHC cell body divided by the stereocilia displacement) of the OHC can be computed by estimating the voltage-displacement ratio for mechano-electric transduction. For small stereocilia displacements, the voltage-displacement ratio is about 10^4 V/cm at low frequencies (Russell, Cody, and Richardson 1986). This leads to a transfer stiffness of 7.5×10^7 dyne/cm^3 for frequencies below the OHC membrane cutoff frequency. If one assumes the OHC membrane acts as a first-order, low-pass filter with a cutoff frequency of 1000 Hz, then the transfer stiffness (K_{OHC}) as a function of frequency (F) in hertz will be

$$K_{OHC} = \frac{7.5 \cdot 10^7}{\dfrac{j\,F}{1000} + 1} \frac{dyne}{cm^3} \qquad (19)$$

In the Neely and Kim (1986) cochlear model, the OHC forces were both frequency- and place dependent. For frequencies near the characteristic frequency in the extreme base, they used a transfer stiffness of 6.15×10^8 dyne/cm^3 for a section length of 100μm. Scaling Eq. 19 for the longer length and evaluating at 40 kHz gives a value for the transfer stiffness of 1.9×10^7 dyne/cm^3, considerably less than that required for the Neely and Kim (1986) model.

Geisler (1991) modeled the OHC force production process as a delayed force proportional to cilia displacement. The required transfer stiffness was similar to the basilar membrane stiffness. In the gerbil base, Olson and Mountain (1991) measured a point stiffness of 9.1×10^3 dyne/cm. This location has a characteristic frequency of 40 kHz. The probe displaced an area of basilar membrane comparable to that used to compute Eq. 19. which leads to a stiffness of 4.5×10^8 dyne/cm^3. At 40 kHz, Eq. 19 predicts that the OHC transfer stiffness will be 1.9×10^6 dyne/cm^3, which is more than two orders of magnitude smaller than required for the Geisler (1991) model.

Kolston et al. (1990) proposed a model in which the OHC feedback force was a complex function of frequency. In their model for the chinchilla, the OHC gain at the 8.5-kHz place was maximum for frequencies around 3 kHz, with a value of 9.4×10^7 dyne/cm^3. They appear to have used a section length of the order of 100 μm. Scaling Eq. 19 for the longer section length and evaluating at 3 kHz gives an estimated value for the actual transfer stiffness of 2.5×10^8 dyne/cm^3, almost three times greater than that required by the Kolston et al. (1990) model. However, if the Kolston et al. (1990) result is scaled up to 40 kHz, then the required gain becomes 4.4×10^8 dyne/cm^3 and the scaled prediction from Eq. 19 becomes 5.3×10^7 dyne/cm^3, almost an order of magnitude too small. These analyses lead to the conclusion that current active models which incorporate local feedback are not capable of replicating the observed basilar membrane data if physiologically plausible OHC forces are used.

5. Afferent Synaptic Physiology

5.1 Background

Data on hair cell synaptic physiology are limited, so most models are based on general concepts of synaptic function that have developed from experiments on other cell types. In this section, some of these concepts are reviewed as well as some of the properties of the auditory nerve firing patterns that presumably arise from synaptic events.

Neurotransmitter release is generally believed to be quantal in nature, involving the release of packets of neurotransmitter from presynaptic vesicles. Depolarization of the presynaptic membrane potential activates a voltage-dependent Ca^{2+} conductance, and the resulting increase in intracellular Ca^{2+} concentration causes vesicles to fuse with the cell membrane and release their contents into the synaptic cleft. After the release, the vesicle is retrieved and refilled with neurotransmitter for subsequent reuse (Zimmermann 1993).

Two pools of vesicles appear to be present near the synapse. One pool, the releasable pool, is the population of vesicles docked at the active zone of the cell membrane and are the vesicles that fuse with the membrane on Ca^{2+} entry. The second pool, the reserve pool, is a group of vesicles which are tethered to the cytoskeleton and are transferred to the releasable pool in response to the physiological activity of the cell.

The released neurotransmitter diffuses across the synaptic cleft and binds with high-affinity receptors on the postsynaptic membrane. The receptor-binding process leads to changes in postsynaptic membrane conductances and hence in postsynaptic membrane potential. If the postsynaptic membrane is sufficiently depolarized, then one or more action potentials will result from the normal spike-generation process. The actual spike pattern generated will depend on the temporal pattern of vesicle release, the kinetics of receptor binding, and subsequent conductance changes, as well as the refractory properties of the spike-generation process.

The chemical signal created by the release of a vesicle must be terminated in some way if the synapse is to be capable of transmitting the temporal information present in the presynaptic vesicle. Diffusion processes are generally too slow to remove neurotransmitter from the cleft at the necessary rates. Typically, neurotransmitter is removed by enzymatic degradation or by active reuptake by the presynaptic membrane. In the latter case, the neurotransmitter can be recycled for future release. Additional mechanisms such as postsynaptic receptor or channel inactivation can also play a role in signal termination.

There is a wealth of information on the activity of auditory nerve fibers, and most attempts to model the IHC afferent synapse have been built upon this data. These nerve fibers are spontaneously active in the absence of sound. The firing pattern is quite random and often is represented as a Poisson process. A more accurate representation is a renewal process, because refractory effects cause the firing probability to be depressed for more than 20 msec after the occurrence of an action potential (Gaumond, Kim, and Molnar 1983). More detailed analyses suggest that even a renewal process is not a perfect model because over long time scales the firing pattern is fractal in nature (Teich 1989).

The nerve fibers are usually classified according to their spontaneous activity. Liberman (1978) found three classes in the cat. He defined the high spontaneous rate class as greater than 17.5 spikes/sec, the medium spontaneous rate class as 0.5–17.5 spikes/sec, and the low spontaneous rate class as less than 0.5 spikes/sec. If the threshold of a fiber is defined as the sound level needed to increase the firing rate by a fixed number of spikes/sec, then the low spontaneous rate fibers generally have higher thresholds than the higher spontaneous rate fibers. Behavioral threshold corresponds to a change in IHC membrane potential of less than 1 mV (Zagaeski 1991; Zagaeski et al. 1994), indicating that the IHC–auditory nerve synapse is remarkably sensitive.

Liberman (1982) and Liberman, Dodds, and Pierce (1990) have demonstrated that fibers from the different classes can be identified morphologically and that all three types can synapse with the same IHC. This last result implies that the differences in threshold and spontaneous activity must result from differences in the synapse or spike generation process and not in differences in sensitivity between different IHCs.

In response to sustained acoustic stimulation, the average firing rate of an auditory nerve fiber increases at the onset of the stimulus and then declines or adapts over time. This is in contrast to the inner hair cell receptor potential, which exhibits no adaptation under similar conditions. There are two components to the adaptation process. There is an initial decline with a time constant of less than 10 msec, usually referred to as rapid adaptation. This time constant is comparable to that of the Ca^{2+} current inactivation discussed in Section 4.3. The rapid phase of adaptation is followed by a slower phase (time constant of 20–90 msec), usually referred to as short-term adaptation. The adaptation process is not complete, and so the average steady-state firing rate in the presence of acoustic stimulation is greater than the spontaneous rate.

The IHC–auditory nerve synapse is quite remarkable in its ability to accurately encode the temporal information present in the IHC receptor potential into changes in firing probability in the auditory nerve fiber. The ability of an auditory nerve fiber to modulate its firing probability in response to a sinusoidal input is usually called phase-locking or synchrony. Several methods have been used to quantify phase-locking but most are approximately equivalent to computing the ratio of the Fourier component in the firing probability pattern at the stimulus frequency to the average firing rate. The frequency limit to phase locking is species dependent but is of the order of 2000 Hz (Ruggero 1992).

In the guinea pig it has been shown that the phase-locking limit corresponds to the cutoff frequency of the IHC membrane. Kidd and Weiss (1990) however have pointed out that the rolloff of phase-locking is steeper than that of the IHC membrane, suggesting that other mechanisms also contribute such as the rate of Ca^{2+} channel activation and the time constant of the Ca^{2+} microdomain at the active zone. In fact the excellent phase-locking ability of the auditory nerve implies that all the events involved in synaptic transmission between the hair cell and nerve must involve time constants of less than 1 msec.

5.2 Neurotransmitter Release Models

Most current models of the IHC synapse focus on neurotransmitter release. Many of the models can trace their origin to the work of Schroeder and Hall (1974). In their model, "quanta" were assumed to be created at a fixed average rate (r) and to disappear by two mechanisms. In one mechanism the probability of disappearance (g) was independent of stimulation and had no effect on the fiber. In the other mechanism, the probability of disappearance (p) was regulated by the mechanical input to the cell and would cause the associated nerve fiber to fire. In this model the number of available quanta (n_v) can be computed by solving

$$\frac{dn_v}{dt} = (r - gn_v) - pn_v \tag{20}$$

In more modern language, n would be the number of vesicles in the releasable pool, the term in parentheses would correspond to the docking process, and p would correspond to the probability of Ca^{2+}-induced vesicle release. In this case, vesicle docking is modeled as a diffusion-like process from an infinite reserve pool. The problem with this type of model is that it has only a single time constant so it cannot model both rapid and short-term adaptation.

This type of model has been extended by Westerman and Smith (1988) by adding a finite reserve pool that in turn was replenished by a diffusion-like process from an infinite pool which presumably represents vesicle synthesis. Vesicle release is assumed to be directly proportional to receptor potential. The addition of another pool results in two time constants so that a more realistic match to the experimental data is possible. The Schroeder and Hall (1974) assumption that nerve firing is directly proportional to vesicle release rate is used here as well. A comparison of the analytical solution to this model to actual nerve firing probabilities led to the conclusion that the permeabilities representing diffusion between pools also had to be stimulus-level dependent. A computational time-domain version of this model has been implemented and tested by Carney (1993) and was found to compare favorably to actual auditory nerve data.

There have been few attempts to include realistic models of the Ca^{2+} influence on vesicle release. Kidd and Weiss (1990) developed and tested a model of the alligator lizard hair cell synapse. No adaptation was included in the Ca^{2+} current but reasonably realistic falloff of phase-locking was produced. Mountain, Wu, and Zagaeski (1991) proposed a model that combine the simple model of Ca^{2+} current inactivation presented in Section 4.3 with a vesicle depletion model similar to that of Schroeder and Hall (1974). Again, nerve firing probability is assumed to be proportional to vesicle release rate. In the most recent version of the Ca^{2+} inactivation synaptic model (D'Angelo 1992), the Ca^{2+} model has been adjusted to reproduce the rapid adaptation time constant and the vesicle depletion model has been adjusted to reproduce the short-term adaptation time constant. This model produces onset and steady-state responses that are very similar to the experimental data (Fig. 4.6). The rapid and short-term time constants also are similar to those measured experimentally at similar stimulus levels.

5.3 Postrelease Models

Most models of the IHC synapse do not include most of the post-vesicle release mechanisms outlined in Section 5.1. One exception is the model of Meddis (1986) in which the concentration of neurotransmitter in the synaptic cleft is explicitly modeled. The cleft is treated as a single compartment in which neurotransmitter is removed by two processes: one process is passive diffusion out ot the cleft and the other is active uptake by the hair cell. The neurotransmitter that is taken up by the hair cell is then repackaged and added to the releasable pool.

Hewitt and Meddis (1991) have developed an extensive test battery for IHC synaptic models and compared the Meddis (1986) model against

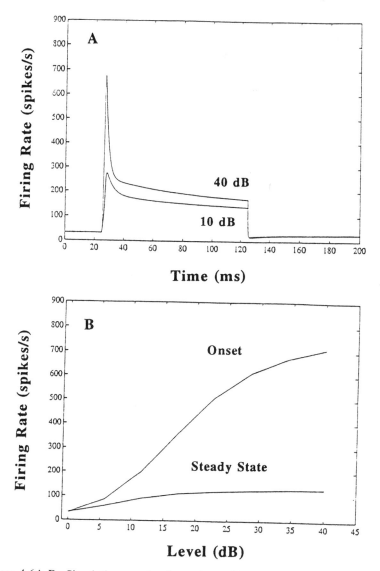

FIGURE 4.6A,B. Simulation results from the Ca^{2+} inactivation model for IHC synaptic transmission. Firing rate in response to a 100-msec tone burst is shown for stimulus levels 10 dB and 40 dB above threshold. Onset and steady-state firing rate as a function of stimulus level compared to threshold.

seven other models. Meddis's model generally performed as well as or better than the other models on all tests. D'Angelo (1992) tested the Ca^{2+} inactivation model with the same battery of tests and found its performance to be similar to the Meddis model.

An alternative approach to modeling postrelease mehanisms has been used by Eggermont (1985), who proposed that short-term adaptation was the result of postsynaptic receptor inactivation and that rapid adaptation was the result of auditory nerve refractoriness. He assumed that free receptor sites could be activated at a very fast rate and then converted at a slower rate to an inactive state. These inactive states were then returned to the free, active state at a rate determined by an enzymatic process. This model performed well in tests of forward masking where a leading tone burst was used to decrease (or mask) the response to a trailing tone burst.

Although few detailed models of the postsynaptic membrane processes exist, several attempts have been made to at least include the properties of the spike-generation process. This is usually done by assuming the vesicle release rate drives a Poisson or renewal process. For example, Diaz (1989) used a Schroeder–Hall style vesicle release model to drive a renewal process that incorporated an absolute refractory period of 1 msec and a relative refractory period with a 1-msec time constant.

More recently, Carney (1993) used the results of Westerman and Smith (1985) to model the refractory properties in more detail. In this model the refractory function consists of an absolute refractory period (τ_0) of 0.75 msec and a relative refractory period with two time constants ($\tau_F = 0.8$ msec; $\tau_s = 25$ msec). The refractory function, $R(t)$, is given by

$$R(t) = 1 - R_F e^{-\frac{(\tau - \tau_0)}{\tau_F}} - R_s e^{-\frac{(\tau - \tau_0)}{\tau_s}} \qquad \text{for } \tau \geq \tau_0$$

$$R(t) = 0 \qquad \text{for } \tau < \tau_0$$

$$(21)$$

where $\tau =$ time since the last spike. The total action potential firing probability function, $F(t)$, then is found by combining the refractory function with a neurotransmitter release model. For example, combining the vesicle depletion model (Eq. 20) with the refractory function yields:

$$F(t) = R(t)pn_v \qquad (22)$$

A Poisson process can then be generated using Eq. 22 for the rate parameter to produce the neural spike train.

6. Efferent Synaptic Physiology

6.1 Background

It has been known for some time that electrical stimulation of the cochlear efferents (MOC system) decreases the sensitivity of auditory nerve fibers (Galambos 1956; Wiederhold and Peake 1966; Wiederhold 1970). The effect of efferent stimulation of sensitivity is maximum for acoustic frequencies near the fiber's characteristic frequency. The efferent effect originally posed a problem for the auditory community because the MOC fibers innervate the OHCs and yet the auditory nerve fibers receive their input largely from the IHCs. This problem was resolved when it was demonstrated that efferent stimulation could alter cochlear mechanics (Mountain 1980), and the efferent system is now thought of as a system for regulating OHC electromotility.

The cochlear efferents innervating the OHCs appear to use acetyl-choline as the neurotransmitter (Eybalin 1993). No detailed models exist for the conductance change associated with efferent synapse on the OHCs. This lack of models is because there has been until recently a lack of intracellular data relating to the efferent system. It is generally believed that efferent activity increases the basolateral conductance because the positive potential in scala media decreases (Fex 1967) and the scala media input conductance increases (Mountain, Geisler, and Hubbard 1980) with electrical stimulation of the efferent system. The extracellular response to sound (cochlear microphonic) increases with efferent stimulation (Konishi and Slepian 1971), which is also consistent with a conductance increase in the basolateral OHC membrane (Wiederhold 1967; Geisler 1974).

The earliest insight into how the OHC efferent synapse might function came from work done in the turtle cochlea. Art and coworkers (Art et al. 1982; Art, Fettiplace, and Fuchs 1984) found that efferent stimulation in the turtle resulted in hyperpolarization of the hair cells. This post synaptic potential appears to arise mainly from an increase in a K^+ conductance, and the time course of the response to a single shock was quite long (100 msec). They also found that the postsynaptic potential consisted of two components, a small early depolarizing component and a slower and much larger hyperpolarizing component. They raised the possibility that the early component may be the primary event tirggered by the efferent stimulation and that it in turn activates the larger, slower component.

Subsequent work using acetylcholine applied to isolated hair cells from a variety of organs including the mammalian cochlea has pointed to the primary event being the entry of Ca^{2+} (Housely and Ashmore 1991; Fuchs and Murrow 1992a,b; Doi and Ohmori 1993). The secondary event, which leads to hyperpolarization, appears to be the activation of a Ca^{2+}-activated K^+ conductance. An alternative hypothesis has been proposed in which the efferents activate a system involving a phosphoinostide-based second messenger cascade (Niedzielski, Ono, and Schacht 1992; Kakehata et al. 1993). The second messenger would act by releasing Ca^{2+} from intracellular storage sites, and the resulting elevation in Ca^{2+} concentration would activate the K^+ conductance.

6.2 Efferent Models

The earliest models of efferent function postulated that efferent stimulation decreased auditory nerve sensitivity by activating a conductance in the basolateral hair cell membrane, which in turn shunted the receptor current away from the afferent synapse (Wiederhold 1967). This approach failed, however, to account for the fact that stimulation of the MOC efferents should affect OHCs and yet the vast majority of auditory nerve fibers receive their input from the IHCs. Geisler (1974) proposed a shunting model in which the conductance increase in the OHCs decreased the endocochlear potential, which in turn decreased the driving force for transduction in the IHCs, thereby decreasing IHC sensitivity. Such models however cannot account for the finding that the efferent effect is greatest for acoustic frequencies near the characteristic frequency.

With the advent of active models of cochlear mechanics, attention turned to modeling efferent effects on OHC motility. Mountain (1986) explored a simple feedback model of cochlear mechanics in which the hypothesized change in OHC membrane conductance from efferent stimulation was modeled as a change in the gain and time constant of feedback to the electromechanical feedback loop. It is not clear, however, what effect an efferent-induced conductance change will have on the response to high-frequency acoustic stimuli, because the OHC membrane impedance would presumably be dominated by the membrane capacitance. One possibility was pointed out by Roddy et al. (1994), who noted that at normal resting potentials the OHCs are operating in a region of the voltage–length curve (see Fig. 4.4) where small changes in membrane potential can cause significant changes in the sensitivity of the voltage-dependent length change. If efferent stimulation hyperpolarizes the OHCs, then the shift in operating point on the voltage–length curve would cause a decrease in sensitivity of electromechanical transduction.

7. Signal Processing Models

Over the years a number of signal processing models have been proposed to mimic the function of the IHC. The best models use some type of nonlinear transfer characteristic is chosen to model the inner hair cell transduction process, and the output of this nonlinearity (V_T) is usually fed to a lowpass filter with a time constant usually chosen to match the time constant (τ_{IHC}) of the IHC (approximately 1 kHz). These models perform reasonably well because the parallel combination of resistance and capacitance in Figure 4.1 acts as a lowpass filter. These filters can be implemented by converting the continuous transfer function to a digital filter using the bilinear transformation (Oppenheim and Schafer 1975). A typical filter can be computed by

$$V_{IHC}[n] = K_A(V_T[n] + V_T[n-1]) + K_B V_{IHC}[n-1]$$

where $$K_A = \frac{1}{2\tau_{IHC}F_s + 1} \qquad K_B = \frac{2\tau_{IHC}F_s - 1}{2\tau_{IHC}F_s + 1}$$ (23)

and F_s is the sample rate.

In the actual hair cell, the hair cell input resistance is nonlinear and time varying. For low-level mechanical inputs, however, the voltage changes are small and the resulting changes in membrane resistance are probably also small, so the nonlinear nature of the membrane filter can be ignored. Also because the voltage changes are small the receptor current will almost exactly mimic the conductance change of the apical membrane. Under these conditions, the signal processing models will produce almost exactly the same results as the more complex biophysically based models. At higher stimulus levels there will, however, be some difference between the two modeling approaches.

The IHC transducer is usually assumed to respond instantaneously to the mechanical input. The transducer could be modeled using Eq. 11 directly, but in fact a variety of nonlinear functions have been used by different authors. One such function is the hyperbolic tangent (Carney 1993), which is mathematically equivalent to the two-state model (see Eq. 10). The hyperbolic tangent function exhibits symmetrical saturation for positive and negative inputs, so if the parameters are adjusted to fit the hyperpolarizing portion of the IHC transducer characteristic then it will saturate too quickly in the depolarizing portion. Ross (1982), in a model designed to replicate auditory nerve activity, used a permeability function that consisted of a permeability which depended on instantaneous basilar membrane displacement in series with a constant permeability. The time-varying permeability was a parabolic function of basilar membrane displacement. Such an approach produces a remarkably good fit to the experimentally determined IHC transfer characteristic. After some appropriate scaling, the series combination proposed by Ross can be considered as the output of the IHC transducer.

$$V_T = V_0 \frac{(x - x_0)^2}{(x - x_0)^2 + K_s} - V_1 \qquad x > x_0$$
$$V_T = -V_1 \qquad x < x_0$$
(24)

where V_0, V_1, and K_s are constants. Figure 4.7A shows the comparison of the results from Eq. 24 to the experimental data. Another type of function that is sometimes used is a rectified logarithmic compressor (cf. Patterson and Holdsworth in press). The following logarithmic compression scheme mimics the IHC transduction process almost as well as the parabolic approach:

$$V_T = V_0 \ln(1 + K_s x) \qquad x > 0$$
$$V_T = -V_0 \ln(1 - K_s x) \qquad x_0 < x < 0$$
$$V_T = -V_0 \ln(1 - K_s x_0) \qquad x < x_0$$
(25)

Figure 4.7B compares the results for Eq. 25 to the experimental data. Both Eqs. 24 and 25 fit the experimental data quite well and have the advantage that they can be computed more quickly than Eq. 11.

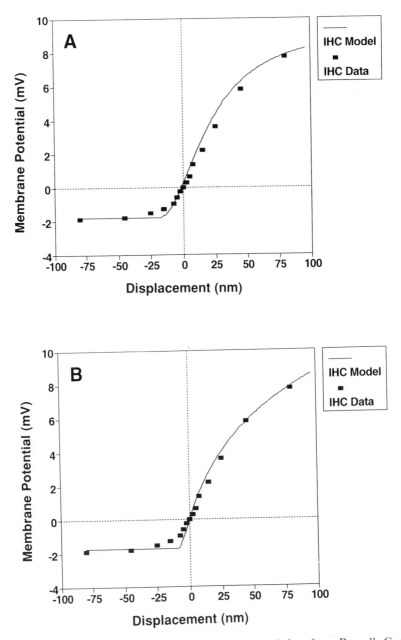

FIGURE 4.7A,B. (A) Comparison of IHC experimental data from Russell, Cody, and Richardson (1986) to the parabolic model (Eq. 24). Model parameters: displacement-offset constant $(x_0) = -19\,\mu m$; saturation constant $(K_s) = 1600\,\mu m^2$; scale factor $(V_0) = 14\,mV$; offset constant $(V_1) = 1.8\,mV$. (B) Comparison of the same experimental data to the logarithmic compression model (Eq. 25). Model parameters: $K_s = 0.0529\,\mu m^{-1}$; $V_0 = 4.7338\,mV$.

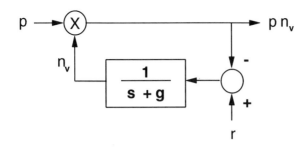

FIGURE 4.8. Equivalent block diagram model for the neurotransmitter depletion model (Eq. 20).

The IHC synaptic models can also be made more computationally efficient. The strategy consists of making simplifying approximations, if necessary, to get the models in a form such that they can be implemented with simple combinations of digital filters, multipliers, and summers. For example, the neurotransmitter depletion model (Eq. 20) is equivalent to the block diagram shown in Figure 4.8. This system acts as an automatic gain control with the time constant $1/g$. The lowpass filter in the gain control loop can be implemented digitally using the bilinear transform as outlined. The same topology can also be used to model the inactivation of the IHC Ca^{2+} as the transition to the inactive state is equivalent to depleting the supply of active channels.

8. Summary and Areas of Future Research

The computational roles of the hair cells and auditory nerve are relatively well understood. The OHCs, through their voltage-dependent length changes, serve to amplify the stimulus to the IHCs. The asymmetry in the IHC tension-gated conductance change serves to half-wave rectify the hair bundle stimulus, and the electrical properties of the IHC membrane form a lowpass filter. This combination of rectification and filtering serves as an envelope detector, which allows the auditory system to process high-frequency acoustic signals. By using envelope detection, the auditory system reduces the frequency response required in the IHC synapse. The adaptation present in synaptic transmission between the IHC and the auditory nerve fibers serves as an automatic gain control that allows the system to preserve the temporal information in the stimulus over a wide range of intensities.

Current models of the various stages of the transduction process produce results that compare well with the experimental data. The models of many hair cell processes, such as the tension-gated channels and voltage-dependent length changes, have become relatively sophisticated and have a firm biophysical basis. The principal weakness of these models lies in details that have not yet been included.

In other cases, model development awaits further experimental study. A number of voltage-dependent conductances have been described in cochlear hair cells, but the kinetics of these conductances have not been studied in detail. As a result, few attempts have been made to model these processes at the cellular level. The fact that second-messenger systems appear to play a key role in the conductances of the OHCs leads to the possibility of a complex combination of regulatory mechanisms that function to modulate cochlear sensitivity. These mechanisms will need to be included in cochlear models that attempt to model efferent effects or long-term changes in cochlear sensitivity. The conductances of the IHC presumably shape the receptor potential, and the Ca^{2+} conductance plays a key role in the IHC synapse. Inclusion of these conductances in IHC models will be necessary if the goal is to accurately predict the detailed temporal properties of the auditory nerve fiber firing patterns.

Additional data are needed to advance the modeling effort concerned with synaptic transmission between the IHC and the auditory nerve. The current models perform reasonably well, but vary considerably in the underlying assumptions used to derive the models. Very little is known about the vesicle release process in the IHC or the nature of the neurotransmitter and postsynaptic receptor. Current models of IHC synaptic transmission successfully mimic some of the features observed in the experimental data, but most fail in one or more ways. One of the notable inadequacies of most models is the difficulty in correctly predicting the response to stimulus decrements (Hewitt and Meddis 1991).

A significant gap in current hair cell models is that few models address interactions between mechanical and electrical properties of hair cells. Freeman and Weiss (1990a,b,c,d) have published an extensive analysis of the hydromechanical forces acting on the hair bundles; however, current models do not take into account the fact that the mechanical properties of the hair bundles appear to be voltage dependent (Crawford and Fettiplace 1985; Assad, Hocohen, and Corey 1989). In addition, Cody and Mountain (1989) have reported that the mechanical input to the IHCs is more complex than would be expected from the basilar membrane displacement. They have suggested that the IHCs may be stimulated by OHC force production directly (Mountain and Cody 1989). To further complicate matters, recent evidence suggests that the OHC itself is a resonant structure and that mechanical stimulation at the resonant frequency can elicit slow contractions (Brundin, Flock, and Canlon 1989; Brundin and Russell 1993). Much work remains to be done so that existing two- and three-dimensional models of OHC electromotility (Dallos, Hallworth, and Evans 1993; Steele et al. 1993; Iwasa 1994) can be elaborated to include OHC resonances as well as slow contractions.

A major difficulty for efforts aimed at systems-level modeling is the lack of consensus as to what constitutes an appropriate micromechanical cochlear model (see Hubbard and Mountain, Chapter 3, this volume). Without a detailed description of the deformations of the reticular lamina and the tectorial membrane, it is impossible to specify the stereocilia displacement for outer hair cells. It is more difficult still to calculate the motion of the stereocilia of inner hair cells. Until these processes are understood, it will be impossible to relate the acoustically evoked responses of hair cells and nerve fibers measured *in vivo* to the rapidly expanding body of data coming from *in vitro* hair cell preparations.

List of Symbols and Terms

Symbol	Quantity	Section in which introduced
A	fraction of channels in state A	4.2
B	fraction of channels in state B	4.2
C	fraction of channels in state C	4.2
C_0	fraction of channels in state C_0	4.4
C_1	fraction of channels in state C_1	4.4
C_2	fraction of channels in state C_2	4.4
C_A	apical membrane capacitance	4.1
C_B	basolateral membrane capacitance	4.1
C_T	total membrane capacitance	4.1
E_1	reversal potential for the first basolateral conductance	4.1
E_{Ca}	calcium reversal potential	4.3
E_N	reversal potential for the nth basolateral conductance	4.1
E_{SM}	scala media potential	4.1
F	frequency	4.5
F_{OHC}	outer hair cell force	4.5
F_S	sample rate	7
$F(t)$	firing probability function	5.3
g	vesicle generation rate	5.2
g_1	first basolateral membrane conductance	4.1
g_A	apical membrane conductance	4.1
g_1	leakage conductance	4.3
g_N	n^{th} basolateral membrane conductance	4.1
g_v	voltage-dependent conductance	4.3
I_A	receptor current	4.5

I_{Ca}	calcium current	4.3
I_v	outer hair cell length change velocity	4.5
K_1	rate constant for transitions from state A to B	4.2
$K_1[Ca^{2+}]$	rate constant for transitions from state C_0 to C_1	4.4
K_{-1}	rate constant for transitions from state C_1 to C_0	4.4
K_2	rate constant for transition from state B to A	4.2
$K_2[Ca^{2+}]$	rate constant for transitions from state C_1 to C_2	4.4
K_{-2}	rate constant for transitions from state C_2 to C_1	4.4
K_3	rate constant for transitions from state B to C	4.2
$K_3[Ca^{2+}]$	rate constant for transitions from state O_2 to O_3	4.4
K_3	rate constant for transitions from state O_3 to O_2	4.4
K_4	rate constant for transitions from state C to B	4.2
K_A	filter coefficient	7
K_B	filter coefficient	7
K_{OHC}	outer hair cell transfer stiffness	4.5
K_S	saturation constant for inner hair cell transducer models	7
L_{max}	maximum outer hair cell length	4.5
L_{OHC}	outer hair cell length	4.5
m	Hodgkin–Huxley gating variable	4.3
n	time index variable	7
n_v	number of available vesicles	5.2
O_2	fraction of channels in state O_2	4.4
O_3	fraction of channels in state O_3	4.4
p	probability of vesicle release	5.2
P_{open}	probability that a channel is open	4.2
P_{short}	probability that a motor molecule is in the short state	4.5
r	rate of vesicle production	5.2
R	gas constant	4.2
R_F	fast refractory coefficient	5.3
R_S	slow refractory coefficient	5.3
$R(t)$	refractory function	5.3
S_0	displacement-sensitivity constant	4.2
S_1	displacement-sensitivity constant	4.2

S_v	voltage-sensitivity constant	4.5
S_x	displacement-sensitivity constant	7
t	time	5.3
T	absolute temperature	4.2
V_0	scale factor for inner hair cell transducer models	7
V_1	offset constant for inner hair cell transducer models	7
V_h	voltage-offset constant	4.3
V_{IHC}	inner hair cell receptor potential	7
V_m	membrane potential	4.1
V_T	inner hair cell transducer output	7
x	cilia displacement	4.2
x_0	displacement-offset constant	4.2
x_1	displacement-offset constant	4.2
Z_{OC}	organ of Corti impedance	4.5
Z_{OHC}	outer hair cell impedance	4.5
α_C	rate constant for transitions from state O_2 to C_2	4.4
β_C	rate constant for transitions from state C_2 to O_2	4.4
α_m	forward rate constant for m	4.3
β_m	reverse rate constant for m	4.3
ΔG_{AB}	energy difference between states A and B	4.2
ΔG_{BC}	energy difference between states B and C	4.2
ΔL	difference between maximum and minimum outer hair cell length	4.5
τ	time since last action potential	5.3
τ_O	absolute refractory period	5.3
τ_F	fast refractory time constant	5.3
τ_{IHC}	inner hair cell membrane time constant	7
τ_S	slow refractory time constant	5.3

References

Art JJ, Crawford AC, Fettiplace R, Fuchs PA (1982) Efferent regulation of hair cells in the turtle cochlea. Proc R Soc Lond B 216:377–384.

Art JJ, Fettiplace R, Fuchs PA (1984) Synaptic hyperpolarization and inhibition of turtle cochlear hair cells. J Physiol 356:525–550.

Ashmore JF (1989) Transducer motor coupling in cochlear outer hair cell. In: Wilson JP, Kemp DT (eds) Cochlear Mechanisms: Structure, Function, and Models. New York: Plenum Press, pp. 107–114.

Ashmore JF, Meech RW (1986) Ionic basis of membrane potential in outer hair cells of the guinea-pig cochlea. Nature 322:368–371.

Assad JA, Hocohen N, Corey DP (1989) Voltage dependence of adaptation and active bundle movement in bullfrog saccular hair cells. Proc Natl Acad Sci USA 86:2918–2922.

Assad JA, Shepherd GMG, Corey DP (1991) Tip-link integrity and mechanical transduction in vertebrate hair cells. Neuron 7:985–994.

Brownell WE, Bader CR, Bertrand D, de Ribaupierre I (1985) Evoked mechanical responses of isolated cochlear outer hair cells. Science 227:194–196.

Brundin L, Russell I (1993) Sound-induced movements and frequency tuning in outer hair cells isolated from the guinea pig cochlea. In: Duifhuis H, Horst JW, van Dijk P, van Netten S (eds) Biophysics of Hair Cell Sensory Systems. Singapore: World Scientific, pp. 182–189.

Brundin L, Flock Å, Canlon B (1989) Sound-induced motility of isolated cochlear outer hair cells is frequency-specific. Nature 342:814–816.

Carney LH (1993) A model for low-frequency auditory-nerve fibers in cat. J Acoust Soc Am 93:401–417.

Cody AR, Mountain DC (1989) Low frequency responses of inner hair cells: evidence for a mechanical origin of peak splitting. Hear Res 41:89–100.

Corey DP, Hudspeth A (1979) Response latency of vertebrate hair cells. Biophys J 26:499–506.

Corey DP, Hudspeth A (1983) Kinetics of the receptor current in bullfrog saccular hair cells. J Neurosci 5:962–976.

Crawford AC, Fettiplace R (1985) The mechanical properties of ciliary bundles of turtle cochlear hair cells. J Physiol 364:359–379.

Crawford AC, Kros C (1989) A fast calcium current with a rapidly inactivating component in isolated inner hair cells of the guinea-pig. J Physiol 420:90P.

Dallos P, Evans BN, Hallworth R (1991) On the nature of the motor element in cochlear outer hair cells. Nature 350:155–157.

Dallos P, Hallworth R, Evans BN (1993) Theory of electrically driven shape changes of cochlear outer hair cells. J Neurophysiol 70:299–323.

Dallos P, Santos-Sacchi J, Flock A (1982) Intracellular recordings from cochlear outer hair cells. Science 218:582–584.

D'Angelo WR (1992) Biophysical models of auditory processing. M.S. thesis, Boston University, Boston, MA.

Davis H (1958) A mechano-electric theory of cochlear action. Ann Otol Rhinol Laryngol 67:789–901.

Diaz JM (1989) "Chopper" firing patterns in the mammalian anteroventral cochlear nucleus: a computer model. M.S. thesis, Boston University, Boston, MA.

Ding JP, Salvi J, Sachs F (1991) Stretch-activated ion channels in guinea pig outer hair cells. Hear Res 56:19–28.

Doi T, Ohmori H (1993) Acetylcholine increases intracellular Ca^{2+} concentration and hyperpolarizes the guinea-pig outer hair cell. Hear Res 67:179–188.

Eggermont JJ (1985) Peripheral auditory adaptation and fatigue: a model-oriented review. Hear Res 18:57–71.

Eybalin M (1993) Neurotransmitters and neuromodulators of the mammalian cochlea. Physiol Rev 73:309–373.

Fex J (1967) Efferent inhibition in the cochlea related to hair-cell dc activity: study of postsynaptic activity of the crossed olivocochlear fibers in the cat. J Acoust Soc Am 41:666–675.

Freeman DM, Weiss TF (1990a) Superposition of hydrodynamic forces on a hair bundle. Hear Res 48:1–16.

Freeman DM, Weiss TF (1990b) Hydrodynamic forces on hair bundles at low frequencies. Hear Res 48:17–30.

Freeman DM, Weiss TF (1990c) Hydrodynamic forces on hair bundles at high frequencies. Hear Res 48:31–36.

Freeman DM, Weiss TF (1990d) Hydrodynamic analysis of a two-dimensional model for micromechanical resonance of free-standing hair bundles. Hear Res 48:1–16.

Fuchs PA, Murrow BW (1992a) A novel cholinergic receptor mediates inhibition of chick cochlea hair cells. Proc R Soc Lond B 248:35–40.

Fuchs PA, Murrow BW (1992b) Cholinergic inhibition of short (outer) hair cells of the chick's cochlea. J Neurosci 12:800–809.

Galambos R (1956) Suppression of auditory activity by stimulation of efferent fibers to cochlea. J Neurophysiol 19:424–437.

Gaumond RP, Kim DO, Molnar CE (1983) Response of cochlear nerve fibers to brief acoustic stiumli: role of discharge-history effects. J Acoust Soc Am 74:1392–1398.

Geisler CD (1974) Model of crossed olivocochlear bundle effects. J Acoust Soc Am 56:1910–1912.

Geisler CD (1991) A cochlear model using feedback from motile outer hair cells. Hear Res 54:105–117.

Gitter AH, Frömter E, Zenner HP (1992) C-Type potassium channels in the lateral cell membrane of guinea-pig outer hair cells. Hear Res 60:13–19.

Hewitt MJ, Meddis R (1991) An evaluation of eight computer models of mammalian inner haircell function. J Acoust Soc Am 90:904–917.

Hodgkin AL, Huxley AF (1952) A quantitative description of membrane current and its application to conduction and excitation in nerve. J Physiol 117:500–544.

Housely GD, Ashmore JF (1991) Direct measurement of the action of acetylcholine on isolated outer hair cells or the guinea pig cochlea. Proc R Soc Lond B 244:161–167.

Howard J, Hudspeth AJ (1988) Compliance of the hair bundle associated with gating of mechanoelectric transduction channels in the bullfrog's saccular hair cell. Neuron 1:189–199.

Howard J, Roberts WM, Hudspeth AJ (1988) Mechanoelectrical transduction by hair cells. Annu Rev Biophys Chem 17:99–124.

Hudspeth AJ (1982) Extracellular current flow and the site of transduction in vertebrate hair cells. J Neurosci 2:1–10.

Hudspeth AJ, Lewis RS (1988a) Kinetic analysis of voltage- and ion-dependent conductances in saccular hair cells of the bull-frog, *Rana catesbeiana*. J Physiol 400:237–274.

Hudspeth AJ, Lewis RS (1988b) A model for electrical resonance and frequency tuning in saccular hair cells of the bull-frog, *Rana catesbeiana*. J Physiol 400:275–297.

Iwasa KH (1994) A membrane motor model for the fast motility of the outer hair cell. J Acoust Soc Am 96:2216–2224.

Iwasa KH, Chadwick RS (1992) Elasticity and active force generation of cochlear outer hair cells. J Acoust Soc Am 92:3169–3173.

Iwasa KH, Li M, Jia M, Kachar B (1991) Stretch sensitivity of the lateral wall of the auditory outer hair cell from the guinea pig. Neurosci Lett 133:171–174.

Jaramillo F, Hudspeth AJ (1991) Localization of the hair cell's transduction channels at the hair bundle's top by iontophoretic application of a channel blocker. Neuron 7:409–420.

Kakehata S, Nakagawa T, Takasaka T, Akaike N (1993) Cellular mechanism of acetylcholine-induced response in dissociated outer hair cells of guinea-pig cochlea. J Physiol 463:227–244.

Kidd RC, Weiss TF (1990) Mechanisms that degrade timing information in the cochlea. Hear Res 49:181–208.

Kim DO (1986) Active and nonlinear cochlear biomechanics and the role of outer-hair-cell subsystem in the mammalian auditory system. Hear Res 22:105–114.

Kolston PJ, Viergever MA, de Boer E, Smoorenburg GF (1990) What type of force does the cochlear amplifier produce? J Acoust Soc Am 88:1794–1801.

Konishi T, Slepian JZ (1971) Effects of the electrical stimulation of the crossed olivocochlear bundle on cochlear potentials recorded with intracochlear electrodes in guinea pigs. J Acoust Soc Am 47:1519–1526.

Kros CJ, Crawford AC (1990) Potassium currents in inner hair cells isolated from the guinea-pig cochlea. J Physiol 421:263–291.

Liberman MC (1978) Auditory-nerve responses from cats raised in a low-noise chamber. J Acoust Soc Am 63:442–455.

Liberman MC (1982) Single-neuron labeling in the cat auditory nerve. Science 216:1239–1241.

Liberman MC, Dodds LW, Pierce S (1990) Afferent and efferent innervation of the cat cochlea: quantitative analysis with light and electron microscopy. J Comp Neurol 301:443–460.

Mason WP (1950) Piezoelectric Crystals and Their Application to Ultrasonics. New York: Van Nostrand.

Meddis R (1986) Simulation of auditory-neural transduction: further studies. J Acoust Soc Am 83:1056–1063.

Mountain DC (1980) Changes in endolymphatic potential and crossed-olivoco-chlear-bundle stimulation alter cochler mechanics. Science 210:71–72.

Mountain DC (1986) Active filtering by hair cells. In: Hall J, Hubbard A, Neely S, Tubis A (eds) Peripheral Auditory Mechanisms. New York: Springer-Verlag, pp. 179–188.

Mountain DC, Cody AR (1989) Mechanical coupling between inner and outer hair cells in the mammalian cochlea. In Wilson JP, Kemp DT (eds) Cochlear Mechanisms—Structure, Function and Models. New York: Plenum Press, pp. 153–160.

Mountain DC, Hubbard AE (1989) Rapid force production in the cochlea. Hear Res 42:195–202.

Mountain DC, Hubbard AE (1994) A piezoelectric model for outer hair cell function. J Acoust Soc Am 95:350–354.

Mountain DC, Geisler CD, Hubbard AE (1980) Stimulation of efferents alters the cochlear microphonic and the sound-induced resistance changes measured in scala media of the guinea pig. Hear Res 3:231–240.

Mountain DC, Wu W, Zagaeski M (1991) A calcium-inactivation model of auditory nerve adapation. Association for Reseach in Otolargngology (ARO) Abstracts 14:153.

Neely ST, Kim DO (1986) A model for active elements in cochlear biomechanics. J Acoust Soc Am 79:1472–1480.

Niedzielski AS, Ono T, Schacht J (1992) Cholinergic regulation of the phospho-inositide second messenger system in the guinea pig organ of Corti. Hear Res 59:250–254.

Olson ES, Mountain DC (1991) *In vitro* measurement of basilar membrane stiffness. J Acoust Soc Am 89:1262–1275.

Olson ES, Mountain DC (1994) Mapping the cochlear partition's sitffness to its cellular architecture. J Acoust Soc Am 95:395–400.

Oppenheim AV Schafter RW (1975) Digital Signal Processing. Englewood Cliffs: Prentice-Hall.

Patterson RD, Holdsworth J (1995) functional model of neural activity patterns and auditory images. In: Ainsworth WA, Evans EF (eds) Advances in Speech, Hearing and Language Processing, Vol. 3. London: JAI Press (in press).

Pickles JO, Corey DP (1992) Mechanoelectrical transduction by hair cells. Trends Neurosci 15:254–259.

Pickles JO, Comis SD, Osborne MP (1984) Cross-links between stereocilia in the guinea pig organ of Corti, and their possible relation to sensory transduction. Hear Res 15:103–112.

Probst R, Lonsbury-Martin BL, Martin GK (1991) A review of otoacoustic emissions. J Acoust Soc Am 89:2027–2067.

Roddy J, Hubbard AE, Mountain DC, Xue S (1994) Effects of electrical biasing on electrically-evoked otoacoustic emissions. Hear Res 73:148–154.

Ross S (1982) A model of the hair cell-primary fiber complex. J Acoust Soc Am 71:926–941.

Ruggero MA (1992) Physiology and coding of sound in the auditory nerve. In: Poper AN, Fay RR (eds) The Springer Handbook of Auditory Research, Vol. 2, The Mammalian Auditory Pathway: Neurophysiology. New York: Springer-Verlag, pp. 34–93.

Rüsch A, Kros CJ, Richardson GP, Russell IJ (1991) Potassium and calcium currents in outer hair cells in organotypic cultures of the neonatal mouse cochlea. J Physiol 434:52P.

Russell IJ, Cody AR, Richardson GP (1986) The responses of inner and outer hair cells in the basal turn of the guinea-pig cochlea and in the mouse cochlea grown in vitro. Hear Res 22:199–216.

Ryugo DK (1992) The auditory nerve: Peripheral innervation, cell body morphology, and central projections. In: Webster DB, Popper AN, Fay RR (eds) The Springer Handbook of Auditory Research, Vol. 1, The Mammalian Auditory Pathway: Neuroanatomy. New York: Springer-Verlag, pp. 23–65.

Sachs F, Lecar H (1991) Stochastic models for mechanical transduction. Biophys J 59:1143–1145.

Santos-Sacchi J (1991) Reversible inhibition of voltage-dependent outer hair cell motility and capacitance. J Neurosci 11:3096–3110.

Santos-Sacchi J, Dilger JP (1988) Whole cell currents and mechanical responses of isolated outer hair cells. Hear Res 35:143–150.

Schroeder MR, Hall JL (1974) Model for mechanical to neural transduction in the auditory receptor. J Acoust Soc Am 55:1055–1060.

Steele CR, Baker G, Tolomeo J, Zetes D (1993) Electro-mechanical models of the outer hair cell. In: Duifhuis H, Horst JW, van Dijk P, van Netten S (eds) Biophysics of Hair Cell Sensory Systems. Singapore: World Scientific, pp. 207–215.

Teich MC (1989) Fractal character of auditory neural spike train. IEEE Trans Biomed Eng 36:150–160.

Warr WB (1992) Organization of olivocochler efferent systems in mammals. In: Webster DB, Popper AN, Fay RR (eds) The Springer Handbook of Auditory Research, Vol. 1, The Mammalian Auditory Pathway: Neuroanatomy. New York: Springer-Verlag, pp. 410–448.

Weiss TF (1982) Bidirectional transduction in vertebrate hair cells: a mechanism for coupling mechanical and electrical vibrations. Hear Res 7:353–360.

Westerman LA, Smith RL (1985) Rapid adaptation depends on the characteristic frequency of auditory nerve fibers. Hear Res 17:197–198.

Westerman LA, Smith RL (1988) A diffusion model of the transient response of the cochlear inner hair cell synapse. J Acoust Soc 83:2266–2276.

Wiederhold ML (1967) A Study of Efferent Ihibition of Auditory Nerve Activity. Ph.D. thesis, Massachusetts Institute of Technology, Cambridge, MA.

Wiederhod ML (1970) Variations in the effects of electrical stimulation of the crossed olivocochlear bundle on cat single auditory nerve fiber responses to tone bursts. J Acoust Soc Am 48:966–977.

Wiederhold ML, Peake WT (1966) Efferent inhibition of auditory nerve responses: dependence on acoustic stimulus parameters. J Acoust Soc Am 40:1427–1430.

Xue S, Mountain DC, Hubbard AE (1995) Electrically-evoked basilar membrane motion. J Acoust Soc Am 97:3030–3041.

Zagaeski M (1991) Information Processing in the Mammalian Auditory Periphery. Doctoral dissertation, Boston University Graduate School, Boston, MA.

Zagaeski M, Cody AR, Russell IJ, Mountain DC (1994) Transfer characteristic of the inner hair cell synapse: steady-state analysis. J Acoust Soc Am 95:3430–3434.

Zimmermann H (1993) Synaptic transmission: cellular and molecular basis. New York: Oxford University Press.

5
Physiological Models for Basic Auditory Percepts

Bertrand Delgutte

On the zigzagging road towards wisdom about the human auditory system we collect knowledge from two entirely different sources of experimental information. First from anatomy and physiology . . . Second, from perception and psychoacoustics . . . Our ever wondering mind tries to combine and explain these findings in terms of some model, law, hypothesis or theory.

J.F. Schouten (1970)

1. Introduction

Explaining auditory perceptual phenomena in terms of physiological mechanisms has a long tradition going back at least to von Helmholtz (1863), and possibly to as early as Pythagoras' experiments on pitch and musical consonance (ca. 530 B.C.; see Cohen and Drabken 1948). In modern practice, such efforts take the form of computational models because these models help generate hypotheses that can be explicitly stated and quantitatively tested for complex systems. Relating physiology to behavior is perhaps the most direct route toward understanding how the auditory system works, because neither physiological nor perceptual data alone provide sufficient information: physiological studies cannot identify the function of the neural structures under investigation, while perceptual studies do not reveal the implementation of these functions. This endeavor is not only an intellectual challenge (Schouten's "ever wondering mind"), it can also have practical value. Perceptual impairments such as difficulties in understanding speech may only yield to surgical and pharmacological cures if the problem is sufficiently well identified at the physiological level. Because any behavior such as speech perception involves a complex physiological system with many interacting components, it becomes essential to identify the roles of these various components in the behavior. Computational models can help determine how a deficit in one or more components affects the overall operation of the system. Such models can also be more generally useful for designing artificial systems for sound reproduction, transmission, and recognition.

This chapter reviews computational models that predict performance in basic psychophysical tasks from the physiological activity of populations of auditory neurons. This review excludes black-box psychophysical models that make little or no use of physiological knowledge, as well as physiological models that have not been explicitly tested against psychophysical data. Models for psychophysical tasks that are treated in other chapters of this volume, such as binaural hearing (Colburn, Chapter 8), and pitch and timbre (Lyon and Shamma, Chapter 6), are also excluded. Primary emphasis is on models based on auditory nerve activity, partly because of the author's expertise and partly because the activity of central auditory neurons has not been described in sufficient detail to allow the formulation of quantitative models, with the exception of models for binaural phenomena. The auditory nerve is a rational starting place for such modeling efforts because all information about acoustic stimuli that reaches the brain is encoded in the patterns of activity of the auditory nerves. This chapter builds upon a previous unpublished review of this area by Colburn (1981).

The first systematic attempt at predicting psychophysical performance from the activity of auditory nerve fibers was that of Siebert (1965, 1968, 1970). Siebert's work was a synthesis of mathematical concepts of signal detection theory (Tanner, Swets, and Green 1956; Green and Swets 1966) with the detailed characterization of the discharge patterns of auditory nerve fibers provided by Kiang et al. (1965). The key idea underlying this work is that the probabilistic behavior of the discharges of auditory nerve fibers (i.e., the fact that the discharges are not identical for repeated presentations of the same stimulus) imposes fundamental limitations on the performance achievable for detecting differences between acoustic stimuli. Making use of the theory of ideal observers, these limits can in principle be determined given a statistical description of the responses for the ensemble of auditory nerve fibers. Most of the models reviewed in this chapter follow the approach outlined by Siebert. These models consist of two components: (1) a peripheral component providing a quantitative description of the discharge patterns of auditory nerve fibers, and (2) an ideal observer (optimum processor) model for a specific psychophysical task based on the information provided by the peripheral component. Specifying an actual neural substrate for the computations performed by the optimum processor is not necessary, although it is desirable and sometimes done in practice.

Siebert's work was an important methodological advance because it provided a rational and rigorous technique for predicting psychophysical performance from patterns of neural activity. As is often the case, the advance came at the price of restrictions in the range of problems that can be addressed. Detection-theoretic models such as Siebert's are most applicable to tasks in which correct and incorrect responses can be objectively determined from knowledge of the stimulus and subject's responses.

These tasks include, for example, detecting a sound signal in masking noise, discriminating two vowels differing in the frequency of the second formant, and determining which of two sound sources is located more to the right. These models are less directly useful for fundamentally subjective tasks in which responses cannot be categorized as being either correct or incorrect. Examples include estimating the loudness or pitch of a stimulus, assigning a phonetic label to a speech utterance, or rating the consonance of musical chords. The emphasis of this chapter is necessarily on the "objective" tasks of detection, discrimination and resolution, even though most biologically significant tasks are probably subjective. Still, simple detection and resolution tasks are relevant to perceptual problems such as speech perception or auditory object recognition because only those stimulus attributes that are easily discriminable can serve as a basis for robust perceptual distinctions.

Although the notion that variability in neural responses limits psychophysical performance is widely recognized, two aspects of detection theory have sometimes been misunderstood. The first one concerns the notion of optimum processor, or ideal observer. It has been argued that the brain cannot be an optimum processor because the required mathematical operations are either too complex or too arbitrary. In practice, optimum processor models can often be implemented using familiar operations. For example, optimal processing of average-rate information for frequency and intensity discrimination is simply a weighted sum of the discharge rates of different fibers (Siebert 1968), which is the most basic operation in neural network models. Optimal processing of temporal information for frequency discrimination is a linear transformation of the autocorrelation function of the spike train (Siebert 1970). Autocorrelation had been proposed as a model of auditory central processing long before optimum processor models were formulated (Licklider 1951). Even if the optimum processor could not be implemented by the brain, comparing psychophysical performance with optimum processing of neural information can still be valuable. The most clear-cut conclusions are reached if optimal processing of a particular response measure (such as average discharge rate) for a particular population of neurons (such as auditory nerve fibers) yields poorer performance than psychophysical observations for a particular task. In such cases, it can be concluded that the task cannot be accomplished solely by processing such information, and that other response measures or other neural populations must also play a role. More often than not, however, predictions based on optimal processing exceed psychophysical performance. Even in such cases, it may still be possible to reach useful conclusions, albeit of a less definitive nature. For example, if the performance of the optimum processor shows the same dependence on stimulus variables as psychophysical observations, the brain may use the same response measure as the model, but may be less efficient. Alternatively, it may be possible to restrict the set of cues available to the central processor or to place additional constraints on the structure of the processor in order to obtain a better match with psycho-physical observations. Such a result would suggest that these constraints may have some physiological validity.

The second source of misunderstanding relating to detection-theoretic models concerns which sources of variability should be included in physiological models of perception. Variability can come from two sources: the stimulus itself, and the neural system. Most psychophysical models have focused on variability in the acoustic stimulus, so-called external noise (Green and Swets 1966). This form of variability must also be considered in physiological models, although in many circumstances its effects may be negligible (Siebert 1968). As we have seen, the probabilistic behavior of neural responses introduces another type of variability, which constitutes a form of internal noise. This intrinsic neural variability has been the main focus of physiological models of auditory perception.

Yet a third type of variability originates from differences in the re-sponses of different neurons to the same stimulus. This cross-neuron variability does not degrade the performance of an ideal central processor, which takes into account the response characteristics of each individual neuron. Indeed, cross-neuron variability can be a major source of infor-mation about the stimulus. For example, variations in the responses of auditory nerve fibers as a function of their characteristic frequency provide information about the frequency of a pure-tone stimulus. Nevertheless, cross-neuron variability can impair the performance of certain nonoptimal central processors. For example, a processor that sums the average dis-charge rates of all the auditory nerve fibers would eliminate place infor-mation about stimulus frequency, and would perform more poorly than a processor that made optimal use of information from a very small number of judiciously chosen fibers. While this example is rather extreme, it does illustrate that the effect of cross-neuron variability on predicted perfor-mance depends on the exact assumptions about the central processor and the psychophysical task.

This chapter begins with a brief summary of the functional organization of the auditory nerve, measures of neural responses, and functional models of auditory nerve activity. The latter constitute a front-end to the psychophysiological models that are the main topic of this chapter. Models for various psychophysical tasks are then described, beginning with very simple tasks such as thresholds in quiet and ending with more complex tasks such as the discrimination of spectral patterns important for speech perception. I conclude with an assessment of which models best predict performance for a wide variety of tasks and how such modeling efforts might be pursued for neural populations within the central auditory system.

2. Physiology and Models of Auditory Nerve Activity

2.1 Functional Organization of the Auditory Nerve

The anatomical plan and functional organization of the auditory nerve is broadly similar in most mammals, including humans (Ryugo 1992). This review is primarily concerned with the cat auditory nerve because most models are based on physiological data for the cat. The auditory nerve consists of a large number of fibers (50,000 in the cat; 30,000 in the human) that connect the cochlea to the cochlear nucleus. Virtually all electrophysiological recordings are from type I fibers, which innervate inner hair cells and constitute 90%–95% of the auditory nerve (Liberman 1982a). Because very little is known about the physiology of type II fibers (Brown 1994), which innervate outer hair cells, these fibers are not considered further.

Type I fibers differ systematically in their characteristic frequency (CF), the frequency for which their pure-tone threshold is lowest (Kiang et al. 1965). There is a precise mapping between the CF of a fiber and its point of innervation along the cochlea (Liberman 1982b). This cochlear frequency map seems to have the same basic mathematical form in most mammals, with species differing mostly in the range of frequencies that is represented in the cochlea (Greenwood 1961, 1990).

Auditory nerve fibers (ANF) also differ in their sensitivity or threshold for pure tones at the CF (Kiang et al. 1965; Liberman 1978). Threshold depends on the spontaneous rate (SR) of discharge, the rate measured when no stimulus is intentionally presented to the ear: high-SR fibers always have the lowest thresholds, while the thresholds of low-SR fibers can be 40–60 dB higher than those of high-SR fibers with the same CF (Liberman 1978). Fibers with different threshold differ not only in SR but also in many other properties, such as dynamic range (Schalk and Sachs 1980; Winter, Robertson, and Yates 1990), strength of two-tone rate suppression (Schmiedt 1982), short-term adaptation (Rhode and Smith 1985; Relkin and Doucet 1991), effects of olivocochlear efferents (Gifford and Guinan 1983), and phase-locking to both pure tones (Johnson 1980) and amplitude-modulated (AM) tones (Joris and Yin 1992). Anatomically, the SR and threshold differences are correlated with differences in the placement and morphology of synaptic terminals on inner hair cells (Liberman 1982a), and in patterns of projections to the cochlear nucleus (Fekete et al. 1982; Liberman 1991).

In summary, the mammalian auditory nerve can be seen as a two-dimensional array of fibers, with the first dimension corresponding to CF (or, alternatively, cochlear place), and the second one to sensitivity or threshold. Any realistic model of the auditory periphery must represent these two fundamental dimensions.

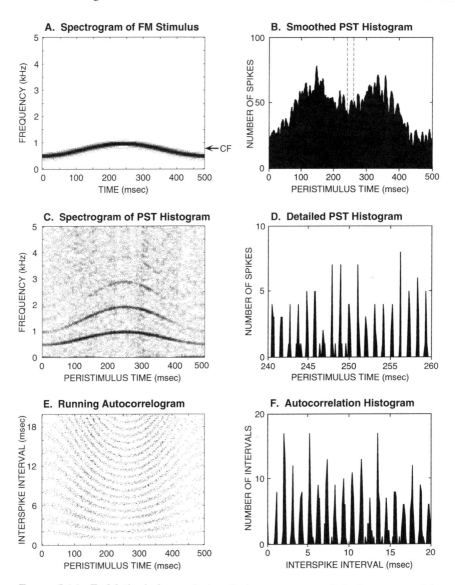

FIGURE 5.1A–F. Methods for analyzing discharge patterns of single neurons. All histograms are based on the responses of an auditory nerve fiber (ANF) (characteristic frequency [CF] = 780 Hz) to a pure-tone stimulus whose frequency was sinusoidally modulated between 500 and 1000 Hz. (A) Spectrogram of the sound stimulus computed with a 20-msec Hamming window. (B) Poststimulus time histogram (PSTH) smoothed with a 10-msec Hamming window to emphasize changes in short-time average discharge rate. (C) Spectrogram of a fine-resolution PSTH (100-μsec bin width) based on the same window as in A. The spectrogram shows a large component at the stimulus frequency, and smaller components at its harmonics. These distortion components are largely due to the half-wave rectifica-

2.2 Measures of Neural Response

Because the discharges of auditory neurons are probabilistic, responses are usually characterized by various histograms that represent statistical estimates of probabilities of discharge (Gerstein and Kiang 1960; Johnson 1978). Each histogram implies certain assumptions about the type of information that is available to the central processor, and psychophysiological models can be classified on the basis of which type of information they utilize. Figure 5.1 shows the major types of histograms for an ANF in response to a frequency-modulated tone stimulus. Each panel in Figure 5.1 is discussed when the corresponding histogram is introduced in the text.

A very general description of then neural response to a sound stimulus is the *instantaneous discharge rate*, representing the probability that a spike occurs as a function of time following the stimulus onset. The instantaneous discharge rate can be estimated from a poststimulus time histogram (PSTH) computed with a bin width sufficiently narrow to reveal the phase-locking of discharges to the stimulus (Fig. 5.1D) (Johnson 1978). PSTHs are computed by averaging responses of a single neuron to multiple stimulus presentations. Under mild assumptions, this operation is mathematically equivalent to averaging responses to a single presentation for multiple neurons having identical statistical characteristics. Such spatial summation is more plausible from a central processing point of view than temporal averaging.

A simple technique for information reduction is to average the instantaneous discharge rate over longer time frames. In effect, this is a form of lowpass filtering, an operation that occurs at most synapses in the central nervous system. For steady-state stimuli such as pure tones, averaging the instantaneous rate over the entire duration of the stimulus may suffice. However, for stimuli with time-varying spectral characteristics such as speech or music, averaging must be done over shorter time frames to preserve information about spectral changes (Fig. 5.1B). Experience with speech analysis and synthesis (Flanagan 1972) suggests that 5- to 40-msec time frames provide sufficient resolution for most purposes. These time

---

tion of the PSTH waveform apparent in D. (D) High-resolution (100-μsec bin width) PSTH that reveals phase-locking of discharges to the stimulus. The 20-msec histogram segment was taken from the center of the stimulus (*dashed lines* in B). (E) Runing autocorrelogram showing how the interspike interval distribution changes with time. Each pixel at coordinates [X, Y] indicates that a spike occurred X msec after stimulus onset and that another one occurred Y msec before. (F) All-order interspike-interval (autocorrelation) histogram for the same stimulus window as in D. Bin width, 100 μsec. The histogram shows peaks at multiples of the 1-msec stimulus period.

frames are consistent with psychophysical measures of temporal resolution (e.g., Moore et al. 1988). In the following, models that make predictions of psychophysical performance based only on the short-time average rates of discharge are called *average-rate models*; models that make use of temporal patterns of discharge within these short time frames will be called *fine-time pattern models*.

Fine-time patterns models can make use of a larger set of response measures than average-rate models. For pure tone stimuli, a convenient measure of the degree of phase-locking of the neural discharges to the stimulus is the synchronization index (Goldberg and Brown 1969; Johnson 1980). This measure is the vector sum of the phase angles of spike arrival times within each stimulus cycle, normalized by the average rate. It is therefore an index of the temporal precision with which spikes occur at a particular phase within the stimulus cycle. The synchronization index is easily generalized to steady-state stimuli consisting of a finite sum of sinusoidal components (Young and Sachs 1979; Javel 1980; Evans 1981). Under appropriate conditions, the relative values of the synchronization indices indicate the relative strengths of the response to each stimulus component[*]. A generalization of these techniques for stimuli with changing spectral characteristics is a *spectrogram* or running spectrum of the PSTH (Fig. 5.1C), which displays how response magnitude varies with both time and frequency (Carney and Geisler 1986). Such spectrograms are obtained by computing Fourier transforms of successive (possibly overlapping) time frames or windows of the PSTH. Time resolution can be traded for frequency resolution by appropriately selecting the window width. In most applications, the 5- to 40-msec windows used for computing short-time average rates are also appropriate for spectrographic analysis. In this case, the short-time average rate is the dc component of the spectrogram.

Alternatives to Fourier analysis of PSTHs are interspike-interval histograms (ISIHs), of which there are two variants: first-order ISIHs (Kiang et al. 1965; Rose et al. 1967), which display the distribution of intervals between *consecutive* spikes, and all-order ISIHs, also known as autocorrelation histograms (Ruggero 1973), which display intervals between nonconsecutive as well as consecutive spikes (Fig. 1F). These measures are appealing from a central processing point of view because, unlike PST and period histograms, they do not require a time reference locked to the stimulus. In addition, interspike intervals can reveal temporal patterns that are not related to the stimulus, such as intrinsic oscillations in cochlear nucleus neurons (Kim, Siranni, and Chang 1990). In practice,

---

[*] This will not always be possible, for example, if distortion products generated by interaction of two components are larger than a third stimulus component at the frequency of the distortion product.

studies that have analyzed responses of ANFs to complex tones using both PST and interval histograms have obtained very similar results (Sachs and Young 1980; Horst, Javel, and Farley 1986). This similarity has been explained theoretically using a nonstationary Poisson model for ANF discharges (Johnson 1978). Because ISIHs do not represent peristimulus time information, they are only appropriate for stimuli with stationary spectral characteristics. A generalization of ISIHs for time-varying stimuli is the short-time (or "running") autocorrelogram (Fig. 5.1E), which displays how the distribution of interspike intervals varies with peristimulus time (Delgutte and Cariani 1992). For a nonstationary Poisson process, the short-time autocorrelogram provides the same information as a spectrogram of the PSTH. Both representations eliminate information about response phase, which is important for certain models of central processing (Loeb, White, and Merzenich 1983; Shamma 1985; Deng, Geisler, and Greenberg 1988; Carney 1994).

Schemes for representing the activity of auditory neurons are only limited by the imagination and theoretical preferences of the investigators, and it is not likely that any single scheme will be appropriate for all stimuli and psychophysical tasks.

2.3 Models of Peripheral Auditory Processing

Models of peripheral auditory processing are an essential component of models that aim at predicting auditory behavior from the activity of ANFs. For this purpose, a functional or black-box model that simulates the key properties of ANF responses without necessarily attempting to simulate the biophysical processes leading to these responses is usually sufficient (Siebert 1965). The following sections identify the response properties that any model should include if it is to be useful for a wide range of stimuli and psychophysical tasks. The block diagram of Figure 5.2 shows how these response properties relate to stages of peripheral auditory processing. More detailed descriptions of these response properties, as well as the physiological mechanisms giving rise to these properties, can be found in an excellent review by Ruggero (1992). Chapter 4 by Mountain and Hubbard (this volume) gives a more detailed treatment of peripheral auditory models from a biophysical perspective.

2.3.1 Frequency Selectivity

The frequency selectivity of ANFs is most readily apparent in tuning curves for pure-tone stimuli (Tasaki 1954; Kiang et al. 1965), but can also be derived from responses to clicks (Kiang et al. 1965), broadband noise (De Boer 1967; Evans and Wilson 1973), tone pairs (Greenberg, Geisler, and Deng 1986), and complex periodic tones (Evans 1981; Horst, Javel, and Farley 1986). Frequency selectivity has traditionally been modeled as

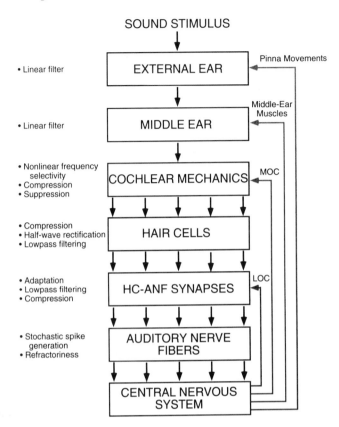

FIGURE 5.2. Stages in the peripheral auditory system and their principal signal processing functions. MOC, medial olivocochlear efferent; LOC, lateral olivocochlear efferents.

a linear bandpass filter patterned after tuning curves (e.g., Siebert 1968). This approach fails to simulate level-dependent changes in frequency selectivity that are apparent in both basilar membrane motion (Rhode 1971; Sellick, Patuzzi, and Johnstone 1982; Robles, Rugger, and Rich 1986) and ANF responses (De Boer and de Jongh 1978). These phenomena have been modeled either by incorporating nonlinear, compressive feedback in the bandpass filter (Kim, Molnar, and Pfeiffer 1973; Hirahara and Komakine 1989; Carney 1993) or by using nonlinear models of cochlear mechanics (Hall 1977; Zwicker 1986; Deng, Geisler, and Greenwood 1988).

2.3.2 Suppression and Combination Tones

Two-tone suppression refers to the decrease in the response to a tone stimulus (the excitor) when a second tone (the suppressor) is introduced.

Two-tone suppression has been observed in responses of ANFs (Sachs and Kiang 1968) and of the basilar membrane (Ruggero, Robles, and Rich 1992). Suppression in ANFs is most easily demonstrated with tones, but is also apparent for bandpass noise (Schalk and Sachs 1980), tones in noise (Kiang and Moxon 1974; Rhode, Geisler, and Kennedy 1979), and speech sounds (Sachs and Young 1979, 1980). From a modeling point of view, it is important to distinguish suppression of the average discharge rate from suppression of the synchronization index at the excitor frequency (Johnson 1974; Greenwood 1986). Synchrony suppression can be produced by very simple models such as an adaptive gain control (Johnson 1974), a compressive nonlinearity (Geisler 1985), or even a half-wave rectifier (Greenwood 1986). In contrast, rate suppression requires a more complex model, particularly for suppressor frequencies below the CF (Kim, Molnar, and Pfeiffer 1973; Hall 1977; Kim 1986). Delgutte (1990a) argued that existing nonlinear models of cochlear mechanics either altogether fail to show rate suppression (Zwicker 1986), or show much smaller rate suppression that the $40-50\,dB$ that are routinely seen in ANF responses (Geisler et al. 1990; Geisler 1992). Some success has been achieved in simulating rate suppression using phenomenological models (Sachs and Abbas 1976; Delgutte 1989; Goldstein 1990), but these models are restricted to either certain classes of stimuli or certain response properties. In this reviewer's opinion, suppression and related nonlinearities remain one of the major challenges in modeling the function of the auditory periphery.

Often associated with suppression in descriptions of cochlear mechanics are combination tones (Goldstein and Kiang 1968). These refer to response components at frequencies that are not present in the stimulus which behave in much the same way as the response to a pure tone at the combination-tone frequency. The most prominent combination tones ($2f_1$-f_2 and f_2-f_1) have lower frequencies than the primaries. Because many natural sounds such as speech, animal vocalizations, and music have a lowpass spectrum, combination tones are often masked by intense low-frequency stimulus components. Thus, combination tones may not be essential to modeling peripheral auditory function for stimulus conditions most often encountered in practice. Of course, combination tones can play an important role in psychophysical experiments on masking and the pitch of complex tones if the stimuli lack low-frequency components (Greenwood 1971; Smoorenburg 1970).

2.3.3 Compression

The discharge rates of most ANFs saturate at moderate to high sound pressure levels (SPL) (Kiang et al. 1965; Sachs and Abbas 1974), although a minority of fibers in the guinea pig show little or no saturation (Winter, Robertson, and Yates 1990). Such nonlinear behavior can be traced to at least three stages of processing in the cochlea (Ruggero 1992). First, basilar membrane motion shows a compressive nonlinearity; for stimulus levels above about 40 dB SPL, basilar membrane displacement grows approximately as a power function of sound pressure with an exponent of $\frac{1}{3}$ (Sellick, Patuzzi, and Johnstone 1982; Robles, Ruggero, and Rich 1986). This basilar membrane nonlinearity is likely to be responsible for a midlevel change in ths slope of rate-level functions for high-threshold ANFs (Sachs and Abbas 1974; Yates, Winter, and Robertson 1990). Second, the relation between hair cell cilia displacement and receptor potential exhibits saturation as well as half-wave rectification (Hudspeth and Corey 1977; Goodman, Smith, and Chamberlain 1982). Yet a third stage of compression is likely to occur at the synapses between hair cells and ANFs. Evidence for this third compression comes from differences in the dynamic ranges of low- and high-threshold ANFs, which potentially innervate the same hair cell (Schalk and Sachs 1980; Winter, Robertson, and Yates 1990). Consistent with these physiological considerations, many models of ANF responses include multiple stages of compressive nonlinearity (e.g., Delgutte 1986; Seneff 1988; Carney 1993; Slaney and Lyon 1993).

2.3.4 Lowpass Filtering

The phase-locking of ANF responses to sinusoidal stimuli falls off rapidly for frequencies above 1 kHz, until it becomes very hard to detect for frequencies above 4–5 kHz (Johnson 1980). Weiss and Rose (1988) have argued that a lowpass filter of at least the fourth order is necessary to model this rapid decrease in phase-locking. Such rapid roll-off is consistent with multiple stages of lowpass filtering, including the capacitance of hair cell membranes, synapses between hair cells and ANFs, and jitter in spike conduction (Weiss and Rose 1988).

2.3.5 Adaptation

In response to stimuli with an abrupt onset such as tone bursts, ANFs show a very rapid rise in discharge rate followed by a more gradual decrease called adaptation (Kiang et al. 1965; Smith and Zwislocki 1975; Smith 1979). Following an adapting stimulus, spontaneous activity as well as responses to transient stimuli are depressed (Smith 1977, 1979; Harris and Dallos 1979). Adaptation consists of different components whose time constants range from less than 5 msec (Westerman and Smith 1984; Yates, Robertson, and Johnstone 1985) to several seconds (Young and Sachs 1973) or even minutes (Kiang et al. 1965). Adaptation is not found in hair cell responses for sinusoidal stimuli, and is therefore thought to arise at the synapses between hair cells and ANFs. Recent models simulate most properties of adaptation (Meddis 1986; Westerman and Smith 1988).

2.3.6 Probabilistic Behavior

The observation that responses of ANFs to different repetitions of the same stimulus are not identical, and the existence of spontaneous activity, suggest that ANF discharges can be modeled as a stochastic point process. Stochastic behavior in neural systems limits the performance for any discrimination or detection task, and is therefore an essential component of models that aim at relating physiology to psychophysics (Siebert 1965). Currently, the most widely used statistical model of ANF discharges is the *nonstationary renewal process* (Siebert and Gray 1963; Gaumond, Molnar, and Kim 1982; Johnson and Swami 1983; Miller and Mark 1992). In this model, the instantaneous discharge rate is the product of a stimulus-dependent component and a refractory component. The stimulus-dependent component represents the excitation of the fiber by the stimulus, and is typically the output of previous stages of the model, including frequency selectivity, nonlinearities, and adaptation. The refractory component represents the decreased probability of discharge following an action potential. The nonstationary Poisson process, which has been widely used to model ANF discharges (Siebert 1968, 1970; Colburn 1973; Johnson 1974, 1978), is a special case of the nonstationary renewal process in which the refractory function is a constant. The Poisson with dead-time model is another special case in which the refractory component is a step function (Teich and Lachs 1979; Young and Barta 1986). The nonstationary renewal process model predicts many properties of both spontaneous and stimulus-driven activity, including the shape of interspike-interval histograms (Gray 1967; Gaumond, Molnar, and Kim 1982) and the ratio of the spike count variance to the mean spike count for short counting intervals (Teich and Khanna 1985; Young and Barta 1986; Delgutte 1987; Miller and Mark 1992). This ratio, known as the Fano factor, is an important variable in models that predict psychophysical performance from neural activity.

One failure of the nonstationary renewal process model is in predicting long-term correlations that have recently been observed in ANF spike trains (Kumar and Johnson 1984; Teich 1989; Teich et al. 1990; Kelly et al. in press). In particular, the Fano factor becomes a power function of counting time for large counting times, an effect that is characteristic of fractal phenomena (Teich 1989). Deviations from the predictions of the renewal process model only become large for counting intervals greater than 500 msec, so that these fractal effects may not be significant in the psychophysically important range of counting times from 5 msec to a few hundred milliseconds. Thus, nonstationary renewal process models might suffice for modeling most psychophysical tasks, at least until the significance of fractal effects is better elucidated.

2.3.7 Feedback

The activity of ANFs is modified by feedback systems, including the middle-ear muscles, the olivocochlear efferent pathways, and, in certain species, pinna movements (Fig. 5.2). I consider here only the medial olivocochlear (MOC) system, whose efferent branch is made up of neurons that project from medial and ventral regions of the superior olivary complex to outer hair cells (Warr and Guinan 1979; Warr 1992). Effects of MOC activity on ANF discharges have been studied through either electrical stimulation of the efferent bundle (Wiederhold and Kiang 1970) or acoustic stimulation of the contralateral ear (Warren and Liberman 1989). With either method, the principal effect of MOC stimulation is to shift the dynamic range of ANFs by as much as 15–30 dB toward higher intensities for stimulus frequencies near the CF (Wiederhold and Kiang 1970; Guinan and Gifford 1988; Warren and Liberman 1989). The contralateral stimulation technique further reveals that these gain changes occur with a time constant of 60 msec and are spatially tuned in the sense that only fibers with CFs close to the frequency of the contralateral tone are affected (Warren and Liberman 1989). To my knowledge, no model of MOC feedback consistent with these basic physiological findings has been developed.

3. Auditory Sensitivity

The most basic perceptual attribute of sounds is their audibility in a quiet background. The pure-tone audiogram, which represents the threshold of audibility as a function of frequency, plays an essential role in the diagnosis of hearing disorders. In most species of vertebrates including cats and humans, the audiogram is roughly U shaped, with good sensitivity in a medium frequency range that differs across species and poorer sensitivity at low and high frequencies (Fay 1988).

When systematic data on tuning curves of ANFs became available, it was noted that profiles of thresholds at CF against CF for the most sensitive ANFs resemble behavioral audiograms for the same species (Kiang et al. 1965; Evans 1975; Dallos et al. 1978; Fay 1978). This resemblance makes sense because ANFs whose CFs are at the frequency of a pure tone will have the lowest thresholds for this tone, at least in normal-hearing subjects. Consequently, these fibers are likely to be used for tone detection. However, a difficulty with such comparisons is that both behavioral and ANF thresholds are defined within an arbitrary criterion (e.g., a certain percent correct detection for behavioral thresholds, and a certain increment in discharge rate for ANF thresholds), so that either threshold can be manipulated by changing the criterion. Signal detection theory allows a more rigorous comparison of behavioral and neural sensitivity by using the same statistical criteria to define both kinds of thresholds (Green and Swets 1966; Geisler, Deng, and Greenberg 1985).

Delgutte (1991) measured the responses of cat ANFs to tones at the CF for stimulus levels near behavioral thresholds. Tone detectability was characterized by the sensitivity index d', which is the ratio of the mean tone-induced increment in discharge rate to the standard deviation of the rate increment over many stimulus presentations. For all ANFs, d' was found to be a power function of sound pressure with an exponent of approximately 1.6. This functional description makes it possible to systematically examine the effect of the criterion on ANF thresholds and to compare ANF thresholds with behavioral thresholds using the same d' units. Figure 5.3 shows both behavioral and ANF thresholds for the cat as a function of frequency. Behavioral thresholds are mean free-field thresholds from 10 separate studies compiled by Fay (1988). Physiological thresholds are mean thresholds at CF based on d' for high-SR fibers. Low- and medium-SR fibers have higher thresholds regardless of criterion, so that they are not likely to be used for tone detection in quiet. A d' criterion of 0.05 was selected to obtain a good fit to behavioral thresholds.

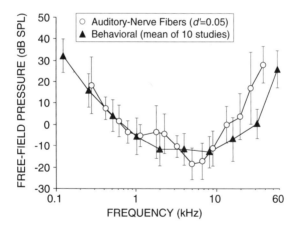

FIGURE 5.3. Physiological and psychophysical pure-tone thresholds as a function of stimulus frequency for the cat. Psychophysical thresholds are the means from 10 different studies summarized by Fay (1988, p. 333). Physiological thresholds are the mean sound pressure levels (SPLs) for which d' is 0.05 for 80 high-SR (spontaneous rate of discharge fibers from 11 cats (Delgutte 1991). Threshold SPLs measured at the tympanic membrane were converted to free-field SPLs by correcting for the transfer function of the cat's pinna and ear canal (Wiener Pfeiffer, and Backus, 1966). Physiological thresholds were also corrected for the effect of opening the bulla using measurements of middle-ear input impedance by Lynch, Peake, and Rosowski (1994). Vertical bars indicate ±1 SD.

This criterion is considerably lower than that of the behavioral measurements, which must be close to unity[†].

Several factors need to be considered in interpreting this difference in criteria between behavioral and ANF thresholds. (1) The tone stimuli were shorter, and therefore less detectable in the physiological experiments than in the behavioral experiments. Specifically, cat behavioral thresholds are at least 5 dB higher for the 50-msec tone stimuli used by Delgutte (1991) than for the 500-msec tones used in most psychophysical studies (Costalupes 1983; Solecki and Gerken 1990). (2) Thresholds for the most sensitive high-SR fibers are 5–10 dB lower than the mean thresholds shown in Figure 5.3 (Liberman 1978). Thus, the physiological thresholds in Figure 5.3 are overestimated by at least 10 dB relative to behavioral thresholds. This 10-dB difference in thresholds translates into a 16-dB difference in d' using the empirical power law relation between sound pressure and physiological d' (Delgutte 1991). Thus, the discrepancy in d' criteria between behavioral thresholds and the thresholds of the most sensitive ANFs is at most a factor of 3 (rather than 20). While this discrepancy is still significant, the central nervous system may further improve detection by integrating information from several ANFs. Specifically, the d' achievable by combining information from N fibers with statistically independent, identically distributed characteristics is equal to the d' for a single fiber times $\sqrt{N}$ (Green 1958; Green and Swets 1966). We conclude that average-rate information from a small number (<10) of the most sensitive ANFs can account for detection of tones in quiet in the cat. This number is only a very small fraction of the fibers that would be available in a critical band.

Figure 5.3 shows good agreement between the frequency dependence of behavioral and physiological thresholds for frequencies up to about 10 kHz. Above 10 kHz, however, physiological thresholds rise faster than behavioral thresholds. Similar observations have been made in several species of mammals (Evans 1975; Dallos et al. 1978). The reason for this high-frequency discrepancy is unknown, and experimental artifacts cannot be ruled out. Specifically, the thresholds of high-CF ANFs are more susceptible to changes in cochlear temperature than those of low-CF fibers (Brown, Smith, and Nuttall 1981; M.C. Liberman, personal communication), so that the rapid rise of ANF thresholds at high frequencies might be caused by a lowering of cochlear temperature resulting from opening the bulla. If the discrepancy between behavioral and physiological audiograms cannot be attributed to an experimental artifact, one would conclude that the central nervous system may process information

---

[†] Although the behavioral studies did not specify their threshold criterion in terms of d', it is likely to be close to 1 because psychophysical procedures are most effective when d' is near 1.

for tone detection more efficiently for high frequencies than for low frequencies.

Because phase-locking of high-SR fibers can be experimentally detected for lower sound levels than increases in discharge rate (Rose et al. 1971; Johnson 1980), it has been suggested that low-frequency tones might be detected from the phase-locking of the discharges of ANFs rather than through increases in average discharge rates. An analysis by Recio, Viemeister, and Powers (1994) showed that in fact when thresholds are defined by statistical criteria, thresholds based on synchronization index are not consistently lower than average-rate thresholds for short-duration (<100 msec) stimuli because the variance of the synchronization index is very large. While this analysis suggests that there would be no advantage in using phase-locking for detecting short-duration tones, it does not rule out this possibility. However, because phase-locking thresholds increase faster with frequency than do rate thresholds, use of phase-locking for signal detection would accentuate the differences between the physiological and behavioral audiograms of Figure 5.3. Thus, physiological results are more consistent with the idea that changes in average discharge rate are used for tone detection by the cat. A similar conclusion was reached for the goldfish, based on an entirely different line of arguments (Fay and Coombs 1983). Thus, similar mechanisms might be used for tone detection in species as different as the cat and the goldfish, suggesting there may be a very general mechanism for sound detection across vertebrates (Fay 1992).

4. Intensity Discrimination

The ability of listeners to distinguish differences in sound intensity is one of the most important aspects of auditory perception. An appropriate loudness is the primary determinant of the subjective quality of a speech communication because insufficient loudness degrades intelligibility, while excessive loudness is uncomfortable. In quiet, normal-hearing listeners can understand speech over a wide range of stimulus levels (100 dB or more). In contrast, hearing-impaired listeners often show *recruitment*, an abnormal growth of loudness that can severely restrict the useful dynamic range for speech communication. A restriction in dynamic range can also be caused by background noise in both normal and hearing-impaired listeners.

Intensity perception also plays a fundamental role in auditory theory. Consider a *place* model in which the stimulus spectrum is represented by the distribution of neural activity across an array of frequency-selective channels. This place-dependent measure of neural activity could be based either on average discharge rates for a *rate-place* model, or on fine time patterns of discharge for a *temporal-place* model. For either model, discriminating between two stimuli differing in spectrum would require detecting changes in neural activity in at least one frequency channel. A special case is that of intensity discrimination, in which an increase in stimulus level causes neural activity to increase in every frequency channel that responds to the stimulus. Thus, for any place model of auditory processing, there is a close connection between the discrimination of spectral patterns and the encoding of stimulus intensity in individual frequency channels (Fletcher 1940; Zwicker 1956, 1970).

4.1 Siebert's Model and Spread of Excitation

A major problem in intensity perception is that the discharge rates of single ANFs increase with stimulus level over a range of only 20–40 dB, while human listeners can distinguish differences in intensity over a range of 120 dB or more. This discrepancy between psychophysical and single-neuron dynamic ranges has been called the dynamic range problem (Evans 1981). The first quantitative attempt to solve this problem was that of Siebert (1968) (see also Whitfield 1967). Siebert's model incorporated the key features of auditory nerve activity that were known at the time (Kiang et al. 1965), including frequency tuning, a logarithmic cochlear frequency map, saturating rate-level functions, and Poisson statistics for ANF discharges. All fibers with the same CF were assumed to have the same threshold. Siebert further assumed that intensity discrimination was based solely on the average discharge rates, and combined information across fibers using the optimum rule for statistically independent Poisson random variables. With these assumptions, the model predicted Weber's law for pure-tone intensity discrimination, that is, the intensity difference limen (DL), expressed in decibels, was constant for sensation levels greater than 15–20 dB. At these stimulus levels, fibers tuned to the stimulus frequency are saturated, so that intensity information is entirely conveyed by unsaturated fibers lying on the edge of the excitation pattern (Fig. 5.4A,B). Due to the logarithmic cochlear map and the triangular shape of the tuning curves, each dB increase in level causes a constant increase in the number of recruited fibers. At the same time, an equal number of fibers becomes saturated, so that the number of fibers conveying intensity information (and therefore model performance) remains constant. Thus, the Siebert model is a pure "spread-of-excitation" model in which intensity is coded by the total number of active fibers rather than by the amount of activity at any one place along the cochlea.

Although Siebert's prediction of Weber's law for pure tones was not inconsistent with the data available at the time, more recent studies have shown that there is a small but significant decrease in intensity DL as stimulus level increases (McGill and Goldberg 1968; Jesteadt, Wier, and Green 1977; Florentine, Buus, and Mason 1987). This "near miss" to Weber's law is observed not only for humans, but also for many species

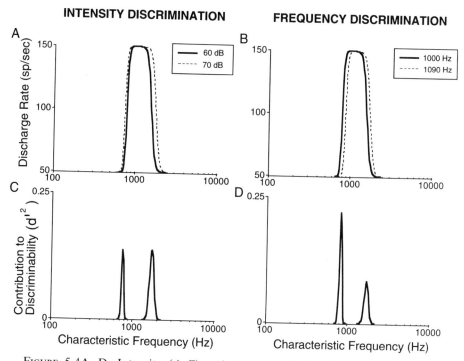

FIGURE 5.4A–D. Intensity (A,C) and frequency (B,D) discrimination for the Siebert (1968) rate-place model. (A) Model discharge rate as a function of CF for a 1-kHz tone stimulus at 60 and 70 dB SPL. The rate profiles for the two intensities only differ in the edges of the pattern where the fibers are not saturated. (B) d'^2 as a function of CF for the discrimination of the two tones from A. This measure gives the contribution of each CF band to discriminability. The total discriminability for the model is proportional to the square root of the area under the curve. (C) Rate profiles for the model for 60-dB tones with frequencies of 1000 and 1090 Hz. (D) d'^2, as a function of CF for the discrimination of the two tones from C. The total area under the d'^2 curve is approximately the same as in C, meaning that the 90-Hz frequency difference in C and D is as discriminable for the model as the 10-dB level difference in A and B.

of mammals (Sinnott, Brown, and Brown 1992). The failure of the Siebert model to predict the near miss may not be a fundamental flaw because a version of the Lachs et al. (1984) model that differed from the Siebert model in that the cohlear filters had a more rounded shape was able to predict the near miss.

Spread-of-excitation models such as those of Siebert (1968) and Lachs et al. (1984) have more fundamental difficulties for situations in which intensity discrimination is accurate with minimal spread of excitation. For example, Weber's law does hold approximately for broadband stimuli such as clicks and noise (Miller 1947; Penner and Viemeister 1973; Raab and Goldberg 1975; Buus 1990). For such stimuli, opportunity for spread of excitation is limited to only very low and very high CFs, where the thresholds of ANFs rise steeply. Spread-of-excitation models were tested in an important psychophysical experiment by Viemeister (1983; see also Viemeister 1974; Moore and Raab 1975). Intensity discrimination was measured for bandpass noise surrounded by band-reject noise intended to prevent spread of excitation. The passband of the noise signal was entirely at high frequencies, for which ANFs show no significant phase locking. Under these conditions, intensity discrimination performance approximately obeyed Weber's law over a broad range of stimulus levels. Moreover, adding the band-reject noise only degraded performance slightly compared to the noiseless case. To the extent that the masking noise was effective in preventing spread of excitation, this experiment shows that spread of excitation is not necessary for accurate intensity discrimination, so that some other mechanism must also be involved. On the other hand, the observation that performance did decrease somewhat in the presence of band-reject masking noise suggests that spread of excitation does contribute to intensity resolution for narrowband signals in quiet.

4.2 Role of High-Threshold Fibers

From the preceding discussion, it is clear that spread of excitation by itself cannot account for intensity discrimination for all stimulus conditions. What is needed is a mechanism for extending the dynamic range in individual frequency channels. One possibility is that the broad distribution of thresholds of ANFs in every CF region (Liberman 1978) makes it possible to encode intensity at high stimulus levels by recruiting high-threshold fibers. This physiological property was not well described in 1968 and therefore not incorporated in the Siebert model. The role of high-threshold fibers was systematically investigated in three models of intensity discrimination (Winslow 1985; Winslow and Sachs 1988; Delgutte 1987; Viemeister 1988; see also Goldstein 1980 and Colburn 1981 for pioneering work in this area). Because these models make similar assumptions and reach similar conclusions, they are discussed together. Model assumptions were broadly similar to those of Siebert (1968), but the emphasis was more on modeling a single frequency channel than on the spatial pattern of neural activity. In this context, a frequency channel represents a group of ANFs that share the same CF, but differ in other properties such as SR, threshold, and dynamic range. The newer models differ from that of Siebert (1968) in that they use a distribution of thresholds and dynamic ranges that closely mimic physiological data (Liberman 1978; Schalk and Sachs 1980). They also differ from the Siebert model in their statistical model for ANF discharges, but we will see later that this is not a major factor. As in Siebert (1968), all three models used the optimum scheme for statistically independent random variables for combining information across fibers.

Results for the Delgutte (1987) model are shown in Figure 5.5A. The intensity discrimination performance for typical high-SR, medium-SR, and low-SR fibers are compared with psychophysical data and with model

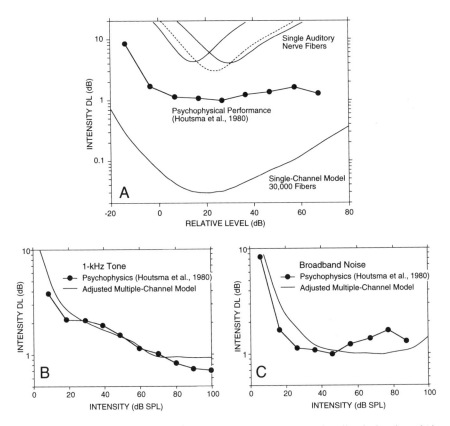

FIGURE 5.5. The Delgutte (1987) rate-place model of intensity discrimination. (A) Performance of a 30,000-fiber, single-channel version of the model compared with human psychophysical performance for broadband noise (Houtsma, Durlach, and Braida 1980) (*filled circles*) and with performance for typical high-SR, medium-SR, and low-SR model fibers (*top*, from left to right). Psychophysical difference limens (DLs) were multiplied by a factor of 2 to compensate for the differences in stimulus duration between physiological data (50 msec) and psychophysical experiments (500 msec). (B) Performance of the multiple-channel version of the model compared with human psychophysical data for intensity discrimination for a 1-kHz tone (Houtsma, Durlach, and Braida 1980) (*filled circles*). In this version of the model, the threshold distribution was adjusted to approximate Weber's law over a wide intensity range in each frequency channel. (C) Same as in B for the discrimination of the intensity of broadband noise.

predictions for an optimal combination of 30,000 fibers (the number of fibers in the human auditory nerve). Psychophysical data are for broad-band noise because spread of excitation (which does not occur in this single-channel version of the model) is likely to be minimal for these stimuli. Single-fiber performance approaches psychophysical performance over a narrow range of levels around a minimum DL, but rapidly degrades as stimulus level increases over the minimum. In contrast, performance for the 30,000-fiber model well exceeds psychophysical performance over the entire range of stimulus levels. Further analysis shows that, for sti-mulus levels exceeding 20 dB SL, model performance is almost entirely determined by high-threshold (low- and medium-SR) fibers. Similar results were obtained for the models of Winslow and Sachs (1988) and Viemeister (1988) and in the physiological study of Young and Barta (1986).

The three models clearly show that psychophysical performance in intensity discrimination can be accounted for by the information available in the average discharge rates of ANFs, even in the absence of spread of excitation. Thus, psychophysical performance is not limited by the infor-mation available in ANFs, but by how the central nervous system makes use of this information. In other words, the key problem in intensity coding is *not* how the wide psychophysical dynamic range can be accounted for from ANFs, which, individually, have a narrow dynamic range, but rather why the central nervous system is so inefficient in processing the intensity information conveyed by the ensemble of ANFs. On the other hand, predictions based on optimal processing of average-rate information clearly deteriorate with increasing stimulus level, while psychophysical performance remains nearly constant when spread of excitation is minimal. Thus, the central nervous system does not behave simply as an inefficient optimum processor: information from high-threshold fibers would have to be processed more efficiently than that from low-threshold fibers to obtain Weber's law.

The question of the relative processing efficiency for fibers differing in thresholds was further investigated by Delgutte (1987). Delgutte adjusted the threshold distribution in his model to approximate Weber's law over a wide range of stimulus levels. In effect, the ratio of the adjusted numbers of fibers to the numbers actually present in the auditory nerve provides a measure of how efficiently each group of fibers is processed. The proportion of low-SR (high-threshold) fibers in the adjusted distribution was about three times greater than the approximately 15% found in the auditory nerve (Liberman 1978). Interestingly, certain types of cells in the cochlear nucleus receive many more inputs from high-threshold ANFs than from low-threshold fibers (Liberman 1991). Such cells may be well suited for intensity coding at high stimulus levels.

While efficient processing of information from high-threshold fibers may provide the basis for intensity discrimination in single frequency channels, spread of excitation across channels is likely to play a role in certain conditions. To examine how these two factors interact, Delgutte (1987) developed a multiple-channel version of his model by adding a bank of bandpass filters mimicking the tuning curves of ANFs. The optimum scheme for Gaussian random variables was used for combining information across as well as within frequency channels. Predictions of this multiple-channel model are compared with psychophysical data for a 1-kHz tone and broadband noise in Figure 5.5B and 5.5C, respectively. The model provides a good fit to the psychophysical data for both stimuli over a broad range of levels: it predicts Weber's law for noise and the near miss for tones. The near-miss behavior results from the increase in the number of active channels with stimulus level and the proportionality of performance to the square root of the number of channels in the optimal combination scheme. In this respect, the Delgutte model is similar to the excitation pattern model of Florentine and Buus (1981) in which Weber's law was assumed for individual frequency channels without specifying a physiological correlate.

4.3 Role of Olivocochlear Efferents

One physiological mechanism that might improve intensity discrimination performance at high stimulus levels and low signal-to-noise ratios is the medial olivocochlear efferent (MOC) system. The possible benefits of the MOC system are most apparent for transient signals in background noise (Winslow and Sachs 1987; Kawase, Delgutte, and Liberman 1993). Continuous background noise compresses the rate-level functions for tone-burst stimuli in two ways (Costalupes, Young, and Gibson 1984): first, the background rate of discharge is raised through stimulation by the noise, and second, the maximum discharge rate is reduced through adaptation by the continuous noise. Activation of MOC efferents through either electrical stimulation (Winslow and Sachs 1987) or acoustic stimulation of the contralateral ear (Kawase, Delgutte, and Liberman 1993) partially counters these effects of background noise: it "decompresses" the rate-level function by decreasing the background discharge rate and increasing the maximum rate for the tone burst.

The effects of MOC stimulation on intensity discrimination in continuous background noise was modeled by Winslow and Sachs (1988). As expected, background noise diminished the ability of ANFs to encode the intensity of tone-burst stimuli. Both the maximum performance and the dynamic range of a single-channel model with a realistic distribution of ANF thresholds was significantly degraded by moderate background noise. Electrical stimulation of MOC efferents restored the maximum performance to its quiet level but had smaller effects on the dynamic range.

The effects of MOC efferents studied by Winslow and Sachs (1988) were induced by electrical stimulation. This situation is very artificial in that the activation of the efferents does not depend on the acoustic inputs to the ears. In natural conditions, efferent activity is likely to be induced through reflex action, possibly modulated by central control. The effect of efferents on ANF activity increases with the intensity of acoustic stimuli delivered to the contralateral ear (Warren and Liberman 1989). This reflex is likely to be even more effective with ipsilateral or binaural stimulation than with contralateral stimulation because roughly two-thirds of MOC neurons best respond to ipsilateral stimuli (Liberman and Brown 1986). In awake animals, this reflex might be modulated by central control to optimize the performance for each perceptual task. For example, MOC neurons that innervate cochlear regions in which the signal-to-noise ratio is particularly low might be selectively activated to implement a form of Wiener filtering. These ideas on the role of the MOC system are consistent with behavioral data showing degraded intensity discrimination in noise for cats in which the MOC fibers have been cut bilaterally (McQuone and May 1993). On the other hand, monolateral section of the olivocochlear efferents in one human patient provided no evidence for an effect on intensity discrimination in noise (Scharf et al. 1994).

4.4 Statistical Issues

The preceding discussion has focused on physiological factors affecting intensity discrimination. Predicted performance in models of intensity discrimination is also affected by the statistical characterization of ANF discharges and schemes for combining information across fibers. These issues are discussed in some detail in the context of intensity discrimination, although they also play a role in models for other psychophysical tasks.

4.4.1 Stochastic Model

Siebert (1968) used a Poisson process to model the statistics of ANF discharges. For Poisson processes, the Fano factor (ratio of the spike count variance to the mean spike count) is always unity. More recent models have modified the Poisson assumptions to yield Fano factors more consistent with physiological data (Teich and Khanna 1985; Young and Barta 1986; Delgutte 1987). Teich and Lachs (1979), Lachs et al. (1984), and Winslow and Sachs (1988) used a Poisson model modified by dead time. Viemeister (1988) and Delgutte (1987) assumed a Gaussian distribution of spike counts with Fano factors empirically derived from physiological data. In either case, variances were two- to threefold smaller than those of the Poisson model at high discharge rates. Because performance expressed as d' is inversely proportional to the square root of the variance, the modified assumptions yield a small improvement in performance over the Poisson model at high intensities. However, these improvements are not sufficient to overcome that failure of single-channel models to predict Weber's law. Thus, the details of the statistical models for ANFs may not be of primary importance for rate-place models of auditory behavior so long as Fano factors are not severely mispredicted[‡].

4.4.2 Cross-Fiber Correlation and External Noise

All the models described so far assumed statistical independence between the discharge characteristics of different ANFs. This assumption is consistent with the physiological data of Johnson and Kiang (1976), who found no evidence for correlation between the discharges of pairs of simultaneously recorded ANFs, even when the two fibers had very similar CFs. Johnson and Kiang (1976) examined spontaneous activity and responses to tones. When the stimulus is itself random (e.g., broadband noise), statistical fluctuations in the stimulus introduce correlations between the discharges of fibers with similar CFs because these fibers share a common excitation (Young and Sachs 1989). Calculations based on a Poisson model suggest that the spike count variance resulting from statistical fluctuations in the stimulus (external noise) is small compared to the variance due to the spike-generating mechanism (internal noise), implying that correlations between pairs of fibers must be small (Siebert 1968). This conclusion is consistent with physiological data showing that the spike count variance is approximately the same for tones and broadband noise that produce the same mean spike count (Delgutte, unpublished observations). Thus, both theory and data concur that statistical fluctuations in the stimulus have minimal effect on discrimination performance for *single* ANFs. However, Erell (1988), pointed out that statistical fluctuations from external noise are correlated for all the fibers whose CFs are within one bandwidth of the cochlear filters, so that the effect of these fluctuations on the performance for a *population* of ANFs may be significant. In other words, while internal noise can be reduced by combining information across fibers, this is much less possible for external noise, which is correlated among fibers with similar CFs. Erell (1988) used Monte Carlo simulations to assess the effect of stimulus fluctuations on the performance of a multiple-channel, rate-place model of intensity and frequency discrimination. His results show that the effect of external noise on model performance is comparable to that of internal noise for broadband noise stimuli. These findings suggest that external noise can

---

[‡] The Fano factor was mispredicted in the Teich and Lachs (1979) model, which assumed that the saturation of the discharge rate of auditory nerve fibers is caused by refractoriness (dead time). This assumption yields infinitesimally small spike count variances at high intensities where mean discharge rates are saturated. This prediction is clearly inconsistent with physiological data, as Lachs et al. (1984) recognized in a later version of their model.

significantly affect the discrimination performance predicted by physiological models for a neural population, even though the effect on each single fiber is small. The effect of stimulus fluctuations should be even greater for narrowband noise than for the broadband noise investigated by Erell (Green 1960; Buus 1990). On the other hand, the effects of external noise are still too small to alter the conclusion that there is sufficient information in the discharge rates of ANFs to account for intensity discrimination. Nevertheless, a more rigorous treatment of external noise is needed in future models.

4.4.3 Combination Schemes

Most intensity-discrimination models used the optimal scheme for statistically independent random variables to combine information across ANFs (Siebert 1968; Delgutte 1987; Viemeister 1988; Winslow and Sachs 1988). This scheme obtains the best possible performance by assigning greater weight to the variables (fibers) that provide the most reliable information. For Poisson or Gaussian random variables, this combination rule takes a particularly simple form, in that the optimal decision variable r is a weighted sum of the spike counts r_i (discharge rates) of the individual nerve fibers (Green 1958; Green and Swets 1966):

$$r = \sum_{i=1}^{N} w_i r_i \qquad (1)$$

The optimal weights are given by

$$w_i = \frac{\Delta \bar{r}_i}{\sigma_i^2} \qquad (2)$$

where $\Delta \bar{r}_i$ is the mean increment in spike count resulting from the change in intensity, and σ_i^2 is the spike count variance. Note that the weights are zero for unstimulated and saturated fibers because $\Delta \bar{r}_i$ is zero for these fibers. With this choice of weights, d' for the optimum combination is given as a function of the d'_i for the individual fibers by

$$d'^2 = \sum_{i=1}^{N} d'^2_i \qquad (3)$$

Lachs et al. (1984) and Viemeister (1988) investigated a simple alternative to the optimal combination scheme in which the decision variable is formed by summing the spike counts of the individual ANFs:

$$r = \sum_{i=1}^{N} r_i \qquad (4)$$

This scheme is interesting because it has been proposed as a possible model for loudness (Fletcher and Munson 1933). It is nonoptimal in that it gives equal weight to saturated and unsaturated fibers, even though the saturated fibers decrease performance because they increase the variance of the decision variable r without contributing to the mean. Accordingly, Viemeister (1988) found that performance of his model for this simple summation scheme degraded more steeply with increasing intensity than that for the optimal combination scheme. Thus the simple summation scheme leads to more severe deviations from psychophysical data than the optimal scheme in the level dependence of DLs. Nevertheless, performance in intensity discrimination could still be accounted for by the simple summation model because the model performance exceeded psychophysical performance over the entire range of levels.

Another suboptimal combination scheme that is often assumed by physiologists is to select the single fiber that has the lowest threshold or, equivalently, the maximum d' (e.g., Kiang et al. 1965; Sinex and Havey 1986). Experience with excitation-pattern models for a wide range of stimuli suggests that this single-channel scheme is not as successful as the optimum scheme in predicting psychophysical data on intensity discrimination (Florentine and Buus 1981). Information for intensity discrimination is usually distributed among a large number of fibers differing in CF and threshold, so that results for the two schemes are likely to differ substantially. On the other hand, Delgutte (unpublished observations) found that tone-on-tone masking patterns predicted by a rate-place model (Delgutte 1989) had similar *shapes* for both combination schemes, although of course the optimal scheme gave better absolute performance than the single-channel scheme. In this masking task, information for detecting the tone signal is usually restricted to a small number of fibers whose CFs are close to (but not exactly at) the signal frequency, so that the two schemes are more likely to give similar results. Thus, the benefits of the optimal scheme over the single-channel scheme depend on the stimulus condition and the psychophysical task.

4.5 Temporal Models

The possibility of using fine-time patterns of discharge for intensity coding is sometimes entertained in the literature, although no explicit temporal model of intensity discrimination has been proposed. One simple possibility would be to replace average discharge rate by a temporal response measure derived from interspike-interval histograms or Fourier analysis of period histograms (e.g., synchronization index). Such a scheme would provide little benefit over rate-place models for predicting Weber's law because these temporal measures saturate at the same intensities as average rate. In fact, the very stability of these temporal measures with respect to changes in stimulus level is advanced as an argument for using these measures as a neural code for the spectrum of complex acoustic stimuli (e.g., Young and Sachs 1979).

One scheme that does show promise for extending the dynamic range for intensity coding beyond the saturation of ANFs is the spatiotemporal coincidence model of Carney (1994). This model is based on the hypothesis that certain cells in the cochlear nucleus act as coincidence detectors in that they only discharge when they receive nearly simultaneous spike inputs from several ANFs. The potential of this model for intensity coding arises from nonlinearities in cochlear mechanics: as intensity increases, basilar membrane motion becomes less sharply tuned, and the phase differences between adjacent points on the basilar membrane decrease. As a result of these changes in the spatial phase pattern, the discharges of ANFs with neighboring CFs become more coincident at high stimulus levels. Carney (1994) implemented a coincidence-detector cell model receiving inputs from ANFs with slightly different CFs. The discharge patterns of the ANF inputs were produced by a peripheral model that simulated essential features of the cochlear nonlinearity (Carney 1993). Results showed that the discharge rate of the coincidence cell for low-frequency tone stimuli increases at least 20 dB beyond the intensity at which the ANF inputs saturate. Thus, spatial coincidence mechanisms can potentially increase the dynamic range of central auditory neurons for intensity coding.

The predictions of the Carney (1994) coincidence model have not been tested against psychophysical data on intensity discrimination. In particular, it does not appear likely that the extension of dynamic range achieved through coincidence would hold for high-frequency tones to which ANFs do not phase lock. Although there is increasing evidence that certain cells in the ventral cochlear nucleus are sensitive to coincidence of their inputs (Carney 1990; Rothman, Young, and Manis 1993; Joris et al. 1994; Palmer et al., 1995), the exact details of the postulated coincidence scheme remain speculative. The significance of the Carney (1994) model is that it provides a physiologically plausible scheme that could be used for intensity coding for certain conditions without relying on the weighted sums of spike counts assumed by rate-place models. This model points to the benefits that can be achieved in physiological models of auditory perception by incorporating knowledge about the physiology of central auditory neurons in addition to constraints imposed by peripheral processing.

5. Loudness

Unlike intensity discrimination, for which listeners' performance can be objectively assessed, loudness is a fundamentally subjective attribute of acoustic stimuli in that there is no rational procedure for deciding whether loudness judgments are correct. Therefore, detection theory cannot, by itself, predict loudness from statistical descriptions of the discharges of ANFs. An additional assumption defining a physiological correlate of loudness is needed. One possibility would be to derive a loudness correlate from a model of intensity discrimination using Fechner's (1860) assumption that all DLs represent equal loudness differences. However, this assumption has been shown to be clearly at variance with direct estimates of loudness (Stevens and Davis 1938). The relationship between loudness and intensity discrimination is still being debated (for review, see Johnson et al. 1993). Until these debates are resolved, most physiological models of loudness have been based on the hypothesis, from Fletcher and Munson (1933; see also von Békésy 1929), that loudness is proportional to the total discharge rate for the ensemble of ANFs. This total discharge rate depends on both the number of excited fibers and on the rate-level functions of individual fibers. Different models have emphasized one or the other of these two components.

The first quantitative test of the Fletcher–Munson hypothesis using ANF data was conducted by Goldstein (1974). The peripheral component of the Goldstein (1974) loudness model was essentially the same as that of the Siebert (1968) model, with saturating rate-level functions, frequency tuning, and uniform distribution of fibers on a logarithmic CF scale. All fibers with the same CF had the same threshold, making the Goldstein model a spread-of-excitation model. Loudness predictions were formed by summing the *driven* discharge rates (i.e., the spontaneous rate was subtracted out) for all the model fibers. The results showed the importance of the peripheral auditory filters in loudness predictions. For one version of the model in which tuning curves lacked low-frequency tails (Kiang and Moxon 1974), predicted loudness for pure tones grew linearly with sound pressure in decibels at high stimulus levels. This prediction contradicts the well-known power law for the growth of loudness (Fletcher and Steinberg 1924; Stevens 1956). However, when the model tuning curves included low-frequency tails, the model did predict the power law with a reasonable exponent. The importance of auditory filter shapes in accounting for the power law was also emphasized by Lachs and Teich (1981) and Lachs et al. (1984) in their spread-of-excitation model.

The results of Goldstein (1974) and Lachs et al. (1984) suggest that, if loudness corresponds to the total discharge rate of the auditory nerve, spread of excitation to CFs well above the stimulus frequency is essential for the power law behavior. As Goldstein (1974) pointed out, this conclusion is inconsistent with psychophysical data on loudness growth functions in the presence of high-frequency masking noise that restricts spread of excitation (Hellman 1974, 1978). These data show that a full range of loudness is obtained even in conditions when spread of excitation is severely restricted, although the rate of growth of loudness is somewhat altered. Thus, pure spread-of-excitation models have the same difficulty with loudness growth as with intensity discrimination.

Pickles (1983) examined whether the Fletcher–Munson hypothesis predicts loudness summation, that is, how loudness grows with the bandwidth of a noise stimulus. Psychophysical data show that, as the bandwidth of a noise of constant sound pressure level is increased, loudness stays nearly constant up to a critical bandwidth, then increases with bandwidth (Zwicker, Flottorp, and Stevens 1957). The Pickles (1983) model was based on physiological data describing how the discharge rates of ANFs grow with the bandwidth of bandpass noise. It was a spread-of-excitation model because it ignored threshold differences among ANFs. As in the Goldstein (1974) model, loudness predictions were derived by summing the driven discharge rates over the fiber ensemble. The model qualitatively predicted the psychophysical phenomenon of loudness summation in that loudness always grew with increasing stimulus bandwidth. This summation results from the compressive nature of rate-level functions: a compressive nonlinearity implies that the total discharge rate is maximum when the stimulus power is spread over a large number of frequency channels. In contrast to psychophysical data, the model showed no indication of a critical bandwidth: there was no range of bandwidths over which predicted loudness remained constant. This result is consistent with the idea that critical bandwidths in loudness summation may be related to central processing rather than to the bandwidths of the peripheral auditory filters (Zwicker and Scharf 1965).

The most extensive attempt at testing the Fletcher–Munson hypothesis to date is that of Jeng (1992). The Jeng model differed from those of Goldstein (1974), Pickles (1983), and Lachs et al. (1984) in that it included a range of ANF thresholds in each CF region and thus was not a spread-of-excitation model. On the other hand, the model lacked realistic rate-level functions (these were essentially step functions in her model). Model predictions for loudness growth of tone stimuli were generally good at low stimulus levels but underestimated psychophysical loudness above 50 dB SPL. More accurate predictions could be obtained by making the proportion of high-threshold ANFs in the model exceed physiological proportions, much as in the Delgutte (1987) model of intensity discrimination. Such manipulations effectively assume that the central processor gives more weight to high-threshold fibers, contrary to the Fletcher–Munson hypothesis that discharge rates from all fibers are summed equally. However, even these threshold manipulations do not overcome the serious difficulties that the Jeng (1992) model has with equal loudness contours. Equal loudness contours for the model were U shaped, roughly paralleling the threshold audiogram for all stimulus levels. This contrasts with psychophysical data, which show a gradual flattening of equal-loudness contours with increasing stimulus level (Fletcher and Munson 1933). In fact, it is not clear what physiological phenomenon could cause this flattening of equal-loudness contours.

In summary, models based on the Fletcher–Munson hypothesis that loudness corresponds to the total discharge rate of the auditory nerve have difficulties predicting some of the psychophysical data (Goldstein 1974; Pickles 1983; Lachs et al. 1984; Jeng 1992). While these models lacked either a realistic distribution of ANF thresholds or appropriate rate-level functions, the discrepancies between model predictions and data are so severe in some cases that it is hard to see how a more realistic model would alter the situation. While the Fletcher–Munson hypothesis may not hold for ANFs, it might be more valid for populations of neurons at more central stages in the auditory nervous system.

6. Frequency Discrimination

Many biologically significant tasks require detecting changes in the frequency of a stimulus component or a cluster of components. Examples include the discrimination of formant frequencies in speech and the detection of directionally-dependent spectral notches produced by the filtering action of the pinna. The simplest instance of this type of task is pure-tone frequency discrimination. Psychophysical data on frequency discrimination are available for a larger number of species than are data on intensity discrimination, and thereby offer opportunities for tighter tests of psychophysiological models (Fay 1988).

We have seen that, for any place model of auditory processing, spectral discrimination is closely related to how intensity is encoded in individual frequency channels. Pure-tone frequency discrimination is no exception, and the rate-place models that were introduced in the context of intensity discrimination also give predictions for frequency discrimination. Pure-tone frequency discrimination is also interesting because it the simplest task for which rate-place and temporal models have been directly compared.

6.1 Rate-Place Models

The first model of frequency discrimination based on the activity of ANFs was that of Siebert (1968). In this spread-of-excitation model, frequency discrimination was based entirely on the unsaturated ANFs lying at the edges of the excitation pattern, particularly the low-CF edge where the discharge rates vary most steeply with CF (see Fig. 5.4C,D). Because the cochlear frequency map was logarithmic, the model predicted a constant Weber fraction $\Delta F/F$ for frequency. The model further predicted that $\Delta F/F$ should be a constant fraction of the Weber fraction for amplitude $\Delta A/A$. This prediction, which holds for any place model (e.g., Zwicker 1956, 1970), arises because, for any ANF, a given change in discharge rate can be produced by changing either the frequency or the intensity of a pure tone, with the ratio between the two variables being determined by the slope of the isorate function (tuning curve). With the cochlear filters used by Siebert (1968) to match the tuning curves of Kiang et al. (1965) for the cat, $\Delta A/A$ was about 15 times $\Delta F/F$.

The prediction that the Weber fraction for amplitude discrimination should be about 15 times the fraction for frequency discrimination does not quite fit human psychophysical data, where the ratio is more nearly 40–50 for 1- to 2-kHz tones (Siebert 1968; Erell 1988). A better match to the human ratio could be obtained by increasing the slopes of the model tuning curves, consistent with evidence that cochlear filters are likely to be more sharply tuned in the human than in the cat (Pickles 1979, 1980; Greenwood 1990). To avoid this difficulty, model predictions based on cat physiology should be compared with psychophysical data for the same species. If we use the intensity discrimination data of Elliott and McGee (1965) and the frequency discrimination data of Elliott, Stein, and Harrison (1960), the ratio is approximately 30 at 2 kHz for the cat, twice that predicted by the Siebert (1968) model. However, intensity discrimination data are sparse for the cat, and recent measurements of frequency DLs in cats (Hienz, Sachs, and Aleszczyk 1993) give considerably higher values than those of Elliott, Stein, and Harrison (1960), so that estimates of this ratio are highly uncertain. Thus, both cat and human data are inconclusive. Developing a model for species such as the chinchilla in which more extensive data are available for both frequency and intensity discrimination might provide a better test of the rate-place model.

The Siebert (1968) model prediction of Weber's law for frequency discrimination is broadly consistent with psychophysical data at intermediate frequencies (0.5–2 kHz in humans, 1–4 kHz in cats), where performance is best (Elliott, Stein, and Harrison 1960; Moore 1973; Wier, Jesteadt, and Green 1977). Psychophysical performance degrades for both lower and higher frequencies, and this degradation is particularly steep in the human at high frequencies (Fig. 5.6). Javel and Mott (1988) examined the frequency DLs for a rate-place model similar to that of Siebert (1968), but which incorporated more realistic tuning curves, a cochlear frequency map, threshold distributions, and cochlear innervation densities. Results for the Javel and Mott model are compared with psychophysical data in Figure 5.6. For frequencies below 2 kHz, the frequency dependence of DLs predicted by the model closely matches psychophysical data for the cat (Elliott, Stein, and Harrison 1960). However, above 2 kHz the model performance is somewhat better than psychophysical performance (by a factor of less than 3). Although the overall fit to the cat data is quite good, the model fails to predict the steep rise in frequency DLs occurring at high frequencies in the human, even if it is modified to reflect the human cochlear map (dashed lines in Fig. 5.6).

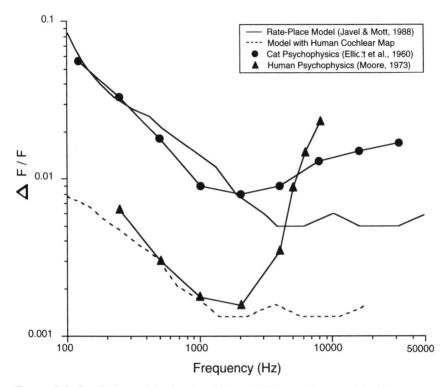

FIGURE 5.6. Predictions of the Javel and Mott (1988) rate-place model of frequency discrimination compared with psychophysical data for cats (*filled circles*) (Elliott, Stein, and Harrison 1960) and humans (*triangles*) (Moore 1973). The vertical ordinate represents the Weber fraction for pure tones. Model perfomance was adjusted so that it would coincide with cat psychophysical data at 2 kHz. The human data are for 50-msec tone bursts (Moore 1973). The *dashed line* shows model predictions with frequencies divided by a factor of 2.8 to reflect the differences in cochlear frequency maps between cat and human (Greenwood 1961, 1990). In addition, model performance was adjusted to match human psychophysical data at 1 kHz.

In summary, while rate-place models provide fairly good predictions of pure-tone frequency discrimination in the cat, they fail to predict the steep rise in frequency DLs at high frequencies for the human. Even in the frequency range where humans have good frequency discrimination, rate-place models may have difficulty predicting the relationship between frequency DLs and intensity DLs, although more detailed data for the cat and other species are needed on this point. These difficulties are not unique to physiological rate-place models, but also arise for other place models such as psychophysical excitation-pattern models (Moore and Glasberg 1986). The steep rise in human DLs at high frequencies is particularly puzzling because it hard to understand why subjects cannot make use of the available rate-place information, even though this information is used relatively more efficiently for other tasks such as intensity discrimination. These difficulties of rate-place models with frequency discrimination suggest an investigation of the obvious alternative, temporal models.

6.2 Temporal Models

The question of whether frequency discrimination is based primarily on a rate-place or on a temporal mechanism has long been debated. At least since the time of Wever (1949), most researchers agree that both mechanisms are involved: temporal mechanisms are thought to be dominant at low frequencies and rate-place mechanisms at high frequencies. However, there is a great deal of disagreement as to the frequency at which the transition between the two mechanisms takes place: estimates vary from 50 Hz to 5 kHz. Siebert (1970) was the first to examine this question quantitatively from the standpoint of the information available in the discharge patterns of ANFs. He incorporated into his 1968 model a statistical description of the phase-locking of ANF discharges to low-frequency tones. Discharges were modeled as a nonstationary Poisson process that provided a good fit to the phase-locking data of Kiang et al. (1965) and Rose et al. (1967). The performance of the optimum processor for frequency discrimination was the sum of two terms, one representing rate-place information and the other representing temporal information. For conditions typical of psychophysical experiments, the temporal term was by far the dominant one. In fact, the temporal discharge patterns of even a single fiber provided sufficient information to account for frequency discrimination in the human (Fig. 5.7). Moreover, the dependence of the temporal term on stimulus duration and level was clearly at variance with psychophysical data. Siebert (1970) concluded that "the brain does not make full, efficient use of the periodicity information coded in the auditory nerve firing patterns."

Goldstein and Srulovicz (1977) argued that the failure of the Siebert (1970) model to account for the dependence of frequency DLs on stimulus variables does not necessarily constitute evidence against processing of fine temporal information, but rather argues against the particular statistics used by the model. Under the plausible assumption that the phase of the tone stimulus is unknown to the observer, this statistic is a linear transformation (a weighted sum) of the autocorrelation function of the spike train, also known as the all-order interspike-interval distribution (Siebert 1970). Goldstein and Srulovicz (1977) proposed that frequency discrimination might instead be based on the *first-order* interval distribution, that is, that the central processor only makes use of intervals between *consecutive* spikes. Using essentially the same Poisson description of ANF discharge patterns as Siebert (1970), they showed that this hypo-

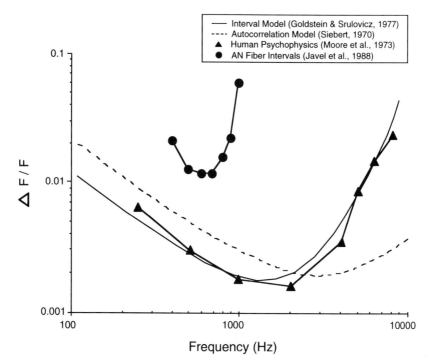

FIGURE 5.7. Predictions of two temporal models of frequency discrimination (Siebert 1970; Goldstein and Srulovicz 1977) compared with human psychophysical data (*triangles*) (Moore 1973) and with frequency DLs measured directly from the first-order interspike intervals of an ANF (*filled circles*) (Javel et al. 1988). Results for the Siebert (1970) autocorrelation model are for a single ANF; results for the Goldstein and Srulovicz (1977) first-order interval model are for nine statistically independent fibers. Model performance was computed using Eq. 3 (autocorrelation) and 6 (first-order intervals) in Goldstein and Srulovicz (1977).

thesis leads to excellent predictions of the frequency dependence of human frequency DLs (Fig. 5.7). The steep rise in DLs at high frequencies was particularly well modeled, and contrasted with the failure of rate-place models to predict this behavior (see Fig. 5.6). On the other hand, direct measurements of frequency DLs based on first-order interspike intervals from single ANFs (Javel and Mott 1988; Javel et al. 1988) rise much more steeply with increasing frequency than do the predictions of the Goldstein and Srulovicz (1977) model (Fig. 5.7). The origins of this discrepancy between physiological data and model predictions are unclear. In any case, this discrepancy suggests that model predictions should be viewed with caution for frequencies above 2 kHz.

Goldstein and Srulovicz (1977) further showed that their model can account for the dependence of the frequency DL on stimulus duration

(Moore 1973). However, this required fitting an additional parameter, the largest interval used by the optimum processor. Further studies of the Goldstein and Srulovicz (1977) model by Wakefield and Nelson (1985) showed that the model can also account for the level dependence of frequency DLs (Wier, Jesteadt, and Green 1977) by introducing yet another free parameter, the synchronization index at threshold. For frequencies below 2 kHz, reasonable fits to psychophysical data could be obtained using fixed values for both free parameters. Above 2 kHz, however, the best-fitting parameters were strongly dependent on frequency. Wakefield and Nelson noted that the model requires extrapolating physiological data beyond the upper frequency limit where phase-locking has been observed (Johnson 1980), and suggested that a rate-place scheme might operate at 4 kHz and above.

While the Goldstein and Srulovicz (1977) first-order interspike-interval model successfully predicts the dependence of human frequency DLs on the frequency, level, and duration of pure-tone stimuli (at least for frequencies below 2 kHz), absolute levels of performance for the model still greatly exceed psychophysical performance when the total number of available ANFs is taken into account. Srulovicz and Goldstein (1983) proposed a further reduction in information that brings performance of their model in better agreement with psychophysical data. They hypothesized that the first-order interspike-interval distribution for each ANF is passed through a central filter matched to the CF of that fiber. This central filter selects frequency components of the interval histogram that are near the CF and its harmonics. The time-average outputs of the central filters plotted against CF form a central spectrum on which basic auditory discriminations are assumed to be based. Because the central spectrum displays a response measure derived from fine time patterns of discharge as a function of CF, it is a *temporal-place* model of auditory processing[§] (Young and Sachs 1979). The Srulovicz and Goldstein (1983) central spectrum model predicts essentially the same dependence of frequency DLs on stimulus variables as the original Goldstein and Srulovicz (1977) model, with absolute levels of performance more in line with psychophysical data. Moreover, the central spectrum provides an appropriate input for the Goldstein (1973) optimum processor model for the pitch of complex tones. Taken together, these models account for an

---

[§] A related but distinct temporal-place model of auditory processing is the average localized synchronized rate (ALSR) proposed by Young and Sachs (1979), and studied by Delgutte (1984), Miller and Sachs (1984), and Miller, Barta, and Sachs (1987). Because this model has not been compared in detail with psychophysical data, it is beyond the scope of this review. Miller, Barta, and Sachs (1987) provide a general argument in favor of temporal-place schemes based on the robustness of cues for frequency discrimination in background noise. Delgutte (1984) pointed out some difficulties of temporal-place models in accounting for psychophysical frequency selectivity.

impressive set of psychophysical data on pitch perception and frequency discrimination. However, physiological evidence for the existence of central (i.e., neural) filters matched to the CF of ANFs is altogether lacking.

In summary, neither rate-place nor temporal (interspike-interval) models can account for all the psychophysical data on pure-tone frequency discrimination. Rate-place models have difficulties predicting the frequency dependence of the DLs at high frequencies, particularly in the human. Interspike-interval models also have difficulty predicting human frequency discrimination at high frequencies (Wakefield and Nelson 1985), and do a poor job for the cat, even though these models are based on physiological data from the cat. A further challenge for interspike-interval models is the inability of cochlear-implant patients to discriminate the frequency of sinusoidal or pulse-train stimuli applied through a single electrode for frequencies above a few hundred Hertz (e.g., Shannon 1983; Townsend et al. 1987). Because such stimuli produce very regular patterns of interspike intervals (Javel et al. 1987; Dynes and Delgutte 1992), they should be easily discriminable for models such as that of Goldstein and Srulovicz (1977). Overall, it seems that *both* rate-place and temporal models better predict psychophysical performance for low frequencies ($<2\,$kHz) than for high frequencies, a result that is at variance with the popular conception that temporal mechanisms dominate at low frequencies and rate-place mechanisms at high frequencies (e.g., Wever 1949). It may be that frequency discrimination at high frequencies is limited by central auditory mechanisms rather than by the information available in the discharge patterns of ANFs.

7. Masking

Most conversations take place in the presence of background noise such as that caused by machinery or competing conversations. While normal-hearing listeners show a remarkable ability to understand speech in noise, hearing-impaired listeners often complain of having difficulties understanding speech in noise even if they do well in quiet. Automatic speech recognition systems also have trouble processing noisy speech. A better understanding of the physiological mechanisms underlying masking may help develop hearing aids and speech processing systems that perform better in noise.

Masking also plays an important role in theories of hearing. Thus far, we have only considered very simple stimuli such as pure tones. Most acoustic stimuli such as speech, music, and environmental sounds have complex spectra whose components interact and potentially mask each other. Thus, a general theory of masking is required for understanding the auditory processing of complex stimuli. Such a theory was first developed by Fletcher and his collaborators (Wegel and Lane 1924; Fletcher and Munson 1937), who hypothesized that masking patterns for tonal signals reflect the spatial pattern of excitation produced by the masker along the cochlea. This hypothesis is based on the assumption that masking is caused by an excitatory mechanism. In this view, masking occurs because the neurons tuned to the signal frequency are excited by the masker, so that the signal level needs to be increased to cause a detectable increment in excitation. Under certain conditions, masking could also be caused by a suppressive mechanism whereby the masker suppresses the response to the signal without exciting neurons tuned to the signal frequency (Sachs and Kiang 1968). Techniques for deriving auditory filters and spatial patterns of excitation from masking data have been considerably refined since the time of Fletcher (Houtgast 1974; Zwicker 1974; Patterson 1976; Glasberg and Moore 1990), but they still rely on the assumption that masking results from an excitatory mechanism. Filters derived by these techniques are important components of auditory models for the processing of speech and music (Moore and Glasberg 1983, 1986; Meddis and Hewitt 1991; Patterson and Holdsworth in press).

In order to test the hypothesis that masking reveals the pattern of excitation produced by the masker, I first analyze mechanisms of masking for single ANFs, then describe models that combine information for signal detection from the ensemble of ANFs.

7.1 Mechanisms of Masking in Single Auditory Nerve Fibers

The response of an auditory neuron to a signal is said to be masked when adding the signal to a masker causes no detectable change in the discharge pattern. If we consider only rate-place information, the masked threshold is the signal level that produces a just detectable increment (or decrement) in average discharge rate over the rate for the masker. There are three physiological mechanisms by which the response of an ANF can be masked: line-busy masking, adaptive masking, and suppressive masking. These three forms of masking are illustrated in Figure 5.8, which shows rate-level functions for a tone signal in the presence of various maskers. Figure 5.8 is based on idealizations of physiological data available in the literature (Javel, Geisler, and Ravindran 1978; Smith 1979; Costalupes, Young, and Gibson 1984). Which form of masking dominates depends on the spectral and temporal relationships between the signal and the masker.

Line-busy masking (see Fig. 5.8A) occurs when the masker is excitatory, that is, when by itself it produces an increment in discharge rate over spontaneous activity. It is most prominent when the signal and the masker occupy the same frequency region (Pickles 1984; Delgutte 1990b), in particular for intensity discrimination, in which case the signal and the

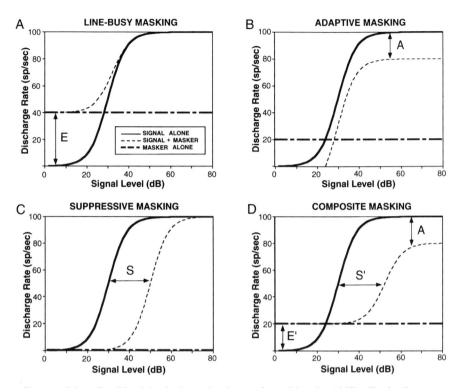

FIGURE 5.8A–D. Physiological mechanisms of masking in ANFs. Each diagram represents idealized rate-level functions of an ANF for a pure-tone stimulus in quiet (*solid line*) and in the presence of a masking stimulus (*dashed lines*). The response to the masker alone is shown by the horizontal *dot-dashed* line. (A) Pure line-busy masking occurring for simultaneously gated signal and masker occupying the same frequency region. *E*, excitation produced by the masker. (B) Pure adaptive masking occurring for a brief signal presented after a long-duration masker. *A*, decrement in discharge rate resulting from adaptation by the masker. (C) Pure suppressive masking occurring for a masker much lower in frequency than the signal. *S*, horizontal shift of the rate-level function resulting from suppression by the masker. (D) Composite masking occurring for a transient signal in a continuous masker occupying a different frequency region from the signal. *E′*, excitation produced by the adapted masker; *S′*, shift in the rate-level function resulting from combined suppression and adaptation.

masker have the same waveform. Line-busy masking results from two factors. The first factor is the compressive nature of rate-level functions (Kiang et al. 1965; Sachs and Abbas 1974). Compression implies that adding a signal of a given intensity results in a smaller increment in discharge rate when the signal is presented over an excitatory masker than when presented alone. The second factor is the increase in variance of the discharge rate with increasing mean rate (Teich and Khanna 1985;

Young and Barta 1986; Delgutte 1987). This increase in variance implies that the same increment in discharge rate will be less detectable when occurring over a high background rate than over a low rate. Taken together, these factors result in that a signal which is detectable in quiet may no longer be so in the presence of an excitatory masker.

Adaptive masking (see Fig. 5.8B) occurs when an excitatory masker has a sufficiently long duration to adapt an ANF, so that the rate in response to subsequent stimuli is decreased (Smith 1977, 1979; Harris and Dallos 1979). Although adaptive masking is an excitatory mechanism, it differs from line-busy masking in that it does not require the signal and the masker to be simultaneous. In fact, adaptation is the basis for forward masking in ANFs, in which case the signal occurs entirely after the masker (Relkin and Turner 1988). Adaptive masking also occurs for transient signals in continuous maskers (Smith and Zwislocki 1975; Costalupes, Young, and Gibson 1984). Measurements of masked thresholds using statistical criteria show that forward masking in ANFs is much smaller than psychophysical forward masking for the same stimuli (Relkin and Pelli 1987; Relkin and Turner 1988). This result suggests that psychophysical forward masking may be primarily caused by central factors rather than adaptive masking in the auditory nerve. This conclusion is consistent with psychophysical data showing forward masking with temporal properties very similar to those for normal subjects in patients with brainstem auditory implants (in which the cochlea and auditory nerve are bypassed) (Shannon and Otto 1990). However, adaptive masking in the auditory nerve might still be important for suprathreshold phenomena (Relkin and Doucet 1991).

Suppressive masking (see Fig. 5.8C) occurs when a masker suppresses the response to the signal through nonlinearities in cochlear mechanics. Unlike line-busy and adaptive masking, it is not caused by an excitatory mechanism because it can occur even if an ANF does not respond to the masker presented along (Sachs and Kiang 1968). As shown in Figure 5.8C, the main effect of suppressive masking is to horizontally shift the rate-level function for the signal toward higher intensities (Javel, Geisler, and Ravindran 1978). Suppressive masking is most prominent when the signal and the masker occupy different frequency regions, particularly when the signal frequency is well above the masker (Sachs and Kiang 1968; Pickles 1984; Delgutte 1990b). Like line-busy masking, but unlike adaptive masking, suppressive masking requires that the signal and the masker be simultaneous.

The three forms of masking are not mutually exclusive, and all three mechanisms may contribute to overall masking in any given situation (see Fig. 5.8D). In an effort to assess the relative contributions of excitatory and suppressive mechanisms to tone-on-tone masking, Delgutte (1990b; see also Pickles 1984; Costalupes, Young, and Gibson 1984) compared masked thresholds of ANFs in two conditions: simultaneous masking and a "nonsimultaneous" condition analogous to the pulsation threshold in psychophysics (Houtgast 1974). This nonsimultaneous threshold, which is not strictly speaking a masked threshold, expresses the excitation produced by the masker in units of the signal level that produces the same excitation. Specifically, the nonsimultaneous threshold is defined as the signal level for which the response to the signal alone exceeds the response to the masker alone by a just detectable increment. The key assumption is that the difference between simultaneous and nonsimultaneous thresholds gives the contribution of suppression to masking. This assumption is justified because (1) excitatory masking is the same in the simultaneous and nonsimultaneous conditions; (2) suppression can only occur when the signal and the masker are simultaneous; and (3) the effect of suppression is to shift the rate-level function for the signal horizontally. Results show that line-busy masking dominates when the signal and the masker are close in frequency, while suppressive masking dominates when the signal frequency is well above the masker frequency, particularly for high masker levels. For example, with a 1-kHz masker at 80 dB SPL, ANFs with CFs in the 5- to 10-kHz region typically showed 40–50 dB of masking that was entirely caused by suppression because these fibers did not respond to the masker alone.

Taken together, physiological studies (Costalupes, Young, and Gibson 1984, Pickles 1984; Sinex and Havey 1986; Delgutte 1990b) show that masking in ANFs is a complex phenomenon involving an interaction of many stimulus-dependent factors. In order to understand how these factors contribute to psychophysical masking in any given situation, it is necessary to use a model that combines information for signal detection from the ensemble of ANFs.

7.2 Rate-Place Model of Masking

In order to assess the role of suppression in masking, Delgutte (1989) incorporated a functional simulation of suppression in his 1987 rate-place model. The model qualitatively simulated most rate suppression phenomena observed in ANFs for tone and noise stimuli, although it somewhat underestimated suppression for suppressor frequencies below the CF. Model masking patterns for a 1-kHz tone are compared in Figure 5.9 with both human psychophysical data (Maiwald 1967) and physiological data from cat ANFs (Delgutte 1990b). Psychophysical data are for a narrowband noise masker to avoid the confounding effects of beats for tonal maskers (Egan and Hake 1950). For both the model and the physiological data, masking patterns were obtained by picking the single fiber with the

---

‖Adaptive masking was minimal in this experiment because both the signal and the masker were brief tone bursts.

1-kHz MASKER

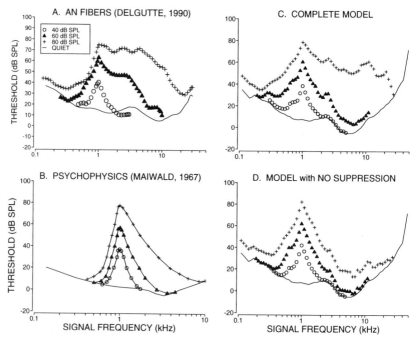

FIGURE 5.9A–D. (A) Physiological masking patterns of ANFs for a 1-kHz tone masker presented at three different stimulus levels (Delgutte 1990b). Each data point represents the minimum masked threshold for the signal frequency among a large sample of ANFs with different CFs and SRs; the *solid line* represents the minimum threshold in quiet. (B) Psychophysical masking patterns for a 1-kHz narrow band of noise (Maiwald 1967). (C) Masking patterns for the Delgutte (1989) rate-place model for the same signals and maskers as in A. As in the physiological data, masking patterns were obtained by selecting the model fiber with the lowest masked threshold for each signal frequency. (D) Masking patterns for a version of the Delgutte (1989) model in which the signal processing elements responsible for suppression were disconnected.

lowest masked threshold for each signal frequency, regardless of CF and SR (Sinex and Havey 1986). This procedure takes into account the possibility that the signal may be detected through fibers that are not tuned to the signal frequency, a phenomenon called "off-frequency listening." The model reproduces three main trends in the physiological and psychophysical data: (1) masking is maximum when the signal and the masker occupy the same frequency region; (2) masking patterns are asymmetrical, in that they extend farther toward high signal frequencies than toward low frequencies; and (3) for signal frequencies above the masker, masking grows faster than linearly with masker level: a 1-dB increase in masker level causes an increase in masked threshold greater than 1 dB. The model and physiological masking patterns, which are based on data for the cat, span a wider frequency range than the human psychophysical masking patterns, consistent with psychophysical evidence that frequency resolution is not as sharp in cats as in humans (Pickles 1979, 1980).

Figure 5.9D shows masking patterns for a modified version of the model in which the elements responsible for suppression were disconnected. Masking patterns without suppression are less asymmetric than for the full model. Furthermore, masking grows more linearly with masker level for signal frequencies well above the masker. This result suggests that the nonlinear growth of masking, which is often called upward spread of masking, is primarily caused by the nonlinear increase in suppression rather than by a nonlinear spread of excitation as assumed by most psychophysicists since Wegel and Lane (1924). This conclusion is supported by the physiological results of Delgutte (1990b), which showed that masking in ANFs grows much more linearly for the nonsimultaneous condition (in which there is no suppressive masking) than for the simultaneous condition.

In addition to tone-on-tone masking patterns (Fig. 5.9), the Delgutte (1989) model qualitatively predicts *unmasking* phenomena whereby a two-tone stimulus can be a less effective masker than one of its components (Houtgast 1974; Shannon 1976). As in psychophysical data, unmasking for the model only occurs for the nonsimultaneous condition. Unmasking is caused mutual suppression between the frequency components of the two-tone masker, and is not seen in simultaneous masking because the signal is also suppressed by the masker in that condition (Houtgast 1974; Delgutte 1989, 1990b). Thus, suppression can result in either masking or unmasking depending on the spectrotemporal configuration of the signal and the masker.

In summary, a rate-place model incorporating suppression can qualitatively account for a wide range of masking phenomena. To some extent, these results are consistent with the assumption of Wegel and Lane (1924) and Fletcher and Munson (1937) that masking is an excitatory phenomenon: excitatory (line-busy) masking is indeed the most important mechanism, particularly when the signal and the masker occupy the same frequency region. However, suppressive masking is also important in certain conditions, and actually dominates for signal frequencies well above the masker. Suppressive masking can be quite large: in some cases, 40–50 dB of masking are obtained even though the masker does not excite the ANFs used for detecting the signal. The large contribution of suppression to masking is at variance with theories that assume a direct correspondence between masking, excitation patterns and loudness (Fletcher and Munson 1937), and challenges the validity of procedures that derive auditory filters and excitation patterns from simultaneous masking data (Zwicker and Feldtkeller 1967; Zwicker 1970; Patterson 1976; Glasberg and Moore 1990). In principle, a better correspondence between masking patterns and excitation patterns could be obtained using nonsimultaneous masking techniques, as suggested by Houtgast (1974), because suppressive masking requires simultaneous presentation of the signal and the masker. In practice, none of the existing nonsimultaneous psychophysical techniques (forward masking and pulsation thresholds) gives as reliable thresholds as simultaneous masking, so that these techniques have essentially been abandoned as procedures for deriving auditory filters and excitation patterns. Currently, the most popular psychophysical procedure for deriving auditory filters is the notched-noise method (Patterson 1976; Moore and Glasberg 1983; Glasberg and Moore 1990). Although physiological data for notchednoise maskers are sparse (Palmer and Evans 1982), suppressive masking is likely to be an important factor in the notched-noise method because this procedure introduces intense noise in suppression areas flanking the signal frequency. A more systematic physiological and modeling investigation of this method using the same stimuli and detection criteria as in psychophysics is needed.

In conclusion, psychophysical procedures that fail to take into account the contribution of suppression to simultaneous masking may only provide a crude approximation to actual cochlear filters and patterns of neural excitation. Models of auditory processing based on such procedures (Zwicker and Feldtkeller 1967; Moore and Glasberg 1983, 1986; Meddis and Hewitt 1991; Patterson and Holdsworth in press) cannot be considered to provide an accurate representation of the response properties of ANFs. Applications of these models to the investigation of neural mechanisms underlying psychophysical phenomena should be viewed with caution. Of course, these models might still be useful for applications such as automatic speech recognition in which detailed simulation of physiological processing may not be essential.

8. Speech Discrimination

There is a large step between the simple psychophysical tasks that I have discussed so far and the much more complex task of speech discrimination. Speech perception depends on temporal properties of the signal (on time scales from a few milliseconds to several seconds) in addition to the spectral properties that are the primary topic of this chapter. The goal of this section is to bridge the gap between the psychophysiological models discussed in this chapter and the computational models reviewed in Chapter 6 (Lyon and Shamma, this volume). More general reviews of the neural encoding of speech are available elsewhere (Sachs 1984; Delgutte in press).

One speech-related task that is a direct extension of the psychophysical phenomena discussed in this chapter is the discrimination between steady-state vowels (e.g., Flanagan 1955; Carlson, Granström, and Black 1979; Kewley-Port and Watson 1994). This task is a special instance of the general problem of spectral discrimination, so that place models can make predictions about performance in vowel discrimination. While many parameters of vowel spectra can be manipulated, there is considerable psychophysical evidence that the frequencies of the first two or three formants play in important role in vowel identification (Carlson, Granström, and Klatt 1979), so that a minimum requirement for models of vowel discrimination is to provide a robust representation of the formant frequencies.

The ability of rate-place models to represent formant frequencies at high sound levels has been questioned because of the saturation of the majority of ANFs (Sachs and Young 1979). On the other hand, modeling results for intensity discrimination suggest that this limitation may not be severe if high-threshold ANFs are taken into account (Delgutte 1987). Indeed, Conley and Keilson (1994) have shown that 5-Hz changes in the second formant frequency (F2) of an [ɛ] vowel would be easily discriminable based on optimal processing of rate-place information (including both low- and high-threshold fibers). This value is considerably smaller than psychophysical F2 DLs for human listeners (Flanagan 1955; Kewley-Port and Watson 1994), suggesting that there is sufficient rate-place information to account for F2 discrimination. However, we have seen that optimum processor models of the type used by Conley and Keilson (1994) fail to predict Weber's law for broadband stimuli (see Fig. 5.5A), so that a scheme more consistent with intensity discrimination is needed.

Figure 5.10B shows the response of the Delgutte (1987,1989) rate-place model to a steady state [ɛ] vowel whose spectrum is shown in Figure 5.10A. These results are based on the version of the model in which the distribution of fiber thresholds was adjusted to approximate Weber's law over a wide dynamic range for individual frequency channels. Following Fechner's (1860) method for deriving sensory scales, the ordinate, labeled *cumulative d'* represents the total number of intensity DLs between absolute threshold and the effective stimulus level for each frequency channel. Because each channel approximately obeys Weber's law, cumulative d' grows nearly linearly with stimulus level in decibels over a wide dynamic range. Cumulative d' can be expressed as a function of the discharge rates of ANFs having the same CF, but differing in thresholds. Figure 5.10B shows that these transformed rate-place patterns provide a good representation of the frequencies of the first two or three formants over a 60-dB range of stimulus levels. Thus, the same rate-place model that successfully predicts many psychophysical phenomena on masking and intensity discrimination also provides a stable representation of speech formants over a wide dynamic range.

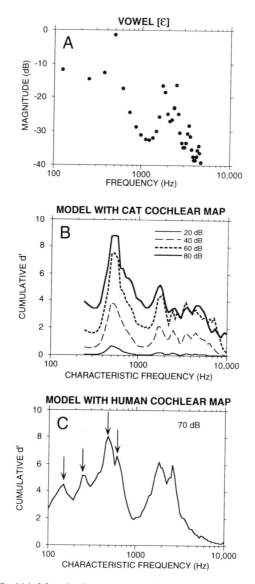

FIGURE 5.10A–C. (A) Magnitude spectrum of a synthetic [ɛ] vowel with a fundamental frequency of 125 Hz. (B) Response of the Delgutte (1989) rate-place model as a function of CF for the vowel at four different stimulus levels. The distribution of ANF thresholds in the model was adjusted to best approximate Weber's law for intensity discrimination within each frequency channel (Delgutte 1987). The ordinate (cumulative d') represents the total number of intensity DLs between absolute threshold and the effective stimulus level in each frequency channel. (C) Response to the 70-dB vowel for a version of the Delgutte (1989) model in which the cat cochlear frequency map was replaced by a human cochlear map using Greenwood's (1961, 1990) formula. *Arrows* indicate peaks at harmonics 1, 2, 4, and 5 of the 125-Hz fundamental.

While maxima in spectral envelope associated with formants are important for speech intelligibility, the fine spectral structure associated with voicing also plays in important role in communication. Voiced sounds such as vowels show a periodicity at the frequency of vibration of the vocal folds. This periodicity, which listeners hear as voice pitch, results in spectral lines at harmonics of the fundamental frequency (Fig. 5.10A). No peaks at harmonics of the 125-Hz fundamental are apparent in the model rate-place profiles of Figure 5.10B. Such harmonic patterns are also lacking in the ANF data of Sachs and Young (1979) and Conley and Keilson (1994), even at low stimulus levels where saturation is not an issue. Both the model and the physiological data are based on the cat, which hears up to much higher frequencies than the human. Figure 5.10C shows results for a version of the Delgutte (1987, 1989) model in which the cat cochlear map was replaced by a human cochlear map using the method of Greenwood (1961, 1990). The change in cochlear map also has the effect of sharpening the tuning of the model cochlear filters, consistent with psychophysical data showing better frequency resolution in humans than in cats (Pickles 1979, 1980). Because the resolution is increased, the rate-place model based on the human cochlear map does show peaks at low-frequency harmonics of the fundamental frequency (arrows in Fig. 5.10C). The number of resolved harmonics in this representation appears to be about five, consistent with psychophysical estimates of frequency resolution in the human (reviewed by Patterson and Moore 1986). Figure 5.10C also shows an enhanced representation of the third formant compared to Figure 5.10B because the relative sensitivity of the human ear in the third formant region is better than that of the cat ear.

A representation with resolved low-frequency harmonics as in Figure 5.10C provides an appropriate input for pattern-recognition, place models of pitch perception (Goldstein 1973; Terhardt 1974), which predict a wide range of pitch phenomena. Thus, a rate-place model based on the human cochlear frequency map may provide sufficient information for the perception of pitch as well as timbre (spectral envelope). Chapter 6 of this volume (Lyon and Shamma) provides more detailed discussions of rate-place and temporal models for pitch and timbre.

9. Summary

9.1 Conclusion

In this chapter, I have systematically reviewed models that predict psychophysical performance in detection and discrimination tasks based on statistical descriptions of the discharge patterns of auditory-nerve fibers and ideal observer models of the central processor. The value of these models is demonstrated by following examples in which tentative conclusions derived from the models run counter to popular ideas about auditory processing.

Pure-tone detection in quiet is likely to be based on the average discharge rates of ANFs, even though synchronization (phase-locking) thresholds are apparently lower than rate thresholds for low frequencies.

The saturation of the rate-level functions of ANFs may not limit intensity-discrimination performance at high intensities. In fact, the central processor is quite inefficient in utilizing average-rate information throughout the intensity range (but particularly at moderate intensities).

Loudness is probably not based on the total discharge rate of the ensemble of ANFs.

Both rate-place and interspike-interval models do a better job of predicting frequency discrimination for low frequencies than for high frequencies.

Masking does not always result from the excitation produced by the masker. Suppressive masking can be quite large, particularly for a signal well above the frequency of an intense masker.

Rate-place models based on the human cochlear frequency map may provide sufficient information for the perception of pitch as well as timbre.

A general result of this review is that rate-place models incorporating a realistic cochlear frequency map, cochlear filters, suppression, rate-level functions, and threshold distribution do a fairly good job of accounting for performance in a wide range of psychophysical tasks. These models predict the main trends in the frequency dependence of thresholds in quiet, masking patterns, and, for the cat, frequency discrimination. The major shortcoming of rate-place models is their failure to predict Weber's law for intensity discrimination when spread of excitation is minimal, but this difficulty can be largely overcome by assuming that information from high-threshold fibers is processed more efficiently than that from low-threshold fibers. This idea is consistent with evidence that certain cell types in the cochlear nucleus receive predominant inputs from high-threshold ANFs (Liberman 1991). Ultimately, information from low- and high-threshold fibers has to be integrated to encode intensity over a wide dynamic range. Winslow, Barta, and Sachs (1987) have proposed a physiologically plausible scheme for how cochlear nucleus cells might perform such integration. However, integration does not have to occur in the cochlear nucleus, so long as the two kinds of inputs remain segregated up to the stage(s) of processing at which integration takes place. The latter view is consistent with the observation that thresholds of central auditory neurons vary over a wide range of levels, although low-threshold neurons seem to remain preponderant (reviewed by Irvine 1992). Some evidence for a topographic representation of intensity in the cat auditory cortex has been presented (Heil, Rajan, and Irvine 1994).

Temporal models can theoretically provide vastly better performance than rate-place models for particular tasks such as frequency discrimination for low-frequency tones or representation of vowel spectra at high sound levels. Although temporal models have not been examined systematically for as many psychophysical phenomena as rate-place models, we have pointed out some general difficulties with this class of models. Specifically, models based on detection of phase-locking are likely to poorly predict the shapes of pure-tone audiograms, and models that make use of temporal patterns *within frequency channels* are likely to have the same difficulties as rate-place models with Weber's law for intensity discrimination because temporal patterns do not change much with intensity. Even in the case of pure-tone frequency discrimination in the human, interspike-interval models seem to have problems for frequencies above 2 kHz.

A fundamental constraint for temporal models is the gradual decrease in the upper frequency limit of phase-locking as one ascends the auditory pathway (reviewed by Eggermont 1993). This degradation in phase-locking suggests that fine-time patterns of discharge would have to be transformed into another type of code at an early stage of central processing, at least for frequencies above 1 kHz. While there is considerable evidence for temporal processing in the cochlear nucleus (Carney 1990; Yin and Chan 1990; Palmer et al. 1995) and superior olivary complex (Goldberg and Brown 1969; Yin and Chan 1990) in that certain cells act as coincidence detectors, such coincidence mechanisms are a far cry from the kind of processing required by interspike-interval models (Licklider 1951; Siebert 1970; Goldstein and Srulovicz 1977). Cross-spatial coincidence models (e.g., Shamma 1985; Deng, Geisler, and Greenberg 1988; Carney 1994) make fewer demands on the central processor than interspike-interval models, but have not been as thoroughly tested for their ability to predict psychophysical data. These models seem to constitute a promising avenue for research into neural mechanisms underlying auditory perception.

Overall, these arguments suggest that rate-place processing may be the basic mode of operation of the monaural auditory system. Temporal cues might still used for special tasks (such as discrimination of low frequencies in the human) that require particularly good performance, or when rate-place cues are degraded by disease or severe listening conditions. At the very least, Siebert's (1970) conclusion that the central processor does not make full, efficient use of the temporal information encoded in the auditory nerve remains very much true. There may be strong physiological constraints on the type of temporal information that can be used efficiently by the central nervous system (e.g., interspike intervals versus cross-neuron coincidence). Unraveling the nature of these constraints and the underlying neural mechanisms is a major challenge for auditory neurophysiologists.

An interesting question is whether performance in any particular task is primarily limited by peripheral or central factors. If we consider *absolute* levels of performance, the limitations must be of central origin because optimal processing of the information available in the auditory nerve generally predicts much better performance than psychophysical observations, even if we consider only rate-place information. On the other hand, the *frequency dependence* of predictions of optimum processor, rate-place models parallels the main trends in psychophysical data for thresholds in quiet, masking patterns, and, for the cat, frequency discrimination. This suggests that the central processor may be uniformly inefficient in the sense that the relative loss in performance resulting from central factors would be the same for every frequency channel. This uniform inefficiency property does not seem to hold for sensitivity because the hypothesis that information from high-threshold fibers is processed more efficiently leads to better predictions for intensity discrimination. Thus, the central nervous system appears to treat differently the two fundamental dimensions (CF and sensitivity) along which ANFs are organized.

9.2 Future Directions

The modeling approach reviewed in this chapter has been applied mostly to very simple stimuli and psychophysical tasks. In principle, the same approach could be used for more complex stimuli and tasks, although the increased importance of central factors might limit the value of models based on auditory nerve activity. For example, temporal processing has been relatively neglected since the pioneering studies of Zwislocki (1960, 1969) on temporal summation of loudness. Physiologically motivated models of temporal integration (Buus and Florentine 1992; Viemeister, Shivapuja, and Recio 1992) and temporal resolution (Zhang, Salvi, and Saunders 1990) are just beginning to be developed. Essentially all models reviewed in this chapter are based on physiological data for the cat, which are sometimes compared with human psychophysical data because appropriate data for the cat are lacking. Developing models for species in which both psychophysical and physiological data are extensive may help explain how vertebrate species with very different peripheral auditory organs can show great similarities in their performance for basic psychophysical tasks (Fay 1992). Yet another possible extension of the present approach would be to develop models for pathological ears. Such models might help understand the behavioral consequences of different types of pathologies and design better hearing aids.

Perhaps the greatest challenge is to develop psychophysiological models for the central nervous system (Colburn 1973; Young 1989; Shofner and Dye 1989). Unlike the auditory nerve in which type I fibers are very homogeneous, central auditory nuclei contain distinct types of cells. Identifying such cell types using morphological, physiological, and cytochemical criteria is a prerequisite for modeling the function of a nucleus. At present, this classification has been most convincingly accomplished for the cochlear nucleus (reviewed by Rhode and Greenberg 1992). Even when relatively homogeneous cell types can be reliably identified, modeling the activity of central auditory neurons raises additional difficulties. All models of auditory nerve activity have assumed that discharge patterns of ANFs are statistically independent for deterministic stimuli, consistent with the data of Johnson and Kiang (1976). This assumption is not tenable for central auditory neurons because their shared inputs and complex interconnections imply that the discharge patterns of these neurons will, in general, be correlated (Eggermont 1993). Indeed, dorsal cochlear nucleus neurons commonly show correlated discharge patterns (Voigt and Young 1980). Such cross-neuron correlation can greatly affect the performance of detection-theoretic models and certainly make modeling more difficult (Winslow 1985). Another difficulty in observing and modeling the behavior of central auditory neurons is that anesthesia and descending (efferent) influences are likely to be increasingly important factors as one ascends the auditory pathway.

Although the challenges are great, so are the potential benefits. A major issue in auditory neuroscience is to identify the functions of the multiple, tonotopically organized cell populations that are found at every stage in the auditory pathway beginning with the cochlear nucleus. By making quantitative predictions about the performance achievable in specific psychophysical tasks given the pattern of activity in a given cell population, detection-theoretic models may help assess the roles played by these populations in different auditory functions. Further benefits may come from incorporating detailed knowledge of the processing performed by central auditory neurons into psychophysiological models. For example, the Carney (1994) coincidence model for intensity coding makes use of time patterns of discharge, but in a way completely different from that assumed by interspike-interval models. Further knowledge of central auditory processing may well render obsolete the traditional distinction between rate-place and temporal models.

Acknowledgments. I express my appreciation to H. Steven Colburn, Owen E. Kelly, Sridhar Kalluri, Nelson Y.S. Kiang, and Mark J. Tramo for their invaluable comments on the manuscript. Peter A. Cariani deserves special mention for reading more drafts than I can count and greatly improving the paper at every level. Deborah T. Flandermeyer and Barbara E. Norris provided expert assistance with the figures. Preparation of this manuscript was supported by NIH grant DC00119.

Abbreviations

ANF   auditory nerve fiber
CF   characteristic frequency
DL   difference limen
d' Psychophysical or physiological sensitivity index
ISIH   Interspike-interval histogram
MOC   medial Olivocochlear system
PSTH   Perior post-stimulus time histogram
SPL   sound pressure level
SR   spontaneous rate of discharge

References

Brown MC (1994) Antidromic responses of single-units from the spiral ganglion. J Neurophysiol 71:1835–1847.

Brown MC, Smith DI, Nuttall AL (1981) The temperature dependency of neural and hair-cell responses evoked by high-frequencies. J Acoust Soc Am 73:1662–1670.

Buus S (1990) Level discrimination of frozen and random noise. J Acoust Soc Am 87:2643–2654.

Buus S, Florentine M (1992) Possible relation of auditory-nerve adaptation to slow improvement in level discrimination with increasing duration. In: Cazals Y, Horner K, Demany L (eds) Auditory Physiology and Perception. Oxford: Pergamon Press, pp. 279–288.

Carlson R, Granström B, Klatt DH (1979) Vowel perception: the relative perceptual salience of selected spectral and waveform manipulations. R Inst Technol Stockholm STL-QPSR3-4:84–104.

Carney LH (1990) Sensitivities of cells in the anteroventral cochlear nucleus of cat to spatiotemporal discharge patterns across primary afferents. J Neurophysiol (Bethesda) 64:437–456.

Carney LH (1993) A model for the responses of low-frequency auditory-nerve fibers in cat. J Acoust Soc Am 93:401–417.

Carney LH (1994) Spatiotemporal encoding of sound level: models for normal encoding and recruitment of loudness. Hear Res 76:31–44.

Carney LH, Geisler CD (1986) A temporal analysis of auditory-nerve fiber responses to spoken stop consonant-vowel syllables. J Acoust Soc Am 79:1896–1914.

Cohen MR, Drabken IE (1948) A Source Book in Greek Science. New York: McGraw-Hill.

Colburn HS (1973) Theory of binaural interaction based on auditory-nerve data. I. General strategy and preliminary results on interaural discrimination. J Acoust Soc Am 54:1458–1470.

Colburn HS (1981) Intensity perception: relation of intensity discrimination to auditory-nerve firing patterns. Internal Memorandum, Research Laboratory of Electronics Massachusetts Institute of Technology, Cambridge, MA.

Conley RA, Keilson SE (1994) Rate representation and discriminability of second formant frequencies of /ε/-like steady-state vowels in cat auditory nerve. Assoc Res Orolaryugol Abstr 17:100.

Costalupes JA (1983) Temporal integration of pure tones in the cat. Hear Res 9:43–54.

Costalupes JA, Young ED, Gibson DJ (1984) Effect of continuous noise backgrounds on rate response of auditory-nerve fibers in cat. J Neurophysiol (Bethesda) 51:1326–1344.

Dallos P, Harris D, Özdamer Ö, Ryan A (1978) Behavioral, compound action potential, and single-unit thresholds: relationships in normal and abnormal ears. J Acoust Soc Am 64:151–157.

De Boer E (1967) Correlation studies applied to the frequency resolution of the cochlea. J Aud Res 7:209–217.

De Boer E, de Jongh HR (1978) On cochlear encoding: potentialities and limitations of the reverse correlation technique. J Acoust Soc Am 63:115–135.

Delgutte B (1984) Speech coding in the auditory nerve. II. Processing schemes for vowel-like sounds. J Acoust Soc Am 75:879–886.

Delgutte B (1986) Analysis of French stop consonants using a model of the peripheral auditory system. In: Perkell JS, Klatt DH (eds) Invariance and Variability in Speech Processes. Hillsdale: Erlbaum, pp. 163–177.

Delgutte B (1987) Peripheral auditory processing of speech information: implications from a physiological study of intensity discrimination. In: Schouten MEH (ed) The Psychophysics of Speech Perception. Dordrecht: Nijhoff, pp. 333–353.

Delgutte B (1989) Physiological mechanisms of masking and intensity discrimination. In: Turner CW (ed) Interactions Between Neurophysiology and Psychoacoustics. New York: Acoustical Society of America, pp. 81–101.

Delgutte B (1990a) Two-tone rate suppression in auditory-nerve fibers: dependence on suppressor frequency and level. Hear Res 49:225–246.

Delgutte B (1990b) Physiological mechanisms of psychophysical masking: observations from auditory-nerve fibers. J Acoust Soc Am 87:791–809.

Delgutte B (1991) Power-law behavior of the discharge rates of auditory-nerve fibers at low sound levels. Assoc Res Otolaryngol Abstr 14:77.

Delgutte B   Neural encoding of speech.   In: Hardcastle W, Laver J (eds) The Handbook of Phonetic Sciences. Oxford: Blackwell (in press).

Delgutte B, Cariani PA (1992) Coding of the pitch of harmonic and inharmonic complex tones in the interspike intervals of auditory-nerve fibers. In: Schouten MEH (ed) The Processing of Speech. Berlin: Mouton-de Gruyter, pp. 37–45.

Deng L, Geisler CD, Greenberg S (1988) A composite model of the auditory periphery for the processing of speech. J Phonet 16:109–123.

Dynes SBC, Delgutte B (1992) Phase-locking of auditory-nerve discharges to sinusoidal electric stimulation of the cochlea. Hear Res 58:79–90.

Egan JP, Hake HW (1950) On the masking pattern of a simple auditory stimulus. J Acoust Soc Am 22:622–630.

Eggermont JJ (1993) Functional aspects of synchrony and correlation in the auditory nervous system. Concepts Neurosci 4:105–129.

Elliott D, McGee TM (1965) Effects of cochlear lesions upon audiograms and intensity discrimination in cats. Ann Otol Rhinol Laryngol 74:386–408.

Elliott D, Stein L, Harrison M (1960) Determination of absolute intensity thresholds and frequency difference thresholds in cats. J Acoust Soc Am 32:380–384.

Erell A (1988) Rate coding model for discrimination of simple tones in the presence of noise. J Acoust Soc Am 84:204–214.

Evans EF (1975) The cochlear nerve and cochlear nucleus. In: Keidel WD, Neff D (eds) Handbook of Sensory Physiology, Vol. V/2. Heidelberg: Springer, pp. 1–109.

Evans EF (1981) The dynamic range problem: place and time coding at the level of the cochlear nerve and nucleus. In: Syka J, Aitkin L (eds) Neuronal Mechanisms of Hearing. New York: Plenum Press, pp. 69–95.

Evans EF, Wilson JP (1973) The frequency selectivity of the cochlea. In: Møller AR (ed) Basic Mechanisms in Hearing. London: Academic Press, pp. 519–554.

Fay RR (1978) Coding of information in single auditory-nerve fibers of the goldfish. J Acoust Soc Am 63:136–146.

Fay RR (1988) Hearing in Vertebrates: A Psychophysics Databook. Winneteka: Hill-Fay.

Fay RR (1992) Structure and function in sound discrimination among vertebrates. In: Webster DB, Fay RR, Popper AN (eds) The Evolutionary Biology of Hearing. New York: Springer-Verlag, pp. 229–263.

Fay RR, Commbs S (1983) Neural mechanisms in sound detection and temporal summation. Hear Res 10:69–92.

Fechner GT (1860) Elemente der Psychophysik. Leipzig: Breitkopf und Härtel.

Fekete DM, Rouiller EM, Liberman MC, Ryugo DK (1982) The central projections of intracellularly labeled auditory-nerve fibers in cats. J Comp Neurol 229:432–450.

Flanagan JL (1955) Difference limen for vowel formant frequency. J Acoust Soc Am 27:613–617.

Flanagan JL (1972) Speech Analysis, Synthesis and Perception. New York: Springer-Verlag.

Fletcher H (1940) Auditory patterns. Rev Mod Phys 12:47–65.

Fletcher H, Munson WA (1933) Loudness, its definition, measurement, and calculation. J Acoust Soc Am 5:82–108.

Fletcher H, Munson WA (1937) Relation between loudness and masking. J Acoust Soc Am 9:1–10.

Fletcher H, Steinberg JC (1924) The dependence of the loudness of a complex sound upon the energy in the various frequency regions of the sound. Phys Rev 24:306–317.

Florentine M, Buus S (1981) An excitation pattern model for intensity discrimination. J Acoust Soc Am 70:1646 1654.

Florentine M, Buus S, Mason CR (1987) Level discrimination as a function of level from 0.25 to 16 kHz. J Acoust Soc Am 81:1528–1541.

Gaumond RP, Molnar CE, Kim DO (1982) Stimulus and recovery dependence of cat cochlear nerve fiber spike discharge probability. J Neurophysiol (Bethesda) 48:856–873.

Geisler CD (1985) Effect of a compressive nonlinearity in a cochlear model. J Acoust Soc Am 78:257–260.

Geisler CD (1992) Two-tone suppression by a saturating feedback model of the cochlear partition. Hear Res 63:203–211.

Geisler CD, Deng L, Greenberg SR (1985) Thresholds for primary auditory fibers using statistically defined criteria. J Acoust Soc Am 77:1102–1109.

Geisler CD, Yates GK, Patuzzi RB, Johnstone BM (1990) Saturation of outer hair cell receptor currents causes two-tone suppression. Hear Res 44:241–256.

Gerstein GL, Kiang NYS (1960) An approach to the quantitative analysis of electrophysiological data from single neurons. Biophys J 1:15–28.

Gifford ML, Guinan JJ Jr (1983) Effects of crossed-olivocochlear-bundle stimulation on cat auditory-nerve fiber responses to tones. J Acoust Soc Am 74:115–123.

Glasberg BR, Moore BCJ (1990) Derivation of auditory filter shapes from notched-noise data. Hear Res 47:103–138.

Goldberg JM, Brown PB (1969) Response of binaural regions of dog superior olivary complex to dichotic tonal stimuli: some physiological mechanisms of sound localization. J Neurophysiol (Bethesda) 32:613–636.

Goldstein JL (1973) An optimum processor theory for the central fomation of the pitch of complex tones. J Acoust Soc Am 54:1496–1516.

Goldstein JL (1974) Is the power law simply related to the driven spike response rate from the whole auditory nerve. In: Moskowitz HR, Scharf B, Stevens SS (eds) Sensation and Measurement. Dordrecht: Reidel, pp. 223–229.

Goldstein JL (1980) On the signal processing potential of high-threshold auditory-nerve fibers. In: van den Brink G, Bilsen FA (eds) Psychophysical, Physiological, and Behavioral Studies in Hearing. Delft: Delft University, pp. 293–299.

Goldstein JL (1990) Modeling rapid waveform compression in the basilar membrane as multiple-bandpass nonlinearity filtering. Hear Res 49:39–60.

Goldstein JL, Kiang NYS (1968) Neural correlates of the aural combination tone 2f1-f2. Proc IEEE 56:981–992.

Goldstein JL, Srulovicz P (1977) Auditory-nerve spike intervals as an adequate basis for aural spectrum analysis. In: Evans EF, Wilson JP (eds) Psychophysics and Physiology of Hearing. London: Academic Press, pp. 337–345.

Goodman DA, Smith RL, Chamberlain SC (1982) Intracellular and extracellular responses in the organ of Corti in the gerbil. Hear Res 7:161–179.

Gray PF (1967) Conditional probability analyses of the spike activity of single neurons. Biophys J 7:759–777.

Green DM (1958) Detection of multiple component signals in noise. J Acoust Soc Am 30:904–911.

Green DM (1960) Auditory detection of a noise signal. J Acoust Soc Am 32:121–131.

Green DM, Swets JA (1966) Signal Detection Theory and Psychophysics. New York: Wiley.

Greenberg SR, Geisler CD, Deng L (1986) Frequency selectivity of single cochlear nerve fibers based on the temporal response pattern of two-tone signals. J Acoust Soc Am 79:1010–1019.

Greenwood DD (1961) Critical bandwidth and the frequency coordinates of the basilar membrane. J Acoust Soc Am 33:1344–1356.

Greenwood DD (1971) Aural combination tones and auditory masking. J Acoust Soc Am 50:502–543.

Greenwood DD (1986) What is "synchrony suppression"? J Acoust Soc Am 79:1857–1872.

Greenwood DD (1990) A cochlear frequency-position function for several species—29 years later. J Acoust Soc Am 87:2592–2605.

Guinan JJ Jr, Gifford ML (1988) Effects of electrical stimulation of efferent olivocochlear neurons on cat auditory nerve fibers. III. Tuning curves and thresholds at CF. Hear Res 37:29–46.

Hall JL (1977) Two-tone suppression in a nonlinear model of the basilar membrane. J Acoust Soc Am 61:802–810.

Harris DM, Dallos P (1979) Forward masking of auditory-nerve fiber responses. J Neurophysiol (Bethesda) 42:1083–1107.

Heil P, Rajan R, Irvine DRF (1994) Topographic representation of tone intensity along the isofrequency axis of cat primary auditory cortex. Hear Res 76:188–202.

Hellman RP (1974) Effect of spread of excitation on the loudness function at 250 Hz. In: Moskowitz HR, Scharf B, Stevens SS (eds) Sensation and Measurement. Dordrecht: Reidel, pp. 241–249.

Hellman RP (1978) Dependence of loudness growth on skirts of excitation patterns. J Acoust Soc Am 63:1114–1119.

Hienz RD, Sachs MB, Aleszczyk C (1993) Frequency discrimination in noise: comparison of cat performances with auditory-nerve models. J Acoust Soc Am 93:462–469.

Hirahara T, Komakine T (1989) A computational cochlear nonlinear preprocessing model with adaptive Q circuits. Proc International Conference on Audio, Speech and Signal Processing 37:496–499.

Horst JW, Javel E, Farley GR (1986) Coding of spectral fine structure in the auditory nerve. I. Fourier analysis of period and interspike interval histograms. J Acoust Soc Am 79:398–416.

Houtgast T (1974) Lateral Suppression in Hearing. Amsterdam: Academische Pers.

Houtsma AJM, Durlach NI, Braida LD (1980) Intensity perception. XI. Experimental results on the relation of intensity perception to loudness matching. J Acoust Soc Am 68:807–813.

Hudspeth AJ, Corey DP (1977) Sensitivity, polarity and conductance change in the response of vertebrate hair cells to controlled mechanical stimuli. Proc Natl Acad Sci USA 76:2407–2411.

Irvine DRF (1992) Physiology of the auditory brainstem. In: Popper AN, Fay RR (eds) The Mammalian Auditory Pathway: Neurophysiology. New York: Springer-Verlag, pp. 153–231.

Javel E (1980) Coding of AM tones in the chinchilla auditory nerve: implications for the pitch of complex tones. J Acoust Soc Am 68:133–146.

Javel E, Mott JB (1988) Physiological and psychophysical correlates of temporal processes in hearing. Hear Res 34:275–294.

Javel E, Geisler CD, Ravindran A (1978) Two-tone suppression in auditory nerve of the cat: rate-intensity and temporal analyses. J Acoust Soc Am 63:1093–1104.

Javel E, Mott JB, Rush NL, Smith DW (1988) Frequency discrimination: evaluation of rate and temporal codes. In: Duifhuis H, Horst JW, Wit HP (eds) Basic Issues in Hearing. London: Academic Press, pp. 224–234.

Javel E, Tong YC, Shepherd RK, Clark GM (1987) Response of cat auditory-nerve fibers to biphasic electrical current pulses. Ann Otol Rhinol Laryngol 96(Suppl 128):26–30.

Jeng PS (1992) Loudness predictions using a physiologically-based auditory model. Doctoral dissertation, City University of New York, New York.

Jesteadt W, Wier CC, Green DM (1977) Intensity discrimination as a function of frequency and sensation level. J Acoust Soc Am 61:160–177.

Johnson DH (1974) The Response of Single Auditory-Nerve Fibers in the Cat to Single Tones: Synchrony and Average Discharge Rate. Ph.D. dissertation, Massachusetts Institute of Technology, Cambridge, MA.

Johnson DH (1978) The relationship of post-stimulus time and interval histograms to the timing characteristics of spike trains. Biophys J 22:412–430.

Johnson DH (1980) The relationship between spike rate and synchrony in responses auditory-nerve fibers to single tones. J Acoust Soc Am 68:1115–1122.

Johnson DH, Kiang NYS (1976) Analysis of discharges recorded simultaneously from pairs of auditory-nerve fibers. Biophys J 16:719–734.

Johnson DH, Swami A (1983) The transmission of signals by auditory-nerve fiber discharge patterns. J Acoust Soc Am 74:493–501.

Johnson JH, Turner CW, Zwislocki JJ, Margolis RH (1993) Just noticeable differences for intensity and their relation to loudness. J Acoust Soc Am 93:983–991.

Joris PX, Yin TCT (1992) Responses to amplitude-modulated tones in the auditory nerve of the cat. J Acoust Soc Am 91:215–232.

Joris PX, Carney LH, Smith PH, Yin TCT (1994) Enhancement of neural synchronization in the anteroventral cochlear nucleus. I. Response to tones at the characteristics frequency. J Neurophysiol (Bethesda) 71:1022–1051.

Kawase T, Delgutte B, Liberman MC (1993) Antimasking effects of the olivocochlear reflex. II. Enhancement of auditory-nerve response to masked tones. J Neurophysiol (Bethesda) 70:2533–2549.

Kelly OE, Johnson DH, Delgutte B, Cariani P (1995) Fractal noise strength in auditory-nerve fiber recordings. J Acoust Soc Am (in press).

Kewley-Port D, Watson CS (1994) Formant-frequency discrimination for isolated English vowels. J Acoust Soc Am 95:485–496.

Kiang NYS, Moxon EC (1974) Tails of tuning curves of auditory-nerve fibers. J Acoust Soc Am 55:620–630.

Kiang NYS, Watanabe T, Thomas EC, Clark LF (1965) Discharge Patterns of Single Fibers in the Cat's Auditory Nerve. Research Monograph #35. Cambridge: MIT Press.

Kim DO (1986) A review of nonlinear and active cochlear models. In: Allen JB, Hall JL, Hubbard A, Neely ST, Tubis A (eds) Peripheral Auditory Mechanisms. Berlin: Springer, pp. 239–247.

Kim DO, Molnar CE, Pfeiffer RR (1973) A system of nonlinear differential equations modeling basilar membrane motion. J Acoust Soc Am 54:1517–1529.

Kim DO, Sirianni JG, Chang SO (1990) Responses of DCN-PVCN neurons and auditory-nerve fibers in unanesthetized decerebrate cats to AM and pure tones: analysis with autocorrelation/power spectrum. Hear Res 45:95–113.

Kumar AR, Johnson DH (1984) The applicability of stationary point process models to discharge patterns of single auditory-nerve fibers. Elec Comp Eng Tech Rep 84-09, Rice University, TX.

Lachs G, Teich MC (1981) A neural counting model incorporating refractoriness and spread of excitation. II. Application to loudness estimation. J Acoust Soc Am 69:774–782.

Lachs G, Al-Shaikh R, Bi Q, Saia RA, Teich MC (1984) A neural counting model based on the physiological characteristics of the peripheral auditory system. V. Application to loudness estimation and intensity discrimination. IEEE Trans SMC-14:819–836.

Liberman MC (1978) Auditory-nerve response from cats raised in a low-noise chamber. J Acoust Soc Am 63:442–455.

Liberman MC (1982a) Single-neuron labeling in the cat auditory nerve. Science 216:1239–1241.

Liberman MC (1982b) The cochlear frequency map for the cat: labeling auditory-nerve fibers of known characteristic frequency. J Acoust Soc Am 72:1441–1449.

Liberman MC (1991) Central projections of auditory-nerve fibers of differing spontaneous rates. I. Antero-ventral cochlear nucleus. J Comp Neurol 313: 240–258.

Liberman MC, Brown MC (1986) Physiology and anatomy of single olivocochlear neurons in the cat. Hear Res 24:17–36.

Licklider JCR (1951) The duplex theory of pitch perception. Experientia 7:128–137.

Loeb GE, White MW, Merzenich MM (1983) Spatial crosscorrelation: a proposed mechanism for acoustic pitch perception. Biol Cybern 47:149–163.

Lynch TJ III, Peake WT, Rosowski JJ (1994) Measurement of the acoustic input impedance of cat ears: 10 Hz to 20 kHz. J Acoust Soc Am 96:2184–2209.

Maiwald D (1967) Beziehung zwischen Schallspektrum, Mitthorschwelle und der Erregung des Gehors. Acustica 18:69–80.

McGill WL, Goldberg JP (1968) Pure tone intensity discrimination and energy detection. J Acoust Soc Am 44:576–581.

McQuone SJ, May BJ (1993) Effects of olivocochlear efferent lesions on intensity discrimination in noise. Assoc Res Otolaryngol 16:51.

Meddis R (1986) Simulation of mechanical to neural transduction in the auditory receptor. J Acoust Soc Am 79:702–711.

Meddis R, Hewitt MJ (1991) Virtual pitch and phase sensitivity of a computer model of the auditory periphery. J Acoust Soc Am 89:2866–2882.

Miller GA (1947) Sensitivity to changes in the intensity of white noise and its relation to masking and loudness. J Acoust Soc Am 19:606–619.

Miller MI, Mark KE (1992) A statistical study of cochlear nerve discharge patterns in response to complex speech stimuli. J Acoust Soc Am 92:202–209.

Miller MI, Sachs MB (1984) Representation of voiced pitch in the discharge patterns of auditory-nerve fibers. Hear Res 14:257–279.

Miller MI, Barta PE, Sachs MB (1987) Strategies for the representation of a tone in background noise in the temporal aspects of the discharge patterns of auditory-nerve fibers. J Acoust Soc Am 81:665–679.

Moore BCJ (1973) Frequency difference limens for short-duration tones. J Acoust Soc Am 54:610–619.

Moore BCJ, Glasberg BR (1983) Formulae describing frequency selectivity as a function of frequency and level, and their use in calculating excitation patterns. Hear Res 28:209–225.

Moore BCJ, Glasberg BR (1986) The role of frequency selectivity in the perception of loudness, pitch and time. In: Moore BCJ (ed) Frequency Selectivity in Hearing. London: Academic Press, pp. 251–308.

Moore BCJ, Glasberg BR, Plack CJ, Biswas AK (1988) The shape of the ear's temporal window. J Acoust Soc Am 83:1102–1116.

Moore BCJ, Raab DH (1975) Intensity discrimination for noise bursts in the presence of a continuous, bandstop background: effects of level, width of the bandstop, and duration. J Acoust Soc Am 57:400–405.

Palmer AR, Evans EF (1982) Intensity coding in the auditory periphery of the cat: responses of cochlear nerve and cochlear nucleus neurons to signals in the presence of bandstop masking noise. Hear Res 7:305–323.

Palmer AR, Winter IM, Jiang G, James N (1995) Across-frequency integration by neurones in the ventral cochlear nucleus. In: Manley GA, Klump GM, Köppl C, Fastl H, Oeckinghous H (eds) Advances in Hearing Research. Singapore: World Scientific, pp 250–263.

Patterson RD (1976) Auditory filter shapes derived with noise stimuli. J Acoust Soc Am 59:640–654.

Patterson RD, Holdsworth J (1995) A functional model of neural activity patterns and auditory images. In: Ainsworth WA (ed) Advances in Speech, Hearing and Language Processing. London: JAI (in press).

Patterson RD, Moore BCJ (1986) Auditory filters and excitation patterns as representations of frequency resolution. In: Moore BCJ (ed) Frequency Selectivity in Hearing. London: Academic Press, pp. 123–177.

Penner MJ, Viemeister NF (1973) Intensity discrimination of clicks: the effects of click bandwidth and background noise. J Acoust Soc Am 54:1184–1188.

Pickles JO (1979) Psychophysical frequency resolution in the cat as determined by simultaneous masking, and its relation to auditory-nerve resolution. J Acoust Soc Am 66:1725–1732.

Pickles JO (1980) Psychophysical frequency resolution in the cat studied with forward masking. In: van den Brink G, Bilsen FA (eds) Psychophysical, Physiological, and Behavioral Studies in Hearing. Delft: Delft University Press, pp. 118–125.

Pickles JO (1983) Auditory-nerve correlates of loudness summation with stimulus bandwidth in normal and pathological cochleae. Hear Res 12:239–250.

Pickles JO (1984) Frequency threshold curves and simultaneous masking functions in single fibres of the guinea pig auditory nerve. Hear Res 14:245–256.

Raab DH, Goldberg IA (1975) Auditory intensity discrimination with bursts of reproducible noise. J Acoust Soc Am 57:437–447.

Recio A, Viemeister NF, Powers L (1994) Detection thresholds based upon rate and synchrony. Assoc Res Otolaryngol Abstr 17:67.

Relkin EM, Doucet JR (1991) Recovery from forward masking in the auditory nerve depends on spontaneous firing rate. Hear Res 55:215–222.

Relkin EM, Pelli DG (1987) Probe tone thresholds in the auditory nerve measured by a two-interval forced-choice procedure. J Acoust Soc Am 82:1679–1691.

Relkin EM, Turner CW (1988) A reexamination of forward masking in the auditory nerve. J Acoust Soc Am 84:584–591.

Rhode WS (1971) Observations of the vibration of the basilar membrane in squirrel monkeys using the Mössbauer technique. J Acoust Soc Am 49:1218–1231.

Rhode WS, Greenberg S (1992) Physiology of the cochlear nuclei. In: Popper AN, Fay RR (eds) The Mammalian Auditory Pathway: Neurophysiology. New York: Springer-Verlag, pp. 94–152.

Rhode WS, Smith PH (1985) Characteristics of tone-pip response patterns in relationship to spontaneous rate in cat auditory nerve fibers. Hear Res 18: 159–168.

Rhode WS, Geisler CD, Kennedy DK (1979) Auditory-nerve fiber responses to wide band noise and tone combinations. J Neurophysiol (Bethesda) 41:692–704.

Robles L, Ruggero MA, Rich NC (1986) Basilar membrane mechanics at the base of the chinchilla cochlea. I: Input-output functions, tuning curves and response phases. J Acoust Soc Am 80:1364–1374.

Rose JE, Brugge JF, Anderson DJ, Hind JE (1967) Phase-locked response to low-frequency tones in single auditory-nerve fibers of the squirrel monkey. J Neurophysiol (Bethesda) 30:769–793.

Rose JE, Hind JE, Anderson DJ, Brugge JF (1971) Some effects of stimulus intensity on response of auditory-nerve fibers in the squirrel monkey. J Neurophysiol (Bethesda) 34:685–699.

Rothman JS, Young ED, Manis PB (1993) Convergence of auditory-nerve fibers onto bushy cells in the ventral cochlear nucleus: implications of a computational model. J Neurophysiol (Bethesda) 70:2562–2583.

Ruggero M (1973) Response to noise of auditory-nerve fibers in the squirrel monkey. J Neurophysiol (Bethesda) 36:569–587.

Ruggero M (1992) Physiology and coding of sound in the auditory nerve. In: Popper AN, Fay RR (eds) The Mammalian Auditory Pathway: Neurophysiology. New York: Springer-Verlag, pp. 34–93.

Ruggero MA, Robles L, Rich NC (1992) Two-tone suppression in the basilar membrane of the cochlea: mechanical basis of auditory-nerve rate suppression. J Neurophysiol (Bethesda) 68:1087–1099.

Ryugo DK (1992) The auditory nerve: peripheral innervation, cell body morphology, and central projections. In: Webster DB, Popper AN, Fay RR (eds) The Mammalian Auditory Pathway: Neuroanatomy. New York: Springer-Verlag, pp. 23–65.

Sachs MB (1984) Speech encoding in the auditory nerve. In: Berlin C (ed) Hearing Science. San Diego: College Hill, pp. 263–308.

Sachs MB, Abbas PJ (1974) Rate versus level functions for auditory-nerve fiber in cats: tone burst stimuli. J Acoust Soc Am 56:1835–1847.

Sachs MB, Abbas PJ (1976) Phenomenological model for two-tone suppression. J Acoust Soc Am 60:1157–1163.

Sachs MB, Kiang NYS (1968) Two-tone inhibition in auditory-nerve fibers. J Acoust Soc Am 43:1120–1128.

Sachs MB, Young ED (1979) Encoding of steady-state vowels in the discharge patterns of auditory-nerve fibers: representation in terms of discharge rate. J Acoust Soc Am 66:1381–1403.

Sachs MB, Young ED (1980) Effects of nonlinearities on speech encoding in the auditory nerve. J Acoust Soc Am 68:858–875.

Schalk T, Sachs MB (1980) Nonlinearities in auditory-nerve fiber response to band limited noise. J Acoust Soc Am 67:903–913.

Scharf B, Magnan J, Collet L, Ulmer E, Chays A (1994) On the role of the olivocochlear bundle in hearing: a case study. Hear Res 75:11–26.

Schmiedt RA (1982) Boundaries of two-tone rate suppression of cochlear-nerve activity. Hear Res 7:335–351.

Schouten JF (1970) The residue revisited. In: Plomp R, Smoorenburg GF (eds) Frequency Analysis and Periodicity Detection in Hearing. Leiden: Sijthoh, pp. 41–58.

Sellick PM, Patuzzi R, Johnstone BM (1982) Measurements of basilar membrane motion in the guinea pig using the Mössbauer technique. J Acoust Soc Am 72:131–141.

Seneff S (1988) A joint synchrony/mean-rate model of auditory speech processing. J Phonet 16:55–76.

Shamma S (1985) Speech processing in the auditory system. II: Lateral inhibition and the central processing of speech evoked activity in the auditory nerve. J Acoust Soc Am 78:1622–1632.

Shannon RV (1976) Two-tone unmasking and suppression in a forward-masking situation. J Acoust Soc Am 59:1460–1470.

Shannon RV (1983) Multichannel electrical stimulation of the auditory nerve in man. I. Basic psychophysics. Hear Res 11:157–189.

Shannon RV, Otto SR (1990) Psychophysical measures from electrical stimulation of the human cochlear nucleus. Hear Res 47:159–168.

Shofner WP, Dye RH (1989) Statistical and receiver operating characteristic analysis of empirical spike-count distribution: quantifying the ability of cochlear nucleus units to signal intensity changes. J Acoust Soc Am 86:2172–2184.

Siebert WM (1965) Some implications of the stochastic behavior of auditory neurons. Kybernetik 2:206–215.

Siebert WM (1968) Stimulus transformations in the peripheral auditory system. In: Kollers PA, Eden M (eds) Recognizing Patterns. Cambridge: MIT Press, pp. 104–133.

Siebert WM (1970) Frequency discrimination in the auditory system: place or periodicity mechanism. Proc IEEE 58:723–730.

Siebert WM, Gray PR (1963) Random process model for the firing pattern of single auditory neurons. MIT Res Lab Electron Q Prog Rep 71:241–245.

Sinex DG, Havey DC (1986) Neural mechanisms of tone-on-tone masking: patterns of discharge rate and discharge synchrony related to rates of spontaneous discharge in the chinchilla auditory nerve. J Neurophysiol (Bethesda) 56:1763–1780.

Sinnott JM, Brown CH, Brown FE (1992) Frequency and intensity discrimination in Mongolian gerbils, African monkeys and humans. Hear Res 59:205–212.

Slaney M, Lyon RF (1993) On the importance of time—a temporal representation of sound. In: Cooke M, Beet S, Crawford M (eds) Visual Representations of Speech Signals. New York: Wiley, pp. 95–116.

Smith RL (1977) Short-term adaptation in single auditory-nerve fibers: some poststimulatory effects. J Neurophysiol (Bethesda) 40:1098–1112.

Smith RL (1979) Adaptation, saturation and physiological masking in single auditory-nerve fibers. J Acoust Soc Am 65:166–178.

Smith RL, Zwislocki JJ (1975) Short-term adaptation and incremental responses of single auditory-nerve fibers. Biol Cybern 17:169–182.

Smoorenburg GF (1970) Pitch perception for two-frequency stimuli. J Acoust Soc Am 48:924–942.

Solecki JM, Gerken GM (1990) Auditory temporal integration in the normal-hearing and hearing-impaired cat. J Acoust Soc Am 88:779–785.

Srulovicz P, Goldstein JL (1983) A central spectrum model: a synthesis of auditory nerve timing and place cues in monaural communication of frequency spectrum. J Acoust Soc Am 73:1266–1276.

Stevens SS (1956) The direct estimation of sensory magnitudes—loudness. Am J Psychol 69:1–25.

Stevens SS, Davis H (1938) Hearing: Its Psychology and Physiology. New York: Wiley.

Tanner WP, Swets JA, Green DM (1956) Some general properties of the hearing mechanism. Univ Michigan Electron Defense Group Tech Rep 30.

Tasaki I (1954) Nerve impulses in individual auditory-nerve fibers of guinea pig. J Neurophysiol (Bethesda) 17:97–122.

Teich MC (1989) Fractal character of the auditory neural spike train. IEEE Trans BME-36:150–160.

Teich MC, Khanna SM (1985) Pulse number distribution for the neural spike train in the cat's auditory nerve. J Acoust Soc Am 77:1110–1128.

Teich MC, Lachs G (1979) A neural counting model incorporating refractoriness and spread of excitation. I. Application to intensity discrimination. J Acoust Soc Am 66:1738–1749.

Teich MC, Johnson DH, Kumar AR, Turcott RG (1990) Rate fluctuations and fractional power-law noise recorded from cells in the lower auditory pathway of the cat. Hear Res 40:41–52.

Terhardt E (1974) Pitch, consonance, and harmony. J Acoust Soc Am 55: 1061–1069.

Townsend B, Cotter N, Van Compernolle D, White RL (1987) Pitch perception by cochlear implant patients. J Acoust Soc Am 82:106–115.

Viemeister NF (1974) Intensity discrimination of noise in the presence of band-reject noise. J Acoust Soc Am 56:1594–1600.

Viemeister NF (1983) Auditory intensity discrimination at high frequencies in the presence of noise. Science 221:1206–1208.

Viemeister NF (1988) Psychophysical aspects of intensity discrimination. in: Edelman GM, Gall WE, Cowan WM (eds) Auditory Function: Neurobiological Bases of Hearing. New York: Wiley, pp. 213–241.

Viemeister NF, Shivapuja BG, Recio A (1992) Physiological correlates of temporal integration. In: Cazals Y, Horner K, Demany L (eds) Auditory Physiology and Perception. Oxford: Pergamon Press, pp. 322–329.

Voigt HF, Young ED (1980) Evidence for inhibitory interactions between neurons in dorsal cochlear nucleus. J Neurophysiol (Bethesda) 44:76–96.

von Békésy G (1929) Zur Theorie des Hörens: Über die eben merkbare Amplituden- und Frequenzänderung eines Tones; Die Theorie der Schwebungen. Z Phyzik 30:721–745.

von Helmholtz HLF (1863) Die Lehre von den Tonempfindungen als physiologische Grundlage für die Theorie der Musik. Braunschweig: Vieweg und Sohn.

Warr WB (1992) Organization of olivocochlear efferent systems in mammals. In: Webster DB, Popper AN, Fay RR (eds) The Mammalian Auditory Pathway: Neuroanatomy. New York: Springer-Verlag, pp. 410–448.

Warr WB, Guinan JJ Jr (1979) Efferent innervation of the organ of Corti: two separate systems. Brain Res 173:152–155.

Wakefield GH, Nelson DA (1985) Extension of a temporal model of frequency discrimination: intensity effects in normal and hearing-impaired listeners. J Acoust Soc Am 77:613–619.

Warren EH, Liberman MC (1989) Effects of contralateral sound on auditory-nerve responses. I. Contributions of cochlear efferents. Hear Res 37:89–104.

Wegel RL, Lane CE (1924) The auditory masking of one pure tone by another and its probable relation to the dynamics of the inner ear. Phys Rev 23:266–285.

Weiss TF, Rose C (1988) Stages of degradation of timing information in the cochlea: a comparison of hair-cell and nerve-fiber responses in the alligator lizard. Hear Res 33:167–174.

Westerman LA, Smith RL (1984) Rapid and short-term adaptation in auditory-nerve responses. Hear Res 15:249–260.

Westerman LA, Smith RL (1988) A diffusion model of the transient response of the cochlear inner hair cell synapse. J Acoust Soc Am 83:2266–2276.

Wever EG (1949) Theory of Hearing. New York: Wiley.

Whitfield IC (1967) The Auditory Pathway. Baltimore: Williams & Wilkins.

Wiederhold ML, Kiang NYS (1970) Effects of electrical stimulation of the crossed olivocochlear bundle on single auditory nerve fibers in cat. J Acoust Soc Am 48:950–965.

Wiener FM, Pfeiffer RR, Backus ASN (1966) On the sound pressure transformation by the head and auditory meatus of the cat. Acta Otolaryngol 61:255–269.

Wier CC, Jesteadt W, Green DM (1977) Frequency discrimination as a function of frequency and sensation level. J Acoust Soc Am 61:178–184.

Winslow RL (1985) A quantitative analysis of rate coding in the auditory nerve. Doctoral dissertation, The Johns Hopkins University, Baltimore, MD.

Winslow RL, Sachs MB (1987) Effect of electrical stimulation of the crossed olivocochlear bundle on auditory nerve response to tones in noise. J Neurophysiol (Bethesda) 57:1002–1021.

Winslow RL, Sachs MB (1988) Single-tone intensity discrimination based on auditory-nerve rate responses in backgrounds of quiet, noise, and with stimulation of the crossed olivocochlear bundle. Hear Res 35:165–190.

Winslow RL, Barta PE, Sachs MB (1987) Rate coding in the auditory nerve. In: Yost WA, Watson CS (eds) Auditory Processing of Complex Sounds. Hillsdale: Erlbaum, pp. 212–224.

Winter IM, Robertson D, Yates GK (1990) Diversity of characteristic frequency rate-intensity functions in guinea pig auditory nerve fibers. Hear Res 45:191–202.

Yates GK, Robertson D, Johnstone BM (1985) Very rapid adaptation in the guinea pig auditory nerve. Hear Res 17:1–12.

Yates GK, Winter IM, Robertson D (1990) Basilar membrane nonlinearity determines auditory-nerve rate-intensity functions and cochlear dynamic range. Hear Res 45:203–220.

Yin TCT, Chan JCK (1990) Interaural time sensitivity in medial superior olive of cat. J Neurophysiol (Bethesda) 64:465–488.

Young EG (1989) Problems and opportunities in extending psychophysical/physiological correlation into the central nervous system. In: Turner CW (ed) Interactions Between Neurophysiology and Psychoacoustics. New York: Acoustical Society of America, pp. 118–140.

Young ED, Barta PE (1986) Rate responses of auditory nerve fibers to tone in noise near masked threshold. J Acoust Soc Am 79:426–442.

Young ED, Sachs MB (1973) Recovery from sound exposure in auditory nerve fibers. J Acoust Soc Am 54:1535–1543.

Young ED, Sachs MB (1979) Representation of steady-state vowels in the temporal aspects of the discharge patterns of populations of auditory-nerve fibers. J Acoust Soc Am 66:1381–1403.

Young ED, Sachs MB (1989) Auditory-nerve fibers do not discharge independently when responding to broadband noise. Assoc Res Otolaryngol Abstr 12:121.

Zhang W, Salvi RJ, Saunders SS (1990) Neural correlates of gap detection in auditory-nerve fibers of the chinchilla. Hear Res 46:181–200.

Zwicker E (1956) Die elementatre Grudlagen zur Bestimmung der Informationskapazität des Gehörs. Acustica 6:365–381.

Zwicker E (1970) Masking and psychological excitation as consequences of the ear's frequency analysis. In: Plomp R, Smoorenburg GF (eds) Frequency Analysis and Periodicity Detection in Hearing. Leiden: Sijthoh, pp. 376–396.

Zwicker E (1974) On a psychoacoustical equivalent of tuning curves. In: Zwicker E, Terhardt E (eds) Facts and Models in Hearing. Berlin: Springer-Verlag, pp. 132–141.

Zwicker E (1986) Suppression and 2f1-f2 difference tones in a nonlinear cochlear preprocessing model. J Acoust Soc Am 80:163–176.

Zwicker E, Feldtkeller R (1967) Das Ohr als Nachrichtenempfänger. Stuttgart: Hirzel Verlag.

Zwicker E, Scharf B (1965) A model of loudness summation. Psychol Rev 72:3–26.

Zwicker E, Flottorp G, Stevens SS (1957) Critical band width in loudness summation. J Acoust Soc Am 29:548–557.

Zwislocki JJ (1960) Theory of temporal auditory summation. J Acoust Soc Am 32:1046–1060.

Zwislocki JJ (1969) Temporal summation of loudness: an analysis. J Acoust Soc Am 46:431–441.

6
Auditory Representations of Timbre and Pitch

Richard Lyon and Shihab Shamma

1. Introduction

Pitch and *timbre* are terms developed to describe musical sounds. By convention, pitch is that perceptual property of a sound that can be used to play a melody, while timbre is a rather more vague perceptual property that distinguishes musical sounds of the same pitch. The pitch of an arbitrary sound is quantified as the frequency in hertz (Hz) of a sinusoid with perceptually matching pitch. Thus, one might say that pitch is a perceptual frequency, and timbre is almost everything else (but not including loudness, position, and perhaps a few other well-defined attributes). In general, a sound source achieves a perceptual unity through pitch, and encodes its identity primarily through timbre. For example, any instrument playing middle C will be heard as a sound source with a pitch of 262 Hz, but the timbre of the sound will tell us whether we are hearing a string excited by a bow, or a tube excited by reed, or a vibrating metal bar, or an electronically synthesized buzz.

But hearing melodies and distinguishing musical instruments are probably not the most important capabilities for animals that evolved excellent hearing. A good theory of pitch and timbre should relate to properties of natural sound sources of importance to an organism, and a good computational model of pitch and timbre should relate to known physiology of the auditory nervous system.

In the following sections, we describe several approaches to modeling pitch and timbre perception computationally. First, in Section 2, we discuss how natural sound sources give rise to sounds that can be usefully described in terms of pitch and timbre. Then, conventional (nonauditory) approaches to measuring and representing pitch and timbre are described in Section 3. In Section 4, we briefly review the fundamental transformations occurring in the cochlea that give rise to the multiplicity of timbre and pitch cues available to the auditory system. In Section 5, various schemes are analyzed by which the early postcochlear auditory stages exctract timbre and pitch representations from the cochlear outputs. A

brief review of central auditory representations of timbre and pitch
it presented in Section 6. We end the chapter with a few comments
regarding the state of our knowledge and the future questions remaining.

2. Origin and Meaning of Timbre and Pitch of a Sound Source

The prime function of the auditory system is to tell the organism about its
world by detecting, locating, and classifying sound sources. In a complex
sound environment, deciding what sound fragments should be grouped
and identified as a single source is perhaps the most difficult and important
low-level processing task. Pitch and timbre in humans are important
percepts that we believe arise from computational strategies for performing this task. (See Chapter 7 by Mellinger and Mont-Reynaud for
more on grouping and sound source segregation.)

Sounds in the environment are produced by vibrating structures such as
rustling leaves, snapping twigs, burbling water, and falling rocks. These
structures must be excited by a driving force to vibrate and emit a sound.
The emitted sound carries information about the driving force, about the
vibrating structure, and about further physical features such as resonating
chambers that modify the sound on its way to our ears.

When the driving function is repetitive, such as in the multiple excita-
tions that the vocal folds produce during voiced speech, the sound consists
of several nearly identical copies of the same emission, possibly overlapping in time, as illustrated in Figure 6.1. The fact that the emissions are

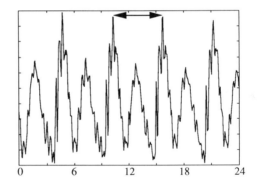

FIGURE 6.1. A periodic speech waveform. The prominent repetitive pattern gives
rise to the percept known as pitch. The pitch period, the interval shown by the
arrows, corresponds closely to the period of a sine wave with the same perceived
pitch. The timbre of the sine wave and the timbre of the speech sound are quite
different, however, corresponding to their very different waveform shapes and
spectra.

so similar is an excellent clue to their common origin, so we might expect organisms to have evolved special mechanisms to recognize repetition. The time intervals between repetitions are also meaningful, as they characterize the excitation function of the sound source. The importance of these properties is demonstrated by the fact that in most cases the rate of repetitive excitation of a source (in repetitions per second) corresponds to what we perceive as its pitch (in hertz). It is critical to recognize here that pitch is only defined for relatively fast repetitions (faster than 20–50 Hz) (Hess 1983). Events occuring at slower rates are perceived anditorily as a distinct sequence of sounds. For example, in a melody (or speech), one recognizes a sequence of notes (or phonemes); however, only within each note (phoneme such as a vowel) can one talk of a pitch percept. (Issues associated with the perception of sound sequences defined over long time intervals are addressed in Chapter 10 by Lewis).

To determine which of many overlapping sound fragments are repetitions, the auditory system must be able to characterize brief fragments along some dimensions other than pitch. Those other dimensions correspond, roughly, to the timbre of the sound, or of the sound source. A simple way to think about the timbre of a repetitive sound is as a characterization of the resonances of the vibrating source and its acoustic enclosure.

In speech, a class of sounds of considerable interest to scientists and other humans, the repetitive excitation comes from the opening and closing of the glottis (vocal folds), while the timbre of a voiced sound is imposed by the resonances of the oral and nasal cavities, as controlled by the tongue, lips, and velum (Flanagan 1972). Thus, for voiced sounds, we might think of the timbre as the smoothed spectrum, or a description of it in terms of resonances and antiresonances, and the pitch as the frequency of excitation. Many nearly periodic, nonspeech sounds can be similarly described.

Taking a more "time-domain" or "temporal" approach, we can also describe timbre on the basis of the time waveform of the repeated emissions of a source, or of a single emission from a nonperiodic source (Fig. 6.1). When the same shape is found repeated in a waveform, a source can be identified and characterized by the repetition rate (pitch) and the vibration pattern (timbre). This simple approach makes sense because in nature the probability of receiving matching repetitive signals from independent sources is very small.

No finite sound can be exactly periodic, so we should be careful to consider effects of onsets, offsets, and limited numbers of repetitions. For example, in studies of musical note perception, it has been found that as few as two periods of the sound can elicit a clear pitch percept, and that the onset of a note has a large impact on its perceived timbre (Bregman 1990). A transient sound with no clear repetition can have clear timbral quality, but usually not a clear pitch (Bregman 1990). Percussive sounds such as drums can have a fairly clear perceptual pitch, based not on a repetitive excitation but on an approximately periodic waveform, at least within some frequency range. And even periodic sounds can have a perceived pitch different from the repetition rate, as is discussed later. These marginal cases provide a wealth of important constraints on theories and computational models of pitch and timbre. For the purposes of this chapter, we focus on the smoothed spectral profile as the physical correlate of timbre. Thus, we shall exclude from our discussion many other secondary physical parameters that are known to influence the perceived timbre of sound, such as the onset and offset characteristics (see Chapter 7 by Mellinger and Mont-Reynaud for more details).

3. Traditional Representations of Timbre and Pitch

For several decades, engineering systems have been developed to process sound signals, be it for speech recognition, music synthesis, or underwater acoustics. In the absence of extensive and reliable biological data, many of these engineering systems were viewed as valuable models of the hearing process. As physiological and psychoacoustical information accumulated, many of these models faded away. Curiously, a few have survived the test of time and have immensely influenced our view of auditory processing despite obvious flaws. The most important example of such algorithms is the short-time Fourier transform (Rabiner and Schafer 1978), more commonly known as the spectrogram (and its related representations such as linear predictive coding [Makhoul 1975] and cepstral coefficients [Rabiner and Schafer 1978]). In this representation, the signal is analyzed by a bank of uniformly spaced and equally tuned filters, and then a measure of the instantaneous output power of the filter is displayed as a function of time (Fig. 6.2). This view of the acoustic signal roughly corresponds to extracting the spectral profile with a resolution that is arbitrarily determined by the bandwidth of the analysis filters.

There are two kinds of problems with the spectrogram as an auditory model. The first concerns implementation details that, in principle, can be readily brought into harmony with known properties of cochlear analysis. For instance, the analysis filters in the cochlea of the ear are not uniformly spaced, but rather logarithmically spaced for frequencies higher than 500 Hz (Plomp 1976). Furthermore, cochlear filters are not of equal bandwidth. Instead, they have approximately constant tuning factors (Qs); that is filters centered at high frequencies are more broadly tuned than those at lower frequencies (Moore 1986). In classical sprectrograms, a fixed window size choice is often a difficult compromise between choices that would be appropriate for low-pitched and high-pitched voices, or between desired low-frequency and high-frequency resolution (Rabiner

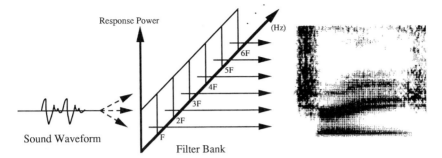

FIGURE 6.2. In creating a spectrogram, the sound waveform is separated into a number of channels equally spaced along a linear frequency axis, and the short-time power in each channel is reported as a function of time.

and Schafer 1978); an approximately constant-Q analysis fits the problem space better.

The second kind of problems with the spectrogram concern the conceptual distortions that it introduced into thinking about auditory processing. The spectrogram has too often been taken as a literal representation of peripheral auditory analysis, with the idea that a one-to-one correspondance exists between the stages of the two. Consequently, for a long time little or no serious consideration was paid to the role played by the detailed shapes and phases of the cochlear filters, and by the fine temporal structure of their outputs, as opposed to simply the output power. As we elaborate in the next section, these factors have turned out to be critical to our understanding of auditory processing.

Similar biases and misconception existed for the representation of pitch. Thus, in examining the time waveform of an acoustic stimulus, the most apparent correlate of pitch is the periodicity of the waveform envelope (see Fig. 6.1). This led to a host of so-called time-domain pitch algorithms that operated on the time waveform of the signal directly (i.e., without any cochlear transformations), and which emphasized the extraction of the signal envelope or the signal's fundamental frequency (Hess 1983). As it turns out, neither of these measures truly corresponds to the pitch percept. For instance, the energy in the fundamental frequency of speech signals is often relatively small, or can be completely filtered out without much effect on the pitch percept. It is also possible to demonstrate that in some signals the periodicity of the envelope does not equal the perceived pitch. This is the case, for example, in an inharmonic series such as 700, 900, and 1100 Hz where the envelope periodicity is 200 Hz whereas multiple pitches are perceived at roughly 175 and 233 Hz, as shown in Figure 6.3 (Schouten, Ritsma, and Cardozo 1962).

Because of these limitations and the sensitivity of the pitch estimates to any additional noise, time-domain approaches have been superceded by

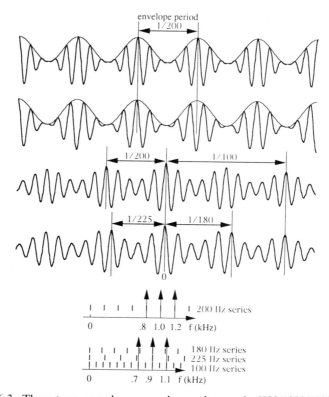

FIGURE 6.3. Three-tone complexes are shown: harmonic (800/1000/1200 Hz, the upper signal) and inharmonic (700/900/1100 Hz, the second signal), along with their respective autocorrelation functions (third and fourth signals). In both cases the complexes can be produced by adding three sine waves, or by amplitude modulating a carrier by a modulating signal. The envelopes, or modulating signal periods, are equal. As shown via the autocorrelation functions, however, the intervals in the fine structure differ between the signals; intervals corresponding to an intergral number of carrier cycles are indicated. The perceived pitch is more closely related to fine-structure intervals, which are also closely related to the frequency-domain pattern of component frequencies, as schematized below the waveforms.

so-called spectral-domain approaches. In these algorithms, the pitch is extracted from the signal's spectrogram, or equivalently, from its short-term autocorrelation function. One example of such algorithms is the estimation of the fundamental from its higher harmonics in the power spectrum (Hess 1983). Pitch estimates thus computed are less sensitive to noise because several periods of the signal are typically processed simultaneously. Furthermore, the estimates are immune to phase distortions because the transformation of the signal to the spectrogram or autocorrelation function discards irreversibly its phase. Consequently, while such approaches have proven more successful and accurate, they have mistakenly reinforced the notion that auditory representations need only be in terms of the magnitude of the transform, as in the case of the spectrogram.

As we elaborate in Section 4, the auditory representation of sound in the early stages of processing includes phase information. For instance, the outputs of the cochlear filters as seen on the auditory nerve do not simply encode the magnitude of the response, but actually encode the time waveform of the response as well. It is therefore by no means obvious how timbre and pitch percepts are derived, because a multiplicity of effective cues exist.

4. Cochlear Transformations of Sound

Sound signals undergo a complex series of transformations in the early stages of anditory processing. Numerous descriptions of these processes exist, ranging from detailed biophysical models to approximate computational algorithms (Lyon 1983; Deng, Geisler, and Greenberg 1988; Shamma et al. 1986; Gitza 1988; Seneff 1988; Patterson and Holdsworth 1991). All models, however, can be reduced to three stages: analysis, transduction, and reduction. In the first stage, the unidimensional sound signal is transformed into a distributed representation along the length of the cochlea. This representation is then converted in the second stage into a pattern of electrical activity on thousands of parallel anditory nerve fibers. Finally, perceptual representations of timbre and pitch are extracted from these patterns in the third stage. In the following, we briefly elaborate on the first two stages of processing; the third stage is discussed in detail in Section 5.

4.1 The Analysis Stage

When sound impinges upon the eardrum of the outer ear, it causes vibrations that are transmitted via the ossicles of the middle ear, which in turn produce pressure waves in the fluids of the cochlea of the inner ear. These pressure waves cause mechanical displacements in the membranes of the cochlea, specifically the basilar membrane. Because of the unique spatially distributed geometry and mechanical properties of the basilar membrane, the vibrations acquire distinctive properties that reflect the structure of the sound stimulus. For instance, the vibrations evoked by a single tone appear as traveling waves that propagate up the cochlea (from base to apex), reaching a maximum amplitude at a particular point before slowing down and decaying rapidly (Fig. 6.4). The abrupt slowdown in the waves makes them appear more "crowded," or equivalently, the spatial frequency of the traveling wave significantly increases near its

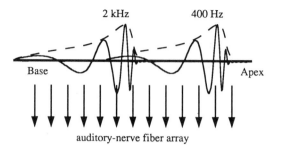

FIGURE 6.4. Schematic of cochlear traveling-wave vibrations evoked by pure tones of two different frequencies. The wavelength of the traveling wave decreases as the wave travels from the base toward the apex, while the amplitude grows and then rapidly decays. Lower frequencies travel further before decaying.

apical end (von Békésy 1960; Rhode 1980). The point at which maximum displacement and subsequent slowdown occurs depends on the frequency of the tone. The higher the frequency, the more basal is the location. This tonotopic order of the responses is an important organizational feature of the entire primary auditory pathway. It is preserved through several point-to-point topographic mappings to the auditory cortex.

One may observe this tonotopic order from a different perspective using acoustic impulses (clicks) rather than continuous tones. In response to an acoustic click, each point along the basilar membrane exhibits decaying oscillations with a different characteristic frequency (CF) (Fig. 6.5). The CF at a particular point is close to the frequency of the tone whose traveling wave maximally excites it. The CF's therefore are also tonotopically ordered with higher frequency oscillations observed closer to the base. For frequencies higher than about 500 Hz, the CFs are logarithmically spaced, that is, octave steps span equal distances along the cochlear partition. Below 500 Hz, the spacing becomes progressively more linear.

Another important feature that covaries with the CF is the decay rate of the responses which exponentially increases toward the basal end of the cochlea. The tonotopic order, combined with the slow decay rates at the low-CF regions of the cochlea, have immediate implications for the representation of repetitive stimuli. For instance, consider a train of acoustic clicks spaced at a regular interval T. If the responses to one click do not decay completely before the arrival of the next, then one would expect an interference response pattern to emerge. At points with CF's that are near integer multiples of the fundamental repetition frequency $1/T$, the interference is constructive and the responses are generally larger than at points in between. Consequently, the repetitive structure of the stimulus is encoded in the time course of the spatial pattern of responses. Conversely, at high CF regions where the decay rate is fast enough that

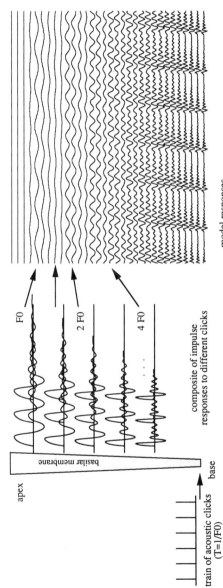

FIGURE 6.5. Single click (A) and click train (B) on the basilar membrane. In response to an acoustic click, each point along the basilar membrane exhibits decaying oscillations with a different characteristic frequency (CF). When the click repeats, the waves interfere constructively and destructively, giving rise to a pattern of activity level as a function of CF.

one click response has died out before the next arrives, the spatial pattern is flat but the click train is well represented in the time pattern. That is, there are two possible sources of information: The first is temporal in that the waveforms of the responses reflect explicitly and unambiguously the interval T. The second source is spatial, in that the pattern of cochlear locations with enhanced responses also reflects unambiguously the interval T.

This duality of temporal and spatial cues is a direct consequence of the fact that the basilar membrane segregates its responses tonotopically, thus creating the spatial cues. As such, the cochlea can be viewed from a more functional, but equivalent, point of view as a parallel bank of bandpass filters. At each point along the membrane, one can measure the displacement as a function of tone frequency, that is, its so-called transfer function. The tonotopic order of the CFs and the associated changes in the decay rate discussed earlier are now expressed, respectively, in terms of the center frequencies and bandwidth of the filters decreasing toward the apex of the cochlea (Fig. 6.6). Once again, above about 500 Hz in humans, the transfer functions appear approximately invariant except for a translation along the log-frequency axis, that is, they maintain a constant relative tuning (or Q factor), and their impulse responses are related to each other by a dilation (Moore 1986).

The foregoing simplified view of the basilar membrane deviates from the real structure in many ways that may be consequential in some applications. In particular, there are nonlinear active cochlear mechanisms that play an important role in enhancing the sensitivity and tuning of the cochlear filters at lower sound levels and which also enhance responsiveness to onsets and changes (Dallos et al. 1990; Ashmore 1993). These phenomena may be less important when dealing with relatively broadband

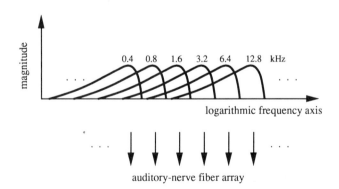

FIGURE 6.6. An illustration of frequency separation in the cochlea as a constant-Q filter bank, mapping log frequencies to place; the magnitude transfer function of the filters is schematized. Q, is tuning factor of a filter.

steady signals at moderate to high levels of intersity (Deng, Geisler, and Greenberg 1988), as in the case of many complex tones used in psychoacoutic experiments. The main impact of these mechanisms on pitch and timbre calcalutions will be to accentuate information at the beginnings (onsets) of sounds, and to suppress information in weak sounds that follow loud sounds (after offsets, as in forward masking).

To summarize, an acoustic signal $x(t)$ entering the ear produces a complex spatiotemporal pattern of displacements $y(t;s)$ along the basilar membrane of the cochlea. To a first approximation, the basilar membrane response is described by the following linear equation:

$$y(t;s) = h(t;s) *_t x(t) \tag{1}$$

where $h(t;s)$ represents the linear impulse response of the cochlear filter at location s along the cochlea ($s = 0$ is the base, and $s > 0$ toward the apex), $y(t;s)$ represents the output of the filter at location s with input $x(t)$, and $*_t$ denotes the convolution operation with respect to time. Computational models sometimes include nonlinear effects as signal-dependent, time-varying gains or impulse responses (Lyon 1983; Slaney and Lyon 1993), but we omit such considerations here to allow a simpler description.

4.2 The Transduction Stage

The mechanical vibrations along the basilar membrane are transduced into electrical activity along a dense, topographically ordered, array of auditory nerve fibers (see Chapter 3 for more details). At each point, membrane displacements cause a local fluid flow that bends small filaments (cilia) that are attached to transduction cells, the inner hair cells (IHCs) (Shamma et al. 1986). The bending of the cilia controls the flow of ionic currents through nonlinear channels into the hair cells. The ionic flow, in turn, generates electrical receptor potentials across the hair cell membranes. The receptor potentials in the IHCs are often regarded as the first internal neural representation of sounds, but they are really only one link in a chain of events involving calcium currents and neuro-transmitter release that finally translates the representation into nerve spikes in cells of the spiral ganglion, whose axons form the auditory nerve (Deng, Geisler, and Greenberg 1988). The receptor signals are then coveyed by the auditory nerve fibers to the central auditory system. In the human auditory system, there are roughly 30,000 fibers, innervating approximately 3000 IHCs along the length of the cochlea (3.5 cm). While a fiber innervates only 1 IHC, several such fibers (typically about 10) may converge onto each IHC.

These three complex transduction stages—the fluid–cilia coupling, the ionic channels, and the receptor potentials—can be surprisingly well modeled by a three-step process (Fig. 6.7): a velocity coupling stage

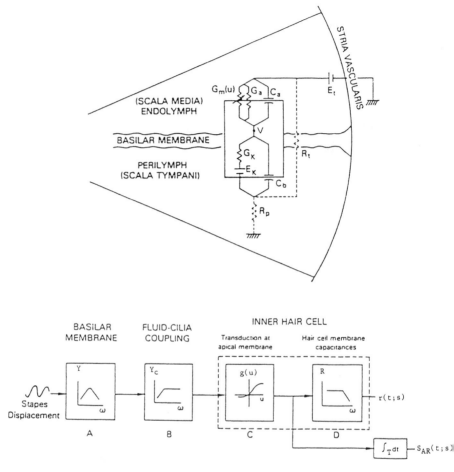

FIGURE 6.7. Inner hair cell stages. *Top*: Schematic diagram of biophysical model of the inner hair cell represents the equivalent electrical circuit of the hair cell and surrounding structures. E_t, endocochlear potential; R_t, R_p, epithelium resistances; E_K, potassium reversal potential; G_K, ionic channel conductance of the hair cell basal membrane; $G_m(y_c)$, mechanically sensitive conductance of the cilia; G_a, leakage conductance of the apical hair cell membranes; y_c, cilia displacement; r, hair cell intracellular potential. *Bottom*: Simplified diagram of the composite cochlear model. Uppercase letters signify the Fourier transforms of their corresponding (lowercase) quantities; ω is frequency.

(modeled by a leaky time derivative), an instantaneous nonlinearity modeling the opening and closing of the ionic channels, and a lowpass filter with a relatively short time constant (<0.2 msec) to describe the capacitive losses and ionic leakage through the hair cell membranes. Detailed considerations of the biophysical bases of these models can be found in Shamma et al. (1986) and Deng, Geisler, and Greenberg (1988). Additional stages involved in encoding receptor potential into spikes are also sometimes modeled, including additional onset-enhancing nonlinear mechanisms that we ignore in this chapter (Lyon 1983; Deng, Geisler, and Greenberg 1988).

In summary, the spatiotemporal patterns of basilar membrane vibrations, $y(t;s)$, are transduced into receptor potentials $r(t;s)$, as follows:

$$r(t;s) = g(y_c(t;s)) *_t w(t) \tag{2}$$

where $y_c(t;s) = y(t;s) *_t c(t)$ is the output of the fluid–cilia coupling, $c(t)$ is the impulse response of the cilia coupling stage, $w(t)$ is the impulse response of the lowpass filter (temporal smoothing window) from the hair cell membrane, and $g(\cdot)$ is an instantaneous nonlinearity that may take various forms depending on the emphasis of the model. For instance, a simple compressive nonlinearity can be

$$g(u) = \frac{1}{1 + e^{-\alpha.u}} \tag{3}$$

where α is the gain at the input of the nonlinearity, and the output varies from 0 to 1. The nonlinearity is sometimes simply modeled as a half-wave rectifier:

$$g(u) = \max(u,0) \tag{4}$$

or as a more complicated and realistic form including variable thresholds θ and saturation levels z:

$$g(u) = \frac{z}{1 + e^{-\alpha(u-\theta)}} \tag{5}$$

This form is also useful in the limit as $\alpha \to \infty$, which results in a threshold-clipping operation:

$$g(u) = \begin{matrix} 0 \text{ if } u < \theta \\ 1 \text{ if } u > \theta \end{matrix} \tag{6}$$

Therefore, the patterns at the output of this stage look similar to the basilar membrane vibrations except for being compressed or distorted by the nonlinearity. Furthermore, temporal fluctuations of $r(t;s)$ are limited to frequencies below 5 kHz because of the lowpass effect of the hair cell membranes. At higher frequencies, the receptor potential conveys primarily the envelope of the basilar membrane vibration at its input. For instance, in the case of a single tone, the receptor potentials are steady, and the frequency is represented by the place of maximal response.

To simplify the presentation in subsequent sections, we shall formally absorb the linear effects of the cilia coupling stage $c(t)$ into the basilar membrane impulse response. Thus, Eq. 2 becomes

$$r(t;s) = g(y(t;s)) *_t w(t) \qquad\qquad (7)$$

where $y_c(t;s)$ in Eq. 2 is replaced by the output of the basilar membrane $y(t;s)$, and the cilia coupling effects are assumed included in $h(t;s)$.

4.3 Stochastic Firings on the Auditory Nerve

The receptor potentials (or enhanced versions thereof) generated at the end of these stages are conveyed via the auditory nerve fibers to the cochlear nucleus, the first station of the central auditory system. This is achieved through a series of transformations in which the receptor potentials are converted into stochastic trains of electrical impulses (firings) on the auditory nerve, as mentioned earlier. Detailed biophysical models of these transformations can be found in Deng, Geisler, and Greenberg (1988) and Westerman and Smith (1984). More abstractly, the stochastic firings can be modeled as nonstationary point processes with instantaneous rates that approximately reflect the underlying receptor potentials (Siebert 1970; Johnson 1980). Recipient neurons in the cochlear nucleus may then reconstruct estimates of the receptor potentials by effectively computing the ensemble averages of activity in locally adjacent fibers (Shamma 1989).

From an information processing point of view, these complex transformations merely convey enhanced receptor potentials to the cochlear nucleus. Consequently, in a functional model they can all be bypassed. Such a simplifying view ignores two types of effects that have figured prominently in several classical models of auditory processing. The first is the adaptive mechanisms operative at the hair cell–auditory nerve junctions that might be important in describing the responses to the onset of sound (Miller and Sachs 1983). To a limited extent, the enhancing action of these mechanisms can be modeled by a linear stage and incorporated into the form of the $w(t)$ filter in Eq. 3.

The second simplification concerns the range of thresholds and spontaneous rates of firings observed in the responses of the auditory nerve. In most fibers (>85%), thresholds are lower than typical conversational sound levels, while their spontaneous rates are relatively high. These fibers exhibit a limited dynamic range between threshold and saturation (approximately 30 decibels [dB]), and are consistently found to be almost totally saturated when driven by broadband sounds (e.g., speech) at moderate levels (60–70 dB sound pressure level [SPL]). The remaining fibers exhibit low spontaneous rates, a wider range of thresholds, and sloping saturations in that they continue at high sound levels to increase their firing rates, but at a very slow pace. Consequently, they are less likely to be saturated at normal sound levels. To model these two populations, it is possible to use two or more nonlinearities $g_i(\cdot)$ with appropriately weighted outputs:

$$r(t;s) = \sum_{i=1,2} g_i(y(t;s)) *_t w(t) \qquad (8)$$

where $g_i(\cdot)$ is the effective instantaneous nonlinearity reflecting the two fiber populations.

4.4 Nonlinear Adaptation

We have ignored nonlinear adaptive effects in basilar membrane mechanics and in the IHC–auditory nerve interface for conceptual and mathematical simplicity. Nevertheless, it is important to consider adaptive effects in computational models of pitch and timbre perception (Lyon 1990). The wide dynamic range of sound signals must be compressed into a range that further auditory computations and representations can handle, without losing contrast between different time and frequency regions of the signal. In the process of adapting, or changing its gain, the auditory periphery also has a big onset-emphasis effect that is salient under some conditions, for example, in the sensitiviy of timbre perception to the parameters of a sound's onset (see Chapter 7 by Mellinger and Mont-Reynaud). Direct evidence for onset emphasis and adaptation effects are most clear in the auditory nerve representation of speech (Miller and Sachs 1983). Adaptation of basilar membrane mechanics to tone and click level is most clearly shown in laser-doppler velocimeter experiments (Ruggero and Rich 1991).

4.5 Examples of Cochlear Outputs

Figure 6.8 shows cochleagrams, or images of cochlear outputs, for a brief voiced speech sound and for a three-tone inharmonic complex. The pitch of the voiced speech is apparent as a clear repetition of a brief

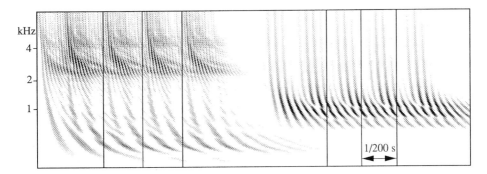

FIGURE 6.8. Detail of cochleagram shows fine temporal and spatial structure in model response to a vowel /i/ (*left*) and to an inharmonic tone complex of 700, 900, and 1100 Hz (*right*). Compare the patterns within the pairs of intervals indicated by the vertical lines.

spatiotemporal pattern; the brief pattern depends on the vowel, and may be thought of as the timbre of the sound. The three-tone complex also has a clear and nearly periodic spatiotemporal structure that corresponds to its several ambiguous pitches and a short-time characteristic pattern that corresponds to its timbre, but as shown in the figure, this inharmonic sound generates patterns that do not quite match.

Figure 6.9 compares a cochleagram with a more conventional representation, the spectrogram, using the same sounds as in Figure 6.8. The cochleagram has been smoothed and resampled at a lower rate in the time dimension to make it more comparable with the spectrograms. Both wideband and narrowband spectrograms are shown for comparison. It

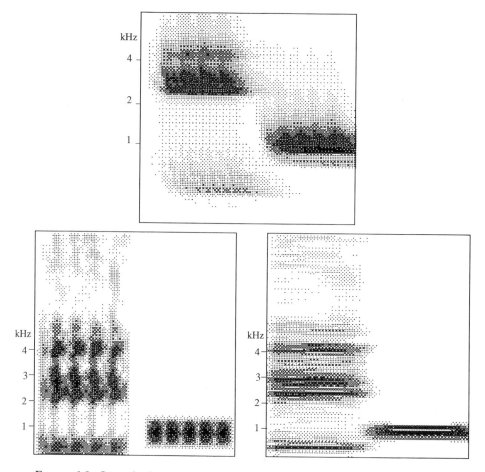

FIGURE 6.9. Smoothed cochleagram (*top*) and wideband and narrowband spectrograms (*bottom left and right*) of vowel /i/ segment followed by an inharmonic tone complex of 700, 900, 1100 Hz.

should be apparent that the spatial axis of the cochleagram is related roughly logarithmically to the frequency scale of the spectrogram. The cochleagram consequently has better spectral resolution at the low end and better temporal resolution at the high end, resulting in an excellent representation of all the features of both sounds compared to the spectrograms. Other advantages of the cochleagram representation depend on the fine temporal structure that has been smoothed away in Figure 6.9, as is explained in Section 5.

5. Representations of Timbre and Pitch in the Early Auditory System

The fundamental question addressed in this section is how the auditory system utilizes the patterns of auditory nerve activity $r(t;s)$ to generate stable and robust representations of timbre and pitch. There are two basic reasons why this question is at all interesting and also a difficult one to answer. The first is that there is a multiplicity of cues available on the auditory nerve, specifically the temporal and spatial cues discussed in Section 4, which make it possible to extract nearly equivalent timbre and pitch estimates in several different ways. The second reason is that the plausability of various algorithms is highly constrained by the biology. The combined impact of these two considerations is to inject considerable uncertainty into the debate on which computational algorithms and biological networks are really operative in the early stages of the auditory system.

Rather than enumerating and reviewing the multitude of such algorithms, we shall instead group them with respect to the nature of the auditory nerve cues that they utilize. This grouping permits us to emphasize the unifying theme that runs within each group and to isolate their inherent strengths and weaknesses independent of the details of the algorithms. Our chosen classification is (i) purely spatial algorithms (exemplified by the analysis in Sachs and Young 1979); (ii) purely temporal algorithms (as in Sinex and Geisler 1983; Gitza 1988); and (iii) mixed spatiotemporal algorithms (as in Shamma 1985b). The terms spatial and temporal here refer specifically to the axis against which the information cues are defined. Thus, spatial refers to the distribution of information along the tonotopic axis, whereas temporal refers to cues defined along the time axis. One fundamental implication of this distinction is that spatial algorithms require an ordered tonotopic axis for their representations. In contrast, temporal algorithms are not affected at all by a shuffling of the tonotopic axis.

The dichotomy between spatial and temporal theories of hearing has had a long history and has taken different forms over the last century. It is, perhaps, best manifested in the spectral versus temporal views of early auditory processing championed respectively by von Helmholtz (1863) and Ohm (1843) on the one hand, and by Wever and Bray (1930) on the other. The "spectral" camp argued that auditory perception is almost purely based on the amplitude or power spectrum of the signal and that therefore the auditory system makes no use of the temporal waveform of the acoustic signal. The "temporal" camp countered that the early auditory system is more like a telephone which simply relays the acoustic waveform to higher centers of the brain, a view supported by recordings of the cochlear microphonic (Wever and Bray 1930).

It is clear, as we discuss in detail in the following sections, that both views had merits. Furthermore, it is safe to assume on the basis of what little is known so far that the brain uses just about all the information available, using a range of different computational strategies. Thus, for understanding typical stimuli from isolated sound sources purely temporal or purely spatial approaches may be adequate. However, for real-world sounds in which features of sources may be interleaved in either or both frequency and time, more complex strategies will almost certainly be required to model human performance quantitatively.

5.1 Purely Spatial Representations

Purely spatial approaches are those in which the relevent timberal or pitch information are represented exclusively in the distribution of activity along the ordered tonotopic axis. This statement has different meanings and implications for timbre and pitch. For timbre, the input from which the percept is derived is the responses of the auditory nerve. Hence, the term spatial implies that the auditory system ignores completely the phase-locked (temporal) response patterns on the nerve. For pitch, the term spatial implies that the input from which the percept is derived is the spectral profile defined along the tonotopic axis of the cochlea, regardless of how this profile comes about. In this sense, it is exactly synonymous with what is traditionally called "spectral" or "pattern recognition" theories of pitch (Moore 1986, 1989).

5.1.1 Spatial Encoding of Timbre

In the purely spatial representations of timbre, no use is made of the fine temporal fluctuations in the hair cell potentials. Rather, the central auditory system is assumed to measure only the relatively long-term average of the response at each channel. In the formulation of Eq. 8, it is equivalent to replacing $w(t)$ by a relatively long time window (Π_T) to measure the average (rather than instantaneous) firing rate of each fiber. The spectral profile $S_{AR}(t;s)$ is then given by

$$S_{AR}(t;s) = g(y(t;s)) *_t \Pi_T = \int_{t-T}^{t} g(y(t;s)) \, dt \qquad (9)$$

where Π_T is taken to be a rectangular window of unit height and duration T. Because of the rectifying nonlinearity, the value of the profile at each point s can be viewed as roughly equivalent to the short-time average power of the time waveform $y(t;s)$ in that channel. Clearly, such a view entails a drastic simplification of how the spectral profile is encoded in the cochleogram, making it much more akin to a traditional spectrogram, but with roughly a logarithmic frequency axis and a frequency resolution determined by the widths of the cochlear filters.

In principle, this average rate representation can be sufficient for the encoding of the spectral envelope, or timbre. Potential problems appear, however, when one considers the limited dynamic range caused by the saturating characteristics of the nonlinearity $g(\cdot)$. As mentioned earlier, in most auditory nerve channels the average rate varies only over a 30- to 40-dB change in stimulus level (Fig. 6.10). Furthermore, despite an apparently wide range of thresholds, physiological experiments in various animals have consistently shown that more than 85% of the nerve fibers are firing at their maximum average rates at relatively moderate sound levels (60 dB SPL) (Sachs and Young 1979). This saturation in turn causes severe compression and deterioration in the clarity of the average rate representation of the spectral envelope. This effect was most elegantly demonstrated in recordings from large populations of auditory nerve fibers responding to speech sounds at various levels, as shown in Figure 6.11 (Sachs and Young 1979).

It is possible of course that an effectively wider dynamic range exists in humans or in awake-behaving animals where feedback from the central nervous system is still operative. However, no evidence yet exists to support this notion. Another possibility is that each of the different types of auditory nerve fibers effectively serves to preserve the average rate representation at a different range of stimulus levels (Winslow, Barta, and Sachs 1987). Such models require two important assumptions: (1) the different types of auditory nerve fibers innervate the dendritic trees of their target neurons in the cochlear nucleus in a specific spatially organized manner, which remains to be anatomically substantiated; and (2) the low spontaneous fibers, which constitute only 15% of auditory nerve fibers, are almost solely responsible for the representation of spectral profiles for moderate and high sounds levels (Viemeister 1988).

5.1.2 Spatial Encoding of Pitch

The term spatial with regards to pitch algorithms refers to algorithms that extract pitch from an explicit representation of the spectral profile along an ordered tonotopic axis, regardless of how this profile is computed. Thus, all pitch algorithms that take as input a spectral profile on a logarithmic frequency axis are considered spatial in character. All these "spectral" algorithms require that the input spectral profile be able to resolve (at least partially) several low-order harmonics, which are the

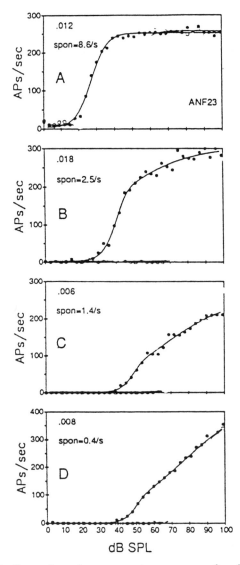

FIGURE 6.10A–D. Saturation of average-rate responses of auditory nerve fibers. Examples of different types of auditory nerve fiber rate-level functions. *Top*: (A,B) illustrate the limited dynamic range of responses (in terms of action potentials per second [APs/sec]) from high and medium spontaneous rate fibers, which together constitute more than 80% of all fibers. *Bottom*: (D) is an example of the remaining low spontaneous fibers, which have a broader dynamic range and a "sloping" saturation.

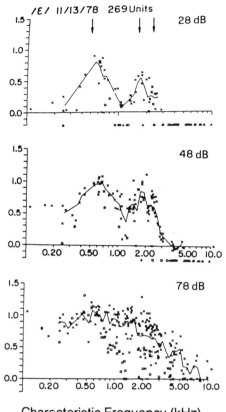

FIGURE 6.11. Profiles of the average firing rates of a large population of cat auditory nerve fibers in response to the vowel stimulus /ɛ/ at three different stimulus levels. The normalized firing rate of each fiber is plotted against the fiber CF. Note the saturation of the profiles and the loss of spectral detail with increasing stimulus level. (Adapted from Sachs and Young 1979.)

most important harmonics for pitch perception (Plomp 1976). In general, the cochlear filters are narrow enough to resolve partially at least up to the fourth or fifth harmonic of a complex periodic sound (Moore 1986). Therefore, spectral profiles derived using purely spatial cues (i.e., the average rate representation discussed previously) may in principle display these low harmonics. Spectral profiles computed by other (not purely spatial) algorithms (as discussed later) have in fact significantly better resolution and may be able to resolve as many as 30 harmonics.

There have been many hypotheses advanced to explain how pitch can be computed from resolved peaks in the spectral profile (Goldstein 1973; Wightman 1973; Terhardt 1979). A common denominator to all is a pattern-matching step in which the resolved component spectrum is compared to "internal" stored templates of various harmonic series in the central auditory system. This can be done explicitly on the spectral profile (Fig. 6.12), or on some transformation of the spectral profile, for example, its autocorrelation function (as in Wightman 1973). The perceived pitch is then taken to be the fundamental of the best matching template. Such schemes have been shown to predict well the perceived pitch values and saliency (Houtsma 1979). There are, however, two often-mentioned shortcomings of these algorithms. The first is biological, in that there is yet no evidence at all for the existance of anything resembling harmonic series templates (Javel and Mott 1988; Schwartz and Tomlinson 1990). The one rather specialized exception is the finding of so-called CF–CF

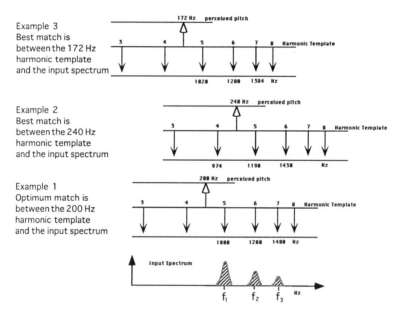

FIGURE 6.12. Pitch perception as a harmonic frequency pattern recognition. This schematic is of the model proposed by Goldstein (1973), which has a probablistic formulation reflecting the random variability in the estimates of the harmonic components at the auditory periphery. *Bottom traces* illustrate the probabilty distribution of frequency measurements of the harmonic complex (1, 1.2, and 1.4 kHz). It can roughly be viewed as the spatial profile of activity induced by a three-tone complex stimulus constituting the fifth, sixth, and seventh harmonics of 200 Hz against the tonotopic axis *Upper traces* are of three examples illustrating best matches of the harmonic frequency template to three different sets of component frequency estimates. The pitch is estimated when the best matching template position correctly aligns its harmonics with those of the stimulus (as in the third trace from the top). Large errors in the pitch estimate would occur if the harmonics are misaligned (as in the upper two traces).

combination-sensitive cells in the auditory cortex of the mustached bat (Suga et al. 1983). These cells respond best when two harmonics of a fundamental frequency (at approximately 30 kHz) are presented simultaneously to the animal.

The second problem for purely spatial pitch algorithms is that pitch percepts can be evoked by high-order harmonics that have been presumed not resolved spatially in the cochlea or elsewhere (Houtsma and Smurzynski 1990). In this case, it is necessary to postulate that higher spectral resolution exists than is generally assumed based on the critical band measures (Houtsma and Smurzynski 1990). This is not at all unrealistic because our ability to detect a change in the frequency of a single tone is an order of magnitude better than that implied by the critical band (1989), which implies that cues for fine frequency resolution are in principle available on the auditory nerve. In fact, spectral extraction algorithms utilizing temporal cues (see following sections) readily achieve this fine spectral resolution.

In summary, it is possible to encode the spectral envelope by purely spatial cues over the entire frequency range, provided saturation effects are not too severe. For the encoding of pitch, purely spatial approaches are only possible in the range where harmonic spectral components are resolved. As we shall see next, using purely temporal cues presents complementary limitations for the encoding of pitch and timbre in that pitch can theoretically be extracted from all places along the fiber array, whereas the spectral profile is available only at the low-CF fibers where phase-locking to the spectral components is present.

5.2 Purely Temporal Representations

In these schemes, the representation of pitch and timbre is derived exclusively from the phase-locked pattern of activity in each auditory nerve fiber. The critical assumption implied here is that the central auditory system makes no cross-channel comparisons and no use at all of the spatial organization or order of the fiber array, that is, the tonotopic axis.

The basic computational operation that all purely temporal algorithms share is some form of period or interval measurement on the responses of each channel, in effect duplicating neurally a frequency analysis similar to that already performed by the cochlea, but operating on the nonlinearly filtered and transduced signals. A variety of ways have been proposed for how this operation might be implemented in the central auditory system. Most involve multiplying a channel's response by progressively delayed versions of itself to form a running autocorrelation function as described in Figure 6.13 (Stevens, 1950).

If such an operation is performed on each spatial channel, the resulting representation has traditionally been called the duplex representation,

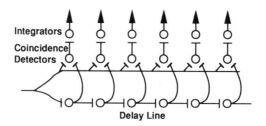

FIGURE 6.13. A simpler neural circuit proposed by Licklider for calculating a short-time autocorrelation function from neural spikes. Each output indicates the degree to which the input matches a delayed copy of itself, with delays being provided progressively by an axonal delay line or other delay mechanism.

and has more recently been referred to as the correlogram (Slaney and Lyon 1992). If the spatial order in this representation is not explicitly used in further processing, then the representation is essentially purely temporal.

Using the auditory nerve signal $r(t;s)$ from Eq. 8, the duplex representation parameterized over time, space, and delay is given by

$$A(t;\tau;s) = \{r(t - \tau;s) \cdot r(t;s)\} *_t \Pi_T = \int_{t-T}^{t} \{r(t - \tau;s) \cdot r(t;s)\} dt \quad (10)$$

where τ is the autocorrelation delay and T reflects the length of the (rectangular) window (Π_T) over which the correlation is computed (Π_T) generalizes to an arbitrary smoothing filter or window by using a slightly more complicated convolution integral). Note that while the hair cell nonlinearity $g(\cdot)$, whether it is compressive or a half-wave rectifier, affects the details of the results, it does not affect the global outcome significantly, as we illustrate later. This is because distorting the waveforms by $g(\cdot)$, for example, half-wave rectifying a locally sinusoidal response or severely compressing it into a square wave, does not affect the periodic character of the waveform (and hence of its autocorrelation function); rather, it adds distortion products that can be tolerated, made use of, or removed depending on the details of the algorithm.

Figure 6.14 shows a block diagram of the correlogram calculation, patterned on Licklider's original duplex theory proposal. Examples of the output of the calculation are shown with the discussion of pitch that follows.

For a more compact representation, the duplex representation may be collapsed across spatial channels by summing or integrating over the s dimension, with an optional nonlinearity, resulting in a short-time, purely temporal profile with no additional dimension:

$$A(t;\tau) = \int_{0}^{s} f(A(t;s;\tau)) ds \quad (11)$$

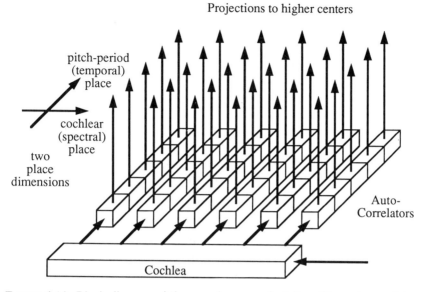

FIGURE 6.14. Block diagram of the correlogram calculation. The autocorrelators are based on delay lines, as shown in Figure 6.13.

If $r(t;s)$ is represented as a set of spikes or impulses, rather than as receptor potentials, then there is a nonzero contribution to the final profile only at delay values that correspond to intervals between spikes within individual channels. This simplification of the autocorrelation operation in terms of impulsive signals is the basis for a variety of interspike-interval and coincidence-detection models of pitch and timbre extraction. The collapsed, purely temporal forms of such discrete models are typically in terms of histograms of intervals or frequencies.

Physiologists have long used interspike-interval histograms (ISIHs) as a way to display the timing structure in auditory nerve responses (Kiang et al. 1965). They are typically constructed using first-order intervals (time from one spike to the next spike on the same neuron), rather than the more general case of all possible intervals within the set of neurons representing a frequency channel, as implied by the autocorrelation formulation. The first-order ISIH representation is quite useful in itself as a representation for timbre and pitch detection (Moore 1986), but including higher order intervals makes the representation more consistent across loudness changes and more suitable for longer pitch periods (Cariani and Delgutte 1992). A single-channel ISIH (either first- or higher order) is sometimes enough to indicate a pitch clearly, but a collection or sum of them across a wide range of frequency channels is needed to obtain information about timbre.

5.2.1 Temporal Encoding of Timbre

To compile an estimate of the spectrum of the stimulus, a purely temporal algorithm would interrogate each channel as to the local periodicities (or frequency components) represented in its temporal waveform, ignoring totally the identity of the underlying cochlear filter, that is, its center frequency (Delgutte 1984; Sinex and Geisler 1983; Gitza 1988; Seneff 1988). For instance, to estimate the amplitude of a 1-kHz component in the stimulus, this component is measured at each channel, and the results are further processed and then compiled in a histogram. One specific example of such an approach is the dominant frequency algorithm (Sinex and Geisler 1983). Here only the largest frequency component from each channel is selected and entered into a histogram of the number of fibers dominated by each frequency. Because of the extensive overlap of the cochlear filters, a frequency component with a large amplitude dominates the periodicity of the waveforms of a proportionately large number of channels, resulting in a large peak in the histogram (Fig. 6.15). A composite of all frequencies would result in a spectrum-like envelope of the histogram. Variants of this algorithm also exist using other response measures such as zero-crossing rates (Allen 1985) and interval measurements over channels with variable thresholds (Gitza 1988).

Another purely temporal representation of timbre can be derived from the duplex representation, or correlogram, after being collapsed across channels as in Eq. 6. This provides a temporally parameterized profile, similar to an autocorrelation function or cepstrum, that can be used directly to characterize timbre nonspectrally (Slaney and Lyon 1992). The delay axis should be as long as one cycle of the finest corresponding

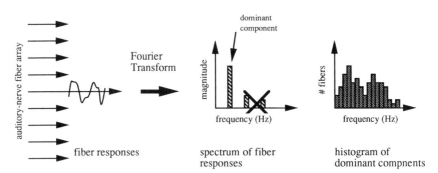

FIGURE 6.15. Schematic representation of the dominant frequency algorithm. The dominant periodicity in the responses of each auditory nerve fiber (independent of the others) is estimated by taking the Fourier transform of the poststimulus histogram and then recording only the frequency of the highest component. A histogram is then made of the dominant frequencies thus measured from all fibers.

spectral detail, which for speech should be typically less than one cycle of the highest voice pitch expected to not confound pitch and timbre. A more complex method of interpreting the duplex representation might possibly get around the need to choose such a window.

All such schemes have been shown to perform admirably well in extracting the acoustic spectrum from the responses of the auditory nerve and over wide dynamic ranges where most of the fibers are saturated in their average rates of firings. Clearly, these schemes cannot be useful for estimating the spectrum of high-frequency components ($>5\,$kHz) where phase-locking is essentially missing and $r(t;s)$ is nearly constant over a cycle. Instead, one has to rely on the spatial schemes discussed earlier.

Are such algorithms biologically plausible? A key operation in all such algorithms is the periodicity or interval measurement of the auditory nerve responses. Implementing this in a biological substrate requires a range of memories to combine waveform or spike values from different points in time. This can take the form of organized delay lines or time constants in biological membranes (as in Fig. 6.13). Such a substrate would have to exist early in the monaural auditory pathway because the phase-locking on which it depends deteriorates significantly in central auditory nuclei (Langner 1992). An analogous substrate for binaural cross-correlation is known in the medial superior olive (Yin and Kuwada 1984), but operates with maximum delays of only around 1 msec, which are too low for monaural pitch and timbre processing below 1 kHz. So far no physiological support of such a scheme exists in the cochlear nucleus, the terminal station of the auditory nerve. In the inferior colliculus, some evidence of organized latencies and other forms of delays have been reported in the bat and the cat (Schreiner and Langer 1988; Simmons, Moss, and Ferragamo 1990). However, if functionally useful, these delays would only be relevent for frequencies less than 800 Hz because no phase-locking to higher frequencies exists beyond the cochlear nucleus (Langner 1992).

5.2.2 Temporal Encoding of Pitch

Because the periodicites in the response waveforms of each channel are directly related to the overall repetitive structure of the stimulus, it is possible to make estimates of the perceived pitches directly from the peaks of the autocorrelation functions of the responses, as expressed for instance in a correlogram, or from the peaks of its collapsed profile (Eq. 11). Compared to the use of this profile as a description of timbre, as discussed earlier, using it for pitch involves longer delays and the information is more in the peak locations than in the shape.

Figure 6.16 shows three examples of steady sounds analyzed into the correlogram representation (the time dimension is not shown because the sounds are steady and there is no good way to show it). In each case, a

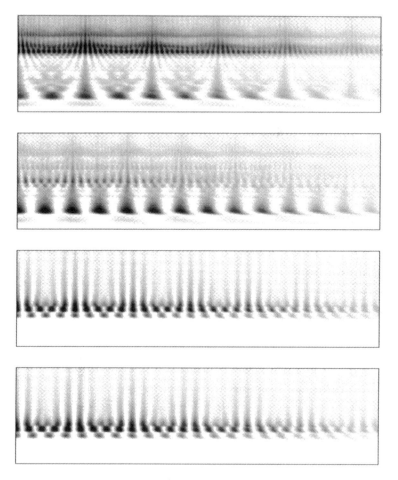

FIGURE 6.16A–D. Four examples (from top to bottom: A, /i/ vowel [ee]; B, /u/, [oo]; C, three-tone harmonic complex; D, three-tone inharmonic complex of steady sounds analyzed into the correlogram representation (the time dimension is not shown as the sounds are steady). In each case, a clear vertical structure or peak in each channel indicates the pitch period, which also shows up as a prominent peak in the summary correlogram.

clear vertical structure or peak in each channel indicates the pitch period, which also shows up as a prominent peak in the summary correlogram.

One example of a correlogram-based pitch detection algorithm is illustrated in Figure 6.17 (Slaney and Lyon 1990). Ignoring for now a step intended to enhance spatial consistency, the correlogram is simply summed across channels as shown in Eq. 6. At each time frame, the resulting profile as a function of delay, or pitch period, is then sharpened by adding in some of the profile from two and three periods of delay,

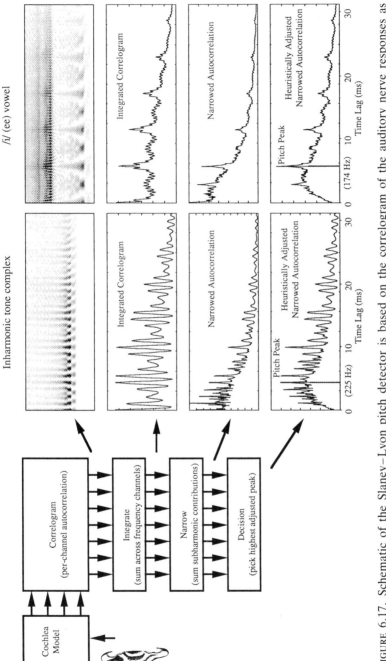

FIGURE 6.17. Schematic of the Slaney–Lyon pitch detector is based on the correlogram of the auditory nerve responses as described in the text.

as in the sharpened or narrowed autocorrelation function (Brown and Puckette 1989), to emphasize more periodic sounds over simple repetitions. This narrowing step is similar to the subharmonic analysis method or the pitch spiral (Patterson 1987). Finally, a peak from the profile is chosen as the pitch period. In a typical computational application, a pitch-tracking algorithm would help decide which of several candidate peaks to follow from frame to frame.

A similar algorithm, but using spike representations and without the narrowing step, has been implemented in a custom analog silicon chip (Lazzaro and Mead 1989) and has been shown to output a fair approximation to the pitch decisions found in psychoacoustic experiments in humans, in response to a range of sound stimuli such as sine waves, complex tones with missing fundamental, comb-filtered noise, and three-tone inharmonic complexes.

This same basic correlogram approach has been used as a method for the display of physiological data (Cariani and Delgutte 1992). By using higher order intervals, their method clearly shows long pitch periods and multiples of the period. Using spikes recorded from auditory nerve fibers at a variety of places or CFs, ISIHs are computed for each fiber and summed into a single autocorrelation-like profile. Stacking these profiles together across time produces a two-dimensional plot analogous to a spectrogram, but with a pitch period axis instead of a frequency axis, as shown in Figure 6.18. Plots from single neurons at various CFs show a clear representation of pitch, as does the sum across CFs.

The summation of intervals method is the basis for several temporal models of pitch perception (Moore 1989). In general, there is a good agreement between the most commonly occuring intervals and the most salient pitches. These approaches differ in their details; for example, one method (Langner 1992) uses oscillators instead of delay lines as a basis for interval measurement.

An alternate approach to the cross-channel aggregation process is to reduce each channel's autocorrelation function to a set of pitch period possibilities and then combine these, rather than combining the auto-correlation values across channels. If pitch value decisions are made independently on each spatial channel, they may still be collapsed by a histogramming or voting scheme into a purely temporal profile, and finally to a single pitch decision.

In summary, it is possible using purely temporal cues to extract per-ceptually accurate representations of pitch over all CFs because phase-locking to the pitch can occur even at high CFs where phase-locking to components near the CF may not exist. In contrast, for timbre representation, it is only possible to extract the spectral envelope in the lower frequency ranges (<3–4 kHz) where phase-locked responses on the auditory nerve are available. Note that this is exactly the opposite situation to that encountered using the purely spatial algorithms.

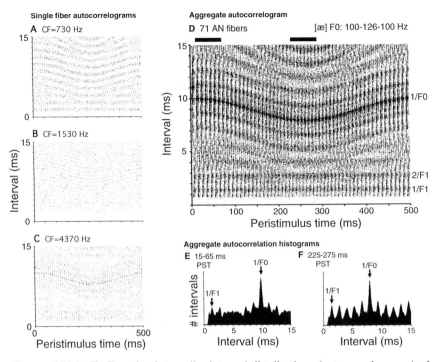

Single fiber autocorrelograms

A CF=730 Hz

B CF=1530 Hz

C CF=4370 Hz

Interval (ms)

Peristimulus time (ms)

Aggregate autocorrelogram

D 71 AN fibers   [æ] F0: 100-126-100 Hz

Interval (ms)

1/F0

2/F1
1/F1

Peristimulus time (ms)

Aggregate autocorrelation histograms

E 15-65 ms   **F** 225-275 ms
PST   PST

intervals

1/F1   1/F0   1/F1   1/F0

Interval (ms)   Interval (ms)

FIGURE 6.18A–F. Running interspike-interval distributions (autocorrelograms) of the auditory nerve for the vowel [ae]. Stimulus, five-format synthetic vowel [ae], as in "had." Formants are at 750, 1450, 2450, 3350, and 3850 Hz; (FO) sinusoidally varies between 100 and 126 Hz. (A–C) Single-fiber autocorrelograms for 100 stimulus repetitions presented at 60 dB SPL (sound pressure level). Each *dot* represents the occurence of one interspike interval of a given length (0–15 msec) ending at a given peristimulus time (0–500 msec). Intervals between both successive and nonsuccessive spikes are included. (A) Unit 28-1, CF = 732 Hz (first formant region), threshold = 2.7 dB SPL, SR = 72 spikes/sec. (B) Unit 27-25, CF = 1531 Hz (second formant region), threshold = 13.4 dB SPL, SR-29.4 spokes/sec. (C) Unit 28-43, CF = 4365, threshold = 1.3 dB SPL, SR = 91 spikes/sec. (D) Aggregate autocorrelogram computed by summing the single-fiber autocorrelograms of 71 fibers across many different CFs. Each *dot* represents 10 or more intervals; *thin line* indicates the fundamental period, 1/FO, as function of peristimulus time. E and F) Aggregate autocorrelation histograms for the two simulus segments are cross sections of autocorrelogram D for the peristimulus regions indicated by the two bars (15–65 msec and 225–275 msec).

5.3 Mixed Spatiotemporal Representations

Because of the evident redundancy of spatial and temporal cues in the response patterns on the auditory nerve, it is possible to conceive of algorithms that utilize various combinations of these cues to generate the timbre and pitch representations. A distinguishing feature of all such algorithms, as we elaborate later, is that they make use of phase-locking indirectly via the spatial cues it creates. There are a few threads common to all spatiotemporal algorithms: (1) the tonotopic axis of the cochlea is necessary for the representation; (2) information conveyed by phase-locking is essential; and (3) use is made of specific properties of the cochlear filters, such as their center frequencies, bandwidths, or overall shapes. Apart from these broad principles, it is difficult to describe these algorithms in general because they can be very different in spirit and detail.

5.3.1 Spatiotemporal Encoding of Timbre

The earliest example of such representations is the so-called average localized synchronous rate (ALSR) (Young and Sachs 1979). The fundamental idea here is to feed the auditory nerve fiber array into a bank of narrowly tuned bandpass filters, where the center frequency of each is set equal to the CF of the fiber at its input (Fig. 6.19). In essence, these

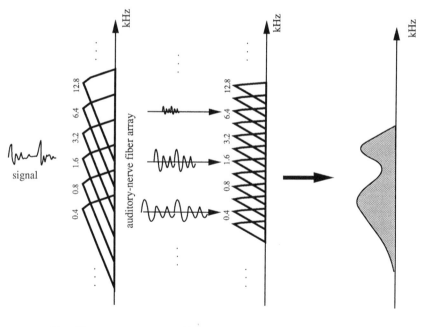

cochlear filters   ALSR bandpass filters   ALSR spectrum

FIGURE 6.19. A schematic of the operations implied by the average localized synchronous rate (ALSR) algorithm used by Young and Sachs (1979). The auditory nerve responses of each fiber are bandpass filtered by a filter whose center frequency is cued to the CF of the fiber. The output of the filter array is then plotted against the tonotopic axis to give the spectrum of the stimulus.

filters are cascaded with the broad cochlear filters to create an overall narrowband frequency analysis, as expressed mathematically by

$$S_{ALSR} = \langle r(t;s) *_t b(t;s) \rangle_T = \langle g(y) *_t b(t;s) *_t w(t) \rangle_T \qquad (12)$$

where $\langle \cdot \rangle_T$ denotes the short-time average power in the interval T, and $b(t;s)$ is the impulse response of the narrow bandpass filter at location s. Therefore, one obtains at the output of this bank a measure of the spectrum of the signal, represented against the spatial (CF) axis of the cochlea. This is immediately apparent in the case of a linear hair cell, that is, $g(y(t;s)) \approx y(t;s) = x(t) *_t h(t;s)$, because S_{ALSR} becomes (ignoring the w term):

$$S_{ALSR} = \langle y(t;s) *_t b(t;s) \rangle_T = \left\langle x(t) *_t \underbrace{h(t;s) *_t b(t;s)}_{\text{effective filter}} \right\rangle_T \qquad (13)$$

Clearly, this algorithm uses the phase-locked responses to estimate the spectral power at each cochlear output, and as such 'fails to generate spectral estimates at high CFs where no phase-locking exists. Nevertheless, a fundamental distinction between it and purely temporal algorithms is that the ALSR filter discards all temporal information outside a band around its CF; that is, it is cued to the cochlear spatial axis. As we discuss next, it is possible via certain spatial opertations to implement this algorithm exactly in the low-CF region and to extend it to the high-CF region.

The LIN algorithm (Shamma 1985b) is a different approach modeled after the function of the (L)ateral (I)nhibitory (N)etworks, which are well known in the visual literature. In the retina, this network enhances the representation of edges and peaks and other regions in the image that are characterized by fast transitions in light intensity (Hartline 1974). In audition, the same network can extract the spectral profile of the stimulus by rapidly detecting edges in the patterns of activity across the auditory nerve fiber array (Shamma 1985a, 1989).

Two reasons motivated the development of the LIN algorithm. The first is the failure of the purely spatial and purely temporal algorithms to utilize the cochlear cues over all CFs. Specifically, purely spatial cues are lost in the lower CF regions because of the hair cell saturating nonlinearity, while purely temporal cues, or phase-locking, are lost in the higher CF regions because of the hair cell lowpass filter. The LIN unifies the representation for all CFs by applying spatial operations that can utilize the temporal cues at low CFs and, at the same time, be compatible with the spatial cues at high CFs. The second important advantage of the LIN is its simplicity and biological plausibility. The LIN does not perform any absolute periodicity or interval measurements and hence requires no delay lines. Instead, as we elaborate later, it uses the spatial cues created by the phase-locked responses to make its spectral estimates.

To see how this comes about, and how the LIN in effect performs the ALSR algorithm in the low-CF regions, we first formulate mathematically the operations that the LIN applies to the auditory nerve patterns. The simplest and most basic function of the LIN is to compute a derivative of the auditory nerve responses with respect to the spatial axis of the cochlea; this is why the LIN is sensitivite to spatial discontinuities in its input patterns:

$$L(t;s) = \partial_s r(t;s) = \partial_s g(y(t;s)) *_t w(t)$$
$$= \{g'(y(t;s)) \cdot \partial_s y(t;s)\} *_t w(t) \qquad (14)$$

where $g'(\cdot)$ is the derivative of the hair cell nonlinearity with respect to its argument and $\partial_s y(t;s)$ is the partial spatial derivative of the basilar membrane patterns. More realistically, this derivative is only an approximation; that is, it is likely accompanied by local smoothing because of the finite spatial extent of the lateral interactions or the convergence of input fibers (Shamma 1989). These and other effects can be formally combined in the output of this stage by an extra spatial term, $v(s)$:

$$L(t;s) = \{g'(y(t;s)) \cdot \partial_s y(t;s)\} *_t w(t) *_s v(s) \qquad (15)$$

where $*_s$ denotes the convolution operation with respect to space and $v(s)$ is a spatial term reflecting the deviation of the LIN from a pure derivative.

The final spectral profile is considered to be the spatial distribution of the short-term average power of the LIN outputs:

$$S_{LIN}(t;s) = \langle L(t;s) \rangle_T \qquad (16)$$

One biologically plausible way to compute this is to half-wave rectify and then integrate the LIN output over some short interval T. Examples of the LIN spectral profile for a speech signal are shown in Figure 6.20.

To see theoretically how this spectral estimate arises from the LIN operations, assume first that the hair cell is linear (i.e., $g'(\cdot) = 1$). Then one can deduce immediately from Eq. 15, by ignoring the effects of the two smoothing terms [$w(t)$ and $v(s)$], that S_{LIN} is given by

---

FIGURE 6.20. Schematic of the lateral inhibitory networks (LIN) processing stages and examples of its outputs. *Top:* the LIN receives a tonotopically ordered input from the auditory nerve. Mutually (or laterally) inhibited neurons effectively perform a derivative operation. The outputs from all LIN neurons against the tonotopic axis provide the spectral estimate. *Bottom:* examples of LIN outputs for the word / magnanimous / (*left*) and a series of vowels (*right*). The peaks reflect the perceptually significant features (e.g., formants) in the stimulus spectrum representing different phonemes. These outputs are computed from heavily saturated inputs, demonstrating that the LIN outputs are not sensitive to the overall level of the stimulus.

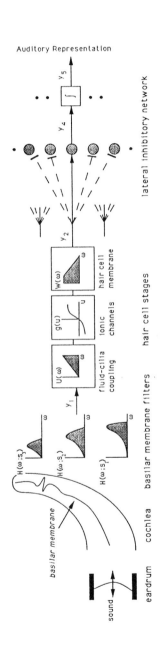

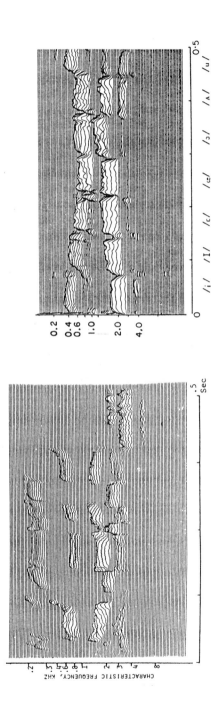

$$S_{LIN}(t;s) = (\partial_s y(t;s))_T = \left\langle x(t) *_t \underbrace{\partial_s h(t;s)}_{\text{effective filter}} \right\rangle_T \qquad (17)$$

In other words, the final LIN spectral profile is effectively derived by passing the input acoustic signal through a bank of filters $\partial_s h(t;s)$. These filters are simply the spatial derivatives of the cochlear filters. Because the cochlear filters are quite asymmetrical with steep roll-offs on their apical side (see Section 4), their derivatives, and hence the effective analysis filters, are relatively narrow (Yang, Wang, and Shamma 1992; Wang and Shamma 1994). The similarity of the LIN and the ALSR estimates is clearly seen when comparing Eqs. 13 and 17. The fundamental difference between the two schemes is in the origin of the narrowband filters. For the LIN, they arise as a direct consequence of the unique shape and characteristics of the cochlear filters and are derived via purely spatial operations.

In the high-CF region, phase-locking to high frequencies deteriorates because of the attenuation of the lowpass filter $w(t)$. The ALSR estimates cannot recover the lost information regardless of the form of the non-linearity $g(\cdot)$, and must give up the narrowband filters there to generate the appropriate estimates (e.g., as in the purely spatial algorithms). In contrast, the LIN estimates are unaffected by the loss of phase-locking because the spatial derivative can be equally and usefully applied to a temporally steady spatial pattern (as in the case of the retina). In other words, the final LIN spectral profiles in the high-CF regions become simply enhanced versions of the profile of the average-rate responses of the auditory nerve.

Finally, what is the effect of a saturating nonlinearity on these spatio-temporal algorithms? In the purely temporal algorithms, the relative amplitude of the different components of a complex spectrum are deduced, in effect, from the numbers of auditory channels they dominate (e.g., Sinex and Geisler 1983). For the ALSR and the LIN, a similar relative measure is preserved because saturating an auditory channel with a compound waveform from two nearby frequency components preserves to first order the relative amplitude of these components in the waveform (Shamma and Morrish 1986; Wang and Shamma 1994). Thus, filtering the saturated waveform with a bank of narrowband fiters produces accurate relative estimates of the spectrum. In the case of the LIN, this can be seen directly from Eq. 15 by assuming the nonlinearity $g(\cdot)$ to be a sigmoidal function that is strongly driven. In this case, $g(\cdot)$ is assumed a Heaviside function and $g'(\cdot)$ is a Dirac delta function. Therefore, $r(t;s)$ becomes:

$$r(t;s) = \{\delta(y(t;s)) \cdot \partial_s y(t;s)\} *_t w(t) *_s v(s) \qquad (18)$$

Because the width (or support) of delta function contains only the origin, then the term in the curly brackets can be interpreted as a sampling process of the quantity $\partial_s y(t;s)$, and the sampling instants are determined by the zero crossings of $y(t;s)$. In other words, saturating the channels is theoretically equivalent to converting the continuous responses into sampled ones, with minimal loss of information. (See Yang, Wang, and Shamma 1992) for a detailed analysis of these equations and their implications.)

As to the biological plausibility of the LIN, the most likely candidate at present to perform these operations is the population of so-called T-type stellate cells of the anteroventral cochlear nucleus (AVCN) (Sachs and Blackburn 1991). These cells exhibit the necessary lateral inhibitory interactions. Furthermore, the spatial pattern of their average-rate responses displays clear and stable representation of the acoustic spectral profile at all stimulus levels. That is, if phase-locked responses are used to convey spectral information, then it is here where the necessary time-to-place transformations must occur. It is uncertain, however, how the inhibition in these cells arises and whether it is fast enough (time constants of <0.5 msec) to carry out the computations implied by the LIN (Shamma 1989).

To summarize, the LIN derives its estimate of the spectral envelope by detecting spatiotemporal discontinuities in the responses on the auditory nerve. These discontinuities are caused by instantaneous mismatches in the time waveforms in different channels because of different frequencies, phases, or amplitudes (Shamma 1985b). Although implemented here as a spatial derivative, the LIN can be similarly described using other spatial operations, for example, a spatial multiplicative correlation between neighboring channels (Deng, Geisler, and Greenberg 1988).

5.3.2 Spatiotemporal Encoding of Pitch

Spatiotemporal representations of pitch are those that utilize directly the fine temporal structure of responses on the auditory nerve and do so via spatial operations that require an ordered tonotopic axis. In contrast, all pitch representations that had a spectral profile as their starting point were defined as purely spatial in character (regardless of how this profile is derived initially from the auditory nerve responses). Similarly, all representations that utilized the phase-locked responses of the auditory nerve in a manner that is insensitive to the tontopoic order were defined as purely temporal. Very few truly spatiotemporal algorithms have been suggested for the encoding and extraction of pitch. This is perhaps because of the success of the purely spatial and purely temporal algorithms and their extensions in accounting for most relevant pitch phenomena and the relative sparsity of physiological data regarding the mechanisms underlying pitch perception in general.

As is the case for timbre, there are various ways in which spatiotemporal representations of pitch can be constructed. One may for instance simply insert a spatial operation into an essentially purely temporal algorithm. For example, in a variation on the correlogram-based pitch detector discussed earlier (Slaney and Lyon 1990), the duplex representation may be smoothed across channels to enhance consistent pitches in nearby channels before being squared and collapsed along its spatial axis. Schemes such as this one, which use spatial coherence in combination with nonlinear filtering, are strictly spatiotemporal but the resulting one-dimensional pitch period profile is qualitatively similar to the purely temporal result. We consider this approach to be only weakly spatiotemporal.

A more radical spatiotemporal scheme involves the use of cross-correlations of phase-locked responses from different cochlear outputs. This approach is outlined here primarily for didactic reasons, that is, to illustrate what is meant by a more strongly spatiotemporal algorithm in the context of pitch.

In all the purely temporal representations (such as the correlogram), pitch was extracted from some form of interval or periodicity measurement on the responses of each auditory channel. To accomplish this, delay lines or intrinsic oscillators are needed early in the monaural auditory pathway. As the existence of such a biological substrate is still in doubt, one may question if such neural delays are actually necessary because the cochlea itself acts as a mechanical delay line. Thus, to compute the autocorrelation function of the response waveform in a given channel, it may be sufficient to cross-correlate the response with those of neighboring channels, which may look substantially similar except for a phase delay caused by the finite velocity of the traveling waves.

Detailed examples of such algorithms can be found in Loeb, White, and Merzenich (1983) and Shamma, Shen, and Gopalaswamy (1989), mostly in the context of binaural hearing problems and the formation of spectral profiles. From a conceptual viewpoint, this type of algorithm is very similar to the LIN for timbre in its use of phase-locked responses indirectly via the spatial features they create when they are mismatched in phase across different channels. As far as we know, however, there has not been a serious theoretical examination of the potential of such a representation in describing the pitch of complex sounds.

Fast software and hardware simulations of these spatiotemporal representations have become available (Mead, Arreguit, and Lazzaro 1991). Specifically, an efficient analog (V)ery (L)arge (S)cale (I)ntegration (VLSI) model of the so-called stereausis network (Shamma et al. 1989) has been constructed with the view of providing an early combined representation for at least three major auditory tasks: perception of timbre, perception of pitch, and binaural localization (Fig. 6.21).

6. Representations of Timbre and Pitch in the Central Auditory System

In this section, we discuss the question of how spectral profiles and pitch maps are further elaborated in the central stages of the auditory system. The importance of this questions stems from the realization that while a spectral profile for instance directly underlies our perception of timbre, this does not in itself reveal the perceptually most significant features of this spectrum, nor does it tell us how to compare (rank or measure the distance between) two spectral profiles. To address these issues, one needs to search for further transformations of the profile representation in higher auditory centers.

An important distinction between early and central auditory representations is the progressive decline of the rate of phase-locking and hence of the temporal cues. For the representation of timbre and pitch, the language of the central auditory system is, consequently, to a large extent spatial. One correlate of this statement is that important stimulus parameters are likely to be spatially organized, much like the tonotopic organization of the cochlea, which is preserved in the auditory pathway up to the auditory cortex. Therefore, the question we briefly address here is what central organizational maps (beyond the tonotopic map) are known to exist, and what computational schemes do these maps imply for the perception of timbre and pitch?

6.1 Representation of Spectral Profiles in the Primary Auditory Cortex

As mentioned earlier, an explicit representation of the spectral profile does exist at the level of the AVCN (Sachs and Blackburn 1991). This pattern is presumably relayed to more central stations, such as the lateral limniscus nuclii, the inferior colliculus, the medial geniculate body, and finally to the primary auditory cortex (AI) and other cortical fields. While much is known about the response characteristics of cells in all these structures, little can be said with certainty about their overall functional organization and the general principles they might apply to the processing and representation of spectral profiles. The one striking exception (to which we shall return later) is the elaborate representation of various stimulus parameters found at all central levels of the mustached bat auditory system (Suga 1984).

In AI, the essentially one-dimensional tonotopic axis of the cochlea is expanded into a two-dimensional sheet, with each frequency represented by an entire sheet of cells, as illustrated by the schematic of Figure 6.22. An immediate question thus arises as to the functional purpose of this

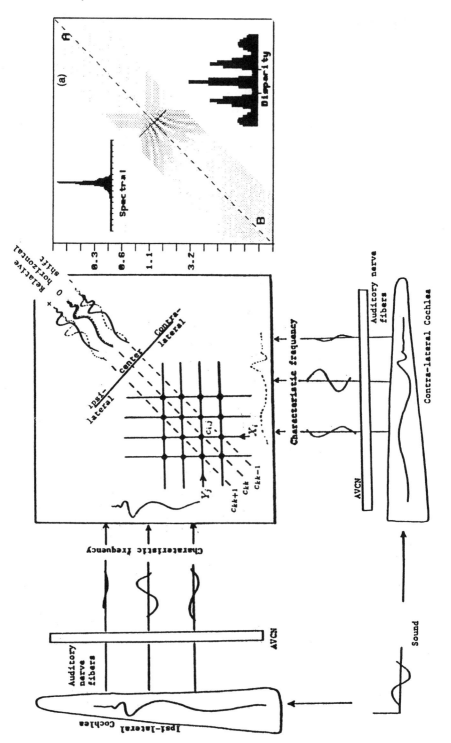

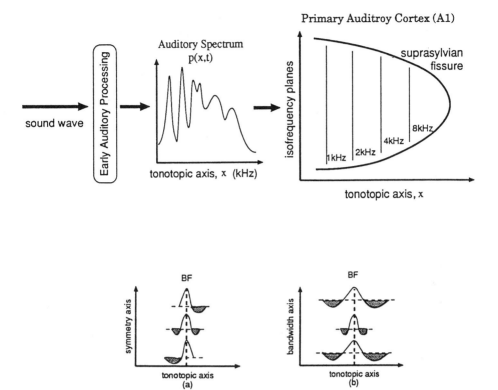

FIGURE 6.22. Schematic of functional organization in primary auditory cortex. *Top* A unidimensional tonotopic axis is expanded to a two-dimensional sheet of cells in AI. *Bottom*: panels representing schematic organization of the response areas in the primary auditory cortex. Left: the response areas vary in their bandwidth along the isofrequency planes of AI. *Right*: Response areas change in their asymmetry along the isofrequency planes.

FIGURE 6.21. Schematic of the stereausis network for cross-correlating the cochlear outputs. The two input arrays can either be from the same cochlea, or from the two cochlei (for binaural inputs), as illustrated in the figure. Along the diagonal, the cells correlate outputs from corresponding CFs. Off the diagonal, outputs of progressively misaligned CF locations are cross correlated. Output patterns computed for a tonal stimulus (1 kHz) are shown to the right of the network schematic. A large peak along the diagonal occurs at the CF location of the tone (1 kHz). Secondary peaks occur along the perpendicular axis around this CF. They are spaced at spatial intervals reflecting the period of the tone (1 msec). This kind of measurement is what is needed for many of the temporally based algorithms of pitch discussed earlier.

expansion and the nature of the spectral features that might be mapped along these isofrequency planes. These questions are addressed by data from physiological mapping experiments in AI, complemented by psychoacoustical studies, which reveal a substantial transformation of the way that the spectral profile is represented centrally.

Specifically, two response features are found that have immediate implications for the representation of spectral shape (and hence timbre). (i) The cell response areas, that is, the excitatory and inhibitory responses they exhibit to a tone of various frequencies and intensities, change their bandwidth in an orderly fashion along the isofrequency planes. Near the center of AI, cells are narrowly tuned. Towards the edges, they become more broadly tuned. This orderly progression occurs at least twice, and it correlates with several other response parameters such as increasing response thresholds toward the edges (Schreiner and Mendelson 1990). The functional implications of this map are uncertain. One intuitively appealing possibility is that response areas of different bandwidths are selective to spectral profiles of different widths. Thus, broad spectral profiles (e.g., broad formant peaks or gross trends such as spectral tilts caused by preemphasis) would drive cells with wide response areas best. Similarly, narrower spectral profiles (e.g., sharp peaks or edges, or fine details of the spectral profile) would be best represented in the responses of cells with more compact response areas. In effect, having a range of response areas at different widths allows us to encode the spectral profile at different scales or levels of detail (resolution). From a mathematical perspective, this is basically equivalent to analyzing the spectral profile into different scales or "bands," much like performing a Fourier transform of the profile. Coarser scales then correspond to the outputs of the "low-resolution" bands, while finer scales correspond to the outputs of the "high-resolution" bands (Shamma Vranic and Versnel in press). Such an analysis of the spectral profile has recently gained significant physiological support by the discovery that cells in AI are indeed tuned to different scales (Shamma, Versnel, and Kowalski in press). Furthermore, these response characteristics seem to be organized topographically across the AI (Versnel, Kowalski, and Shamma in press). A similar organization has been thoroughly studied for the receptive fields of the visual cortex (de Valois and de Valois 1988), where it is known as the spatial frequency analysis theory.

(ii) The response areas exhibit systematic changes in the symmetry of their inhibitory response areas. For instance, cells in the center of AI have sharply tuned excitatory responses around a best frequency (BF), flanked by symmetric inhibitory response areas. Toward the edges, the inhibitory response areas become significantly more asymmetrical, with inhibition dominated by frequencies either higher or lower than the BF. This trend is repeated at least twice across the length of the isofrequency plane (Shamma et al. 1993). It is intuitively clear that response areas with different symmetries would respond best to input profiles that match their symmetry. For instance, an odd-symmetrical response area would respond best if the input profile had the same local odd-symmetry, and worst if it had the opposite odd-symmetry. As such, one can state that a range of response areas of different areas is capable of encoding a measure of the relative evenness and oddness of a local region in the profile. From an opposite perspective, it can be shown mathematically that the local symmetry of a pattern can be changed by manipulating only the phase of its Fourier transform (Shamma, Versnel, and Kowalski 1994; Shamma Vranic and Versnel in press). Therefore, the axis of response area symmetries in effect is able to encode the phase of the profile transform, thus providing a complementary description to that of the magnitude along the scale axis discussed previously.

Mathematical models of these ideas abound in the vision perception literature and have proven to be extremely valuable both in interpreting psychophysical and physiological experiments and in engineering applications of these ideas in image processing (de Valois and de Valois 1988). No such attempts have as yet been applied in explaining higher auditory percepts such as timbre, pitch, or binaural phenomena. Interestingly, something akin to these ideas has been known for several decades as homomorphic processing, and it involves computations of Fourier-like coefficients of the acoustic spectrum called cepstral coefficients (Rabiner and Schafter 1978). While quite different in their detail, the cepstral coefficients in a global sense encode similar types of information about the shape of the spectral profile, and they have been universally found to be very useful in such engineering applications as automatic speech recognition systems.

Finally, there is an alternative representation of the spectral profile that has been hypothesized to exist in AI (Suga and Manabe 1982; Phillips and Orman 1984). It is based on the fact that cortical (and many other central auditory cells) often exhibit strongly nonmonotonic responses as a function of stimulus intensity. In a sense, one can view such a cell's response as being selective to (or encoding) a particular intensity. Consequently, a population of such cells, tuned to different frequencies and intensities, may provide an explicit representation of the spectral profile by their spatial pattern of activity. This scheme is not a true transformation of the spectral features represented, but rather is strictly a change in the means of the representation from the rate code, as in the T cells of the AVCN (Sachs and Blackburn 1991), to a place code. The most compelling example of such a representation is seen in the (D)oppler (S)hifted (C)onstant (F)requency (DSCF) area of AI in the mustache bat (Suga and Manabe 1982). In this animal, there is a hyperrepresentation of the frequency 61 kHz in a large isofrequency plane that is used for target ranging. The amplitude at this frequency seems to be mapped explicitly in AI by an array of intensity-tuned cells. However, an extension of this representation to multicomponent stimuli has not yet been found in any species.

6.2 Representation of Pitch in the Central Auditory System

In all pitch algorithms described earlier in Section 5, the extracted pitch is assumed implicitly or explicitly to be finally representable as a spatial map. Thus, in the purely temporal and the spatiotemporal algorithms, phase-locked information on the auditory nerve is used before it deteriorates in its journey through multiple synapses to higher centers of the brain. In fact, many studies have confirmed that synchrony to the repetitive features of a stimulus, be it the waveform of a tone or its AM modulations, becomes progressively worse toward the cortex. For instance, although maximum synchronized rates in the cochlear nucleus cells can be as high as in the auditory nerve (4 kHz), they rarely exceed 800–1000 Hz in the inferior colliculus (Langner 1992), and are less than 100 Hz in the anterior auditory cortical field (Schreiner and Urbas 1988). Therefore, it seems inescapable that pitch be represented by a spatial map in higher auditory centers. So, does such a map exist?

It is a remarkable aspect of pitch that, despite its fundamental and ubiquitous role in auditory perception, only two types of reports exist of physiological evidence of a spatial pitch map, and neither has been independently confirmed. The first are obtained in human subjects using nuclear magnetic resonance (NMR) scans of the primary auditory cortex (e.g., Langner 1993). Basically, the data show that low-BF cells in AI can be activated equally by a tone at the BF of these cells, or by higher order harmonics of this tone. As such, it is inferred that the tonotopic axis of the AI, or at least its lower BFs, essentially represents the frequency of the "missing fundamental" in addition to the pitch of a simple tone.

Attempts at confirming these results using higher resolution single and multiunit recordings in animals have unequivocally failed (Schwartz and Tomlinson 1990). Thus, the NMR results must be seen either as experimental artifacts or that coding of pitch in humans is different than in primates and other mammals in some basic way. The second evidence of spatial pitch maps was reported by Schreiner and Langner (1988) in the inferior colliculus of the cat. Using amplitude modulated tones, they found an explicit mapping of the best modulation frequencies (BMFs) along the isofrequency planes. The spatial map is somewhat difficult to tease out because the range of BMF's in each isofrequency plane depended on the CF. Nevertheless, this detailed report supports the hypothesis that pitch is extracted as early as the inferior colliculus. Furthermore the fine structure of the maps is in harmony with many of the predictions that follow from the purely temporal algorithms discussed earlier. These results have not yet been duplicated in other mammals or by other groups. More problematic, however, is the lack of reliable evidence of the existance of these maps in the cortex.

The absence so far of spatial maps of pitch in the cortex may be because the maps sought are not at all as straightforward as we imagine. For instance, it is conceivable that a spatial map of pitch can be derived from the cortical representation of the spectral profile discussed in the preceeding section (Wang and Shamma in press). In this case, no simple explicit mapping of the BMFs would be found. Rather, pitch will be represented in terms of more complicated spatially distributed patterns of activity in the cortex. Of course, it is also possible that pitch maps may not exist beyond the inferior colliculus. This possibility is counterintuitive given the results of ablation studies showing that bilateral cortical leasions in the auditory cortex severely impair the perception of pitch of complex sounds (Whitfield 1980), but does not affect the fine discrimination of frequency and intensity of simple tones (Neff, Diamond, and Casseday 1975).

7. Future Directions

Timbre and pitch are two fundamental attributes of any sound signal. Only together with other attributes, such as loudness, localization in space, and onset and offset characteristics, however, does a sound achieve its unitary identity and be perceived as eminating from a particular source. As such, pitch and timbre serve the parallel purposes of characterizing a sound signal and segregating its source from the multitude of other sources that are likely to be active in any natural environment. Research to date on pitch and timbre perception has focused heavily on the first aspect of the problem, that of describing the auditory representations of timbre and pitch of single sound sources. The issues, approaches, and representations discussed in this chapter all essentially fall in this category. Therefore, a key future direction in the physiology, phychoacoustics, and modeling of pitch and timbre is the understanding of how pitch and timbre cues are extracted from sound mixtures and applied to the separation and understanding of such mixtures. These computational studies must remain within the constraints of known or plausible physiology to be useful.

List of Abbreviations

Acronym		Section in Which Introduced
Q	tuning factor of a filter	3
CF	characteristic frequency	4.1

BM	basilar membrane	4.1
IHC	inner hair cells	4.2
ALSR	average localized synchronous rate	5.3
LIN	lateral inhibitory networks	5.3
AVCN	anteroventral cochlear necleus	5.3.1
AI	primary auditory cortex	6.1
BF	best frequency	6.1

Mathematical Symbols

t	time axis	
s	spatial axis along the length of the cochlea	
$h(t;s)$	impulse response of the basilar membrane at location s	4.1
$x(t)$	input acoustic signal	4.1
$y(t;s)$	output of the cochlear filter from acoustic input $x(t)$	4.1
$*t$	convolution operation with respect to time	4.1
$c(t)$	impulse response of the fluid–cilia coupling stage	4.2
$y_c(t;s)$	cilia displacements at location s from cochlear response $y(t,s)$	4.2
$g(\cdot)$	the transfer characteristics of hair cell nonlinearity	4.2
a	gain of the input of the nonlinearity	4.2
$r(t;s)$	hair cell intracellular receptor potential	4.2
Π_T	integration window of length T	5.1
$S_{AR(t;s)}$	auditory spectral profile from average-rate hypothesis	5.1
$A(t;\tau;s)$	duplex representation parameterized over time (t), delay (τ), space (s)	5.2
S_{ALSR}	auditory spectral profile from ALSR algorithm	5.3.1
$b(t;s)$	impulse response of the central narrowband filters in the ALSR algorithm	5.3.1
$L(t;s)$	the spatial derivative of $r(t;s)$: $\partial_s r(t;s)$	
$S_{LIN}(t;s)$	auditory spectral profile from the LIN algorithm	5.3.1

References

Allen J (1985) Cochlear modelling. IEEE ASSP 2(1):3–29.
Ashmore J (1993) The ear's fast cellular motor. Curr Biol 3:38–40.
Bregman AS (1990) Auditory Scene Analysis. Cambridge: MIT Press.

Brown J, Puckette M (1989) Calculation of a "narrowed" autocorrelation function. J Acoust Soc Am 85:1595–1601.

Calhoun B and Schreiner C (1993) Spatial frequency fitters in cat primary auditory cortex. Abstract (581.8) in Soc. Neuroscience Meeting Washington D.C.

Cariani P, Delgutte B (1992) Coding of the pitch of harmonic and inharmonic complex tones in the interspike intervals of auditory-nerve fibers. In: Schouten M (ed) The Processing of Speech. Berlin: Mouton–de Gruyter.

Dallos P, Geisler C, Mathews J, Ruggero M, Steele C (eds) (1990) Mechanics and Biophysics of Hearing. Berlin: Springer.

Delgutte B (1984) Speech coding in auditory nerve: processing schemes for vowel-like sounds. J Acoust Soc Am 75:879–886.

Deng L, Geisler D, Greenberg S (1988) A composite model of the auditory periphery for the processing of speech. J Phonetics 16:3–18.

de Valois R, de Valois K (1988) Spatial Vision. New York: Oxford University Press.

Flanagan J (1972) Speech Analysis, Synthesis, and Perception, 2nd Ed. Berlin: Springer.

Gitza O (1988) Temporal non-place information in the auditory-nerve firing patterns as a front-end for speech recognition in noisy environments. J Phonetics 16:109–124.

Goldstein J (1973) An optimum processor theory for the central formation of the pitch of complex tones. J Acoust Soc Am 54:1496–1516.

Hartline H (1974) Studies on Excitation and Inhibition in the Retina. New York: Rockefeller University Press.

Hess W (1983) Pitch Determination of Speech Signals. Berlin: Springer.

Houtsma A (1979) Musical pitch of two-tone complexes and predictions by modern pitch theories. J Acoust Soc Am 66(1):87–99.

Houtsma A, Smurzynski J (1990) Pitch identification and discrimination for complex tones with many harmonics. J Acoust Soc Am 87:304–310.

Javel E, Mott J (1988) Physiological and psychophysical correlates of temporal processing in hearing. Hear Res 34:275–294.

Johnson DH (1980) The relationship between spike rate and synchrony in the responses of auditory-nerve fibers to single tones. J Acoust Soc Am 68:1125–1122.

Kiang N, Watanabe T, Thomas E, Clark L (1965) Discharge patterns of single fibers in the cat's auditory nerve. Research Monographs No. 35 Cambridge: MIT Press.

Langner G (1992) Periodicity coding in the auditory system. Hear Res 60:115–142.

Langner G (1993) Spatial representation of periodicity pitch in the human auditory cortex. In: Abstracts, Society of Neuroscience Meeting, 1993, Washington, D.C., Abstr. 581.11.

Lazzaro J, Mead C (1989) Silicon modeling of pitch perception. Proc Natl Acad Sci. USA 86:9587–9601.

Licklider J (1951) Aduplex theory of pitch perception. Experientia 7:128–134.

Loeb G, White M, Merzenich M (1983) Spatial-correlation: a proposed mechanism for acoustic pitch perception. Biol Cybern 47:149–163.

Lyon R (1983) A computational model of binaural localization and separation. IEEE Proc International Conference on Acoustics, Speech, and Signal Processing, Boston, MA April.

Lyon R (1990) Automatic gain control in cochlear mechanics. In: P. Dallos (ed) The Mechanics and Biophysics of Hearing. Springer Verlag.

Makhoul J (1975) Linear prediction: a tutorial review. Proc IEEE 63:561–580.

Moore B (ed) (1986) Frequency Selectivity in Hearing. London: Academic Press.

Moore B (1989) An Introduction to the Psychology of Hearing, 3rd Ed. London: Academic Press.

Mead CA, Arreguit X, Lazzaro J (1991) Analog VLSI Model of Binaural Hearing. IEEE Trans Neural Networks 2:230–236.

Miller M, Sachs M (1983) Representation of stop consonants in the disharge patterns of auditory-nerve fibers. J Acoust Soc Am 74:502–517.

Neff W, Diamond I, Cassiday J (1975) Behavioral studies of auditory discrimination: central nervous system. In: Keidel W, Neff W (eds) Handbook of Sensory Physiology, Vol. 5. New York: Springer-Verlag.

Ohm G (1843) Uber die defintion des tones, nebst daran geknupfter theorie der sirene und ahnlicher tonbildender vorichtungen. Ann Phys Chem 59:513–565.

Patterson R (1987) A pulse ribbon model of monaural phase perception. J Acoust Soc Am 82:1560–1586.

Patterson R, Holdsworth J (1991) A functional model of neural activity patterns and auditory images. In: Ainsworth WA (ed) Advances in Speech, Hearing, and Language Processing, Vol. 3. London: JAI Press.

Phillips D, Orman S (1984) Responses of single neurons in posterior field of cat auditory cortex to tonal stimulation. J Neurophysiol (Bethesda) 51:147–163.

Plomp R (1976) Aspects of Tone Sensation. London: Academic Press.

Rabiner L, Schafer R (1978) Digital Processing of Speech Signals Englewood Cliffs: Prentice Hall.

Rhode W (1980) Cochlear partition vibration: recent views. J Acoust Soc Am 67:1696–1703.

Ruggero R, Rich N (1983) Two-tone distortion in the basilar membrane of the cochlea. Nature 349:413–414.

Sachs M, Blackburn C (1991) Processing of complex sounds in the cochlear nucleus. In: Altschuler R, Bobbin R, Clopton B, Hoffman D (eds) Neurophysiology of Hearing: The Central Auditory System. New York: Raven Press.

Sachs M, Young E (1979) Encoding of steady state vowels in the auditory nerve: representation in terms of discharge rate. J Acoust Soc Am 66:470–479.

Schreiner C, Langner G (1988) Periodicity coding in the inferior colliculus of the cat. II. Topographical organization. J Neurophysiol (Bethesda) 60:1823–1840.

Schreiner C, Mendelson J (1990) Functional topography of cat primary auditory cortex: distribution of integrated excitation. J Neurophysiol (Bethesda) 64:1442–1459.

Schreiner C, Urbas J (1988) Representation of amplitude modulation in the auditory cortex of the cat: I. The anterior field (AAF). Hear Res 21:227–241.

Schwartz D, Tomlinson R (1990) Spectral response patterns of auditory cortex neurons to harmonic complex tones in alert monkey (*Macaca mulatta*). J Neurophysiol (BEthesda) 64:282–299.

Seneff S (1988) A joint synchrony/mean-rate model of auditory processing. J Phonetics 16:55–76.

Shamma S (1985a) Speech processing in the auditory system: I. Representation of speech sounds in the responses of the auditory nerve. J Acoust Soc Am 78:1612–1621.

Shamma S (1985b) Speech processing in the auditory system: II. Lateral inhibition and the central processing of speech-evoked activity in the auditory nerve. J Acoust Soc Am 78:1622–1632.

Shamma S (1989) Spatial and temporal processing in central auditory networks. In: Koch C, Segev I (eds) Methods in Neuronal Modelling. Cambridge: MIT Press.

Shamma S, Morrish K (1986) Synchrony suppression in complex stimulus responses of a biophysical model of the cochlea. J Acoust Soc Am 81: 1486–1498.

Shamma S, Shen N, Gopalaswamy P (1989) Stereausis: binaural processing without neural delays. J Acoust Soc Am 86:989–1006.

Shamma S, Versnel H, Kowalski N (1995) Ripple analysis in the ferret primary auditory cortex: I. Respse characteristics of single units to sinusoidally rippled spectra. Auditory Neurosciency 1(2). (in press).

Shamma S, Vrainc S, Versnel H (1995) Representation of spectral profiles in the auditory system: Theory, Physiology, and Psychoacoustics. In: Mauley G, Klump G, Koppl C, Fastl H, Oechinghaus H (eds) Advances in Hearing Research, World Scientific Publishers.

Shamma S, Chadwick R, Wilbur J, Rinzel J (1986) A biophysical model of cochlear processing: intensity dependence of pure tone responses. J Acoust Soc Am 80:133–145.

Shamma S, Fleshman J, Wiser P, Versnel H (1993) Organization of response areas in ferret primary auditory cortex. J Neurophysiol (Bethosda) 69(2): 367–383.

Shamma S, Versnel H (1995) Ripple Analysis in the ferret primary auditory cortex. II. Prediction of single unit responses to arbitrary spectral profiles. Auditory Neuroscience 1(2) (in press).

Schouten JF, Ritsma RL, Cardozo BL (1962) Pitch of the residue. J Acoust Soc Am 34:1418–1424.

Siebert W (1970) Frequency discrimination in the auditory system: place or periodicity mechanisms? Proc IEEE 58:723–730.

Simmons J, Moss C, Ferragamo M (1990) Convergence of temporal and spectral information into acoustic images of complex sonar targets perceived by the echolocating bats. J Comp Physiol A 166:449–470.

Sinex D, Geisler D (1983) Responses of auditory-nerve fibers to consonant-vowel syllables. J Acoust Soc Am 73:602–615.

Slaney M, Lyon R (1990) A perceptual pitch detector. In: Proceedings, IEEE International Conference on Acoustics, Speech, and Signal Processing, Albuquerque, NM. April.

Slaney M, Lyon R (1992) On the importance of time—a temporal representation of sound. In: Cooke M, Beet S, Crauford M (eds) Visual Representations of Speech Signals. Chichester: Wiley.

Stevens K (1950) Autocorrelation analysis of speech sounds. J Acoust Soc Am 22:769–771.

Suga N (1984) Neural mechanisms of complex-sound processing for echolocation. Trends Neurosci 6:20–27.

Suga N, Manabe T (1982) Neural basis of amplitude spectrum representation in auditory cortex of the mustached bat. J Neurophysiol (Bethesda) 47:225–255.

Suga N, O'Neill W, Kujirai K, Manabe T (1983) Specificity of combination-sensitive neurons for processing of complex biosonar signals in auditory cortex of the mustached bat. J Neurophysiol (Bethesda) 49:1573–1626.

Terhardt E (1979) Calculating virtual pitch Hear Res 1:155–182.

Versnel H, Kowalski N, Shamma S (1995) Ripple analysis in the ferret primary auditory cortex: III. Topographic and columnar organization of ripple response characteristics. Auditory Neuroscience 1(2) (in press).

Viemeister N (1988) Psychophysical aspects of auditory intensity coding. In: Edelman G, Gall W, Cowan W (eds) Auditory Function. New York: Wiley.

von Békésy G (1960) Experiments in Hearing. New York: McGraw-Hill.

von Helmholtz H (1863) Die Lehre von der tonempfindungen als physiologische Grundlage fur die Theorie der Musik, 1st Ed. Braunschweig: Vieweg.

Wang K, Shamma S (1994) Self-normalization and noise robustness in early auditory representations. IEEE Trans Audio Speech 2(3):421–435.

Wang K, Shamma S (1995) Modelling auditory functions in the primary auditory cortex. IEEE Trans Audio Speech (in press).

Westerman L, Smith R (1984) Rapid and short-term adaptation in auditory-nerve responses. Hear Res 15:249–260.

Wever E, Bray C (1930) Action currents in the auditory nerve in response to acoustical stimulation. Proc Natl Acad Sci USA 16:344–350.

Whitfield I (1980) Auditory cortex and the pitch of complex tones. J Acoust Soc Am 67:644–647.

Wightman F (1973) A pattern transformation model of pitch. J Acoust Soc Am 54:397–406.

Winslow R, Barta P, Sachs M (1987) Rate coding in the auditory nerve. In: Watson C, Yost W (eds) Auditory Processing of Complex Sounds. Hillsdale: Erlbaum.

Yang X, Wang K, Shamma S (1992) Auditory representations of acoustic signals. IEEE Trans Inform Theory 38:824–839.

Yin T, Kuawada S (1984) Neuronal mechanisms of binaural interactions. In: Edelman G, Gall W, Cowan W (eds) Dynamic Aspects of Neocortical Function. Neurosciences Institute Publications. New York: Wiley, pp. 263–314.

Young E, Sachs M (1979) Representation of steady-state vowels in the temporal aspects of the discharge patterns of populations of auditory-nerve fibers. J Acoust Soc Am 66:1381–1403.

7
Scene Analysis

DAVID K. MELLINGER AND BERNARD M. MONT-REYNAUD

1. Introduction

One person talks to another in a crowded, noisy room; a soloist performs a concerto with an orchestra; a car screeches to a halt in the street outside: in each of these situations, the auditory system is faced with the problem of separating several different sources of sound from the complex, composite signal that reaches the ears.

As part of hearing, sound sources are identified and separated without apparent effort, despite the fact that many different processes interact to produce the sound signal reaching the ears and that the acoustic events produced by the different sources may overlap in time, in frequency, or in other characteristics important to hearing. These confusing factors make it difficult not only to find distinct sources in a sound signal, but also to reliably associate acoustic events with specific sources. Experiments have revealed some of the processes involved in the human auditory system, but most of its workings remain to be modeled. Our aim in this chapter is to describe some of the mechanisms used in hearing sound sources and in separating them when they occur simultaneoulsy.

1.1 Domains for Sound Separation

The sounds encountered by the auditory system may be broadly classified into speech, music, and environmental sounds. However, it seems plausible that many basic mechanisms of auditory source separation operate similarly in the various sound domains. We believe that such problems as sound source localization and source separation are best addressed at the auditory level, rather than separately for each domain. A generic modeling approach cutting across the various sound domains is desirable for intellectual economy. If successful, it becomes motivated by its application.

On the other hand, we are not assuming that the various domains are to be treated completely the same way. Not only is there much domain-specific information to be used, but the importance and relevance of generic auditory cues will vary from one domain to another. In the rest of this section, we briefly review some of the main cues.

Salient characteristics of musical sounds are pitch and rhythm. Pitch is beneficial for source formation and source separation in that the simple integer ratios between the frequency of the fundamental of a musical note and of its harmonics provide a basis for grouping the frequencies associated with a note. But it is also detrimental in that harmonically related notes (unisons, or notes an octave or a perfect fifth apart) may have enough coinciding harmonics to make pitch detection more ambiguous and source separation more difficult. Rhythmic structure, similarly, provides order that can be useful in grouping notes into separate sources, but it can also lead to false associations, as when several instruments play notes at the same instant, inducing the auditory system to believe that they all come from the same source. Work in source separation in the musical domain has typically been aimed at the transcription of polyphonic sound (Moorer 1975; Chafe and Jaffe 1986). In this arena, several harmonically related voices combine in the signal, providing a challenge for source identification and separation; background noise is typically quite low.

Source separation in speech research is usually driven instead by a need to separate out a particular speaker from background noises and environmental sounds, although some of the work (Parsons 1976; Weintraub 1985; Cooke 1991) does focus on multiple voices. The speech problem is characterized by the mixture of voiced, harmonic vowels and unvoiced, wideband consonants and also by the rapid rate of change of speech sounds (Borden and Harris 1984; Moore 1989). Again, each of these characteristics can aid or impede the progress of event separation. Voiced sounds in speech offer the same positive and negative factors as pitch in music, with the added boon that the pitches of vowels from two simultaneous talkers are less likely to be harmonically related than those from two simultaneous instruments. Consonantal sounds have sharp attacks and decays, providing strong temporal cues for event detection, but convey little spectral information for grouping. The rapid rate of change gives grouping mechanisms much material to work with, making the task easier, but because cues do not persist for long their characteristics are not easy to detect. This points out the need for an auditory periphery model (ear model or signal processing front-end) that is very finely tuned in its ability to extract transient information accurately.

Other sounds, loosely lumped together into the category of environmental sounds, have not been extensively studied as signals to be explicitly separated though they are an important part of everyday life. Although there has been at least one effort to characterize certain basic physical processes of sound production (Warren and Verbrugge 1984), no systematic attempts at categorization have been made, leaving formation and separation problems for environmental sound sources essentially unaddressed.

1.2 Definition: Auditory Scene Analysis

Our overall framework is called auditory scene analysis (Bregman 1990) and covers the entire hearing process beginning with the reception of a sound signal, through several stages of auditory processing, and culminating in the formation and separation of auditory sources. Here some terms are introduced, the central problems of auditory processing are defined, and some key auditory processes are described.

The terms *feature*, *event*, and *source* are used to refer to different levels of organization of sounds in the auditory system. Because these terms overlap somewhat, definitions may help straighten out to which general levels of organization they are referring. A feature is a part of the sound signal occurring at (or centered at) a specific time and frequency. It includes such things as an onset of sound energy of a particular frequency at a particular time, or a change in frequency in the harmonic of a pitched sound. Feature filtering in the auditory system is strongly data driven, meaning that the context of higher level objects plays very little role in it. While a feature is instantaneous, an event extends over a range of time, and perhaps also over multiple frequencies. We call an *event* an auditory phenomenon of relatively short duration that exhibits constancy or continuity through time. It has an onset and an offset and represents the lowest time-extensive perceptual entity. In music, a single note is normally an event, although some musical notes whose characteristics change with time could be treated as multiple events. In speech, events are usually syllables, although consonant clusters may sometimes be multiple events by virtue of their multiple onsets and offsets.

Features and events are auditory phenomena, points or regions in time–frequency space. In contrast, an auditory source (or stream) is a perceptual object, more permanent than an event, to which an explanation is attached. Bregman (1990) defines it as "our perceptual grouping of the parts of the . . . spectrogram that go together [acting] as a center for our description." There are two important points here. The first is that a source is a grouping of lower level phenomena—all the tones from the violin part of a sonata, for example. The second is that it unifies our mental description of the auditory field, providing a perceptual handle for explanatory processes. We attach an origin, or source, to each sound we hear, using new sources for sounds that have not been assigned one yet. We usually have some explanation for how the sound is generated, be it vocalization, a vibrating string, one object hitting another, or some other process. Electronically synthesized sounds do not necessarily fit any of these common physical models. We have many ways of explaining such sounds, sometimes by reference to other sounds whose physical models are known, sometimes simply by learning that sounds that are produced in certain contexts or that have certain characteristics are electronic.

Formation processes explain representations at one descriptive level by relating them to corresponding higher level representations. Thus, event formation accounts for a number of lower level features by grouping them into an event, while source formation provides an explanation for one or more events by assigning them to a source. Separation processes are closely associated with formation processes: while formation emphasizes the grouping effect within the objects at one level, separation focuses on the partitioning into distinct groups. The two processes occur in tandem.

Following Bregman, we call sequential formation a grouping process that associates entities over time; simultaneous formation is the complementary grouping process, a term applied to entities that happen concurrently. These distinctions between sequential and simultaneous grouping can be applied at the levels of features, events, or sources.

1.3 An Overview of Scene Analysis in the Auditory System

We present here an overall view of the stages involved in auditory scene analysis. Because this view is undoubtedly wrong in an least some particulars, we present it not as an exact model but rather as a way of placing auditory computations in context and as a means of motivating discussion about auditory processes and their relationships.

Figure 7.1 shows the overall structure of our view. A more detailed description of the system, as algorithms and implementations of parts of it, is supplied in Sections 2–4. This structure comprises a process modeling filtering and transduction in the ear and cochlea, a series of filters responsive to various features, a process for forming features into events, and a process for grouping events into sources.

1.3.1 Ear Filtering and Transduction

The first stage of any auditory model must be an ear model, that is, a functional unit that models the filtering done by the outer ear (and perhaps the head) and middle ears and the transduction of sound to neural impulses in the cochlea. The ear model includes a process representing basilar membrane action in which sound waves propagate from high frequencies to low, triggering responses from hair cells at specific frequencies. It also contains mechanisms for gain control and intensity encoding, as the 120 decibels (dB) of dynamic range perceivable by the auditory system (Moore 1989) must be reduced to the much smaller dynamic range of firing rates of neurons. The ear model also confronts

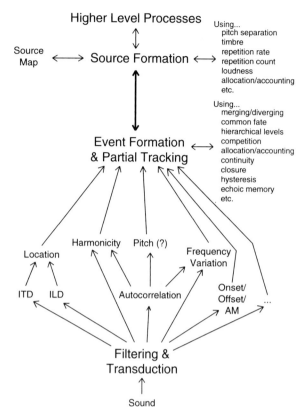

FIGURE 7.1. Auditory framework for scene analysis. Sound enters the ear, is transduced into neural firings in the cochlea, and is filtered by low-level feature-detectors. Grouping mechanisms form features into events, and then events into auditory scenes.

the filtering trade-off in which finer time resolution leads to coarser frequency resolution and vice versa.

1.3.2 Feature Filtering

Next come a series of filtering processes, each of which extracts or filters a different type of feature that may be present in the data. These filters output a number of feature maps, representations in time and frequency and perhaps other dimensions of these features. Some of the features found at this level include:

Amplitude onsets and offsets: changes in the amplitude of a frequency channel over some short period of time

Frequency modulation: changes at some rate in the frequency of a sound component (partial)

Harmonicity: how well the sound energy at a particular frequencies fits into a harmonic series

Amplitude variation, for changes in sound energy or neural firing intensity over relatively long periods of time: these are distinguished from the short-time amplitude onset because of the prevalence of the latter in neurons of the auditory system (Rhode and Smith 1986; Pickles 1988)

Interaural time and level differences: for sound localization

1.3.3 Event Formation

After these features have been filtered, some directly from auditory nerve fiber firings and some from previous feature maps, the features are analyzed to parcel out the energy represented by neural firings into various sound events. This event formation process uses different features in different ways and must obey a number of rules. For example, one rule is that if there is a sound from a particular event at one frequency at one time, then sound at the same or a nearby frequency a few milliseconds later is likely to belong to the same event. Another rule is that onsets at different frequencies at or near the same time probably belong to the same event. Information about event formation need not be completely data driven, that is, need not come entirely from the low-level feature filters just described: it can also include knowledge about patterns of learned or recently heard sounds, including timbre, short-time speech patterns, and other time–frequency schemas.

1.3.4 Source Formation

Finally, the source formation process assigns events to separate sources. Up to this point, all the processing has occurred on fairly short time scales. Features are represented at points in time–frequency space, and events typically range from a few tens of milliseconds to a few seconds. Sources last arbitrarily long—a fan you hear may have been running all day, for example—and act as grouping points for all the events that compose them.

The source formation process is where the auditory evidence (the events) meets with an explanation as a known world object that usually has some kind of permanence in the environment. Explanations are often based on domain-specific knowledge, although purely auditory explanations of very low specificity are available by default: something is heard as a sound but we do not know what it is. In all cases, the explanation (or interpretation or high-level pattern) makes some claims to a specific subset of events.

To complete the separation of the incoming sound signal into its constituent sources, we need accounting principles to state that any sound energy manifestations (evidence) must be predicted by some source (explanation).

1.4 Scope

This chapter considers the conceptual framework of auditory processing just overviewed. It includes:

A review of the introspective, psychoacoustic, and neurophysiological evidence for the response in the auditory system to various features useful for scene analysis (Section 2)

A description of the principles and rules used by the event formation mechanism, including integration of features and of nonauditory information (Section 3)

Implementations of filters for some of these features, and of a subset of these principles sufficient to handle some musical event-separation problems (Section 4)

Scene analysis is a complex, poorly understood process in the auditory system, and attempts to model it are in their infancy. The material covered in Sections 2 and 3 reveals some of the complexity of the process; a deeper look may be seen in Bregman (1990). The implementations developed to date, including the one discussed in Section 4, are incomplete, although it is hoped that they point the way toward a more complete model.

The perceptual, integrative approach used in the study of scene analysis may prove to be an insightful method for considering higher level thought processes. The framework of computation—the types of processes and representations—uncovered in the study of perception may extend to other parts of the brain and provide some clues for researchers modeling higher levels of thinking.

2. Features for Auditory Scene Analysis

This section covers the features that may be useful for data-driven simultaneous separation and includes psychoacoustic and neurophysiological evidence for the filtering of these features in the auditory system.

2.1 Feature Maps

Before discussing the various feature-extraction processes, it is first necessary to mention a representation of stimuli that has turned up in numerous places in the auditory system. A feature map is a spatial arrangement of cells, of one or more dimensions, that encodes some perceptual feature(s) (in the auditory system, some acoustic feature[s]). A given cell (usually a neuron) in a map characteristically responds best to one combination of values of the features encoded by the map. For instance, the earliest feature map in the the auditory system is the set of hair cells on the basilar membrane. Here, one feature, frequency, is spatially encoded, with high frequencies at the basal end and low frequencies at the apical end. Ignoring phase-locking for the moment, each cell has a characteristic frequency at which its threshold is lowest.

Feature maps are analog and continuous. Analog in this sense means just that the values in a map could be represented as real values rather than, say, Boolean truth values. Continuous means that the axes of a feature map are smoothly varying domains, such that a small change in position in this domain represents only a small change in the underlying properties. In addition, the range is continuous in that a small change in the intensity value at a particular position in a feature map corresponds to a small change in the represented feature.

2.1.1 Feature Maps in the Brain

Feature maps are widespread in perceptual systems. In the visual system, several maps are organized as two-dimensional images, called retinotopic because the dimensions are associated with the retinal image. In the auditory system, many maps are organized tonotopically (according to tone frequency) with one dimension (place) representing a place along the cochlea. (However, Shepard [1989] pointed out that some feature maps may not be represented by tonotopic maps in the auditory system.)

Other dimensions of variation are embedded in feature maps. Suga (1990) has found several such maps in the auditory cortex of the mustached bat (*Pteronotus parnellii*), many specialized for eocholocation. "FM-FM" maps compare pulsed pitches emitted by the bat with returning echoes and have axes of echo delay, for target range, and of amplitude, for target size. Another map has axes of frequency (tonotopy) and frequency-change rate, for Doppler-shifted target velocity measurement. Schreiner and Mendelson (1990) found a map in the cat encoding "spectral tilt," where an amplitude change from one frequency to another induced firing, and Shamma found strong evidence for a similar map in the ferret (Wang and Shamma 1995). Amplitude modulation maps in the cat (Schreiner and Langner 1988) encode short-time amplitude fluctuations, perhaps as a first step toward sound localization. Maps encoding the spatial location of a sound have been found in several animals; Yin and Chan (1988) provide a good survey. The diagram in Figure 7.2 (from Knudsen 1981) shows graphically how the directions in space around the head, azimuth and elevation, are smoothly mapped onto two spatial dimensions of a structure in the barn owl (*Tyto alba*) brain, the magnocellularis lateralis pars dorsalis.

2.1.2 Locality of Processing

The computations that produce feature maps are primarily local. A local computation from one feaure map to another uses only points in a small neighborhood of the location to be computed. Thus, to compute a value

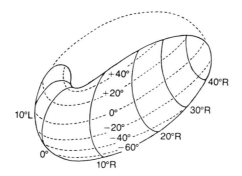

FIGURE 7.2. Two-dimensional space map in the magnocellularis lateralis pars dorsalis (MLD) of the owl. Sounds at a given angle in space (relative to the owl's head) will cause neural activity centered at a single position in this map. The horizontal curving axis of the map is azimuth, the vertical axis for elevation. (From "The hearing of the barn owl," by E.I. Knudsen. Copyright © 1981 by Scientific American, Inc. All rights reserved.)

at a particular position in a map representing some feature, a local process uses only values near the corresponding position in the input map(s). Local computations between feature maps have come to dominate the field of machine vision (see, for example, Adelson and Bergen 1986). Local computations, as well as feature maps representing their results, are often called topographic.

Some hearing processes are apparently nonlocal. For instance, the part of pitch perception that is place-based must integrate information from across the spectrum, because harmonics from many widely spaced frequencies can contribute to pitch perception. It may be that there is some neural mapping that rearranges the tonotopic order of auditory nerve fibers in such a way that local computations can perform these nonlocal operations.

The axes of a feature map other than time and frequency are typically the variation of some parameter of the filter that computes the map. In a map that represents frequency modulation, for example, the parameter might be the rate of modulation; positive for upward frequency modulation, negative for downward. In a map for amplitude onsets, the dimension of variation might be the characteristic duration of onset.

2.1.3 Map Computation

The basic structure of a feature map model of audition is to perform computations from one feature map to another. The primary map from which all others are computed is the time × log-frequency pattern of firing of cochlear nerve fibers, with time measured relative to some reference time t_0. All other maps are computed from this one, either directly or indirectly via intermediate maps, generally using local computational operators. These computed maps extract features of various types for use by the grouping process to be described later.

2.1.4 Organizing Maps in Space

One question that arises about the auditory system is how maps representing different features are arranged. If each map is kept in a different location in the auditory pathway, then there arises the problem of bringing together the data that belong together. The data for a specific frequency, say, would be scattered over several areas if several maps have frequency as a dimension, and such information must be brought together for many types of computation. This recombination problem is not easily solved. If, on the other hand, maps are overlapping in space, then the different cues could be computed next to each other. The only problem with this plan is that there is simply not enough space in the three dimensions of the brain to represent all the various maps.

Fortunately, a computer implementation of a feature-map model such as used in Section 4 need not suffer from this dilemma, thanks to the random access nature of computer memories. Maps are simply put in separate locations in memory and associate values at, say, a common frequency by accessing the maps with the right index for that frequency in each map. To what degree this is a violation of physiological compatibility is a question that may be answered as more is learned about the auditory system.

2.2 Partials

In a spectrogram or neural spectrogram of a pitched sound, frequency lines called *partials* immediately stand out. They often appear in parallel groups; when their respective frequencies form a harmonic series, at least approximately, the partials are called *harmonics*. An important task of any event formation and detection system is to identify partials and group together those that belong to the same event. The algorithm for event formation given in Section 4 describes a representation for partials and uses them as a primary means to track changes in the spectrum of a sound over time.

2.3 Features for Event Formation

The remainder of this section reviews various low-level features that can be used for sound event formation and separation, providing psychoacoustic and neurophysiological evidence for the presence in the auditory system of mechanisms responsive to the features. The features to be described at this level are what Bregman calls cues for simultaneous integration because they are short-time cues that can be thought of as operating at slices in the same short interval (Bregman 1990). These features are amplitude onset, frequency modulation, harmonicity, amplitude modulation, and spatial location.

In this part we address raw features, such as amplitude onset, frequency modulation, or location, that later provide a basis for the detection of common fate. In Figure 7.3, common amplitude onset of several partials can be seen at about 340 msec, where a number of partials appear at different frequencies. The score for this sequence is Figure 7.4. Common frequency modulation can be seen around 1500 msec, where several partials have a parallel decline in frequency. The grouping of partials by common characteristics of features is addressed later, under event formation.

This section is a review of evidence for several of the features used in event formation and, later, source formation. This evidence is mainly of two forms. Psychoacoustic evidence demonstrates that the feature

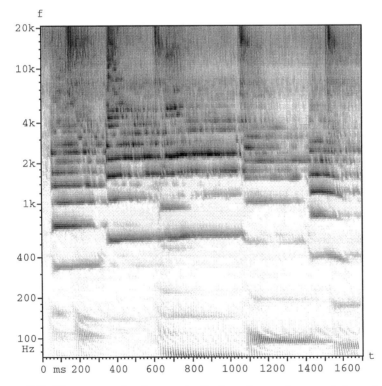

FIGURE 7.3. Time versus log-frequency display of saxophone and drum tones. Saxophone notes starting at 40, 340, 1050, and 1420 msec show up each as a series of harmonically spaced partials. Note the frequency change in the second and last notes. High-pitched sounds between 10 kHz and 20 kHz are snare drum notes.

FIGURE 7.4. Score for saxophone part of the previous figure showing the pitches of the four notes.

in question is an important part of the scene analysis process. Neuro-physiological evidence shows that there exist neurons in the auditory system that respond to a certain feature. Such evidence suggests that these features are filtered fairly early in the auditory pathway, enabling them to be used as information for an event or source formation process. In this review, we occasionally refer to introspective demonstrations that are widely known or are especially telling, even if they have not been put through the rigors of a psychoacoustic test.

2.4 Onset

One of the features acting most strongly for grouping of related partials is common amplitude onset, or simply common onset; Hartmann (1988) places it first in a list of stream-segregation features. In nearly all musical instrument tones, the start of a note is marked by the rapid rise of all strong partials within a period of about 40 msec (Grey 1975). This common onset gives the auditory system a strong cue that the partials belong to a single source. In many types of ensemble playing, performers take great pains to have a common onset. This does not greatly inhibit the ear's ability to hear the separate instruments, but if one instrument leads or lags it stands out noticeably.

Pierce (1983) presented an illustration of the power of onset asynchrony to pull apart a single source to make several. He played a short tone made up of a number of harmonics of the fundamental frequency at 220 Hz, shown schematically as the note at the left in in Figure 7.5. He

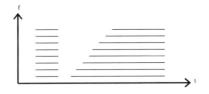

FIGURE 7.5. Tone pattern in Pierce's (1983) demonstration. Each horizontal line represents a sinusoidal tone. At *left*, all sinusoids blend to make a single, harmonic note. At *right*, successive tones begin at approximately 0.5-sec intervals and are briefly heard separately before blending with the other, existing tones.

then played the same set of harmonics, but introduced each of them 1 sec after the previous one. As each harmonic enters, it stands out briefly as a separate tone before merging with the existing mass of partials. Because each new partial is part of a harmonic series with the existing ones, the only evidence present for the grouping process in the auditory system is that the partial has a different onset from the others. This makes it stand out initially as a separate event. Why it later merges with the other partials is discussed in Section 3.

2.4.1 Psychoacoustic Evidence

Gordon (1984) studied the differences between various different sound characteristics that fall under the general name of "onset." The physical onset time of a note is the instant a sound is physically present. With sufficient intensity and a suitable amplitude envelope, perceptual onset, the instant when a note is first audible, occurs at the same point in time. The perceptual attack time (PAT) of a tone is its moment of perceived attack relative to the physical onset; the PAT can be later than the onset in, for example, reed and bowed string tones. Because of these differences in perceived attack of different instruments, it is important in measuring perceptual onset time differences to define a "standard instrument" for comparison. Gordon measured PAT by having subjects adjust timings of repeating sequences A—B—A—B... from two instruments, and discovered that (1) subjects could be consistent in their judgment to 1%–2% of the beat interval, (2) perceptual fusion of two instruments occurred when onsets were less than 2 msec apart, and (3) a physical measure corresponding to perceptual data does exist. An important characteristic of onset synchrony is that it forms an important part of the timbre of a musical instrument. Grey's (1975) multidimensional scaling study of instrument timbres identified three dimensions of important timbral variation; one dimension showed that the auditory system is quite capable of detecting small differences in the onset (and offset) times of the different partials.

Experiments of Bregman and others have confirmed the effect of onset synchrony in grouping partials together. A classic experiment (Bregman and Pinker 1978) varied the characteristics of three sine tones to change the way they are grouped into streams by the auditory system. Figure 7.6 illustrates the common onset portion of the experiment, in which sine tones A, B, and C are presented repeatedly. The degree of onset synchrony between partials B and C, with a threshold of about 30 msec, strongly affected the perception of whether partial B stood by itself or combined with C to make a richer timbral tone complex BC.

Rasch (1979) measured onset asynchrony among different instruments in an ensemble and found that the asynchrony (RMS) varied from 27 to 49 msec. With Bregman and Pinker's 30-msec difference and Gordon's

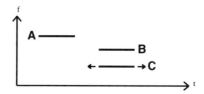

FIGURE 7.6. Tones for the onset part of Bregman–Pinker (1978) experiment. Each line represents a sinusoidal tone. When tone C starts at the same instant as B, it blends with B to produce a different timbre BC; the streams heard, with repetition, are A—A—A—A and BC—BC—BC—BC. When C starts with a sufficiently different onset time than B, it is heard as a separate note; B is then heard in the same stream as A because of pitch nearness, and the streams are A—B—A—B and, at half the rate, C—C—C—C.

(1984) instrument tone attack times of as long as 30 msec, this suggested that the onset asynchrony is important in enabling the auditory system to hear separate instruments in an ensemble. Darwin (1984) added extra energy near the first formant of a vowel sound, finding that the vowel quality changed if the extra energy was within 32 msec of the vowel onset but did not change at longer time offsets.

Neurophysiological evidence

Evidence for neurons that fire upon onset of a particular frequency is fairly extensive. Pickles (1988) described a set of cells in the cochlear nucleus responsive only to onset. Young et al. (1988) pointed out that precisely timed onset responses are necessary for sound lateralization and suggested that some cochlear nucleus cells may be specialized for this purpose. Schreiner and Urbas (1986) reported that some neurons in the auditory cortex respond best to signals which rise or fall with a 10-msec characteristic time.

Offset

Common offset provides a related but much less important cue for grouping the parts of a spectrum produced by one source. Offsets are less salient psychoacoustically than onsets. An example demonstrating this may be heard when the Pierce (1983) example (Fig. 7.5) is played backward. As the harmonics cease one by one, a slight change in the timbre of the tone complex is heard, but no tone stands out and the perceptual effect is generally much weaker. Pickles (1988) discussed evidence for the filtering of offsets in the auditory system; offset cells are less common than onset ones, but are indeed found.

2.5 Frequency Modulation

Frequency modulation (FM) refers here to the change in frequency of a partial. Filtering of frequency modulation may be based on the output of some pitch extraction process. *Pitch* is a subjective measure (or combination of several subjective measures) of the height of a tone, as on a musical scale. It is a perceptual phenomenon, and is related to the periodicity or fundamental frequency of a harmonic or quasi-harmonic sound. It may be contrasted with frequency, which is a physical measure of the repetition rate of a pure tone.

Many pitch detectors have been proposed, some of them inspired by engineering and others by physiology. One way or another, they combine information from across the spectrum to determine the fundamental frequency or frequencies that are present in the sound. As Bregman (1990) pointed out, however, the pitch percept is influenced by the source-segregation process. Here, we focus on early mechanisms that act independently of pitch perception, which could be part of the input, directly or indirectly via a source-separation process, for a pitch mechanism.

2.5.1 Psychoacoustic Evidence

Chowning (1980) noticed that synthesized singing-voice partials—a vowel sound—became more like a single voice when identical slight vibrato was added to the synthetic partials. This suggests an auditory mechanism that uses the vibrato as a grouping feature. Lacking this feature, the synthesized partials sound like a collection of buzzy separate sounds. With it, the partials crystallize into a single voice in a sudden and striking way.

McAdams (1984) found that partials with constant frequency ratios, as are generated by most natural vibrating sources, are grouped together by the auditory system more easily than are those with constant frequency differences. McAdams (1989) further investigated the grouping of singing-voice sounds, discovering that vowels with vibrato group together more than vibrato-less vowels, even when the vibrato between partials is out of phase. It may be that the difficulty of the listening task is not in separating the vowels from one another, but rather in having the partials form a source at all.

Rasch (1978) played notes with and without 4% vibrato against a background of a masking tone. He found that tones could be decreased in level by 17.5 dB and still be equally audible against the background when vibrato was added. This amount of vibrato, typical of musical instruments, makes notes stand out even when they are much less intense than a masking tone.

Carlyon and Stubbs (1989) experimented to determine whether changes in the frequency of one partial of a harmonic or inharmonic complex were heard because of direct frequency change or because the partial became mistuned with respect to the rest of the complex. They found that harmonic complexes were more easily heard than inharmonic ones against a noise background, and that FM of a partial is also more easily heard in the harmonic case. Their conclusion could be seen as suggesting that response to FM is solely dependent on harmonicity, but their data show some effect of FM, although not nearly as much as of harmonicity. The strong perceptual effect of common FM in Chowning's synthesis example (Chowning 1980) suggests that FM must play some part. Perhaps the distinction between source *separation* and source *formation* (see Section 1.2) is important here, with FM having the stronger effect on the latter. This suggestion is supported by McAdams's experiment mentioned earlier; further experimentation would be interesting.

Kay and Mattews (1972) found that playing a modulated tone of one FM depth caused subjects to become less sensitive to FM at a shallower depth, provided that the shallower modulation was at approximately the same rate as the deeper. The experiment was controlled for pure-frequency adaptation. Perhaps the salience of frequency-varying sound results entirely from the passage of a moving partial into new frequency channels. Gardner and Wilson (1979) tested this hypothesis by playing a series of tones that swept downward in frequency until the subjects had become adapted to such tones, then played an upward-sweeping tone through the same frequency band. The adaptation affected only the threshold for downsweeps, not for upsweeps, suggesting that there are mechanisms sensitive to the direction of frequency movement.

Neurophysiological evidence

Evidence for auditory neurons sensitive to FM comes from several sources. The earliest is probably Whitfield and Evans (1965), who recorded units in the auditory cortex. They found that many units were more responsive to FM tones than to any steady tone, and that some responded only to FM. They also found that FM-responsive units tended to fire at a particular point in the modulation sinusoid, for example, when the tone was rising. They also found units responsive to wide frequency sweeps but not to steady-state tones or to narrow FM and units more responsive to sweeps in one direction than the other.

Møller (1977) found some units in the cochlear nucleus for which an "optimal rate of frequency change exists" in eliciting a neural response. Mendelson and Cynader (1985) also found units in the cat auditory cortex responsive to the direction and rate of a frequency sweep and suggested a columnar organization to the auditory cortex, a map of FM sweep rate.

2.6 Harmonicity

Harmonicity refers to the degree to which a partial falls into a harmonic series with other partials. It is important in the source separation process because many sound sources in nature have harmonic or nearly harmonic partial structures resulting from their origin in a vibrating medium. Because different vibrational sources usually vibrate at different frequencies, their partials from separate harmonic series, enabling a scene analysis mechanism to distinguish them.

The relationship between harmonicity and perception of harmony, or pleasantness, is a complex one that is touched on only briefly here. The existence of beautiful pieces of music that rely on partials with nonharmonic spacing (i.e., spacing other than integer multiples of the fundamental), such as *Stria* by John Chowning and *Inharmonique* by Jean-Claude Risset, demonstrates that perception of pleasantness is not completely tied to harmonicity. Bregman (1990) put forth the interesting idea that we have two separate mechanisms for hearing harmony, one that gives rise to a perception of consonance or dissonance, and a second that produces fusion of partials into a sound source. This is closely related to the observation of Moore, Peters, and Glasberg (1985) that low-frequency mistuned partials "appear to stand out from the complex tone as a whole, [while] for high harmonics, the mistuning is detected as a kind of 'beat' or roughness." Shepard (1982) has investigated the role of harmonic relations in musical structure, and Balzano (1980) suggested a possible underlying group-theoretic basis for 12-tone harmonic relations.

2.6.1 Psychoacoustic Evidence

The strength of harmonicity as a grouping cue is well enough established psychoacoustically that it is usually used as a basis against which experimenters measure competing forces for segregation. For instance, McAdams's thesis work (1984) (see Section 2.5.1) started out with harmonic tones on which he imposed (in one experiment) different types of frequency modulation. One type of modulation, constant-difference modulation, introduces inharmonicity into a tone. The tone breaks into two sources when a sufficient modulation depth is reached. The experiments of Moore, Glasberg, and Peters (1985) echo this result. Cohen (1984) showed that pitch perception is not completely tied to harmonicity, in that subjects were able to assign consistent pitch values to tones with small departures from harmonicity. Also, studies of perception of multiple simultaneous vowels have typically used the harmonicities present in the vowels as the main, or only, grouping cue (Parsons 1976; Weintraub 1985; Assman and Summerfield 1989).

Efforts at modeling pitch detection (Licklider 1951; Schroeder 1968; Terhardt 1972; Goldstein 1973; Wightman 1973; Meddis and Hewitt 1991) have implicitly or explicitly assumed that the harmonicity of a sound contributes to its grouping, so that a single pitch is heard instead of an ungrouped collection of harmonics.

Neurophysiological evidence

So far, "harmonicity" neurons have not been found in the auditory system, that is, neurons that integrate information from widely spaced frequencies and respond only if the frequencies present are in a harmonic series. In the signal processing literature, the closely related notion of *virtual pitch* has been considered as a means to map spectral frequencies to possible pitches (Wise, Caprio, and Parks 1976).

It is possible that associations between harmonics are discovered in the time domain, based on phase-locking. The synchrony of neural firings on the basilar membrane to the peaks in a periodic sound signal could be used at higher levels to detect harmonicity at frequencies up to 4–5 kHz, because units responding to frequencies that are multiples of a fundamental will stay in a constant phase relationship with one another.

2.7 Amplitude Modulation

Common amplitude modulation is the parallel variation in intensity of a number of partials. It derives its importance as a grouping cue from the fact that most physical processes that change the intensity of one partial will change the intensity of all the others at the same time.

2.7.1 Psychoacoustic Evidence

Békésy (1963) found that sine waves at 750 and 800 Hz, when presented to opposite ears, could be made to form a single "sound image" by the imposition of coherent amplitude modulation (AM) of 5–10 Hz. The strongest line of evidence for the salience of common amplitude modulation comes from the series of comodulation masking release (CMR) experiments begun by Hall, Haggard, and Fernandes (1984). The fundamental discovery was that noise bands that vary in amplitude coherently, with the same modulating function, are not as effective at masking other within-band signals as those that are modulated incoherently. The suggestion is that a scene analysis mechanism can use AM evidence from flanking noise bands to group the noise bands together, leaving the signal more perceptible because it is a separate source.

Many experiments have followed this initial discovery; see Moore (1990) for a review. Cohen and Schubert (1987) measured the frequency range over which a flanking noise band contributes to CMR, finding it to be about two-thirds octave below the signal frequency and one-third octave above. In a sine-tone study similar to Figure 7.6, Bregman et al. (1985) found that common amplitude modulation did function to integrate two tones.

Schooneveldt and Moore (1987) found a sharply tuned component of CMR when the noise bands are near the signal band, and suggested that the auditory system listens to the target signal in the silent periods of beats between the noise carrier frequencies. This is supported by the fact that CMR fails for frequency-modulated, instead of amplitude-modulated, sounds (Schooneveldt and Moore 1988). However, Schooneveldt and Moore's finding of dichotic CMR that cannot depend on within-channel cancellation suggests that a mechanism similar to one useful for scene analysis may also be present.

Neurophysiological evidence

Schreiner and coworkers have performed an extensive series of recordings of neurons responsive to amplitude change at several places in the auditory system. These neurons could be part of a mechanism that uses amplitude modulation as a cue for source separation, although there is no evidence as yet for this function. Schreiner and Urbas (1986) found that units in the auditory cortex could be characterized by a best modulation frequency (BMF) as determined by phase-locking of neural firings to the signal, and that BMF varied in a consistent way with frequency. They also found (Schreiner and Urbas 1988) that units varied in sharpness of tuning, some being tuned to relatively exact AM rates and some to relatively broad ones.

Schreiner and Langner (1988) found a concentric arrangement of units across the inferior colliculus ordered by BMF. Sharpness of tuning also varied in an orderly concentric map, but with a different center than the BMF map. Thus each unit embodies a unique BMF–sharpness pair.

2.8 Location

The auditory system performs a significantly difficult task in deciding from which spatial location a sound originates. Spatial location is a significant cue for sound separation, although not an indispensable one. To convince yourself of the latter, simply listen to two people talking at once with one ear covered. Mechanisms for spatial localization are covered in Chapter 8.

2.8.1 Psychoacoustic Evidence

A series of experiments with a common theme support the idea that spatial location is an important grouping cue. An experimenter plays a signal and a masking noise diotically and measures the signal level threshold necessary to hear it in the noise. He or she then introduces a delay into either the signal or the masker in one ear and measures the threshold again. The threshold is usually lower in the second case; the difference in thresholds is called the masking level difference (MLD), which was discovered by Jeffress et al. (1956). Several explanations have been advanced for it (Durlach 1964; Hafter 1971; Jeffress 1972), the surviving one being that the delay in one ear makes the auditory system localize the signal and masker at different places. Thus, the signal is better separated as a source and can be heard more clearly. Jeffress (1972) proposed a cross-correlation process, implementable as neural delay lines, that could provide a means for source separation based on spatial location. This model has recently been extended and implemented on a computer by Lindemann (1986), who used it as a spatial locali- zation process. He added lateral inhibition and a contribution from a monaural filter to produce a method that appears to work well for localizing monophonic signals and perhaps polyphonic ones as well.

Durlach's equalization and cancellation theory (Durlach 1963, 1964; Metz, von Bismark, and Durlach 1968) suggests that MLD results from a process that shifts the mental representation of the stimulus in one ear in time or amplitude until it matches the stimulus in the other ear as well as possible. The matching part of the sound, which is the noise in the MLD effect, can then be subtracted (canceled) from each ear, leaving behind the target signal. Such a mechanism, if present, could be considered a part of the scene-analysis process that separates sources by spatial location.

Bregman (1990) found that that sounds perceived to be in separate streams are difficult to relate rhythmically. Van Noorden (1975) played identical tones to the two ears, describing them as being lateralized to opposite sides of the head, and discovered that subjects found it difficult to tell if the tones were evenly spaced in time, that is, if the right-ear tones were exactly halfway between the left-ear tones.

Neurophysiological evidence

Again, Chapter 8 reviews neurophysiological evidence for the extraction of spatial location cues. How such evidence might be used for auditory scene analysis is left to Section 4.

3. Event and Source Formation

The previous two sections covered the feature filtering stage of the auditory model. This section outlines the higher level processes in which feature information is used to decide what events are present in the sound signal and to group these events into sources. Most of this section is an overview of the factors that influence these processes, while the next one contains an implementation of an event-formation algorithm that incorporates some of these factors. This section is but a brief overview of the important aspects of event and source formation, as that subject is broad and deep. Much of Bregman's 700+ page book (Bregman 1990) is devoted to it, and only the surface of the subject is touched here. Nonetheless, the hope is to describe enough of these processes that the implementation in Section 4 will be comprehensible.

This section begins with a review of some of the characteristics of source perception: how sources are organized in time and in level of detail, how different perceptual organizations compete with one another, and what constraints they must obey. It continues with the processes of event formation and then source formation, describing some of the factors that influence each. The section draws throughout on the works of Moore (1989) and especially of Bregman (1990).

3.1 Characteristics of Event and Source Formation

The terms formation and separation refer to the same processes but with different emphases. Formation, either of events or of sources, means the associating of different parts of the time–frequency image to make a single object. This is the complement of separation, which emphasizes the portion of the process that distinguish parts of the image as belonging to different sources. Neither formation nor separation can exist without the other, but formation emphasizes the unifying part of the process while separation emphasizes the discriminating part.

3.1.1 Simultaneous versus Sequential

Another important distinction is that between simultaneous and sequential grouping, also called (Bregman 1990) vertical and horizontal grouping, respectively. Simultaneous grouping is the process by which we hear different, simultaneously sounding frequencies, potentially widely spaced parts, as arising from the same source. Harmonicity is a kind of simultaneous grouping. Simultaneous grouping is associated closely with event formation: although events last over time and require features filtered at different times to be integrated, much of the task of event formation can be thought of as finding which of the partials that exist at any instant belong together.

Sequential integration is the part of the scene analysis process that forms groups over time. In a piece of music with several parts, one or more melodies and harmonies, each part may be thought of as a separate source.

Sequential and simultaneous organization happen concurrently, influencing each other's progress. Simultaneous organization clearly happens first in most musical listening, for we hear quite easily such simultaneous cues as common onset, common offset, and harmonically spaced partials, even when several instrument notes are competing for attention. However, Bregman and Rudnicky (1975) showed that a sequential grouping of notes into separate streams can affect simultaneous perception of the timbre of the notes.

3.1.2 Hierarchical Perception

Although grouping of events into sources is a hierarchical process, sources themselves have hierarchical structure as well. Think of hearing an orchestra. At the largest grouping scale, the entire orchestra is a single source. At successively finer scales of analysis, one can hear the string section as a source, the first violin section as a source, or perhaps even an individual violinist as a source. Each of these levels of grouping is contained within the previous; the organization of sources is hierarchical.

This hierarchical organization is partly directed by attention. That is, one can focus attention on a particular level of the hierarchy to make apparent the source present at that level. The levels are not always as clear-cut as in the example just given, as when a piece of music has many interweaving elements that come into and go out of existence rapidly. Indeed, probably everyone has had the experience of hearing a piece of music differently after repeated listening. Often the difference results from perceptual reorganization of the sources, with musical events placed into different streams after the deeper analysis enabled by familiarity.

3.1.3 Competing Organizations

Several possible descriptions, or organizations, of a sound signal are possible. A visual analogy is the Necker cube illusion, in which the visual system can impose either of two organizations on a set of line segments, producing percepts with two different front corners. Several different factors may tend to favor one organization over another; which one wins is determined by the number, types, and strengths of the factors. Many of Bregman's experiments have exploited this competition between grouping factors. For instance, the Bregman and Pinker experiment (Section 2.4.1) exploited competition between a cue for common onset that led to grouping two tones together, and a cue for nearness in frequency that led to grouping a different pair together. Similarly, the Bregman–Rudnicky experiment mentioned in Section 3.1.1 exploited competition between onset and harmonicity as grouping cues.

3.1.4 Allocation and Accounting

Two complementary principles from Gestaltist vision work are important in auditory scene analysis. The first, known in auditory scene analysis research as the principle of *exclusive allocation*, says that "a sensory element should not be used in more than one description at a time" (Bregman 1990). Any element of the incoming sound, say a unit of energy at a particular time and frequency, should be assigned to only one source; it cannot do double duty. The complementary principle of accounting requires that all incoming sound energy be assigned to one source or another. If a sound is heard that cannot be assigned to any existing source, then it becomes a source by itself. Allocation and accounting happen at all the levels of organization mentioned in Section 1. All features perceived must be accounted for by being associated with events, and each unit of intensity of a feature is generally associated with but a single source (but note the exceptions following). Likewise, all of the events produced by an event-formation process must be accounted for as part of one, and generally only one, source.

Evidence for the principle of exclusive allocation comes from several experiments that supplement common sense. Some of them have been seen already, as in the Bregman–Pinker experiment (see Section 2.4.1). There, a tone could belong to only one of the streams at a time; one cannot hear it simultaneously as part of both possible streams.

A number of cases have been found in which allocation fails, causing one stimulus to contribute to two or more events or streams. This phenomenon, known as duplex perception (Ciocca and Bregman 1989), was first pointed out in sound by Rand (1974). Gardner and Darwin (1986) found it present in a study of vowel harmonics in which they modulated the frequency of a single harmonic of a vowel sound. The modulated harmonic stood out from the rest of the complex as would be expected from its frequency-modulation cue, but the timbre of the vowel did not change as it did when the harmonic was eliminated entirely. A similar effect is found in McAdams's computer-music oboe sound (Reynolds 1983), in which the even harmonics contribute to both a soprano-like voice and an oboe sound.

The effect is seen in harmonicity and pitch perception as well. Moore, Glasberg, and Peters (1985) mistuned one harmonic of a harmonic complex by 3% or more and found that it stands out from the complex as a separate entity while still changing the perceived residue pitch slightly. Although the principle of exclusive allocation is sometimes violated, duplex perception seems to be the exception rather than the rule. Violation seems to happen more with speech sounds than other types of sounds (Moore 1989), which may reflect the operation of a separate speech processing module.

3.2 Natural Constraints

The principles of accounting and exclusive allocation in perceptual systems derive from constraints on the nature of sounds to which the auditory system is exposed. Several other such constraints affect the ability to hear and separate sources.

3.2.1 Common Fate

One general class of constraints, called common fate by gestaltists, holds that different parts of the perceptual field—of vision, audition, and probably other senses—that are associated with a source usually have common properties. These common properties are reflected in the auditory system, which has mechanisms to use these commonalities as cues for scene analysis. Many of these have already been discussed: common onset and offset, common frequency and amplitude modulation, and harmonicity or common fundamental frequency. (Note: The latter commonality, not involving motion or change, would not fit the Gestaltists' meaning of common fate.)

Continuity

A general class of constraints, also reflected in auditory scene alaysis mechanisms, is that of continuity or similarity over time. This constraint arises from the fact that most sounds tend not to change in character rapidly. A clarinet is not likely to suddenly sound like an automobile. A piano is not likely to drop its third harmonic halfway through a note. The principle of continuity, like accounting and allocation, applies at many levels of organization, from the continuity of partials in a note to the continuity of subject matter in speech. It is explored further under event formation (Section 3.3) and source formation (Section 3.4).

Closure

Another important principle from the Gestalt school is closure, the idea that incomplete sensory stimuli can sometimes be closed or made complete by the perceptual addition of nonexistent parts. An example of this may be heard in speech interrupted by brief noise bursts in which the speech is perfectly intelligible even though some phonemes are lost. Indeed, subjects generally find it difficult to tell where the interrupting noises occurred; perceptual completion works so well that we are not even aware of the place of its effect (Miller and Licklider 1950; Dirks and Bower 1970).

   Closure is a part of the auditory scene analysis mechanism. It helps preserve continuity of partials across interruptions, and helps at a higher level to preserve the continuity of events in streams.

3.3 Event Formation

The auditory system brings together information about a sound signal from various perceptual mechanisms to decide which events are present. This process, event formation, associates sound from across the time–frequency sensory field, grouping it into events as they occur for use by higher levels of the auditory system.

   Event formation is primarily a spectral organization, placing sound energy into the correct groups at each instant. The process clearly uses knowledge from the recent past, as for instance in Pierce's demonstration (see Fig. 7.5) where information about a recent onset contributes to the perceptual organization for a while after the onset occurs. Event formation even uses "knowledge from the future," as demonstrated by Warren (1982): the decision of when an event is perceived to end is made later than the event actually ends.

3.3.1 Grouping Partials

How does the auditory system perform event formation? One possible way is by tracking partials and keeping some kind of evaluation of how well they associate with each other. In this conception, partials are placed in collections and remain there as long as evidence for their belonging is strong enough. For instance, if a partial has FM at a rate that matches that of the other partials in its group, then this match constitutes evidence that the partial belongs to the group. Evidence can also go against keeping a partial in group: if a partial begins at a different time than the members of an existing group, then this onset mismatch is evidence that the partial does not belong to the group.

Partials usually obey the common-fate constraints mentioned earlier in that common onset, offset, frequency modulation, and amplitude modulation are all evidence that several partials belong to the same event. Partials also obey the continuity constraint not changing frequency rapidly. Bregman and Dannenbring (1973) studied this effect, finding that a sound with a continuous frequency trajectory was more likely to be grouped as a single source than one with a discontinuous trajectory; this was true even if the sound had a gap so long as the frequency trajecty suggested by sound before and after the gap was continuous.

In accumulating evidence for the creation and maintenance of groups of partials, the absence of cues can be as important as their presence. For instance, the absence of harmonicity for a given partial tends to make it stand out as a separate source (Moore, Peters, and Glasberg 1985). The absence of any sound at all drives the decision about when an event ends in Warren's experiment discussed in Section 3.3.

Contribution strength

Each cue has a different contribution strength for its partial's presence in a group, a strength that depends on the context of the event. Common onset seems to be one of the stronger cues, as Hartmann noted (Hartmann 1988). Parsons (1976), Hartmann (1988), and Cooke (1991) all noted the importance of harmonicity as a grouping cue. Common FM may be one of the weaker cues (Carlyon 1991), but it is important musically in that other cues are often missing and it is the only one available to hear out multiple voices. Cue strength may vary depending on sound characteristics, as when an onset in one frequency channel is perceptually unimportant because it arises from a partial that is varying in frequency.

Context for cue strength can come from learned patterns. For example, in listening to piano music, one listens for the percussive onset of each note. Violin notes have attacks that are much more spread out in time, providing less of an onset cue; in this case, the harmonicity and pitch separation cues may be acting most strongly.

Hysteresis

A final charactreristic of groups of partials is that they display hysteresis; that is, a group tends to stay together in the absence of evidence that it should split apart, even for some period of time after the arrival of evidence for splitting. Conversely, different events will fail to merge for a brief duration despite evidence that they should do so. An example is Pierce's example in Section 2.4. Each new partial stands alone as a separate event for a brief period of time, roughly half a second, before merging with the existing partials; evidence must accumulate for the new partial's membership in the complex before it fuses. Bregman (1978) also observed this effect, noting that sounds tend to be heard as a single stream when they start but to split apart after several seconds of accumulating evidence.

3.4 Source Formation

Source formation is the process in which events are grouped over time into coherent souces, each group perceived as a complete stream. Like event formation, it is a complex dynamic operation depending on many factors beyond the scope of this chapter. A few of these factors are listed here as a brief overview of the principles that could be used in a system that goes beyond the event-formation stage implemented in the next section. These factors are generally cues that furnish the auditory system with evidence for or against an event belonging to a source. Event and source formation are closely related processes, neither existing without the other.

3.4.1 Pitch Separation

Two successive notes near each other in frequency tend to be placed by the auditory system into a common source, and conversely (Fucks 1962; Dowling 1978). This fact is salient in melody: Ortmann (1926) studied a large number of melodies, tabulating the pitch intervals between adjacent notes. He found that the intervals used in melodies tended to be small, with approximately a reciprocal ($1/df$) relationship between interval in semitones and frequency of occurrence.

J.S. Bach used pitch separation in a technique called virtual polyphony. Erickson (1982) analyzed the C major solo violin sonata, in which a single instrument plays interleaved notes that are widely separated into three disjoint ranges of pitch. If the piece is played rapidly enough, the notes in each range are heard as a separate stream. In other words, one instrument can become several sources by virtue of pitch separation.

Pitch separation has been used extensively as a grouping cue in psychoacoustic experiments. It was used in Bregman and Pinker's classic experiment (Section 2.4.1) to induce tones to group together in a single stream. Another example is the experiment of Bregman and Rudnicky (1975) diagrammed in Figure 7.7, in which the frequency separation of tones A and B separates them from the sequence of the X tones. Both pitch similarity and spectral similarity may be used as grouping cues (Bregman 1990). Deutsch's (1975) striking scale illusion shows how strong the pitch similarity cue can be.

3.4.2 Timbre

The timbre, or tone quality, of an instrument is one of the most important characteristics by which we associate notes with separate instruments. Wessel (1979) demonstrated convincingly the salience of timbre for source grouping, as shown schematically in Figure 7.8. There, **x** represents one timbre and **o** a different one. When the sequence is played slowly, the notes sound like three repeated ascending tones that change in timbre. At a high enough repetition rate, the notes of each timbre are heard in separate streams, with **x** notes and **o** notes in two separate descending streams.

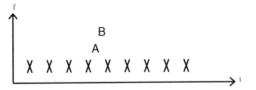

FIGURE 7.7. The Bregman–Rudnicky (1975) experiment. When ony X—A—B—X is presented, it is heard as a four-note melodic stream. When the entire sequence of X's is heard, A and B stand alone as a separate auditory stream.

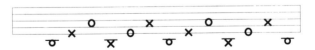

FIGURE 7.8. Wessel's (1979) timbre illusion. X stands for a tone with one timbre and O for one with a different timbre. At slow speeds, all the tones are heard sequentially and make a rising sequence. At fast speeds, the X's and O's separate out to make two distinct streams, each descending.

Bregman (1990) also reported an effect in notes of various pitch with bright and dull timbres. Whenever the timbres were sufficiently similar, that is, when the "formant peaks" in the spectrum were at least as close is frequency as the pitches, then subjects heard the timbral grouping more easily than the pitch grouping.

3.4.3 Rate and Number of Repetitions

The effect of repetition rate on event grouping, like the effect of harmonicity on partial grouping, is so well established that it is frequently used as the independent variable in streaming experiments. Fast sequences of tones split apart into multiple sources more easily (see, for instance, Bregman and Pinker [1978] or Bregman and Dannenbring [1973]).

The number of times that a stimulus is repeated influences the perception of sources in it. Bregman (1978) cited van Noorden as the first to notice this effect and described his experiments with it: van Noorden made up a sequence of notes that played once sounds like a single four-note melody but which played repeatedly sounds like two separate streams of notes. Anstis and Saida (1985) also found a similar result when testing subjects on a sequence of notes alternating between just two pitches.

3.4.4 Loudness

Sound pressure level has two effects on source perception. The first is that low signal intensities promote fusion. Hartmann (1988) reviewed some of the experiments that use this effect. Another effect of loudness on source formation, somewhat disputed, is that sounds of equal or nearly equal loudness tend to be grouped together. Van Noorden (1977) played a repeating sequence of tones, with alternate tones at different loudnesses. Near 35 dB sound pressure level (SPL), at sufficiently high repetition rates, he found that only 3 dB of loudness difference was enough to cause fission of the sequence into two sources.

Bregman (1990) pointed out that van Noorden had his subjects *trying* to hear separate streams, which undoubtedly gave a lower threshold than would be found for involuntary fission. Indeed, the loudness difference needed for involuntary fission would almost certainly introduce forward and backward masking, greatly complicating the task of finding out whether loudness differences alone were responsible for fission. Bregman concluded that loudness differences are not as powerful cues for stream formation as, say, frequency or timbre differences.

Shepard (personal communication) has demonstrated stream separation based on a loudness difference, which leads to the paradoxical effect that a decrease in amplitude can cause an increase in salience.

3.5 *Neural Speculation*

The principles just presented made no attempt to model in detail the action of the auditory system. Here is presented speculation as to how the auditory system may represent the knowledge that different parts of a sound signal belong to different sources and perhaps integrate information from other senses.

3.5.1 Labeling and Filtering

Two different ways that sources may be separated in the auditory system spring to mind; these may be called the *filtering* and *labeling* hypotheses. The filtering hypothesis says that different sources are separated into different neural channels, so that each source gets its own set of neurons for propagating information to the brain. (This approach is the one used in the maps of the event-formation algorithm in Section 4.8.) The labeling hypothesis says that the same set of neurons transmits information about different sources to the brain, but that this information is somehow labeled to identify which parts of it belong to which sources. (Computer simulations of this labeling might show each event with a different hue.)

Both filtering and labeling have obvious strengths and weaknesses. It is easy to see how filtered events could be transmitted neurally, while a method by which a neural signal could label a set of firings as being distinct from one another seems difficult to imagine. On the other hand, the auditory system extracts many different kinds of patterns from a sound, including at least the properties covered in Sections 2 and 3, and replicating the data transmission circuits in the auditory system once for each possible source would seem to be an immense neural burden.

3.5.2 Cortical Oscillation

A potential answer to this dilemma has appeared within the past 7 years. If this theory is applicable, it would imply that the labeling hypothesis is correct. The basic idea is that synchronized neural firing is the label by which different features of a sound signal are identified as belonging to the same source. In this representation, somewhat like time-division multiplexing, different sources share the same neurons to compute and transmit information, but use these neurons at different times to do so. Because neurons typically fire repeatedly at a rate as great as several hundred spikes per second, different sources could be represented by having each one claim a distinct phase or frequency of repetitious firing as its mode of identification.

Although questions abound, the most prominent asking how different neural processes that perform parts of the source formation process, such as feature detection, agree on the phase or frequency to use, evidence for such a mechanism in other senses continues to grow, and it is possible to imagine the mechanism applying to audition as well.

Theoretical underpinning sof the cortical oscillation theory were advanced by von der Malsburg and Schneider (1986), who suggested that synchronous oscillation (firing) of cortical neurons could bind together sense impressions in vision and perhaps other senses as well. They referred back to the idea of Hebb (1949), who suggested that features were organized and processed by means of neural assemblies (von der Malsburg 1986). Freeman (1975) had long recorded synchronicity of firings in olfactory neurons and suggested that it might be binding together separate features in that sensory pathway.

Singer and others, while probing neurons in the visual system, noticed that units fairly far apart from each other were firing in synchrony at a rate of roughly 40 Hz (Barinaga 1990). Gray and Singer (1989) then recorded synchronous firing in different columns of the visual cortex in cats in response to certain stimuli. Gray et al. (1989) found the same result for properties of the entire visual field, leading them to conclude that oscillation synchrony may be a mechanism for grouping spatially separated features.

One factor complicating the understanding of auditory synchronization that is not present in vision is the presence of phase-locking of neural firing to the waveform. It may be difficult to tell whether a given example of synchrony is caused by feature binding or merely phase-locking. Indeed, it may be that phase-locking acts as a trigger for feature binding; perhaps this is why harmonic sounds group together so well. Certainly common onset of energy at different parts of the spectrum, triggering as it does a cascade of neural firing in a short time, could initiate synchronicity in the firing of feature filters. Jenison et al. (1991) pointed out the extent of synchrony of auditory nerve fibers. In the absence of competing periodicities, a speech formant at one frequency triggers synchrony across fibers of a wide range of characteristic frequencies.

The only auditory work in this area has been on the grassfrog. Johannesma et al. (1986) recorded up to four neurons simultaneously, finding that their firing became synchronous with certain stimuli (but not at a rate that implied mere phase-locking to the waveform). They suggested a search for neural assemblies in the auditory system, ones that would "contribute to the generation of figure out of ground, percept out of stimulus, [or] sense out of fact."

These questions and more will have to be answered by neurophysiological experiments.

3.6 Summary of Event and Source Formation Mechanisms

Event and source formation are part of a hierarchical structuring that the auditory system imposes on incoming sound stimuli. In constructing sound percepts, the auditory system must perform allocation and accounting, ensuring that a rough correspondence is maintained between the sound energy present at each frequency and the perceptual objects that explain the sound.

The auditory system uses several principles to make the correct association between sound stimuli and auditory objects. Common fate, including common onset and offset, common frequency modulation, harmonicity, and common amplitude modulation, enables the auditory system to make groupings of different frequency components. Continuity ensures that brief interruptions or changes do not terminate partials, events, or sources. Closure "fills in" obscured parts of perceptual entities. Information from other senses can also contribute to auditory grouping decisions. Hysteresis ensures that groups tend to stay together for a short period of time after the appearance of contradicting evidence.

Several cues affect the grouping of events into sources. In tonal sequences, events near in pitch tend to be grouped together, as do events with a similar timbre. Faster occurrence tends to split sequences of events apart into separate streams, as does repeated presentation. Quiet sounds tend to be grouped into a common source, while the effect of different loudness levels on grouping processes is unclear.

3.7 Integration

The issues that appear at the boundary between the auditory scene analysis model and the rest of the system can be quite elusive. The input to auditory processes is rather well defined: the sound pressure waves reaching the listener, or more precisely, the pressure field over time around the listener's head. Less well defined is the output of an auditory scene analysis model. Experimentally and introspectively, no method is known to isolate such an output: with physiological data, one does not know where to look or what to look for; with psychological data, the responses to auditory stimuli are not precluded from involving nonauditory processes as well. The mechanisms of cross-modal effects (such as effects of lip movement on phoneme recognition) must be considered in building a complete auditory model, and more pervasive than these are the roles of cognitive functions, training, and domain knowledge. It is possible to design specific experiments that minimize or eliminate the effect of nonauditory processes, but in a general model, it is a challenge to draw a clean and plausible boundary line between auditory and nonauditory processes.

An alternative approach (Fodor 1983) tries to avoid this difficulty by eliminating the notion of generic auditory analysis. As described by Massaro (1987), this view assumes that "speech is special in that a specialized speech module is directly responsible for phoneme recognition, rather than operating on preliminary input from the auditory system." Such a speech-specific alternative may have some appeal for certain speech-specific tasks, but the goal here is the broader one of modeling auditory perception. It is very hard to understand how an architecture based on a notion such as Fodor's could be robust in the presence of multiple sources. Would speech, music, and other domains of auditory experience each require their own specialized ways to handle source separation, or is it just speech that is special? In either case, the mechanism seems inefficient, a most unlikely design even with quasi-unlimited processing resources. What is desired is a generic solution to the generic aspects of auditory processing, yet a solution that is capable of interfacing with domain-specific processes.

The position that follows from this discussion is that a comprehensive model or framework of auditory processing, instead of specifying a single well-defined output representation, must focus on a multilevel auditory representation and describe some sort of approach to the integration problem.

There are substantial integration issues within the auditory system alone. Beyond this, a full integration model should also support cross-modal effects and the role that cognitive functions (domain-specific knowledge) may have in source formation and source separation. Although we do not address all these issues here, it is important to grasp the nature of this soft inner boundary of auditory scene analysis.

4. Example Implementation

The preceding two sections presented, respectively, a number of low-level features used by the auditory system for scene analysis with evidence for the computation of them in the auditory system, and a number of principles that apply to higher level grouping processes. Now we turn to a computer implementation of some of these auditory processes, showing first algorithms to filter two of these features, onset and frequency modulation, with examples of their performance. These cue filters are presented as examples of the sort of processing meant by "physiologically compatible" filters for auditory scene analysis. Following these filters is an algorithm for integrating information from the cue filters. This process uses the output of the cue filters, and implements many of the principles covered in Section 3.

These filters and algorithms are designed for and tested on musical sounds, although many of the methods work on other types of sounds as well. The descriptions here are necessarily brief; a detailed description is available in Mellinger (1991). The filters of Sections 4.3–4.5 model processes that happen very early in the auditory system and are thus unlikely to be specialized for any particular domain of sound. The output of such filters may be useful for speech recognition systems, and indeed similar filters have proven useful for recognition of animal calls (Mellinger and Clark 1993). However, the set of cue filters is particularly applicable to musical sound; a system aimed at speech processing, say, would probably include some mechanism for handing formants, stops, releases, etc.

Two noteworthy implementations of auditory models have appeared recently. Cooke (1991) and Cooke and Crawford (1993) presented a model of the auditory system oriented toward source separation. Like the model discussed here, Cooke's model includes layered auditory processing, with a cochlear layer computing a place-based representation, filtering elements that respond to various kinds of features in the sound signal, and a grouping mechanism that tracks harmonics over time. His "synchrony strand" representation captures both relatively low-frequency harmonics and higher frequency partials in a single unified process. This makes it especially applicable to speech processing. He also identified many of the types of features useful for source separation, and suggested a grouping mechanism that brings all the information together to make decisions about what sources are present. Cooke tested his model on a set of sounds made up of sentences containing no stop consonants, like "I'll willingly marry Marilyn," plus various kinds of interfering sounds. Cooke's model performs well on this test set, reliably tracking speech and interfering sounds.

Brown's implementation (Brown 1992; Brown and Cooke 1993) is somewhat similar to the one presented here; it uses feature maps as a basic representation, building up higher levels of knowledge by computing from one feature map to another. In addition to cue filters for onset and frequency modulation similar to the ones to be presented here, Brown's model also detects periodicites in the sound as an aid to making source separation decisons. It also includes an integration mechanism for groups of partials believed to belong to a common source.

The implementations just described, as well as the one described here, rest to some extent on guesses about the nature of processing in the auditory system. The extent of our understanding of auditory processes requires more guesswork the farther the modeling is from the periphery. The model covered here is limited in that it covers only some of the cues for scene analysis mentioned in Sections 2 and 3. However, the framework is believed sound; more cues could be incorporated by developing filters for them similar to those in Sections 4.3–4.5 and using the output of the filters in the integration model of Seciton 4.6.

4.1 Sampling

The implementation here of feature maps, using arrays, implicitly includes the idea of sampling. The auditory system may or may not contain such sampling, so a possible breakdown of physiological compatibility occurs here. For instance, the role or presence of time sampling in various auditory processes in unclear. As any signal of finite bandwidth can be accurately represented by a sampled signal, however, information is not necessarily lost. The sampling interval in a computational model must reflect the auditory system's sensitivity along each dimension of variation. Topographic maps in the auditory system are common, and each position in such a map represents a sample of the parameters that vary along the dimensions of the map.

In the implementational model in this section, the dimensions of each feature map are sampled finely enough to capture all the important information and represent all the important processes. In thinking about feature maps, it is usually safe, because of their analog, continuous nature, to think of them as continuous representations.

4.2 The Cochlear Model

The first step in the implementation of auditory processes is a model of the filtering and transduction performed by the ear, including the cochlea. Several ear/cochlea models have appeared (Patterson 1987; Seneff 1988; Meddis and Hewitt 1991), some including mechanisms for filtering pitch and frequency modulation (Patterson 1987); Chapters 4 and 5 include a more detailed account. The ear/cochlea model used for the experiments here is the Lyon model, as implemented by Lyon and Slaney (Lyon 1982, 1984, 1986; Lyon and Mead 1988; Slaney 1988, 1990). This one was chosen because of its four-stage automatic gain control, which compresses its output signal to a dynamic range that is better suited to neural representation. Although some other models are more accurate at characterizing the dynamics of neural populations (e.g., Cooke's [1991] State Partition Model), none of these covers as many of the various mechanisms present in the ear for gain control as Lyon's.

Instead of the place dimension of the cochlea (which is approximately but not exactly proportional to the logarithm of frequency), we use the exact measure *height*, defined by $h = \log f$, where f represents frequency. On the height scale, a harmonic relationship such as $1:3$ appears as a constant difference in vertical position throughout the map. The output of the cochlear model, a function of time and height, is represented here as $c(t,h)$, where t represents time.

The Lyon–Slaney ear model also computes a correlogram, a feature map with the three dimensions of time, height, and autocorrelation delay d. This map is computed at a given time step in each height channel independently by autocorrelating the cochleagram output over a short time window w:

$$C(t,h,d) = \int_{t}^{t+w} c(t_1,h)\, c(t_1 + d,h)\, dt_1$$

4.2.1 Feature Maps in a Model

In a computational model, a feature map is represented as an n-dimensional array of real-valued numbers, where each number, usually positive, represents the intensity of some feature present in a signal. The number of dimensions n varies from one to three. A typical feature map uses place or height as one of its dimensions, and may have other dimensions, such as time or delay; the value at a particular position typically represents an average rate of neural firing for that position.

In the nervous system, time is usually its own representation, which is to say that it is not coded explicitly (spatially) but implicitly (temporally) in the instantaneous state of the system, that is, as a real-time signal. In our computational realization, time is measured relative to some designated starting time and is simplified by spatializing it. Many of our maps are two-dimensional maps of height against time, analogous to spectrograms with a $\log f$ vertical spacing.

4.3 Filtering Amplitude Onset

An amplitude onset is a rapid rise in sound amplitude over a short period of time; see Section 2.4 for a discussion of the salience of onsets in scene analysis. An onset is represented in a cochleagram by a rapid rise in neural firing rate in a channel. Such onsets can be seen in Figure 7.9, which is a half-second slice of time from a piano performance of a toccata by Frescobaldi (Fig. 7.10). Onsets are easily seen at the start of each note.

The method employed here for filtering onsets is a simple cross-correlation operator. The cross-correlation integral used here operates on a single frequency channel of the cochleagram, a single horizontal slice in a cochleagram. The cross-correlation is defined by

$$o(t,h) = \int_{t}^{t+w} k(t_1)\, c(t + t_1,h)\, dt_1$$

Here, $c(t,h)$ is the cochleagram, $k(t)$ is a correlation kernel, $o(t,h)$ is the output of the onset filter, and w is the window length.

Cross-correlation is physiologically compatible in that for fixed k its output may be thought of as a sum of products of its inputs (or delayed inputs), and such weighted-sum operations are within the functional capability of a neuron. The function $k(t)$, called the kernel, must be appropriately chosen to produce the desired result. The kernel $k(t)$ is the representation as a correlation kernel of a time-differentiation operator (Adelson and Bergen 1986), and looks like Figure 7.11. (The kernel is one dimensional; the vertical dimension shows its value as it extends over the horizontal dimension.)

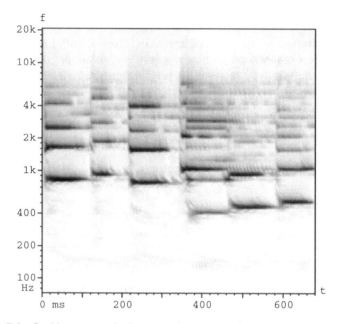

FIGURE 7.9. Cochleagram of piano performance of Frescobaldi's toccata #6 showing onsets at about 0, 120, 220, 340, 360, 470, and 580 msec. Each note appears as a series of harmonics.

FIGURE 7.10. Score for Figure 7.9.

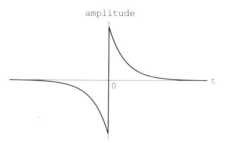

FIGURE 7.11. One-dimensional onset kernel is cross correlated with each height channel of the cochleagram to produce the onset filter output.

The kernel shape is motivated by the action of cells in the cochlear nucleus that respond only at tone onset; Pickles (1988) suggested an excitatory input and a delayed inhibitory input. The cross-correlation kernel corresponding to this has an upward spike that responds strongly to energy in the cochleagram, say at a time t_0, and a downward spike immediately preceding the upward one which inhibits response when there is energy in the channel before the upward spike.

A sample onset map, the result of this per-channel cross-correlation, is shown in Figure 7.12. Here, strong responses occur where there is a rapid rise in energy in a channel. The onset filter described here responds to amplitude onsets but not specially to common amplitude onsets; it does not look for associations between different channels of the cochleagram.

4.3.1 Onset Kernel Function

Several characteristics of this onset filter, or operator, must be tuned for best response. One is the nature of the kernel function. Two kernels that have been tested both have a negative, inhibitory region followed by a symmetrical positive, excitatory region. One was a rectangular funciton and the other an exponentially decaying function; the exponential produces the sharper response for onset filtering.

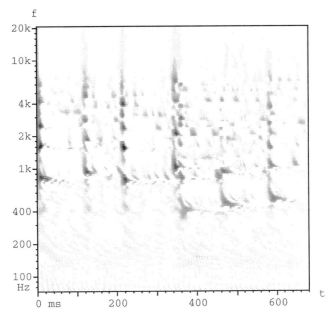

FIGURE 7.12. Onset-filter output map for the Frescobaldi piano excerpt. Note the correspondence of dark areas (high filter output) with the onset times of Figure 7.9.

Another tuning parameter of the onset kernel is its response time. How quickly should the exponential curve decay away from its central ± spike? Latency of onset responses in the cat auditory system vary from about 2 msec in the cochlear nucleus (Rhode and Smith 1986) to 20 msec in the auditory cortex area AI (Pickles 1988). Although a certain latency in a neuron does not strictly imply that the unit uses a characteristic decay of that amount of time, it does set an upper bound on the time duration of the unit's processing. The characteristic ($1/e$) decay times used here vary from 2.3 to 18.1 msec.

Experiments with this onset filter have shown the value of using a range of widths. Different musical instruments have different attack times (Grey 1975; Gordon 1984), and so respond best to different kernels. Each characteristic response time is represented by a different kernel, so there are several kernels, one per response time. The onset feature map thus ranges over four different values of the characteristic decay time. Shorter kernels produce output that is more time accurate but also noisier, and the converse is true for longer kernels.

4.3.2 An Artifact

This onset filter captures important onsets in the sound signal reasonably well, but also picks out some features that do not correspond to perceptual onsets. This happens when a sound is varying in frequency, moving from one cochlear channel to another. As a partial moves from one cochlear channel to another, the new channel has a sudden increase in intensity, causing a strong onset filter response.

This spurious response may be misleading to a higher level scene analysis mechanism because it indicates an onset when in fact none is heard by the human auditory system. Conversely, it may actually help in cases where several partials from one source are moving in parallel, triggering common onset responses that enable a grouping mechanism to make the correct associations among partials.

4.3.3 Filtering of Amplitude Modulation and Offsets

By extending the kernel of this onset filter in time, it can be used as an amplitude modulation filter for the sort of changes found in speech sounds and tremolo in musical ones (see Section 2.7). Indeed, there is no qualitative difference between an amplitude onset and amplitude modulation. They are distinguished here merely by the time scales involved, with amplitude changes happening within the time period found by Grey (1975), about 40 msec, called onsets, and slower rate changes called amplitude modulation.

Also, by inverting the onset kernel in time, the filter becomes an offset filter (see Section 2.4.1). Offset synchrony in scene analysis is not as important as that of onset synchrony, and filtering of it is not pursued further here. Table 7.1 summarizes the characteristics of the onset kernel as has been discussed.

4.4 Filtering Frequency Modulation

The second event-formation feature to be filtered in this computaitonal auditory model is frequency modulation (FM); see Section 2.5 for a discussion of its importance for scene analysis. FM is the movement of partial(s) in frequency, and can be seen the cochleagram in Figure 7.13. Here, two simultaneous sounds move in pitch with different vibrato envelopes. The lower sound, with four harmonics, has a slower and shallower vibrato than the upper sound, with six harmonics.

A source separator is ultimately interested in common FM, that is, modulation in the frequency of several partials in the same direction at the same rate. Such partials have the same slope in the image at one

TABLE 7.1. Summary of onset kernel characteristics.

Kernel shape:	zero-sum $-/+$ spike
Kernel function:	exponential decay
Characteristic decay time:	2.3–18.1 msec
Normalization:	for equal response

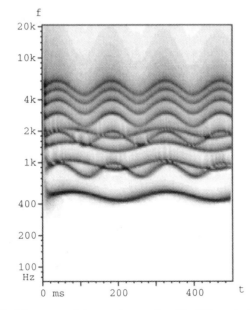

FIGURE 7.13. Cochleagram of two notes, each with different vibrato. Each note appears as a series of harmonics that change in parallel over time.

instant, at one vertical slice. The goal is to find partials with such identical slopes.

4.4.1 Frequency Modulation in the Cochleagram

The method to be described here for filtering a certain slope of frequency-modulated sound operates on the cochleagram, and is similar to some methods that have been used for filtering in computational machine vision (Adelson and Bergen 1986). A two-dimensional cross-correlator that operates on rectangular areas of the cochleagram is used.

The kernel of this cross-correlation operator is chosen to filter FM at a certain rate, for example, at two octaves/sec. The kernel is tilted from the horizontal to an angle corresponding to its specific rate of FM, so that kernels with steeper axes filter higher rates of FM. The kernel in Figure 7.14, when cross-correlated with the cochleagram of Figure 7.13, responds most strongly when it is centered on a partial that is varying in frequency at the kernel's characteristic rate, as is seen in the result map of Figure 7.15. Several factors combine to make this happen. One factor is the use of height, or logarithm of frequency, as the scale in the cochleagram. This ensures that partials that are moving up or down in frequency at a certain rate measured on a logarithmic scale such as octaves/second, semitones/second, or cents/second, will have the same slope. Note that the scale used for the frequency axis here is strictly logarithmic, rather than the scale of equal spacing in the cochlea that is skewed toward linear spacing at lower frequencies.

Using a log-frequency scale, the cross-correlation equation becomes

$$F(h,t) = \int \int_{-\infty}^{\infty} k(h_1,t_1)c(h + h_1, t + t_1)\,dh_1\,dt_1$$

where h is height ($\log f$), t is time, $c(h,t)$ is a two-dimensional function, $k(h,t)$ is a two-dimensional kernel, and $F(h,t)$ is a the output of the FM filter.

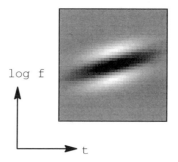

FIGURE 7.14. Two-dimensional kernel for frequency modulaton (FM) filtering. This is cross correlated with a cochleagram to produce the FM filter output.

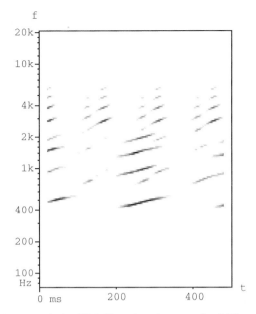

FIGURE 7.15. Output of the FM filter for the sound of Figure 7.13. This was computed with the kernel of the previous figure; strong responses are seen where partials are rising in frequency.

The kernel is the Cartesian product of two one-dimensional kernels. In the time (horizontal) dimension, it is Gaussian to localize its effect to a short time period, to pick out instances of frequency change on a short time scale. In the height dimension, the kernel is center surround, so that it responds to FM only at its characteristic rate. Partials changing at other rates intersect negative regions of the kernel and produce zero or small values.

4.4.2 Center-Surround Functions

Also important is the choice of the center-surround function that is multiplied by the Gaussian function to make the kernel. The usual choice in machine vision work (Adelson and Bergen 1986; Heeger 1991) has been a Gabor function, which is a sinusoid windowed by a Gaussian. Here, the interest is not in the grating shapes best detected by Gabor funcitons, but rather in single partials with a specific FM rate. For this purpose, a center-surround function like the one in Figure 7.16 with a single excitatory lobe and surrounding inhibitory lobes works best.

The center-surround function f should sum to zero or a negative number: $\int_{-\infty}^{\infty} f(x)\, dx = 0$. If it sums to a positive number, then a crossing partial will elicit a positive response, causing smearing in the output image.

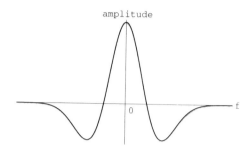

FIGURE 7.16. A center-surround function, the second derivative of a Gaussian. A vertical section through a FM kernel has this profile.

4.4.3 Tuning the FM Kernel

Several parameters influence the operation of the FM kernel. A spread parameter in the time direction determines how wide the Gaussian function is for the kernel. The trade-off between short and long kernels mirrors the familiar time–frequency trade-off (Oppenheim and Schafer 1975; Rabiner and Gold 1975) of the discrete Fourier transform. A short kernel uses information from a relatively small duration of the signal, responding more quickly and filtering out relatively ephemeral frequency sweeps in the sound. A long kernel, while responding more slowly, is more accurate in filtering only FM at its characteristic rate.

Whitfield and Evans (1965) reported, in measuring neural responses to frequency-varying tones in the cat primary auditory cortex, that "as the modulation rate increased, the phase of firing of the unit appeared to lag progressively with respect to the modulation waveform." This suggests a delay time that is approximately constant, as would be produced by an FM filter with a certain size of kernel. (The delays in question, about 20 msec, are generally much longer than the latency of response to tone onset.) Whitfield and Evans also reported a fall-off in unit response at a sinusoidal FM rate of 15 Hz. This rate matches the change in a perception of beating to one of roughness, perhaps because following the FM is not possible at higher rates. The kernel length used here, 43 msec, corresponds to the steepest slope portion of the descending part of a 15-Hz modulator.

The kernel's spread in the frequency direction is chosen so that the positive region (the dark axis in the image of Fig. 7.14) covers the width of a partial in the cochleagram. Pure-tone (sinusoidal) components in the cochleagram produced by Lyon's ear/cochlea model have a $1/e$ size of about four cochlear channels because the spread of activation on the basilar membrane; this is equivalent to 2.2 semitones. Accordingly, the center-surround spread in frequency is set such that the positive part of the function is about 2 semitones high.

The slope of the kernel's axis determines the rate of FM it filters, or, equivalently, the maximum point of its spatial frequency in time–height space. To do grouping by common FM, the filter must find FM at any of the rates that commonly occur in sound. FM in music can happen in smooth steps, as in glissandi and notes with vibrato, or in discrete steps, as in discrete scale runs and trills. The highest rate of FM found in examining a number of pieces of music has been about 4–5 octaves/sec, occurring in rapid whole-note trills and in deep, fast vibrato. Accordingly, the limits of the range of FM filtering have been set at ±5 octaves/sec.

Another question is how many different rates are needed within this range, and what their spacing should be. Using the tunings of parameters just given, the outputs of filters at successive FM rates begin to overlap when the rates are spread about 2 octaves apart, and overlap somewhat more when the rates are 1.25 octaves/sec apart. Consequently, FM rates distributed at 1.25 octaves/sec apart are used.

The best spacing of FM rates depends on the size and especially the time spread of the kernels. As mentioned, a longer lasting kernel is more specific about the FM rate it filters than a short-lasting one. If the rates were more highly specific, each one would cover less of the desired range, so there would need to be more, closer spaced kernels than described in the last paragraph.

Table 7.2 summarizes the characteristics of the FM kernel discussed previously.

4.5 Frequency Modulation in the Correlogram

A second method for filtering FM in a sound works on the three-dimensional correlation feature map, the correlogram, computed by the Lyon–Slaney ear model program. This method supplements the cochleagram method and performs well in different regimes. It is based on the fact that small changes in frequency show up as relatively large changes in the correlogram image.

In the correlogram, a change in the frequency of a partial, upward for example, has two effects. First, the intensity spots corresponding to that partial move upward in the image because they are increasing in frequency. The amount that they move up is the same as in the cochleagram, so there would be no improvement in sensitivity over the method of Section

TABLE 7.2. Summary of FM kernel characteristics.

FM rates:	5, 3.75, 2.5, . . . , −5 octaves/sec
Frequency shape:	zero-sum center surround
Time shape:	Gaussian
Frequency spread:	2.2 semitones
Time spread:	43 msec characteristic decay

4.4.1 if that were the only effect. The spots also group more closely to the left, because an increase in frequency corresponds to a decrease in period, and this decrease is represented by a correspondingly smaller lag value in the correlogram. The space between spots on any horizontal line (any frequency channel) in a correlogram frame is just the period.

This change can be filtered by a correlation-like operator similar to the one described earlier by using a center-surround kernel designed to detect the motion of spots in the correlogram. The correlogram is a three-dimensional array, so that the correlation operation (and kernel) must be three dimensional as well. This presents quite a computational burden. The dimensionality can be reduced to two by noticing that each correlogram spot is constrained to move along a slanting line determined by the spot's frequency and lag value. By taking a two-dimensional slice that includes this line over time, the correlation dot-product operator may be reduced to two dimensions.

Here the method is not described in any detail; the reader is referred to a full account (Mellinger 1991), which covers the factors involved in designing kernels for the operation. A comparison of the cochleagram- and correlogram-based methods is given there as well; this may be summarized with the statement that the correlogram method works best at filtering slow, shallow frequency modulation, and the cochleagram method at faster, deeper modulation.

An example of correlogram-based filtering may be seen in Figures 7.17 and 7.18. First is shown a sound in which the even partials have a slight amount of modulation and the odd partials none; second is shown the corresponding correlogram-based filter output, with strong responses where the partials are decreasing in frequency.

4.6 Algorithm for Event Formation

This remainder of this section presents an algorithm for event formation that incorporates some of the principles put forth in Section 3. The algorithm is limited in scope, as its main goal is to demonstrate the implementation of grouping principles outlined in Section 3 and to show that the computational cue filters produce results that can be used for event formation. The algorithm also shows how some principles for event formation can be implemented. The algorithm does no source formation; that is, once events are detected, there is no attempt to place them in sequential streams of the type mentioned in Section 3.4. That process could operate on the output of this one, but it is more complex than the treatment here. This algorithm is not an attempt to model neural mechanisms, because little is known about them at this level.

This algorithm is also limited in that it uses for input only information from the cochleagram plus the features covered earlier, onset and

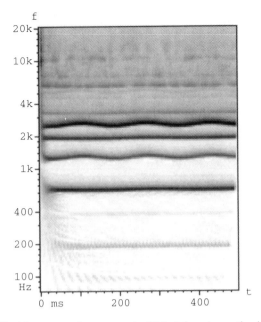

FIGURE 7.17. Cochleagram of an excerpt of McAdams's synthesized sound shows the harmonics of an oboe-like sound; the even and odd harmonics have different amounts of vibrato.

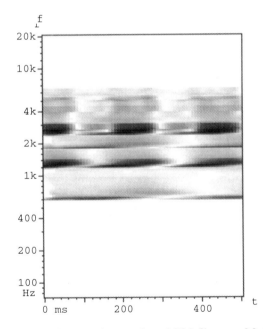

FIGURE 7.18. Output of the correlogram-based FM filter on McAdams's sound. This output map is for an FM rate of 0.5 octave/sec, and shows dark areas where the partials are varying at or near that rate.

frequency modulation. The other features mentioned in Section 2— harmonicity, amplitude modulation, and location in space—are ignored. Any complete event formation algorithm would have to incorporate all these cues and more to have a valid claim of being a model of human auditory scene analysis. It should be emphasized again that this algorithm was designed for use with musical sounds.

4.7 Overview of Operation

The algorithm operates by keeping track over time of two important descriptions of the sound: the set of partials and the set of events. A partial is simply a peak in the cochleagram over some number of time steps. Partials are found by noticing when they start, as revealed by the onset feature map that is one of the inputs to the algorithm, and then tracking them over time until they terminate. This tracking uses cues from the cochleagram, to track partial peaks (spectral peaks) over time and to notice when they disappear, and from the FM feature map, to track changes in the frequency of partials and to keep partials separate when they cross one another.

Events are formed of groups of related partials. Partials are put in close relation by having a common onset when they first start up, or by having common FM as they continue in time. Events can capture and lose partials over time, although this is relatively rare. The input feature maps can contribute evidence both for and against the association of a particular partial with a particular event, as well as contributing evidence both for and against the tracking of a particular partial to particular frequency channels.

The output of the algorithm is the set of events that were found during the processing. Each such output event is stored in a separate feature map, which is the same size as the cochleagram map representing the input. Each such output map has a zero value everywhere but at those time–frequency points where a partial for this event existed; at such a point, the value is the corresponding value from the cochleagram, showing the analog of intensity of neural firing rate for that $T \times F$ point. At this level, events can be detected; each one found by the algorithm is produced as a distinct output.

4.7.1 Cycle of Processing

The process moves through the duration of the input sound signal, one step at a time. A step in this case is a single time step at the sampling rate of the input maps; in the examples in this section, the feature maps have been downsampled to a rate of 441 Hz, so each time step is $1/441 =$ 2.27 msec. This unit of time is called a time slice.

At the start of the processing cycle for each time slice, there may be some existing partials and events. (This is of course not true at the very beginning of the sound, where no partials or events exist yet, but it is true in the general case.) Existing partials are updated by finding what frequency they move to in the new time slice. They are also checked to see whether their peak cochleagram firing intensities have fallen below a minimum threshold for existence and are terminated if so.

Next, onset information is processed. The onset map is checked to see if there are any noticeably strong onsets; onsets that may result from FM of an existing partial are ignored. If there are onsets, the peak values of the onset map are found and turned into new partials. New partials that begin at the same time slice, or nearly the same slice, are grouped in one event on the evidence of common onset.

Next, the algorithm processes FM information. For each event, the FM values of the partials that compose the event are compared. If a partial has FM sufficiently different from that of the other partials in its event, it is separated from the event and placed into its own newly created event. Also, if partials in two different events have sufficiently similar FM, then the two events merge, assembling their partials into one unified event.

Finally, information about whatever events currently exist, if any, is recorded in the output feature maps. For each partial, the current time slice and the partial's current frequency are used to index a $T \times F$ point in the output map for whichever event to which the partial belongs, and this point is marked to provide a record of which partials were part of the event.

4.7.2 Affinity Groups

Partials and events, two of the most important data structures of this algorithm, have already been introduced. Another important one links these two types: the affinity group. An affinity group is simply a set of partials that are closely enough related, or have a high enough affinity for each other, that they are considered to belong to the same event. At any time, each event has its own affinity group that consists of those partials that are currently members of the event. Partials can move into and out of an affinity group as a consequence of the algorithm steps just outlined, but the event-affinity group relation is fixed and one to one. A partial can belong to at most one affinity group, an expression of the principle of exclusive allocation, but it is not required to do so: a partial can be floating, without being attached to any affinity group, for a brief period just after creation (and in a few other cases to be described) until enough information accumulates to assign it to an affinity group or to form its own affinity group if none of the existing ones is acceptable.

4.7.3 Affinity Function

Closely related to the affinity groups is the affinity function $aff(p_1,p_2)$, which keeps track of how strong the belief is at any moment that two partials p_1 and p_2 belong to the same event. $aff(p_1,p_2) = 1$ represents certainty that p_1 and p_2 belong to the same event and $aff(p_1,p_2) = 0$ represents certainty that they should be separate. This function aff is symmetrical; that is, $aff(p_1,p_2) = aff(p_2,p_1)$. The values of the aff function for a partial p are normally initialized to 0.5 when p is created (i.e., $aff(p,p_i) = 0.5$ for all partials p_i).

The values of aff are changed by various processes over time. For instance, two partials p_1 and p_2 that exhibit common FM at a given time will cause the value $aff(p_1,p_2)$ to be raised, while those that undergo frequency change in different directions will have their aff value lowered. Similarly, two partials beginning at close to the same instant, that is, with a common onset, will have their aff value incremented to near one.

The affinity function provides a general means for tracking partials over time, one of the principal problems that must be solved by any scene analysis system. Methods for filtering other types of scene analysis cues, such as harmonicity or spatial location, could be incorporated into this model by using the outputs of the filters over time the change the value of the aff function in ways appropriate to the filter type.

4.7.4 Neighbor Groups

Neighbor groups are groups of partials that are near or at the same frequency. They are created when two partials merge and when two diverge. The principal function of a neighbor group is to keep track of which events may be associated with the partials in the group. Partials in a neighbor group can swap their events around relatively easily. For example, suppose that two partials p_1 and p_2, with events e_x and e_y, respectively, come together and merge. They form a neighbor group. A short while later suppose the peak of the merged group splits; a reasonable assumption is that p_1 and p_2 are diverging again, perhaps because they just crossed each other or touched briefly without crossing.

The problem now is that it is not clear which of the two split partials (i.e., which peak) belongs with which event. The solution adopted in this algorithm is to make a guess based on similarity of frequency modulation, but to allow easy correction in case the guess is wrong.

The newly split partials are kept in a neighbor group. At each time slice, both partials are checked to see if either one has accumulated enough evidence to be certain that the guess was right; that is, if the mean aff value between the partial and the other partials in its event is sufficiently high. If so, the partial breaks out of the neighbor group and becomes solidly attached to its event again. (If there had been exactly two partials in the neighbor group, as in the previous example, then the neighbor group now has only one member and effectively ceases to exist.)

4.8 Results

How well does this algorithm work, given that the cue filters from section 4 are providing the input feature maps? This section presents the results of running it on the output of the cue filters earlier in this section.

The sound from the Frescobaldi toccata (see Fig. 7.9) was interesting for this test because two of the notes happen nearly simultaneously, at 340 and 360 msec. The grouping algorithm found each piano note as a separate event, although it did not always find all the partials for each note. The durations of the notes are usually about right, though some of the partials are cut off short.

A second sound is the mixture of two notes that have different rates of frequency modulation (see Fig. 7.13). In this sound, the two events heard when you listen to the sound are the two notes. These events are difficult for the program to separate correctly for several reasons. They both start at the same time, leaving no common onset cue to distinguish them. At onset, the partials are muddled, not becoming clear for 60 or 70 msec. The partials cross one another repeatedly and sometimes touch without crossing, making partial tracking difficult. There are numerous small features, such as destructive interference of partials at 280 msec, that also make tracking difficult.

The feature maps resulting from running the algorithm on this sound are shown in Figure 7.19 and 7.20. At the beginning of the sound, all the partials are grouped together because they have a common onset. After about 180 msec, enough FM information has been collected to pull apart the two notes, assigning them to separate events. Both events share the same history up to the breakpoint.

A third sound is the McAdams oboe sound (Reynolds 1983) of Figure 7.17. The goal here is to separate out the even and odd harmonics because of their independent FM. This sound is difficult because of the subtlety of FM in it, providing little for the event formation algorithm to work with. The output maps produced by the algorithm are shown in Figures 7.21 and 7.22. Again, at the onset of the sound, the even and odd harmonics are grouped together because of their common onset. Again, after a period of time, the two sets of harmonics split apart into two events, exactly as happens perceptually when listening to the sound.

4.8.1 Limitations of This Algorithm

The model presented here fails to work well in some musically significant situations, namely those in which the onset and FM cues used by the

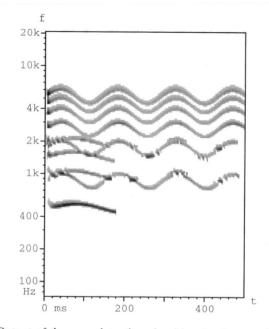

FIGURE 7.19. Output of the even-detection algorithm for the sound of Figure 7.13 (the two-note mixture). This is event one produced by the process; note that the partials of the displayed note do not separate until after about 180 msec, when enough evidence accumulates to force the splitting decision.

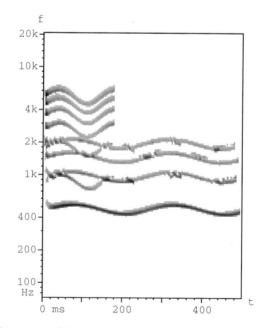

FIGURE 7.20. Event two of the two-note mixture. The first 180 msec of this figure is shared with Figure 7.19, because until that time of splitting there was only one event. The part of the cochleagram common to both events is shown in both event 1 and event 2.

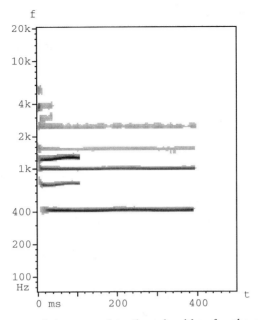

FIGURE 7.21. Output of the event-detection algorithm for the sound of Figure 7.17. This event contains the odd harmonics.

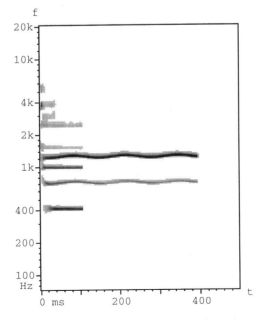

FIGURE 7.22. A second event produced by the event-detection algorithm for the same sound. This event contains the even harmonics.

algorithm are weak or misleading, as in the examples in the previous section. One example of this is when different instruments, or different tones of a multivoiced instrument like the piano, are played nearly simultaneously, as can happen in playing a chord. There may be an onset cue at the start of the chord as the voices commence at slightly different times, but because of the absence of continuing cues, the model will soon group all the partials together. This problem could be cured, at least partially, with a harmonicity cue detector, which could be built using a "harmonic comb" correlation filter than responds to groups of harmonically related partials in a cochleagram.

Another type of failure occurs when too few harmonics are resolved by the cochlear filter. When harmonics are closely spaced, the amplitude of each cochlear channel drops and the partial tracker usually does not detect a partial. This problem becomes especially severe when there are many sources, as in orchestral or choral music.

Intense noise can also make the model perform poorly. String instrument tones are relatively noisy; a rapid succession of such tones, as in a string quartet (or larger group), is very difficult for the model to separate, or even to filter out reasonable onsets.

4.8.2 Quantitative Evaluation

Cooke (1991) described a quantitative evaluation method for computational scene analysis models. A set of test signals with speech sounds, "interfering" sounds, and combinations of these is used. A given scene analysis system must separate the various sounds by grouping their harmonics and/or as a single source. The fraction of times this is done correctly over the entire set is the rating for the system.

If such a set of test sounds were developed for musical scene analysis, it could be a viable way to test the present method. Unfortunately, such a set has not been gathered, principally because of the difficulty of collecting a wide enough variety of musical sounds to form a reasonably complete test set. The set would need to cover variation along at least the dimensions of musical instrument timbre, musical style, tempo, number of simultaneous instruments, reverberation environment, noise level, and noise type, as well as a host of other factors. Qualitative evaluation and analysis like that used in the previous section will have to suffice until such a comprehensive testbed is developed.

5. Summary and Conclusions

5.1 Summary

Sound enters the ear and is transduced, after some gain control processes, into neural firing roughly representing sound amplitude at each frequency. The map of cochlear neuron firings, or cochleagram, is the first of several feature maps—arrays in some number of dimensions—that represent the presence or intensity of some feature. The spectral representation provided by the cochleagram serves as the input to a number of elementary processes that filter features in the sound, including onset, frequency modulation, harmonicity, amplitude modulation, and spatial location. These features, in turn, drive an event-formation process that integrates information from the feature filter output maps and the cochlear nerve firings to decide over time which parts of the spectrum are associated with which events. Event formation influences and is influenced by another process, source formation, that places events into separate sources such that all events that are in an auditory stream originate from the same source. Information about what events are in what sources is passed to higher levels of the system.

This chapter has covered these topics:

An overall discussion of auditory scene analysis, its problems, its methods, and its boundary areas.

A survey of the psychoacoustic and neurophysiological literature for information about how humans and other animals may respond to features and use them for source separation.

A representation for features that may work for several levels of the auditory model and for other tasks like pattern recognition.

Methods for onset filtering and frequency modulation filtering, along with knowledge about how to tune them for musical sounds.

An algorithm using some of the principles of scene analysis to make event decisions, based on integrating cochlear information with onset and frequency-modulation data.

An implemented auditory model providing a framework in which further computational experiments may be carried out.

5.2 Future Work

The performance of our model is far from matching that of human audition. People understand speech even in a room crowded with other talkers; no current source-separation system comes even close to achieving this level of performance at the separation task. We do not yet know even what all of the cues are that might be used for source separation, much less how these cues might be found in a sound signal and filtered out. What is the role of pattern recognition, for example, of phoneme or word recognition, of musical instrument timbre recognition, or of recognition of footsteps and telephone bells? How does the auditory system deconvolve source characteristics and the effects of room acoustics? How do we filter out noise? How do short- and long-term acoustic memory affect source separation? How do linguistic, semantic, and cross-modal clues come into play?

Apart from such longer term research issues, there are quite a few relatively straightforward areas that need immediate work. One possibility is to improve the feature filters. The onset filter could be expanded to capture slower amplitude modulation as well as onsets. Most likely the simple decaying exponential spike used here would need some parameter of variation. Some kind of nonlinearity would likely help with this and possibly with onset filtering as well. The frequency modulation filter could perhaps be improved by nonlinearities, possibly lateral inhibition, and expanded to cover wider band partials than the musical ones concentrated on here.

Harmonicity is probably one of the most important cues for source separation, but other feature filters need to be added to the model. The relationship of pitch detection to harmonicity (or periodicity) detection is still unclear. Regarding pitch, we must distinguish pitch cues (low-level features) from pitch decisions (perceptual effects), and the issue is whether the separation cues interact with one or both of these; additional information from psychoacoustic, neurophysiological, and computer modeling experiments could shed some light here. Cooke's (1991) amplitude modulation techniques may be important too, as well as his method of filtering and representing harmonic structure. Location has been neglected in the field of auditory modeling for scene analysis: no one has yet tried to incorporate localization cues into a sophisticated multicue separation system.

The partial tracking and event-formation algorithm of Section 4 also needs improvement. As noted there, it incorporates only a few of the principles and factors that affect event formation, enough to show the utility of the implemented feature filters. A more complete algorithm would use all of these and probably other as-yet-undiscovered cues as well. In addition, the implemented model does not address source formation. Thus it cannot, for instance, decide which notes in ensemble music belong to which instruments; it can at best notice that single notes exist and determine their spectral content over time. A complete model would use the cues and principles outlined in Seciton 3 and in Bregman's book to track sources over time.

Another direction of work would be to investigate types of sounds other than music. Speech and environmental sounds each have specific characteristics that may require some adaptation and tuning of the source-separation system. For practical applications involving user-assisted source separation, a resynthesis system could be designed to regenerate sound from the output of the event formation algorithm. Such a system would need a modified front-end: either an ear model that records the amounts of gain applied so it could be reversed on output, or a linear front-end like the constant-Q Fourier transform (Schwede 1983). Because the output of the event formation algorithm is essentially binary—what parts of the spectrum are and are not included in each event—the linearity of stages past the front-end need not matter; the initial time–frequency representation is sufficient to regenerate the sound. Most likely a sinusoidal partial tracker (Serra 1988) would improve reconstruction quality.

Another area of work is to examine other auditory processes than source separation in the framework described in the auditory model presented here. For instance, the features filtered here could be used as input to a pattern recognizer, one that identifies time–frequency patterns on the basis of what features are present in what frequency and time relationships.

A final interesting line of work would be to follow up on the cortical oscillation theory of Section 3.5.2. Could one develop a physiologically compatible algorithm for event formation and source formation incorporating synchronous neural firing as an object representation? How would it work?

5.3 Conclusion

This chapter presented some auditory system evidence, a rough model, and some implementational techniques for scene analysis and suggested directions for further work. These lines of development increase and will continue to increase our knowledge of the auditory system and its source-separation methods. In addition, we may soon have a fairly complete source-separation system that can be used for practical applications, including speech separation from interfering sounds, detection of sounds in some kinds of noise, and music transcription.

List of Symbols

Symbol	Quantity	Section in Which Introduced
$C(t,h)$	cochleagram value	4.2
$C(t,h,d)$	correlogram value	4.2
d	delay, or lag	4.2
f	frequency	4.2
$F(t,h)$	FM filter value	4.4.1
h	height, or $\log(f)$	4.2
$o(t,h)$	onset filter value	4.3
t	time	4.2
w	window length	4.2

References

Adelson EH, Bergen JR (1986) The extraction of spatio-temporal energy in human and machine vision. In: Proceedings, Workshop on Motion: Representation and Analysis, pp. 151–155. Los Alamitos, CA: IEEE Computer Society Press.

Anstis S, Saida S (1985) Adaptation to auditory streaming of frequency-modulated tones. J Exp Psychol Hum Percept Perform 11(3):257–271.

Assman PF, Summerfield Q (1989) Modelling the perception of concurrent vowels: vowels with different fundamental frequencies. J Acoust Soc Am 88:680–697.

Balzano GJ (1980) The group-theoretic description of twelvefold and microtonal pitch systems. Comp Music J 4:66–84.

Barinaga M (1990) The mind revealed? Science 249:856–858.

Békésy G von (1963) Three experiments concerned with pitch perception. J Acoust Soc Am 35(4):602–606.

Borden GJ, Harris KS (1984) Speech Science Primer: Physiology, Acoustics, and Perception of Speech. Baltimore: Williams & Wilkins.

Bregman AS (1978) Auditory streaming is cumulative. J Exp Psychol Hum Percept Perform 4(3):380–387.

Bregman AS (1990) Auditory Scene Analysis. Cambridge: MIT Press.

Bregman AS, Dannenbring G (1973) The effect of continuity on auditory stream segregation. Percept & Psychophys 13(2):308–312.

Bregman AS, Pinker S (1978) Auditory streaming and the building of timbre. Can J Psychol 32(1):19–31.

Bregman AS, Rudnicky A (1975) Auditory segregation: stream or steams? J Exp Psychol Hum Percept Perform 1(3):263–267.

Bregman AS, Abramson J, Doehring P, Darwin CJ (1985) Spectral integration based on common amplitude modulation. Percept Psychophys 37:483–493.

Brown GJ (1992) Computational Auditory Scene Analysis. Ph.D. thesis, University of Sheffield, England.

Brown GJ, Cooke M (1993) Physiologically-motivated signal representations for computational auditory modeling. In: Cooke M, Beet SW, Crawford M (eds) Visual Representations of Speech Signals. New York: Wiley, pp. 181–188.

Carlyon RP (1991) Discriminating between coherent and incoherent frequency modulation of complex tones. J Acoust Soc Am 89(1):329–340.

Carloyon RP, Stubbs RJ (1989) Detecting single-cycle frequency modulation imposed on sinusoidal, harmonic, and inharmonic carriers. J Acoust Soc Am 85(6):2563–2574.

Chafe C, Jaffe DA (1986) Source separation and note identification in polyphonic music. Proc IEEE Int Conf Acoust Speech Sig Proc 2:25.6.1–25.6.4.

Chowning JM (1980) Computer synthesis of the singing voice. In: Sound Generation in Winds, Strings, Computers. Stockholm: Royal Swedish Academy of Music, Publ. No. 29, pp. 4–13.

Ciocca W, Bregman AS (1989) The effects of auditory streaming on duplex perception. Percept Psychophys 46(1):39–48.

Cohen EA (1984) Some effects of inharmonic partials on interval perception. Music Percept 1(3):323–349.

Cohen MF, Schubert ED (1987) Influence of place synchrony on detection of a sinusoid. J Acoust Soc Am 81(2):452–458.

Cooke MP (1991) Modelling Auditory Processing and Organisation. Ph.D. thesis, University of Sheffield, Sheffield.

Cooke MP, Crawfod MD (1993) Tracking spectral dominances in an auditory model. In: Cooke MP, Beet SW, Crawford MD (eds) Visual Representations of Speech Signals. New York: Wiley, pp. 197–204.

Darwin CJ (1984) Perceiving vowels in the presence of another sound: constraints on formant perception. J Acoust Soc Am 76(6):1636–1647.

Deutsch D (1975) Two-channel listening to musical scales. J Acoust Soc Am 57(5):1156–1160.

Dirks DD, Bower D (1970) Effect of forward and backward masking on speech intelligibility. J Acoust Soc Am 47(4):1003–1008.

Dowling WJ (1978) Scale and contour: two components of a theory of memory for melodies. Psychol Rev 85(4):341–354.

Durlach NI (1963) Equalization and cancellation theory of binaural masking-level differences. J Acoust Soc Am 35(8):1206–1218.

Durlach NI (1964) Note on binaural masking-level differences at high frequencies. J Acoust Soc Am 36(3):576–581.

Erickson R (1982) New music and psychology. In: Deutsch D (ed) The Psychology of Music. London: Academic Press, pp. 517–536.

Fodor JA (1983) Modularity of Mind. Cambridge: MIT Press.

Freeman WJ (1975) Mass Action in the Nervous System. London: Academic Press.

Fuchs W (1962) Mathematical analysis of formal structure of music. IRE Trans Inform Theory, IT 8:225–228.

Gardner RB, Darwin CJ (1986) Grouping of vowel harmonics by frequency modulation: absence of effects on phonemic categorization. Percept Psychophys 40(3):183–187.

Gardner RB, Wilson JP (1979) Evidence for direction-specific channels in the processing of frequency modulation. J Acoust Soc Am 66(3):704–709.

Goldstein JL (1973) An optimum processor theory for the central formation of he pitch of complex tones. J Acoust Soc Am 54(6):1496–1516.

Gordon JW (1984) Perception of Attack Transients in Musical Tones. Ph.D. thesis, Dept. of Music, Stanford University, Palo Alto, CA.

Gray CM, Singer W (1989) Stimulus-specific neuronal oscillations in orientation columns of cat visual cortex. Proc Natl Acad Sci USA 86:1698–1702.

Gray CM, König P, Engel AK, Singer W (1989) Oscillatory responses in cat visual cortex exhibit inter-columnar synchronization which reflects global stimulus properties. Nature 338:334–337.

Grey JM (1975) An Exploration of Musical Timbre. Ph.D. thesis, Dept. of Music, Stanford University, Palo Alto, CA.

Hafter ER (1971) Quantitive evaluation of a lateralization model of masking-level differences. J Acoust Soc Am 50(4):1116–1122.

Hall JW, Haggard MP, Fernandes MA (1984) Detection in noise by spectro-termporal pattern analysis. J Acoust Soc Am 76:50–56.

Hartmann WM (1988) Pitch perception and the segregation and integration of auditory entities. In Edelman GM, Gall WE, Cowan WM (eds) Auditory Function: Neurobiological Bases of Hearing. New York: Wiley, pp. 623–645.

Hebb DO (1949) The Organization of Behavior. New York: Wiley.

Heeger DJ (1991) Nonlinear model of neural responses in cat visual cortex. In: Landy MS, Movshon JA (eds) Computational Models of Visual Processing. Cambridge: MIT Press.

Jeffress LA (1972) Binaural signal detection: vector theory. In: Tobias JV (ed) Foundations of Modern Auditory Theory, Vol. II. London: Academic Press, pp. 351–368.

Jeffress LA, Blodgett HC, Sandel TT, Wood III CL (1956) Masking of tonal signals. J Acoust Soc Am 28:416–426.

Jenison RL, Greenberg S, Kluender KR, Rhode WS (1991) A composite model of the auditory periphery for the processing of speech based on the filter response functions of single auditory-nerve fibers. J Acoust Soc Am 90:773–786.

Johannesma P, Aertsen A, van den Boogaard H, Eggermont J, Epping W (1986) From synchrony to harmony: ideas on the function of neural assemblies and on the interpretation of neural synchrony. In: Palm G, Aertsen A (eds) Brain Theory. Berlin: Springer, pp. 25–47.

Kay RH, Matthews DR (1972) On the existence in human auditory pathways of channels selectively tuned to the modulation present in frequency-modulated tones. J Physiol 225:657–677.

Knudsen EI (1981) The hearing of the barn owl. Sci Am 245(6):113–125.

Licklider JCR (1951) A duplex theory of pitch perception. Experientia 7:128–133.

Lindemann W (1986) Extension of a binaural cross-correlation model by contralateral inhibition. I. Simulation of lateralization for stationary signals. J Acoust Soc Am 80:1608–1622.

Lyon RF (1982) A computational model of filtering, detection, and compression in the cochlea. Proc IEEE Int Conf Acoust Speech Sig Proc 2:1282–1285.

Lyon RF (1984) Computational models of neural auditory processing. Proc IEEE Int Conf Acoust Speech Sig Proc 36.1.1–36.1.4.

Lyon RF (1986) Experiments with a computational model of the cochlea. Proc IEEE Int Conf Acoust Speech Sig Proc: 1975–1978.

Lyon RF, Mead CA (1988) Cochlear hydrodynamics demystified. Tech Rept CSTR 88–4, California Institute of Technology, Pasadena.

Massaro DW (1987) Speech Perception by Ear and Eye: A Paradigm for Psychological Inquiry. Hillsdale: Erlbaum.

McAdams S (1984) Spectral Fusion, Spectral Parsing, and the Formation of Auditory Images. Ph.D. thesis, Stanford University, Palo Alto, CA.

McAdams S (1989) Segregation of concurrent sounds I: Effects of frequency modulation coherence. J Acoust Soc Am 86(6):2148–2159.

Meddis R, Hewitt M (1991) Virtual pitch and phase sensitivity of a computer model of the auditory periphery: I. Pitch identification. J Acoust Soc Am 89(6):2866–2882.

Mellinger DK (1991) Event Formation and Separation in Musical Sound. Ph.D. thesis, Dept of Music, Stanford University, Palo Alto, CA.

Mellinger DK, Clark CW (1993) A method for filtering bioacoustic transients by spectrogram image convolution. Proc IEEE Oceans '93, pp. 122–127.

Mendelson JR, Cynader MS (1985) Sensitivity of cat auditory primary cortex (AI) neurons to the direction and rate of frequency modulation. Brain Res 327:331–335.

Metz PJ, von Bismark G, Durlach NI (1968) Further results on binaural unmasking and the EC model. II. Noise bandwidth and interaural phase. J Acoust Soc Am 43(5):1085–1091.

Miller GA, Licklider JCR (1950) The intelligibility of interrupted speech. J Acoust Soc Am 22(2):167–173.

Møller AR (1977) Coding of time-varying sounds in the cochlear nucleus. Audiology 17:446–468.

Moore BCJ (1989) An Introducion to the Psychology of Hearing, 3rd Ed. London: Academic Press.

Moore BCJ (1990) Co-modulation masking release: spectro-termporal pattern analysis in hearing. Br J Audiol 24:131–137.

Moore BCJ, Glasberg BR, Peters RW (1985) Relative dominance of individual partials in determining the pitch of complex tones. J Acoust Soc Am 77(5): 1853–1860.

Moore BCJ, Peters RW, Glasberg BR (1985) Thresholds for the detection of inharmonicity in complex tones. J Acoust Soc Am 77(5):1861–1867.

Moorer JA (1975) On the Segmentation and Analysis of Continuous Musical Sound by Digital Computer. Ph.D. thesis, Dept. of Music, Stanford University, Palo Alto, CA.

Oppenheim AV, Schafer RW (1975) Digital Signal Processing. Englewood Cliffs: Prentice-Hall.

Ortmannn O (1926) On the melodic relativity of tones. Psychol Monogr 35(1): 1–47.

Parsons TW (1976) Separation of speech from interfering noise by means of harmonic selection. J Acoust Soc Am 60(4):911–918.

Patterson RD (1987) A pulse ribbon model of peripheral auditory processing. In: Yost WA, Watson CS (eds) Auditory Processing of Complex Sounds. Hillsdale, NJ: Erlbaum, pp. 167–179.

Pickles JO (1988) An Introduction to the Physiology of Hearning. London: Academic Press.

Pierce JR (1983) The Science of Musical Sound. New York: Freeman.

Rabiner LR, Gold B (1975) Theory and Application of Digital Signal Processing. Englewood Cliffs: Prentice-Hall.

Rand TC (1974) Dichotic release from masking for speech. J Acoust Soc Am 55(3):678–680.

Rasch RA (1978) The perception of simultaneous notes such as in polyphonic music. Acustica 40:21–33.

Rasch RA (1979) Synchronization in performed ensemble music. Acustica 43: 121–131.

Reynolds R (1983) Archipelago. New York: C. F. Peters.

Rhode WS, Smith PH (1986) Encoding timing and intensity in the ventral cochlear nucleus of the cat. J Neurophysiol (Bethesda) 56(2):261–286.

Schooneveldt GP, Moore BCJ (1987) Comodulation masking release (CMR): effects of signal frequency, flanking-band frequency, masker bandwidth, flanking-band level, and monotic versus dichotic presentation of the flanking band. J Acoust Soc Am 82(6):1944–1956.

Schooneveldt GP, Moore BCJ (1988) Failure to obtain comodulation masking release with frequency-modulated maskers. J Acoust Soc Am 83(6):2290–2292.

Schreiner CE, Langner G (1988) Coding of temporal patterns in the central auditory nervous system. In: Edelman GM, Gall WE, Cowan WM (eds) Auditory Function. New York: Wiley, pp. 337–361.

Schreiner CE, Mendelson JR (1990) Functional topography of cat primary auditory cortex: distribution of integrated excitation. J Neurophysiol (Bethesda) 64(5): 1442–1459.

Schreiner CE, Urbas JV (1986) Representation of amplitude modulation in the auditory cortex of the cat. I. Anterior auditory field. Hear Res 21:227–241.

Schreiner CE, Urbas JV (1988) Representation of amplitude modualtion in the auditory cortex of the cat. II. Comparison between cortical fields. Hear Res 32:49–64.

Schroeder MR (1968) Period histogram and product spectrum: new methods for fundamental-frequency measurement. J Acoust Soc Am 43(4):829–834.

Schwede GW (1983) An algorithm and architecture for constant-Q spectrum analysis. Proc IEEE Int Conf Acoust Speech Sig Proc 3:1384–1387.

Seneff S (1988) A joint-synchrony/mean-rate model of auditory speech processing. J Phonet 16:55–76.

Serra X (1988) An Environment for the Analysis, Transformation, and Resynthesis of Music Souds. Ph.D. thesis, Dept. of Music, Stanford University, Palo Alto, CA.

Shepard RN (1982) Geometrical approximations to the structure of musical pitch. Psychol Rev 89:305–333.

Shepard RN (1989) Internal representation of universal regularities: a challenge for connectionism. In: Nadel L, et al. (eds) Neural Connections, Mental Computation. Cambridge: MIT Press, pp. 104–134.

Slaney M (1988) Lyon's cochlear model. Technical Report 13, Apple Computer. Available from the Apple Corporate Library, Cupertino, CA 95014.

Slaney M (1990) Interactive signal processing documents. IEEE ASSP Mag 7(2):8–20.

Suga N (1990) Cortical computational maps for auditory imaging. Neural Networkds 3:3–21.

Terhardt E (1972) Zur Tonhöhenwahrnehmung von Klängen II: Ein Funktionsschema. Acustica 26:187–199.

van Noorden LPAS (1975) Temporal Coherence in the Perception of Time Sequences. Ph.D. thesis, Technische Hogeschool Eindhoven, Netherlands.

van Noorden LPAS (1977) Minimum differences of level and frequency for perceptural fission of tone sequences ABAB. J Acoust Soc Am 61(4):1041–1045.

von der Malsburg C (1986) Am I thinking assemblies? In: Palm G, Aertsen A (eds) Brain Theory. Berlin: Springer, pp. 161–176.

von der Malsburg C, Schneider W (1986) A neural cocktail-party processor. Biol Cybern 54:29–40.

Wang K, Shamma S (1995) Auditory analysis of spectro-temporal information in acoustic signals. IEEE Engineering in Medicine and Biol 14(2):186–194.

Warren RM (1982) Auditory Perception: A New Synthesis. New York: Pergamon Press.

Warren WH Jr, Verbrugge RR (1984) Auditory perception of breaking and bouncing events. J Exp Psychol Hum Percept Perform 10(5):704–712.

Weintraub M (1985) A Theory and Computational Model of Auditory Monaural Sound Separation. Ph.D. thesis, Stanford University, Palo Alto, CA.

Wessel DL (1979) Timbre space as a musical control structure. Comp Music J 3(2):45–52.

Whitfield IC, Evans EF (1965) Responses of auditory cortical neurons to stimuli of changing frequency. J Neurophysiol (Bethesda) 28:655–672.

Wightman FL (1973) The pattern-transformation model of pitch. J Acoust Soc Am 54(2):407–416.

Wise JD, Caprio JR, Parks TW (1976) Maximum likelihood pitch estimation. IEEE Trans Acoust Speech Sig Proc 24(5):418–423.

Yin TC, Chan JCK (1988) Neural mechanisms underlying interaural time senstivity to tones and noise. In: Edelman GM, Gall WE, Cowan WM (eds) Auditory Function: Neurobiological Bases of Hearing. New York: Wiley, pp. 385–430.

Young ED, Shofner WP, White JA, Robert J-M, Voigt HF (1988) Response properties of cochlear nucleus neurons in relationship to physiological mechanisms. In: Edelman GM, Gall WE, Cowan WM (eds) Auditory Function: Neurobiological Bases of Hearing. New York: Wiley, pp. 277–312.

8
Computational Models of Binaural Processing

H. STEVEN COLBURN

1. Introduction

Computational models in this chapter are defined to include models that lead to explicit, quantitative predictions for the phenomena that are being modeled. They may be posed purely in terms of the information that is available for the task, in which case the computed predictions are evaluated using information-theoretical or other statistical communication theory techniques, or they may be posed in terms of mechanisms or algorithms. Both types of computational models are included in this chapter. We do not include models that have been suggested but not evaluated or models which are not sufficiently explicit to allow precise predictions.

Binaural processing is reflected in our everyday experiences, and one of the goals of computational modeling is to understand the mechanisms by which we take advantage of the differences between the signals reaching or ears. Of the many subjectively important phenomena that critically involve binaural processing, one of the most obvious is sound source localization, which has been known for many years (Rayleigh 1907) to involve processing of interaural differences. Other day-to-day listening tasks that are substantially assisted by the use of two ears include reverberation suppression and selective listening. (This is often referred to as the "cocktail party effect," which describes our ability to hear out a desired sound source in an environment containing multiple sources). In general, the perceptual effect of listening with two ears is the creation of a three-dimensional world of sound in which specific sound events can be placed and analyzed with minimal confusion with other sound events.

Scientific investigations of binaural processing include both physiological and psychophysical studies, and some of the computational models that we consider are models of neural activity (see Section 3) and some are models of binaural psychophysics (see Section 4). Some of the psychophysical models include physiological elements explicitly; these models are also included in Section 4. To orient the reader and to establish the basic phenomena, psychophysical data that are important for understanding the models discussed here are summarized in Section 2. No similar summary for physiological or anatomical data is provided because it is more convenient to describe these data at the time when the models are presented. These data have been reviewed in previous volumes of this series (see Irvine 1992; Schwartz 1992).

This chapter summarizes attempts to understand the way that binaural information is represented and the way that it is processed by the auditory system. The conceptual goals of this chapter are similar to those of the chapter by Colburn and Durlach (1978), which was written in 1974. In the past 20 years, there has been a significant increase in the availability of computational resources, and developments in models of binaural processing reflect this increase. In addition, information about neural activity in binaural cells is significantly more extensive and is progressing rapidly, which also changes the nature of the models and style of the modeling. However, many of the models that were discussed in the previous chapter are still used and are still of interest, and we refer the reader to that chapter (Colburn and Durlach 1978) for more information about some of the models that are treated very lightly here. A recent review of binaural psychophysical models, emphasizing those based on interaural cross-correlation, has been prepared by Stern and Trahiotis (1995). For readers interested in an encompassing presentation of binaural phenomena, particularly those aspects related to free-field listening, the important book by Blauert (1983) is recommended.

2. Summary of Binaural Psychophysical Data

Basic empirical phenomena that are considered to be directly related to models of binaural processing are outlined in this section. A more inclusive summary of psychophysical data can be found in other volumes in this series (e.g., Wightman and Kistler 1993). As one would expect, modeling efforts tend to address relatively limited sets of phenomena. This is inevitable (and appropriate) when even restricted sets of data are not satisfactorily understood. However, our understanding must ultimately encompass all our empirical knowledge, and models should be evaluated with the broadest empirical perspective. For the purposes of this chapter, we introduce only data that have had primary impact on the formulation of the computational models that we discuss. Although most data are from experiments with headphones, free-field results are outlined first because they are part of our everyday experience and because they are the phenomena of ultimate concern to the processor.

2.1 Free-Field Experiments

Most free-field experiments are related to sound localization, which is simply the determination of the direction and distance of the source of the sound energy arriving at the listener. Our attention is restricted to direction judgments, as most studies have been, although information about distance judgments is available (see Blauert 1983). We separate the relatively straightforward localization task with a single source in anechoic space from the messy but interesting condition in which the listener is localizing sounds in a complex environment, which might involve multiple sources, echoes, and reverberation. Other free-field phenomena, such as the improvements in speech intelligibility achieved when the target speech and interfering sounds are separated in space, are equally important but less investigated and thus are treated briefly in a final subsection.

2.1.1 Localization of Single Sound Sources in Anechoic Space

The auditory localization of sources in typical environments is an ability that is taken for granted by most of us. It is striking when we make mistakes; the underlying processes occur naturally and without an awareness of the process. From the point of view of our daily lives, we are seldom aware of or influenced by our limitations. In anechoic space with most stimuli, listeners with normal hearing can resolve about 1° of horizontal displacement and about 5° of vertical displacement in the direction of maximal performance, straight ahead (Mills 1958). This resolution limit is quantified by the minimum audible angle (MAA), the smallest angular change in the direction of a sound source that can be reliably distinguished from a given reference angle.

Horizontal resolution deteriorates as one moves off to the side. For wideband stimuli, the horizontal MAA increases from 1° to about 5° at the sides (Hausler, Colburn, and Marr 1983); for tonal stimuli, cues are ambiguous (because of front–back symmetries), and performance depends on frequency and the details of the experiment (e.g., Mills 1958). The deterioration in performance off to the sides is partly caused by a decrease in the sensitivity to changes in the interaural time difference (ITD) and interaural intensity difference (IID) as the reference interaural differences increase from zero and partly by the decrease in the physical changes in the ITD and IID for a given change in angle. The latter effect is a consequence of the physical relationship between the angle and the interaural differences at the eardrum. For example, the dependence of the ITD τ on the azimuthal angle θ of the source can be approximated at low frequencies by the relationship

$$\tau = k \sin(\theta) \qquad \text{for } -\pi/2 < \theta < \pi/2 \qquad (1)$$

where θ is defined such that straight ahead corresponds to 0. (See Kuhn 1987 for a careful analysis of the acoustics of sound localization.) This implies that changes in these parameters are related by

$$\Delta\tau = k \cos(\theta) \Delta\theta \tag{2}$$

which, for $\Delta\tau$ sensitivity held constant at $(\Delta\tau)_0$, would imply that

$$\Delta\theta = \frac{(\Delta\tau)_0/k}{\cos(\theta)} \tag{3}$$

The fact that the angle increment goes to infinity at 90° within this approximation reflects the fact that the interaural time delay does not change when a source moves through the maximum value at 90° symmetrically from front to back. A similar ambiguity applies to interaural level differences. For actual listeners, complete symmetry would not hold, but the general problem remains and one might expect that front–back discrimination would be based on changes in the received spectrum similar to the changes that occur from changes in elevation directly in front of the observer.

The general hypothesis, that localization judgments are based on a combination of information from monaurally available spectra and interaural differences in time and intensity, has been suggested and supported with a variety of evidence for many years (Butler 1969; Searle et al. 1976; Hausler, Colburn, and Marr 1983). Which source of information is used in a particular circumstance depends on a number of factors, notably including the nature of the signal, the *a priori* information about the signal and the environment, the direction of the source, and the details of the experiments. Performance also depends on the modulation of the received signals by the motion of the head, on reverberation patterns, and on the compatibility and consistency of the received signals with the listener's experience.

When one considers experiments that are more complicated than simple discrimination, it is clear that sound localization depends on more than the basic sensitivity to the information available. The two primary examples of these types of experiments are absolute identification experiments and pointing experiments. In both cases, the stimuli are chosen from a set of possibilities that may cover a significant range of possible directions, and the stimulus set is an important parameter of the experiment. Because the listener must choose among a number of possible locations, a memory of the cues (attributes of the received signals that are related to the source direction) associated with the possible directions is required in addition to the sensitivity to changes in the cues. Larger stimulus sets also tend to contain more conditions in which a single cue has the same value for more than one stimulus. Formally speaking, absolute identification requires subjects to map a fixed set of n stimulus conditions (that are known to the subject) to a set of n responses such that each stimulus has one correct response and $n - 1$ available incorrect responses. On the other hand, a pointing experiment requires subjects to "point" to the location of the perceived sound, by reporting a set of coordinates, by adjusting a reference source to have the same direction, or by physically pointing (either in space or in a representation). In practice, the stimulus sets may be the same for both types of experiments so that the experiments differ formally in the allowable responses and in the knowledge of the subjects about the set of possible stimuli. (Practically speaking, very large stimulus sets would probably be handled the same way by subjects in both types of experiments.)

Pointing experiments on sound localization of isolated sources in anechoic space were reviewed extensively by Middlebrooks and Green (1991) and by Wightman and Kistler (1993). They showed that performance using broadband stimuli is comparable to what one might expect from the discrimination performance; namely, responses show little bias (shift in mean responses from the actual stimulus) and the scatter in responses from the mean response is of the order of a few degrees, with the smallest deviations in the horizontal plane straight ahead.

An important attribute of sound localization performance is the number of "confusions." Confusions are said to occur when responses are far (relative to normal scatter) from the stimulus position. Typically, the distribution of responses from a given stimulus would be multimodal when confusions are present. Confusions are interpreted as the response to ambiguities in some of the attributes of the physical stimulus. The most significant confusions are related to symmetries about the interaural axis, a problem that Mills (1958) described as the "cone of confusion," which is reflected by front–back and up–down confusions for a given lateral shift that corresponds to roughly constant interaural differences (Makous and Middlebrooks 1990; Wenzel et al. 1993). With broadband stimuli, there are few confusions between front and back (about 2%–10% confusions in Makous and Middlebrooks's data) and between up and down, but confusions become increasingly likely as the stimulus bandwidth or duration decreases.

In absolute identification experiments on sound localization, subject responses are determined by choosing one of a fixed set of possible source directions. In their analysis of direction identification performance, Searle et al. (1976) concluded that errors tend to increase with the total range of angles spanned by the stimuli. A direct measurement of an analogous range effect in headphone lateralization experiments was reported by Koehnke and Durlach (1989). Both reports included speculation that the range effect results from factors similar to those that apply to absolute identification of intensity relative to intensity discrimination, notably memory effects. Memory effects have not been discussed in association with pointing studies.

2.1.2 Localization in Complex Environments

When there are multiple sources present and the task of the subject is to localize the sources separately, performance depends on the extent of spectral overlap of the signals. When the signals have essentially nonover-lapping spectra of comparable intensities, performance is not dramatically affected by the presence of multiple sources. When the spectra are over-lapping, the nature of the signals, their temporal sequence, and their statistical relationship become critical issues and several special cases must be distinguished. A presentation of these results can be found in Blauert's (1983) book.

The localization of a single source in reverberant space shares some attributes of multiple sources in that the echoes constitute additional sources; however, the echoes are specifically related to the original wave-form. This situation has been the focus of many studies with both free-field and analogous headphone stimuli. These studies are often referred to under the general title of "the precedence effect," which originally referred to the observation that the direction perceived in response to a direct sound and its echo corresponds, for appropriate delays between the direct and echoed components, to the direction of the sound arriving earlier, or preceding. For short delays, the perceived direction is inter-mediate between the direct and delayed sound directions; for intermediate delays (in the precedence region), the perceived direction is toward the earlier (direct) sound direction; for larger delays, two sounds are per-ceived, each with its own direction. The delay that separates the pre-cedence condition from the separately perceived echo condition is called the "echo threshold." The term precedence effect has also been used for a variety of phenomena that are presumed to be related to the original observations, including a lack of sensitivity to the interaural parameters of an "echo sound" (Zurek 1980) and aspects of the suppression of physiological responses to delayed components of transient stimuli (Litovsky and Yin 1993). Perceptual studies of precedence effects have been reviewed by Blauert (1983) and by Zurek (1987). Recent free-field studies (Rakerd and Hartmann 1985, 1986; Clifton 1987; Blauert and Col 1989; Clifton and Freyman 1989) have described interesting long-term effects that are dependent on the history of stimulation and listeners' expectations.

2.1.3 Other Free-Field Phenomena

Although various other phenomena have been addressed in free-field studies, few of these phenomena have been specifically addressed by computational models. Most free-field studies are consistent with expecta-tions from headphone experiments, and modeling efforts have addressed the headphone data, which are more plentiful and generally easier to describe because the physical acoustics are so complicated in the free-field case. Studies for which physical acoustics are an essential part of the phenomena include sound localization as just discussed, intelligibility studies, and auditory scene analysis. Intelligibility studies involve judgments about the content of a target speech source in the presence of other interfering speech or noise sources. Our abilities in these tasks are commonly referred to as the "cocktail party effect." For information about recent work on intelligibility studies in the free field, see the series of papers by Bronkhorst and Plomp (1988, 1989, 1992). Auditory scene analysis (Bregman 1990) refers to the perception of the acoustic world as multiple sound sources located in an acoustic environment and to the relationship of these perceptions to the waveforms reaching the two ears. The complexity of this analysis can be appreciated by observing that the signals from all sound sources are combined acoustically and must be parsed (reseparated) by internal processing.

2.2 Headphone Experiments

When headphones are used to present stimuli, it is easy to control the stimuli to the two ears independently. This allows independent measurements of the effects of and sensitivity to the interaural time differences (ITDs), interaural intensity differences (IIDs), and other stimulus attributes. Headphone stimulation also allows direct exploration of several phenomena that are taken for granted in the free-field case. These phenomena include binaural fusion, the process by which a single perceived auditory object is generated from separate stimuli at the two ears, and time–intensity trading, the process by which the influences of ITDs and IIDs are combined to determine the lateral position of an image (typically inside the head with headphone stimulation).

With respect to binaural fusion, it is remarkable that there is almost no awareness of the reception of separate sounds by our two ears. As E. Colin Cherry (1961) noted, we have "two ears [but perceive] one world." There are of course limits to the differences that can be accommodated into a fused percept, and the investigation of these limits has had an important role in the development of computational models of binaural processing. It is clear (as one would expect) that fused images are generated by common frequency components at the two ears with relatively stable interaural relations, and that the images of individual frequency components can often be tracked across the head individually. Initial descriptions of binaural processing as a cross-correlation mechanism (Sayers and Cherry 1957) grew out of these considerations.

The fact that the ears can be stimulated separately raises the possibility that stimuli may be presented that are extremely unlikely to arise in free-field listening and are, in that sense, unnatural stimuli. This issue is particularly important for binaural studies because the interaural relationships of sounds are normally constrained by the physical environment. In spite of the fact that experimental stimuli sometimes have very complex internal images and perceptions, our language and representations for these perceptions are relatively crude. This should be kept in mind as we consider both subjective data generated from unusual stimulus configurations and objective performance, which may vary over time as the subject learns to appreciate the complexity of the image space or as the system changes to process new classes of stimuli. The plasticity of the binaural system has not been studied extensively; important studies include those of Held (1955) and Florentine (1976).

2.2.1. Simulations of Free-Field Conditions

As noted, one would expect that the acoustic issues in the transformation between free-field stimuli and stimuli presented over headphones should be almost trivial. If one simply constructs the waveforms in the ear canals to be the same as the free-field case, then the same perceptions should result. Accordingly, there have been numerous attempts to recreate the sensations of free-field listening using headphone stimuli (Davis 1980; Wightman and Kistler 1989; Ericson and McKinley 1992; Foster and Wenzel 1992; Kulkarni, Woods, and Colburn 1992; Wenzel 1992). The most remarkable aspects of these attempts is the difficulty one has in achieving complete success. It is clearly possible to present sound images with variable and more or less appropriate perceived directions; it is difficult to generate stimuli that are perceived as located outside the head; and it is particularly difficult to simulate images in front of the observer. Several reviews of these experiments are available (Durlach et al. 1992; Wenzel et al. 1993) and this material is not reviewed further here. It is clear that the issues that are involved include the nature of the source material, careful measurement and matching of the source-to-ear transfer functions (including individual variations), the influence of head movement on the received signal, the presence of realistic reverberation, visual cues, and the psychological set of the observers. It is likely that different effects limit perceptions in different circumstances, and of course the list may be incomplete (e.g., bone-conducted vibrations).

2.2.2 Lateral Position of Sound Images

If one presents the same stimuli to the two ears and varies the interaural time delay (ITD) or the interaural intensity difference (IID), the sound image moves inside the head (as one would expect from the physical acoustics of a moving source in free field). Generally, the auditory image (which is normally fused for common inputs) is located toward the ear that receives the leading or the more intense stimulus. If combinations of ITD and IID are imposed, image lateralities can be matched to obtain equivalent combinations of these differences. This equivalence of the effect on lateral position of ITD and IID is called time–intensity trading. The basic observations of these phenomena are presented at some length in Durlach and Colburn (1978); accordingly, this discussion is relatively sketchy, focusing on descriptions that are basic for understanding the computational models and on recent observations.

The relationship between lateral position and ITD depends on the as frequency content of the stimulus. Tonal stimuli below about 1500 Hz have perceived lateral positions that are sensitive to ITD (or, equivalently for continuous tones, interaural phase delay); the image moves toward the side of the leading ear for small delays, and, because the stimulus is periodic, the image location is ambiguous for delays near half-period (or π-phase) delays. We are not sensitive to phase shifts in tonal stimuli for frequencies greater than about 1500 Hz, an observation that is interpreted as being consistent with the decrease in phase-locking to the stimulus fine structure observed in mammalian auditory nerve fibers as frequency increases. In the cat, for example, the synchronization to stimulus fine structure decreases above about 800 Hz, and no phase-synchronization is seen above a few kilohertz (Johnson 1980). [The barn owl, in contrast, shows phase-locking for frequencies as high as 9 kHz (Sullivan and Konishi 1984).]

For wideband stimuli, the perceived image moves monotonically toward the leading side as delay increases from zero, and the image starts to diffuse as the delay increases beyond about 1 msec. This is interpreted to indicate that the range of internal delays available for processing is on the order of a millisecond. For continuous noises at larger delays, the perception approaches that of independent noises: either separate images located at each ear or a cloud that fills the head when delays are greater than about 15 msec. For wideband transient stimuli, the image breaks into two successive events for large delays.

The relation between lateral position and ITD is particularly interesting for bandpass stimuli, as shown in a recent series of papers (Bernstein and Trahiotis 1982, 1985; Trahiotis and Bernstein 1986; Stern, Zeiberg, and Trahiotis 1988; Trahiotis and Stern 1989, 1994). The phase ambiguity for tonal stimuli noted above results in perceptions that have ambiguous locations for narrowband stimuli with low center frequencies. As bandwidth increases, this ambiguity is resolved by the requirement of consistency over frequency, which is also reflected in the increasing prominence of the stimulus envelopes (that are related by the ITD). Thus, for low frequencies and delays greater than a half period, the image moves across the head as the bandwidth increases from narrowband (in which case the phase ambiguity results in a lateral position corresponding to a smaller delay of the opposite polarity) to wideband (in which case the stimulus position matches the side of the leading ear). These results lead to a consideration of the processing of the ongoing envelope as distinct from the ongoing fine structure. In the frequency domain, these stimuli raise issues related to the consistency of ITD across frequency channels. These issues are reconsidered below in discussions of the models.

A related issue is the mapping of ITD to lateral position as a function of stimulus frequency. This topic has been addressed in several studies (Sayers [1964] and Yost [1981] for tone bursts; Shackleton, Bowsher, and Meddis [1991] for narrowband-filtered clicks). These studies concluded that a fixed ITD does not result in a fixed lateral position when different low frequencies are compared.

The lateral position of a sound image is also influenced by changes in the IID (interaural intensity difference). The IID is well defined for cases in which the signals differ only by an amplitude scale factor; however, when the relationship between the signals is more complicated or when signal parameters vary in time, the definition of the interaural difference in intensity is not obvious. Consider, for example, the case of amplitude-modulated tones presented binaurally. If one defines the IID as a very short time measure, which in the limit might be the ratio of the envelopes, then an ITD in the stimulus would also create a time-varying IID. For the purposes of this chapter, we adopt the convention that, unless specifically stated to the contrary, the IID is defined to be the difference between the average intensities measured over the whole stimulus duration. For cases in which the stimuli at the two ears are simply scaled versions of each other, perceptions are relatively straightforward. The lateral positon of the image moves toward the ear with the more intense stimulus and approaches the monaural stimulation case monotonically.

The characterization of the stimulus waveform becomes more complex when the stimuli at the two ears are not related simply by an ITD or an IID. In the more general case, stimuli can have relationships that are combinations of ITDs and IIDs, that are dynamically changing, that vary over frequency, or that have statistically independent components. In many of these cases, the auditory image is more complicated than a simple compact, fused image; this complexity is not discussed here. When the ITD and IID are both varied, the lateral position is determined by their combined effect and is described as time–intensity trading as noted above. For small values of ITD and IID, the interaural differences combine approximately linearly in microseconds and decibels. For larger increments, the effects are nonlinear and effects cannot be simply described. Other attempts to understand the process of lateralization and signal processing in general have led to stimulus situations in which the early part of a stimulus has different parameters than the later parts (Zurek 1987; Shinn-Cunningham, Zurek, and Durlach 1993), transient stimuli in which the phase delays and group delays are different (Henning 1983), stimuli in which the ongoing portion has a different delay than the onsets (Buell, Trahiotis, and Bernstein 1991; Zurek 1993), and stimuli in which the fine structure and the ongoing envelope of the waveform are delayed differently (Trahiotis and Bernstein 1986; Stern, Zeiberg, and Trahiotis 1988; Trahiotis and Stern 1989). These experiments lead to the following conceptualizations: phase delay is more important than group delay for low-frequency transients; both onset and ongoing ITDs affect the lateral position of longer stimuli, and the low-frequency components of a stimulus tend to determine the perceived lateral position. Further discussion of these issues can be found in the cited references.

2.2.3 Interaural Discrimination

It is probably obvious from the preceding discussion that the processing of interaural time and intensity differences is a central issue in the understanding and analysis of binaural processing. One of the basic constraints for computational models of binaural processing is the ability to predict the sensitivity to changes in interaural time and intensity differences. As we have noted, interaural differences are fundamentally related to the localization of sound sources and are present in most binaural stimulus conditions.

Sensitivity to interaural time and intensity differences is usually measured in discrimination experiments, and results are summarized as just-noticeable differences (jnds). A jnd in some parameter is defined as the smallest change that can be reliably distinguished from no change. A rough statement of human ability in these tasks (Durlach and Colburn 1978) is that the interaural time jnd is about 10 μsec and the interaural intensity jnd is about 0.5 decibels (dB). These are values near the best that human subjects achieve; the actual value obtained in a given experiment will depend on the subjects, their training, and the psychophysical technique used for the measurement. For example, in a study that reports relatively quick measurements of abilities in relatively naive subjects, the average interaural time jnd is about 40 μsec for subjects with normal hearing. The jnd varies with frequency for tonal stimuli, and human subjects show essentially no sensitivity to interaural phase shifts for frequencies above about 1.5 kHz.

For wider bandwidths, ITD discrimination performance is possible for all frequencies, although performance based on stimuli limited to high frequencies is not so good as low-frequency performance. Performance with high-frequency stimuli is presumably based on synchrony to the envelopes of the filtered stimulus waveforms. As one would expect, jnds also depend on the level of the stimulus, although the dependence is relatively mild for levels more than about 25 dB above threshold. Also, as one would expect, performance in angle discrimination is consistent with interaural time and intensity jnds in conditions for which interaural difference information is more useful than monaural spectral information. In general, interaural differences are more useful for horizontal discrimination except for reference directions toward the sides and for vertical discrimination with reference positions toward the sides.

The abilities of subjects to discriminate simultaneous changes in ITD and IID have also been investigated. Experiments from this general category are discussed in Durlach and Colburn (1978) as "combination jnds." These experiments include studies of the discriminability of combinations of fixed values of ITD and IID (Gilliom and Sorkin 1972), coherent masking paradigms in which the masker and target waveforms are derived from a common waveform (e.g., Hafter and Carrier 1970; McFadden, Jeffress, and Lakey 1972), and studies of the discriminability of time–intensity traded stimuli (Hafter and Carrier 1972; Ruotolo, Stern, and Colburn 1979). In general, these studies are consistent with the notions that lateral position is a prominent perceptual cue but not the only cue for interaural parameter discrimination. In particular, when the lateral position cue is eliminated, other aspects of the auditory image provide information. As one would expect from the complexity of the image space, there are large training effects, and performance continues to improve after many thousands of trials (Ruotolo, Stern, and Colburn 1979).

The temporal integration of binaural information for sequences of transient stimuli has been studied in a series of experiments by Hafter and his colleagues (Hafter, Buell, and Richards 1988). They measured the interaural time jnd for trains of click stimuli as a function of the number of clicks and the interclick interval. They found the interesting result that later clicks contribute less and less to performance (relative to an optimum integration of information). That is, there is a clear failure to integrate information. It is also interesting that an interruption in the sequence by a brief noise burst or an anomalous gap appears to "restart" the processor, and clicks after the interruption contribute as if they were initial clicks.

Finally, we note two other categories of discrimination studies that are influencing the development of computational models. Motivated by the precedence effect (see Section 2.1.2), studies of the interference of early and late components of paired stimuli have demonstrated that discrimination of interaural parameters of the later stimulus is more affected by the initial stimulus than is the earlier by the later (see discussion in Zurek 1987). Second, interference effects of stimulus components at different frequencies have also been investigated (McFadden and Pasanen 1976; Davis 1985; Zurek 1985; Dye 1990; Trahiotis and Bernstein 1990; Buell and Hafter 1991; Woods and Colburn 1992). Some of these results have been interpreted in terms of auditory scene analysis, as we discuss below, although not all the data are consistent with this interpretation.

2.2.4 Temporal Sluggishness

The ability of the binaural system to follow temporal fluctuations in the interaural parameters has been investigated by Grantham and colleagues (Grantham and Wightman 1978, 1979; Grantham 1982, 1984). They have demonstrated that the binaural system is remarkably "sluggish" in the sense that fluctuations in ITD, IID, or interaural cross-correlation more rapid than about 5 Hz cannot be discriminated from statistically decor-related stimuli (which have incoherent interaural fluctuations). A decor-related reference stimulus must be used because fluctuations that cannot be tracked (i.e., followed) lead to a broadening of the perceived image. (This is to be expected, because any fluctuations in interaural parameters lead to a decrease in the long-term cross-correlation.) This "binaural sluggishness" has implications for the processing of interaural information and has been incorporated into some models of binaural processing, as we discuss later.

2.2.5 Binaural Masked Detection

Conditions in which a target waveform is detected in the presence of binaural masking waveforms are referred to as binaural masked detection. The interesting cases for binaural modeling are those in which interaural differences are useful for the detection of the target. This includes cases in which the interaural parameters of the target and masker waveforms have fixed but different interaural relationships and cases in which the target is monotic (presented to one ear) and the masker is diotic (presented identically to both ears). In these cases, the presence of the target changes the interaural parameters. For example, when the masker is a Gaussian random noise and the target is a tone, the presence of the target causes random fluctuations in the ITD and IID about the reference values (which are determined by the masker).

Binaural detection experiments have played a major role in the development of binaural models. This is presumably related to both the importance of hearing in noisy environments and the richness of the binaural detection data. The parameter space is extensive, and the effects of changes in the parameters are large. For example, in the classic binaural detection condition N_0S_π of a low-frequency tone in wideband noise, when the noise is interaurally identical and the tone contains an interaural polarity reversal (a phase shift of 180° between the tones at the two ears), the threshold of detection is approximately 12 dB lower than the threshold obtained for a similar condition with the tone in phase at the two ears. This binaural detection advantage is a function of frequency and decreases from about 12 dB to about 4 dB as the frequency increases up to 1.5 kHz, after which it remains approximately constant. The evidence is very strong that the binaural advantage comes from the ability to compare the signals at the two ears and to exploit the dependence of the distribution of the interaural differences in the target. For wideband noise maskers, the presence of the tonal target near threshold is perceived as an additional tonelike image located away from the noise image (often located ambiguously at one or both ears). For narrowband noise maskers, the presence of the tonal target near threshold is perceived as a broadening of the image of the noise. Binaural detection data are discussed at length in the chapter by Durlach and Colburn (1978).

Important new information about binaural detection and binaural processing in general is being provided by experiments with reproducible noise waveforms (Gilkey, Robinson, and Hanna 1985; Isabelle and Colburn 1991). In these experiments, performance is measured for individual masking noise waveforms in the context of a randomly presented set of reproducible noise waveforms. Results from these experiments appear to have important implications for models of binaural processing; as we discuss below in the context of binaural models, essentially none of the available models that are able to describe the usual detection data is able to describe the dependence of performance on the individual noise waveform.

2.2.6 Binaural Pitch

Another category of experimental results that have had a significant impact on the development of binaural computational models are binaural pitch phenomena. Generally speaking, binaural pitches correspond to situations in which pitches can be perceived with reasonable reliability only when both ears receive stimuli. That is, if the stimulus is turned off in either ear, the pitch perception disappears. For this reason, these phenomena are often called "binaural-creation-of-pitch phenomena." There are several different kinds of binaural pitches, and a summary, including representative data for most, is found in Durlach and Colburn (1978). For our purposes in the current chapter, it is sufficient to review the situations in which the pitches are generated and to note that these pitches are usually perceived as having specific locations in the (internal) perceptual space; that is, trained subjects report the locations inside the head for these pitch perceptions. The lateral positions of the pitch images are of interest in models for lateral position and binaural pitch. Even though trained subjects can give consistent reports of the pitches and their positions, some of these pitch phenomena are relatively weak percepts and are difficult to hear. Generally, pitches of this type are easier to hear with continuous than with burst versions of the stimuli, especially for listeners who are unfamiliar with these percepts.

One of the strongest (and earliest reported) pitches is the Huggins pitch (Cramer and Huggins 1958), which is generated by a stimulus in which one ear receives a wideband Gaussian noise and the other ear receives an all-pass filtered version of this noise. The all-pass filter for this effect has zero phase shift (and therefore does nothing to the signal) for all frequencies except those in a narrow band in which the phase is a function of frequency, shifting by 2π over that band. Because the noise phase is random with respect to frequency, the noises at each ear are statistically identical and individually contain no information about the phase shift. If the band is in the low-frequency range, roughly the same range (below 1.5 kHz) that is required for sensitivity to the phase of sinusoidal stimuli, a pitch that matches the frequency of the center of the transition band is heard with little difficulty. At higher frequencies, pitches can be perceived and matched by most subjects with practice and good experimental circumstances (Klein and Hartmann 1981).

Pitches that appear to be related to the Huggins pitch can be generated with other stimuli which have interaural differences that change rapidly or have distinctive values in localized frequency regions. A notable example is binaural edge pitch (Klein and Hartmann 1981; Yost, Harder, and Dye 1987). Even stronger pitches can be generated with stimuli that have multiple, locally distinctive (interaurally) frequency regions that are harmonically related (Bilsen 1977). A much more subtle but very interesting pitch is called dichotic repetition (DRP) pitch (Bilsen and Goldstein 1974; Bilsen 1977; Raatgever 1980). The word dichotic indicates that the ears receive different stimuli. In this case, one ear receives a noise signal and the other ear receives a delayed version of this signal with delay of the order of 10 msec. If the delayed version is presented with no polarity shift (DRP+), the perceived pitch is matched to a frequency that is equal to the inverse of the delay. If the delayed version is presented with a polarity reversal (DRP−), the perceived pitch is ambiguous and can be matched to two pitches, one above and one below the frequency that is equal to the inverse of the delay.

The locations within the head of the auditory images associated with these pitches are relatively well defined and have been measured in several cases (Raatgever and Bilsen 1977, 1986). The locations are simply related to the stimulus configurations and are successfully predicted by several computational models of binaural processing, as noted later in the discussions of the individual models.

The last example of binaural pitch is the perception of a pitch at the fundamental when selected harmonics are presented separately to the two ears (Houtsma and Goldstein 1972). This percept is expected and strong if the ears receive multiple harmonics; however, listeners can also perceive pitches in the case in which each ear receives only a single harmonic of a pair of successive harmonics, even when the pair is chosen randomly for each presentation.

Finally, for several binaural pitch experiments results are most easily understood by postulating a dominant frequency region for pitch (Raatgever 1980; Raatgever and Bilsen 1986). This frequency dominance region seems to require a mechanism that emphasizes stimulus frequency components near 800 Hz. As noted, in some models this emphasis is explicitly added and in others it comes as a natural by-product of other assumptions in the models.

2.3 Summary Comments

The psychoacoustic phenomena summarized here were selected to include those with the most direct impact on computational models. The free-field data have been influential primarily in terms of the overall properties of the data with relatively little quantitative modeling. The headphone data, which are fundamentally concerned with the processing of inter-aural differences, particularly ITD and IID, have been used to specify parameter values and to test computational models. Specifically, quantitative predictions for lateral position judgments, discrimination of ITD and IID, binaural masked detection, and pitch experiments have been compared in detail to experimental results. This chapter does not show quantitative fits between models and data because of space limitations. Our discussion of models is more concerned with the conceptual structure of the models, and only general conclusions from comparisons with data are described.

3. Models of Binaural Interaction in Brainstem Neurons

In this section we consider explicit models for binaural processing by neurons in brainstem nuclei. Although influences of contralateral stimulation are observed in the primary auditory neurons (presumably caused by the activity of olivocochlear efferent fibers), neurons in the superior olivary complex (SOC) are the first neurons in the ascending auditory pathway exhibiting binaural interaction with a time scale and sensitivity comparable to the binaural interaction observed in behavioral measurements. Several cell types from either the SOC or from the inferior colliculus (IC) have been modeled explicitly and are discussed in this section. As a guide to terminology and a primitive description of the neuroanatomy of the brainstem, Figure 8.1 shows a cartoon of the ascending auditory pathway, starting with the primary auditory nerve fibers.

In our discussion of these models we use the simplified categorization of Goldberg and Brown (1969). According to this scheme, cells types are identified by a two-letter code in which the first letter represents the predominant response to ipsilateral stimuli and the second letter represents the predominant response to contralateral stimuli. These categories are discussed more fully in Irvine (1992); we consider only EE-type, EI-type, and IE-type cells. The EE-type cells, which are excited by either ear alone, have been identified as candidates for the representation of ongoing interaural time delay information in the SOC. The EI-type and IE-type cells, which are excited by one ear and show inhibitory effect when the other ear is stimulated, have been identified as candidates for the representation of interaural level differences (Boudreau and Tsuchitani 1968) as well as interaural onset time differences (Caird and Klinke 1983) in the SOC and the IC. In addition, Kuwada and Batra (1991) and Joris and Yin (1990) have shown that EI-type cells in the lateral superior olive (LSO) are also capable of representing ongoing timing information.

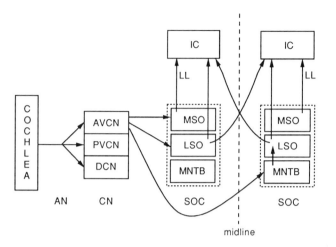

FIGURE 8.1. Organization of major nuclei in the ascending auditory pathway from the auditory nerve to the inferior colliculus. Abbreviations used within the figure and within the text are AN, auditory nerve; CN, cochlear nucleus; AVCN, anteroventral cochlear nucleus; PVCN, posteroventral cochlear nucleus; DCN, dorsal cochlear nucleus; SOC, superior olivary complex; MNTB, medial nucleus of trapezoid body; MSO, medial superior olive; LSO, lateral superior olive; LL, lateral lemniscus; IC, inferior colliculus.

It should be noted that most data from the brainstem have been obtained from experiments in which barbiturate anesthetics were used, and that Kuwada, Batra, and Stanford (1989) have demonstrated that sodium pentobarbital (a barbiturate) has significant effects on the patterns of responses observed in neurons of the IC of the rabbit. Specifically, the anesthesia appears to enhance the effects of inhibition. Because all the models have been evaluated with respect to data from anesthetized animals, it is important to keep this limitation in mind. Data from Kuwada and Batra (1991) suggest that SOC recordings in the unanesthetized rabbit are similar to those from anesthetized dog and cat that were used to compare to the models, although details may be different.

We also note that the discussions in this part of the chapter are focused on the assumptions of the models and their consistency with electrophysiological data. In Section 3.1, models of EE cells in the SOC are discussed; in Section 3.2, models of EI cells in the SOC are discussed and in Section 3.3 models of binaural cells in the IC. The implications of these neural models for the modeling of psychophysical data are considered briefly in Section 3.4. Quantitative models that attempt to integrate physiological results into models of psychophysical data are considered in Section 4.

3.1 Models of EE Cells in the SOC

3.1.1 Jeffress Model

The most prominent and important model of binaural interaction was suggested by Jeffress (1948, 1958) as a mechanism for sensitivity to interaural time delays. In his classic 1948 paper, Jeffress anticipated much that has been measured with modern electrophysiological techniques and suggested an interaural timing mechanism based on a network of cells, each of which would respond to the coincidence of input action potentials from fibers on each side. A diagram of this model is shown in Figure 8.2.

A critical aspect of this model is that each cell in the network is situated such that it is maximally excited by a different interaural time delay

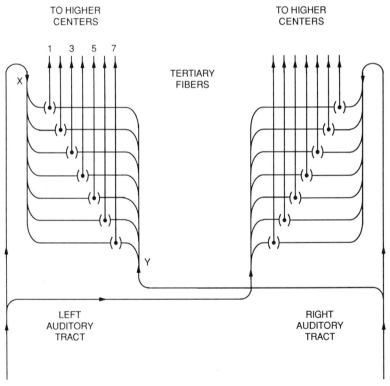

FIGURE 8.2. Neural network proposed by Jeffress for localization of low-frequency tones. The tertiary neurons act as coincidence detectors: a tertiary neuron is more likely to respond when firing from left and right secondary fibers reach the cell body at times that are closer to simultaneity. Interaural time differences in the firings of the input fibers are thus converted to differences in the spatial excitation pattern of the output fibers. (Reproduced from Jeffress 1948 with permission.)

(ITD) of the stimulus. In essence, each cell is associated with a different internal ITD, and a particular external ITD stimulates a distribution of activity along the network that is maximal for the cell corresponding to that delay. As we shall see, this structure, a coincidence network with a distribution of internal delays that translate external delay to a position for the maximum in the network, remains a viable model for the activity of EE cells in the medial superior olive (MSO) and a mechanism for sensitivity to interaural time sensitivity at low frequencies. Most of the modeling work since 1948 can be viewed as essentially refining this basic notion provided by Jeffress.

Although the basic notion of the Jeffress model is consistent with data from many recordings from EE cells, the model is incompletely specified. To predict the behavior of actual cells or to specify the information provided to the central auditory system (i.e., the information available for judgments in psychophysical situations), the descriptions of the coincidence mechanism and of the input firing patterns must be explicitly specified. In the following paragraphs, we summarize the models that have given an explicit computational formulation to the Jeffress model. Before reviewing these quantitative modeling efforts, we review physiological data that have explicitly addressed issues raised by the Jeffress formulation.

The first location in the ascending auditory pathway in mammals at which there is clear evidence of binaural interaction is at the level of the superior olive (Irvine 1992; Schwartz 1992). Similar evidence from birds can be found in Carr and Konishi (1990). Guinan and his colleagues (Guinan, Guinan, and Norris 1972; Guinan, Norris, and Guinan 1972) reported that cells with EE characteristics are found particularly in the MSO, and the primary cells in this nucleus are believed to be of EE type. It is well known that single-cell recordings from cells in the MSO are very difficult to obtain, and reports from cells that have been unambiguously localized to the MSO are very limited. We consider two studies of cells in the MSO that are directly related to the Jeffress model, the study of dog MSO by Goldberg and Brown (1968, 1969) and the study of cat MSO by Yin and Chan (1990). Both studies show the obvious properties predicted by the Jeffress model: that the cells should be excitable by either ear, the cells should have a single maximum in the firing rate versus delay curve for each period of a sinusoidal stimulus, and the delay for which a maximum rate of firing is found should be independent of the stimulus. (Goldberg and Brown measured the dependence on interaural level differences and Yin and Chan measured the dependence on frequency. Except for dependencies that could arise from cochlear effects, the data are consistent with the model.) One of the fundamental properties of the model suggested by Jeffress is that the excitation from each ear would arrive "in phase" for delays that correspond to the maximum firing rate. This notion was tested explicitly in both studies by a comparison between the interaural delay resulting in the maximum rate of response and the delay that is predicted from the phases of monaural responses.

Variations in the interaural level difference did not affect the location of the peaks in the period histogram in the small number of cases measured by Goldberg and Brown (1969). Evidence from these cells leads to the conclusion that the temporal structure of the responses is not influenced significantly by the interaural intensity differences. These data are not consistent with the latency hypothesis (Jeffress 1948). According to this hypothesis, increases in intensity cause decreases in the latency of responses at each ear so that IIDs are converted internally to ITDs. It is possible, however, that the main effects of interaural intensity differences on the temporal patterns appear in neurons that are stimulated away from their best frequencies. This possibility is consistent with the observations of Anderson et al. (1971) that the phase of the period histogram for tonal stimulation of the auditory nerve depends on the level of the tone for frequencies off the characteristic frequency and not at the characteristic frequency.

The latency hypothesis was formulated on the basis of responses to click stimuli, which have not been studied in cells that were histologically verified to be in the MSO. However, in cells deduced to be located in the MSO from the characteristics of the gross potential, Hall (1965) observed significant effects of intensity in firing patterns in response to binaural click stimuli. Hall's observations are discussed below in connection with his model for psychophysical data. Auditory-nerve fiber responses to clicks show level-dependent latencies in the sense that earlier peaks (separated by the period corresponding to the characteristic frequency) grow relative to later peaks as the level increases. The timing of the responses within individual peaks does not show significant effects of level.

Both Goldberg and Brown (1969) and Yin and Chan (1990) noted that the rate of firing of MSO cells in response to binaural inputs at a "bad phase" was often below the rate of response to a single monaural input at the same level. This was interpreted by Goldberg and Brown to indicate that there must be inhibitory inputs. Yin and Chan noted the consistency of this interpretation with the presence of inhibitory inputs to MSO cells. As we see in the models described below, inhibitory inputs are not necessary for the observed behavior, although there is plentiful evidence of inhibitory inputs to the principal cells in the MSO. (Inhibition is discussed further in Section 3.1.6.)

3.1.2 Interaural Delay Mechanisms

There are two important aspects to the Jeffress model. The first is the coincidence detector cell, which has been modeled explicitly, as we discuss, and the second is the mechanism for the generation of the interaural delays. The interaural delays are sketched in the Jeffress model as arising from differences in the pathlengths of the neurons that innervate the coincidence cells. Most models of coincidence networks have focused on this mechanism, either explicitly or implicitly, and there are anatomical and physiological data that are consistent with this mechanism (see, for example, Young and Rubel 1983; Konishi et al. 1988; Smith, Joris, and Yin 1993). Another possible mechanism for the generation of interaural delays was suggested by Schroeder (1977). Specifically, Schroeder noted that if neurons from different locations along the cochlear partition or, equivalently, with different characteristic frequencies are stimulated by a common stimulus, there will be an effective time delay between them. Thus, if neurons from opposite sides and with slightly different characteristic frequencies are compared, there will be an interaural delay imposed without the need for delay lines in addition to the cochlear delay lines that are incorporated into the tuning characteristics.

Although it has been noted that the best frequencies of the monaural responses of MSO cells are approximately equal (Guinan, Guinan, and Norris 1972; Yin and Chan 1990), Shamma, Shen, and Gopalaswamy (1989) pointed out that the amount of mistuning required to generate the observed values of delay (generally less than a millisecond) is very small and could be available with no noticeable mismatch of tuning between the input fibers. Of course, both delay mechanisms may operate for a given cell, and there may be little functional difference between these two alternatives. Shamma, Shen, and Gopalaswamy (1989) described a network that implements a representation of binaural information based on mistuning; however, because the representations are compared with psychoacoustic data and not with neural firing patterns, this model is discussed in Section 4.3 and not further here. An analysis of quantitative models of firing patterns including the mistuning delay was reported recently by Bonham and Lewis (1993); because this work was reported after this review was completed, it is not discussed here.

3.1.3 Shot-Noise Model

Colburn, Han, and Culotta (1990) presented an explicit model of activity of the MSO cell that included *only* excitatory inputs and that generates output patterns that are consistent with the patterns recorded by Goldberg and Brown and by Yin and Chan as has been described. Their model was a relatively abstract, functional model that was based on an internal variable which was identified as the model membrane potential. Input firings from either side generate a jump in this potential with an exponential decay to the resting potential, analogous to excitatory postsynaptic potentials (EPSPs) in a real neuron. We refer to this as the "shot-noise" model following standard engineering terminology for a filtered sequence of pulses distributed as a Poisson process (Davenport and Root 1958). A threshold parameter established the potential at which an output firing would be generated in the model cell. The membrane potential was reset to zero following an output firing. This model cell has only two parameters, the time constant of the EPSP and the threshold. (Of course, the model that describes the firing patterns of the inputs to the cell is also important and contains several critical parameters.) The shot-noise model provides an excellent fit to the physiological data with few parameters.

A sketch illustrating the operation of the shot-noise model is shown in Figure 8.3. The inputs are described by a Siebert–Gaumond model (Siebert 1970; Gaumond, Molnar, and Kim 1982) that is characterized by the driving function and a refractory function. The driving function for the sinusoidal stimuli of concern here was an exponentiated sine wave. The refractory function included a relative refractory interval of 1 msec. The output patterns of this model showed excellent agreement with data from specific neurons reported in the literature. In particular, the model demonstrates the fallacy of the argument that a reduction in the firing rate when stimulation is added to a second ear demonstrates the presence

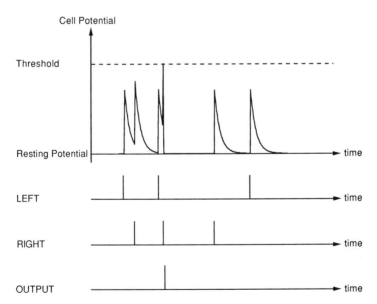

FIGURE 8.3. Shot-noise model of an MSO cell (Colburn, Han, and Culotta 1990). Input firings are indicated by *vertical lines* on the time axes labeled *LEFT* and *RIGHT*. The top graph shows the hypothetical cell potential in response to the input patterns shown. Cell potential jumps in response to inputs from either ear and decays exponentially to the resting potential. Outputs occur when the summed cell potential crosses threshold. Because input firings must occur close together to cause the cell to cross threshold, this model cell operates as a kind of coincidence detector.

of inhibitory inputs. This model does not provide evidence that there are no inhibitory inputs; it simply shows that inhibitory inputs are not required for the observed response. Within the model, the binaural rate can be reduced below the monaural rate because the instantaneous firing rate goes below spontaneous during the half-cycle of the stimulus period, and at bad phase events on each side are aligned with this below-spontaneous rate. For monaural stimulation, higher rates are obtained because one side is always stimulated by spontaneous activity, in additon to "monaural coincidences" as discussed in Colburn, Han, and Culotta (1990).

3.1.4 Point-Neuron Model

A point-neuron model that is consistent with the models described above and formulated in terms of several distinct types of conductance channels has been developed by Han and Colburn (1993) and shown to provide the same level of description of the MSO data as the shot-noise model (Colburn, Han, and Culotta 1990). The point-neuron model is similar to the point-neuron models of MacGregor (1987); it has four conductance channels: one responds to excitatory inputs, one responds to inhibitory inputs, one represents the delayed potassium channel that opens in response to an output action potential, and one represents the constant, residual conductance of the membrane. An equivalent circuit for the model is shown in Figure 8.4

Although the model allows for inhibitory inputs, most of the simulations used excitatory inputs alone. The point-neuron model is more explicitly descriptive of the physiological mechanisms than the functional model, but the cost, from a modeling point of view, is a significant increase in the number of parameters. Even though the model for the input fibers is maintained to be the same as the earlier model, the conductance functions in response to input firings and the conductance change following an action potential are variable, as are the constant potentials associated with each channel and the resting conductance and capacitance. In additon, one can choose, as before, the threshold potential above which an action potential is generated.

Some of these parameters can be approximated on the basis of our general knowledge of electrophysiology, but there remains a lot of freedom to manipulate parameters to fit observed data. The parameter that has the greatest impact on the results and that has to be smaller than our physiological estimate is the effective duration of the excitatory conductance change. This conductance change must be a small fraction of the period of the stimulus waveform. This model leads to a questioning of the simple membrane description in which the membrane properties are fixed until an action potential is generated. A mechanism that might allow a longer duration change in the excitatory channel without losing the ability to synchhronize to the inputs has been suggested by the recent work of

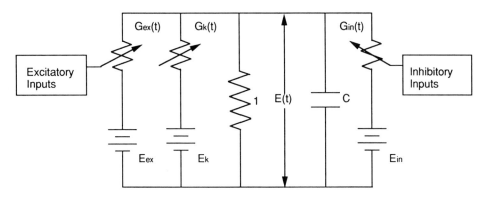

FIGURE 8.4. Point-neuron circuit model for MSO cell (Han and Colburn 1993). The excitatory and inhibitory conductances, G_{ex} and G_{in}, are controlled by the excitatory and inhibitory inputs. Because ipsilateral and contralateral inputs have the same effect on the model cell, they are not distinguished in the circuit diagram. The potassium channel conductance $G_k(t)$ is triggered by the output firings that are generated whenever the cell potential $E(t)$ crosses the threshold. Note that because all conductances and the capacitance were scaled by dividing by the constant conductance, the normalized constant conductance is unity and the normalized capacitance C is equal to the resting membrane time constant. The batteries in series with each dynamic conductance represent the equilibrium potential associated with each ionic species, normalized by subtracting the resting potential. (Reproduced from Han and Colburn 1993 with permission).

Stutman and Carney (1993). They noted that a nonlinear, low-threshold channel of the type reported by Manis and Marx (1991) and by Smith and Banks (1992) could result in an effective sharpening of the timecourse of the excitatory change in membrane potential.

3.1.5 Rate-Combinaiton Model

Although our discussion has been restricted to mammalian data, there are also data from birds that are consistent with the coincidence detector idea. In particular, both anatomical and physiological data from the nucleus laminaris (NL) of the chick (Young and Rubel 1983) and the owl (Sullivan and Konishi 1986; Carr and Konishi 1988, 1990) are similar to MSO data in mammals, and a computational model of these data has been developed by Grün et al. (1990). This model, illustrated by the sketch in Figure 8.5, represents the inputs from each side in terms of firing probability functions (i.e., rate funcitons), which are combined by addition together with a constant inhibition term (a negative quantity) to form a "generator potential." This generator potential is then transformed by a sigmoid nonlinearity to represent the firing probability (period histogram) of the NL neuron. The outputs of this model have been compared

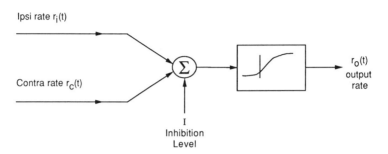

Ipsi rate $r_i(t)$

Contra rate $r_c(t)$

Σ

$r_o(t)$
output
rate

I
Inhibition
Level

FIGURE 8.5. Rate-combination model (Grün et al. 1990). The instantaneous firing rate of the model cell $r_0(t)$ is derived from the sum of two excitatory input rates, an ipsilateral ipstantaneous firing rate $r_i(t)$, and a contralateral instantaneous firing rate $r_c(t)$, and a constant inhibitory level I. The effective excitation represented by the sum of the three inputs is rectified through a memory-less, saturating rectifier to generate the instantaneous firing rate of the cell.

to the period histograms of NL neurons with good agreement. One of the interesting aspects of this model is that the inhibitory input is not synchronized with the fine structure of the stimulus but is used to shift the operating point with respect to the nonlinear transformation. This can be considered functionally as an adjustment to the parameters of the model, an adjustment that would depend on the level of inhibitory activity.

3.1.6 Inhibiton in the MSO

The significance of the inhibitory inputs to the principal cells in the MSO remains uncertain. There is considerable evidence for the presence of inhibition in the anatomy and the neuropharmacology of the MSO (Adams and Mugnaini 1990; Cant and Hyson 1992; Schwartz 1992) and in the physiological data from slice preparations (Grothe and Sanes 1993). The MSO apparently receives inhibitory inputs from the medial and lateral nuclei of the trapezoid body. The role of inhibiton *in vivo* has not been explored experimentally. The modeling work just described suggests that the inhibitory inputs are not necessary for the rates and temporal structure of the MSO response to tonal stimuli, although they may be important for other types of stimuli (see Section 3.3) or for overall modulation and control of the sensitivity of the neurons, as in the model of Grün et al. (1990), for example.

3.2 Models of EI Cells in the SOC

3.2.1 Background

Boudreau and Tsuchitani (1968) reported cells in the lateral superior olive (LSO) that showed excitatory responses to inputs at the ipsilateral ear and inhibitory responses to contralateral inputs. They pointed out that these cells provide a mechanism for sensitivity to the interaural intensity of the input stimuli. These cells individually provide a measure of binaural interaction. Bourdreau and Tsuchitani plotted the rate of firing of LSO cells from their study versus interaural intensity difference with the average intensity (averaged in decibels) as a parameter and showed that the cell firing rates can be approximated as depending only on the interaural intensity difference with small effects of the overall intensity.

These responses are similar to other measurement of the behavior of LSO neurons reported by Guinan, Guinan, and Norris (1972) and by Caird and Klinke (1983). At least partly because of these measurements, the LSO has often been referred to as the location for the measurements of interaural intensity difference (IID); however, Kuwada and Batra (1991) have shown that EI cells in the SOC may show sensitivity to the interaural phase delay (IPD) of low-frequency tones that is comparable to the sensitivity of the EE cells in the SOC, as discussed in Seciton 3.1. The dependence of the firing rate of these cells on the interaural phase of the low-frequency tonal stimulus shows a characteristic delay for which the firing rate is a minimum for all frequencies, suggesting an EI mechanism following an internal delay. A similar dependence on ongoing ITD for amplitude-modulated stimuli has been demonstrated for a wide range of characteristic frequencies by Joris and Yin (1990). Caird and Klinke (1983) noted that their EI cells (which were presumed to be in the LSO) were sensitive to the interaural onset delay of the stimuli (and not sensitive to the interaural phase of the carrier waveform or to the fine structure of the waveform). To some extent, of course, one would expect a dependence on the onset time differences in almost any mechanism that responds to level differences as an onset delay implies that one stimulus is turned on before the other and an IID is inevitably generated during the onset intervals.

3.2.2 Shot-Noise Model

Colburn and Moss (1981) evaluated a simple model of EI cell activity that was suggested by Guinan, Guinan, and Norris (1972) and that is closely related to the model of Molnar and Pfeiffer (1968) for spontaneous activity in the cochlear nucleus. The basic notion is similar to the shot-noise model described in Section 3.1.2 in which the membrane potential of the model neuron is characterized as the internal variable of the model. Input firings from ipsilateral inputs cause a depolarization pulse to be added to the membrane potential, and inputs from the contralateral inputs cause a hyperpolarizaiton pulse to be added to the membrane potential. These individual pulses are assumed to cause a net result that is approximated as the sum of the individual pulses (for subthreshold depolarizations). The cell is assumed to generate an output firing when the depolarization exceeds a threshold amount. Whenever an output firing is generated, the membrane potential is reset to the resting potential.

The operation of this model can be visualized with Figure 8.3 if inputs from the inhibitory side generate negative pulses. The behavior of this cell model was evaluated assuming that the excitatory and inhibitory potential changes were of the same shape and with the same amplitude, specifically pulses that decay exponentially from an initial amplitude. The advantage of exponential pulses is mathematical simplicity: only in this case is the past history of the inputs irrelevant to the future behavior when the current value of the potential is known. Further, because there is a possibility that the inhibitory input could drive the potential to an arbitrarily hyperpolarized state, it is necessary to limit the amount of hyperpolarizaiton that is allowed. Colburn and Moss assumed that the potential reached a barrier at the potential corresponding to the same amount below the resting potential as the threshold is above the resting potential. With this assumption there are two parameters to this model: the threshold, relative to the size of the potential changes generated by the inputs, and the time constant of the decay of the input pulses. The input patterns were taken to be Poisson processes because the inputs from each side were assumed to result from the superposition of a significant number of separate input fibers, each of which was not Poisson. The superposition of a number of statistically independent random point processes are for many purposes well approximated by a Poisson process (Snyder and Miller 1991).

Using mathematical techniques similar to those of Molnar and Pfeiffer (1968), Colburn and Moss demonstrated that the overall patterns of response of the model cell were similar to those measured in the LSO by Boudreau and Tsuchitani (1968) and that the statistics of the model cell response, as represented by the shapes of the interval histograms, were similar to the values measured in the LSO by Guinan, Guinan, and Norris (1972). Specifically, the histograms of the model and the real cell agreed in measures of the mean, the mode, and the width. Also, the model cell showed an appropriate dependence of the rate of firing on the interaural intensity difference and on the overall intensity.

3.2.3 Point-Neuron Model

As one might expect, the shot-noise model just described can be reformulated as a point-neuron model with little change in predictions (Diranieh 1992). The expected dependence on onset time delay and the phase dependence reported for low-frequency sinusoidal stimulation (see Section 3.2.1) were also demonstrated in the Diranieh model. In EI cells, the dependence on the interaural delay shows that the minima of the responses occur at the delay that for the EE cells would generate maxima at all frequencies. One of the most interesting aspects of this study is that the interval histograms for binaural stimulation had a different shape than the interval histograms for monaural stimulation even when the rates of firing were the same for both conditions (binaural and monaural stimulation). This conclusion applies to the shot-noise model (Section 3.2.2) as well as the point-neuron model.

3.2.4 Point-Process Description

Johnson and his colleagues described, in an elegant series of papers (Johnson et al. 1986; Zacksenhouse, Johnson, and Tsuchitani 1992), the statistics of the firing patterns from EI neurons in the LSO. These reports focused on the statistical properties of the firing patterns and, at least in the published work to date, consider only patterns that are generated by high-frequency tones with constant intensities. These studies did not address the cellular mechanisms within the LSO explicitly; rather, they described the statistics of the output patterns in a time-stationary case with the levels of the ipsilateral (excitatory) and contralateral (inhibitory) stimuli as parameters. They used the definitions of Snyder and Miller (1991) related to conditionally Poisson processes for their description and specified the process by the probability of firing in a small time increment. This probability, when normalized by the time increment, is the instantaneous rate of firing, which is a product of the stimulus-dependent driving funciton $\mu_0(\cdot)$ and the history-dependent recovery function $r(\cdot)$ that specifies the refractory properties of the neuron. The sequential dependence of the intervals between firings for high-frequency stimulation is a major focus of this work. Because the model is fundamentally a mathematical description of the patterns and is not formulated as a mechanism, we describe the model with illustrative equations rather than with figures.

Initially (Johnson et al. 1986), only monaural stimulation of the ipsilateral (excitatory) ear was considered, and it is shown that the history effects can be approximated by a first-order interval dependence (essentialy, a tendency for long intervals to be followed by short intervals and vice versa). The interval dependence is incorporated into the mathematical representation by shifting the recovery function an amount that depends on the size of the previous interval. More specifically, the spike train is described by the following expression for the instantaneous rate of firing as a function of τ_{n+1}, the length of the $n + 1$ interval, given τ_n, the length of the nth interval:

$$\tilde{\mu}(\tau_{n+1}; \tau_n) = \mu_0(E)r[\tau_{n+1} - s(\tau_n; E)] \qquad (4)$$

where, as defined, $\mu_0(E)$ is the stationary driving function that depends on the excitatory level E and $r[\cdot]$ is the recovery function with an argument that is a shifted version of the current interval. The amount of shift is determined by the funciton s(·), which decreases with the size of the previous interval (and also depends on the level of excitation). Note that the dependence of the function s(·) on τ_n means that, for a long previous interval, the shift is smaller and the recovery is faster. Thus, the probability distribution of intervals following a longer interval is shifted toward shorter intervals, and the size of the shift depends on the size of the previous interval. The process is thus characterized by a model with three functions: the driving function, the unit recovery function, and the shifting function. This characterizaiton works for neurons with low rates of firing such that the shifts are not bounded by the apparent absolute dead time of the pattern.

In Zacksenhouse, Johnson, and Tsuchitanni (1992), the description is extended in two ways. First, the effects of contralateral (inhibitory) stimulation are represented as a *scaling* of the recovery function. Second, neurons with high rates of firing (such that the expected shift of the function to shorter intervals is blocked by the absolute dead time of the cell) can be described by a "suppress-and-rebound" effect. This is represented in the following equation for the instantaneous rate of firing:

$$\tilde{\mu}(\tau_{n+1};\tau_n) = \mu_0(E)\, S(E,I)\, \{r[\tau_{n+1} - d - s(\tau_n; E)]$$
$$+ g(\tau_{n+1};\tau_n)\}u(\tau_{n+1} - d) \tag{5}$$

where the previous expression has been modified by the scaling function $S(E,I)$ and where the rest of the recovery function has been rewritten to include the suppress-and-rebound factor $g(·)$ with the deadtime d. The function $u(·)$ is the unit step function that is zero for negative values of its argument and unity for positive arguments. The operation of this mechanism of course depends on the factor $g(·)$, which is not described here. The net effect of the suppress-and-rebound term is to keep the probability zero during the dead time but to let the unit catch up with an increased probability of response shortly after the dead time.

These papers provide an excellent description of the statistical properties of the LSO firing patterns and the relation of these patterns to the input stimuli for the high-frequency, steady-input case. They provide insights into the structure of the point processes that are unexpected, notably the simple shifting of the recovery functions. Although the mechanisms for generating the output patterns from the inputs to the LSO are left unspecified, an interesting discussion of the relationship of these results to mechanisms within the neuron are presented in the discussions of the published papers.

3.2.5 Multiple-Component Membrane Model

Johnson and his colleagues have also presented a membrane-oriented model of single LSO neurons (Johnson, Zertunche, and Pelton 1991; Johnson and Williams 1992). This study makes use of available software

packages to simulat the behavior of model neurons. A significant hypothesis from their work is that a critical role is played by a calcium-
dependent potassium channel. A discussion of the implications of the
point-process modeling for single-neuron modeling is contained within
the discussion section of the paper by Zacksenhouse, Johnson, and
Tsuchitani (1992).

3.2.6 Rate-Combination Model

A model of LSO activity based on combinations of average rates of firing
was presented in a pair of papers by Reed and Blum (1990) and Blum and
Reed (1991). The first paper is primarily concerned with the structure of
the connections between the LSO cells and their inputs from the anteroventral cochlear nucleus (AVCN) and the medial nucleus of the trapezoid
body (MNTB) and with the ability of these cells to encode interaural
intensity information. Because there are no explicit comparisons with
physiological data, this paper is discussed in Section 4.3 with other
physiologically based models of psychophysical performance.

The second paper is concerned with the response properties of individual LSO cells and with the development of the connections between
these cells. The model LSO cells are specified by the following assumptions: input fibers are specified by their rate-intensity functions, which are
modeled by four-parameter saturating functions with parameters that
specifiy spontaneous rate, threshold, slope (roughly), and saturated rate.
The excitatory inputs describe the activity of cells in the AVCN, and the
inhibitory inputs describe the activity of cells in the contralateral MNTB.
A set of synaptic rules are used to determine the number and thresholds
of excitatory and inhibitory inputs for each cell. The output rate is a
nonlinear saturating function of the sum of the excitatory rates minus the
inhibitory rates. The output saturating function is similar to the other
saturating functions with the spontaneous rate and the threshold set equal
to zero. In summary, the responses of the model cells are determined by
the function that specifies the input rates, the assumed synaptic connections from input cells to the model cell, and the function that specifies the
output rate from the sums of the input rates.

The output of these model cells is consistent with the output of actual
LSO cells. The rate of firing of the model cells increases with increases in
the excitatory (ipsilateral) intensity and decreases with increases in the
inhibitory (contralateral) intensity. Blum and Reed pointed out that the
model, like the actual cells, shows the property that the rate of firing of
the neurons decreases with increases in overall level when the IID is held
fixed. This property is a consequence of the saturation properties of the
rate functions. Specifically, if the rate of firing of the excitatory fiber is
higher than that of the inhibitory inputs, then the excitatory rate will
saturate before the inhibitory rate; thus, as the overall level increases, the

rate of inhibitory inputs will continue to increase while the excitatory input rate will saturate and the net increase in inhibition over excitation will cause an overall decrease in the output firing rate. This property would be expected for any models with this kind of saturation, including the Colburn and Moss model discussed earlier. The Blum and Reed model is also used to discuss alternative innervation schemes and the development of synaptic connections.

3.3 Models of Binaural Neurons in the Inferior Colliculus

Much more is known about the response patterns of neurons in the inferior colliculus (IC) than about those from the superior olivary nuclei, presumably because the technical problems are less challenging in recording from this nucleus. The variety of response types is broader, and this chapter does not attempt to survey the large number of studies of these neurons (see Irvine 1992). Even though this is a critically important location for binaural interaction, few quantitative models have been developed for these data, in part because the inputs to these neurons, which include the SOC neurons, have not been adequately characterized. We consider two explicit computational models (the only models of which we are aware). It should be noted, however, that many of the empirical resutls from the IC are very similar to those observed in the SOC and thus many of the models described in Sections 3.1 and 3.2 could be considered to be models of the IC data as well. However, they have not yet been evaluated from this point of view.

The first model that has been explicitly specified and evaluated by comparison to physiological measurements from the IC is the model reported by Sujaku, Kuwada, and Yin (1980). The basic structure is a coincidence model of the type we have discussed with the addition of a crossed collateral presynaptic inhibitory input on each side. Four delays can be independently specified for the four inputs (two inhibitory and two excitatory). This model is illustrated in Figure 8.6.

The model explicitly specifies the synaptic interaction in terms of a neurotransmitter substance with first-order dynamics and postsynaptic potentials that are characterized as decaying exponentials with an amplitude that depends on the strengths of the input transmitter levels (after the presynaptic inhibiton is included). The inputs to the model are of the same general type as the models described. Specifically, they were modeled as nonhomogenous Poisson processes with a dead time equal to one-tenth of a period. The driving function of the Poisson process was taken to be proportional to the amplitude of a half-wave-rectified signal. Simulated results from this model are compared to actual data from the cat IC for cases in which the interaural time delay is varied statically and

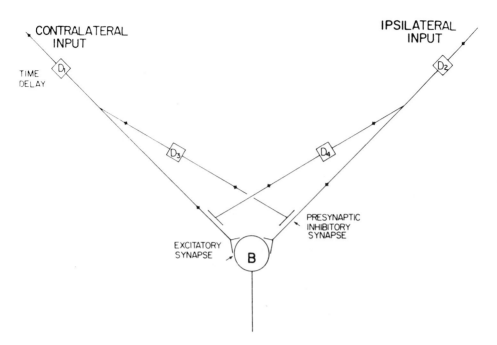

FIGURE 8.6. Point-neuron model for IC cell with crossed collateral presynaptic inhibition (Sujaku, Kuwada, and Yin 1980). Inputs are patterns of neural firings. The blocks labeled D_i are time delays. The inhibitory processes and the excitatory effects on each side are described by synaptic models that include adaptation and saturation. (Refer to original report for specific characterizations.) (Reproduced from Sujaku, Kuwada, and Yin 1980 with permission.)

with a binaural best paradigm (Yin and Kuwada 1983). Model results are compatible with the observed neural responses in typical cases for which responses are cyclic and insensitive to the direction and speed of the phase change in the dynamic (binaural beat) case. In additon, with appropriate choices of parameters, the model describes the observed ITD dependence of a subset of IC neurons that show sensitivity to the speed and direction of the time-varying delay in the binaural-beat stimulus case.

An interesting set of results that imply the operaiton of a long-lasting inhibition at the level of the IC (or peripheral to it) can be seen in the responses to transients in the data of Carney and Yin (1989). It is remarkable that an ipsilateral click is able to inhibit the response to a contralateral click that is applied as much as 100 msec later. This inhibiton can be seen in both directions, although it is usually not symmetrical. As the authors suggest, this long-lasting inhibiton could be the result of single or multiple inhibitory postsynaptic potentials. Recent data with more complex stimuli (Litovsky and Yin 1993; Yin and Litovsky 1993) show inhibitory effects that appear to be related to those observed by Carney and Yin (1989). In Litovsky and Yin's experiments, the stimuli are motivated by the precedence effect in psychoacoustics and show, at the level of a single IC neuron, interactions that appear to be related to phenomena observed with human listeners. Specifically, the response of a cell to a pair of binaural clicks with different ITDs (or to a pair of clicks from separated speakers) is predictable from the response to a single binaural click (or a click from a single speaker) that has a time delay (or chosen according to the precedence effect. For the loudspeaker location case, the single speaker would have a position between the two speakers for time delays within a millisecond and near the first speaker for delays in the range of 2–10 msec.)

Colburn and Ibrahim (1993) reported results of simulations with an explicit computational model that appears to be consistent with many of the transient responses of IC neurons as just discussed (Carney and Yin 1989; Litovsky and Yin 1993; Yin and Litovsky 1993). It is also consistent with the inhibition of transient responses observed in the MSO by Rupert, Moushegian, and Whitcomb (1966). This model hypothesizes that the cell receives excitatory and inhibitory inputs. The excitatory inputs are similar to those in the Han and Colburn (1993) model described in Section 3.1.4, and the inhibitory inputs are relatively slow and long lasting relative to the period of low-frequency stimulus components (tens of milliseconds). The relatively complex patterns of responses appear to result from the relative timing, strength, and duration of the excitatory and inhibitory inputs and from the refractory properties of the model cell. Although the net effects of inhibition are represented by Colburn and Ibrahim (1993) as a single equivalent inhibitory channel for each side with long time constants, more complicated structures could lead to equivalent behavior with shorter individual time constants. Although this model is consistent with many of the data, it fails to predict the observed difference for some cells between (1) the ITD for maximum response to a single click and (2) the ITD for the initial click that generates the maximum inhibition on the following click. With long-lasting inhibition from each inhibitory channel, the optimal inhibitory delay cannot be generated by the relative delay of the inhibitory inputs to this cell. Thus, the inhibitory inputs must arise from the output of another cell that is itself sensitive to ITD.

Finally, neurophysiological data from the owl (Konishi et al. 1988) suggest a sequence of processing steps that includes activity of the owl IC. These suggested mechanisms have not been published as a computational or quantitative model, although there are no obvious barriers to such a model. According to the available data, there are separate pathways from the lower nuclei for interaural time and interaural intensity information. These pathways come together in the external shell of the IC, and the frequency-based representation is integrated into a spatially organized representation in which time, intensity, and frequency coherence are used to resolve ambiguities that would be associated with individual cues at individual frequencies (Knudsen 1984).

3.4 Relationship of Neural Models to Psychophysical Data

The neural models described in the previous subsections generate patterns of activity that show sensitivity to changes in interaural time delay and interaural intensity differences in the stimulus waveforms. Because the neurons being modeled are also generally tuned to particular ranges of frequencies, one can realize that the neural models could be developed into psychoacoustic models for the extraction of ITD and IID values as functions of frequency and time. The availability of the neural information is generally consistent with the broad categories of human and animal abilities, including binaural fusion, separate lateralization of separate stimulus components, sensitivity to ITD and IID, detection based on interaural differences, pitch from binaural differences, and other general phenomena. The development of computational models of these psychoacoustic abilities requires considerably additional work.

The development of a quantitative model for predictions for objective experiments generally requires a specification of internal variability and assumptions about how information is combined over neurons, over time, and over frequency. (Objective experiments are those for which listeners' responses can be determined as correct or incorrect according to a mathematical function [the arguments of which are the stimulus and the responses].) We draw particular attention to the specification of variability that arises internally, which is often called internal noise. A critical attribute of models that make predictions for objective sensitivity measures (such as predictions for interaural jnds or detection thresholds) is the presence and characteristics of internal noise. Without some internal noise or variability, statistical decision theory predicts perfect performance whenever a stimulus causes any change in the mean internal representation; thus, the specific incorporation of randomness or noise into a model is necessary for sensitivity predictions. As the neural activity is stochastic at the auditory nerve level, simulations that include this randomness could be used to estimate performance in objective as well as subjective experiments because repeated simulations with random seeds in the generation of the input patterns can be used to estimate the variability.

Most of the models described here have not been evaluated in terms of the variability of the outputs, nor have they been related to objective psychophysical data. This is not surprising when one considers the complexity of the stimuli and the number of neurons that would have to be incorporated into the simulation. It is a difficult task to derive quantitative predictions for the wide variety of data that have been measured, and there has been little work of this type. Models that have been developed on this basis are described in Section 4.3.

4. Models of Binaural Psychophysics

4.1 Background

As Jeffress (1948) has pointed out, his coincidence network model provides a mechanism for the extraction of information about interaural time difference (ITD). In addition, this network also provides a mechanism for the estimation of the interaural correlation function (ICF) as noted by Licklider (1959). These related conceptions are incorporated, either explicitly or implicitly, into essentially every model of binaural processing.

The relationship between these two conceptions of the Jeffress network (the ITD estimator and the ICF estimator) can be appreciated by considering a simple model of the network in which (1) the peripheral neural coding is such that the neural firing patterns at the periphery are represented by (nonhomogeneous) Poisson processes with rates $r_L(t)$ and $r_R(t)$, (2) there is a set of internal delays such that the inputs to the coincidence detectors have a relative delay of τ_m between left and right inputs, and (3) the output is generated with probability $d(x)$ whenever the inputs to the coincidence device are separated by x, and (4) the function $d(x)$ is nonzero only for x near zero. With these assumptions, then one can show (Colburn 1977) that the expected value of the number of outputs L_m over an interval $(0, T)$ for the mth coincidence device can be approximated [for sufficiently narrow $d(x)$] by

$$E[L_m] = \int_0^T r_L(t)r_R(t - \tau_m)\,\Delta\,dt \qquad (6)$$

where Δ is equal to $\int d(x)\,dx$ [the area under $d(x)$]. The close relationship between the expression for $E[L_m]$ and the cross-correlation function of the input firing rates, combined with the fact that the firing rates are roughly equal to filtered and rectified versions of the stimulus waveforms, makes it clear how the coincidence device could be used to estimate the ICF. The set of outputs of the network considered as a function of τ_m provide an estimate of the ICF.

Independently of the mechanism by which the ICF is estimated, the ITD can be related to the ICF in many circumstances. For conditions in which the stimulus ITD is constant, the value of interaural delay for which the ICF is a maximum is often used in engineering systems as an estimate of the ITD. Of course, when the stimulus ITD varies in time, the relationship between the ICF (which incorporates an average over time) and the ITD is less clear and depends on the duration over which the ICF is averaged (i.e., the duration over which the outputs of the coincidence detector are counted or smoothed). One of the historically important cases in which the ITD varies in time is binaural masked detection. In a representative condition usually notated as the $N_\tau S_{\tau,\pi}$ case, the masking noise is delayed by τ and the target stimulus is both delayed by τ and reversed in polarity. In this case, the total stimulus ITD varies randomly around the reference value of τ with a variance that depends on the stimulus signal-to-noise ratio (Zurek 1991). The special case with $\tau = 0$ is the classical binaural detection case, usually notated as $N_0 S_\pi$; all this discussion applies to that case as well. As the ITD varies over time, the pattern of outputs of the coincidence network vary with time; thus, a local (in time) average of the outputs of the coincidence network would lead to a local estimate of the ITD, whereas a long-time average of the outputs would lead to an estimate of the ICF. For the $N_\tau S_{\tau,\pi}$ detection condition with the target level near threshold, the long-term ICF would be maximum at the internal delay matching the external noise delay τ independent of the presence of the target and the presence of the target would be seen in the shape of the ICF, Specifically, the ICF would have a smaller maximum correlation and a greater width along the τ_m axis.

In connection with the comments about the relationship between the ICF and ITD estimation, the special case in which the effective input stimuli are narrowband may be interesting because most current conceptions of binaural models assume that interaural comparisons are made between frequency-matched, bandpass-filtered versions of the stimulus waveforms. If the stimuli are sufficiently narrowband, they can be conceptualized as sinusoidal stimuli with slowly varying amplitudes and phases (the narrower the bandwidth, the slower the variation). For times that are short compared to these variations, the short-time ICF would simply be the cross-correlation function of two sinusoids, and the only interaural information would be contained within the interaural phase difference represented by the ITD and in the level differences, which do not appear in the normal ICF. If the amplitudes are ignored, complete information about the interaural temporal relations of the stimuli is contained within the sequence of ITD values.

The interaural intensity differences (IIDs) in the stimulus (defined for the narrowband stimuli considered here as the ratio of the envelopes) are generally not contained within the ICF or within the sequence of ITD values, at least not directly. (Because the amplitude envelope and the minimum-phase function of a narrowband stimulus are constrained by Hilbert transform relations [Oppenheim and Schafer 1989], the ITDs and IIDs should also be constrained in some way, but the nature of the constraints has not been explored.) Indirect mechanisms for the incorporation of IID information into the ITD and ICF have been suggested, including (1) the latency hypothesis (Jeffress 1948), according to which the latency of neural responses decreases with increases in the level of the stimulus; (2) the assumption of a separate mechanism for the estimation of IID and the assumption that both time and intensity information are separately available to an ideal processor; and (3) the assumption that the separately estimated IID is combined with the ITD information, possibly by weighting the ICF and possibly by a forced combination to a combined variable that is assumed to be available for decision making.

With respect to the latency hypothesis, several authors have argued against the hypothesis on the basis of psychophysical data, including von Békésy (1930) and Sayers and Cherry (1957). From the physiological data, the issues are complicated. Auditory-nerve responses to sinusoidal stimuli show negligible shifts of phase with level at the characteristic frequency (CF) of the fiber (over most of the stimulus range) but show shifts at frequencies away from CF with direction of shift that depends on whether the stimulus frequency is above or below the CF (Anderson et al. 1971). Auditory nerve responses to impulsive (acoustic click) stimuli show a level dependence of the temporal pattern, but primarily as a change in the relative size of the response at peaks in the PST histogram that are separated by multiples of the period of the CF; shifts are not seen in the timing of the response within a given peak. In terms of separate mechanisms, physiological mechanisms for the estimation of IID have been suggested, usually in terms of networks of EI or IE cells, and specific models for this purpose are discussed in Section 4.3. At this point, there is no definitive answer as to the relative importance of the various possibilities listed here.

In the rest of Section 4, we separate the discussion into "black-box" models and "pink-box" models. In general, black-box models operate mathematically on the stimulus waveforms to predict psychoacoustic performance and pink-box models incorporate descriptions of physiological data explicitly. This distinction is somewhat artificial within this chapter. As we shall see, most of the black-box models that we discuss are based on the ITD or the ICF of the stimulus, and, in terms of our foregoing discussion, could be restated to include a coincidence network with explicit physiological descriptions, transforming them to what we would include as pink-box models. On the other hand, we have included within the pink-box section models that imply a physiological mechanism (usually by analogy with a model that has presented a physiologically explicit version) even when the model is actually unconstrained by actual physiological data.

We have decided to use this separation, even though it is problematic, for three reasons. First, most of the models in our black-box category were originally presented and analyzed in the literature without explicitly incorporating physiological information. Second, in some cases, physiological detail is a distraction to the basic structure of the information processing and provides no additional insight. Third, we think this separation allows contiguous discussion of closely related models. Some compromises in the categorization are being made to allow the grouping of similar models together even when members of the group could be put into the other category. (Our overview and summary of all the models is included as Section 5.)

4.2 Block-Box Psychophysical Models

Models that relate performance to the input stimuli without attempting to incorporate what is known about internal processing are called black-box models. The advantages of black-box models are their ability to abstract those aspects of the stimulus that are of fundamental importance for the tasks considered as well as their ability to demonstrate directly the relationships among different perceptual phenomena. They are generally efficient means of representing our state of understanding of the phenomena. The most important criterion for a black-box model is the number of free parameters relative to the quantity and complexity of the data that they are able to predict or describe.

In our previous review chapter (Colburn and Durlach 1978), we divided black-box models into four categories: count-comparison models, interaural difference models, noise-reduction models, and interaural correlation models. The interested reader should consult that chapter for a review of modeling results and ideas up to 1974, the year that chapter was written. In this section, some of the material from that chapter is repeated for historical perspective and continuity of ideas. We do not include any information on count-comparison models because there has been no recent work in that category. The other three categories of models are considered in the next few paragraphs, although much of the recent work related to interaural correlation models was done in a physiologically explicit context and is treated in Section 4.3 in the discussion of pink-box models.

4.2.1 Interaural Difference Models

As described above, Jeffress (1948) noted that the coincidence network he suggested could be used to estimate the interaural time delay (ITD) simply by looking for the delay that leads to the maximum output in the network. Webster (1951) noted that the ITD is a source of information about the presence of the target in a binaural detection experiment (when the interaural relationships of the masker and the target are different). Webster's (1951) hypothesis, that the distribution of ITD sample values is the basis for detection, was the first of a sequence of models (Hafter 1971; Yost 1972) that are based on the assumption that estimates of the ITD, the IID (interaural intensity difference), or combinations of these interaural differences are used for detection judgments. The most prominent combination is a fixed linear combination with a frequency-dependent combination weight that results in a variable analogous to the lateral position. We refer to these models collectively as interaural difference models. The motivating ideas of these models in the current context are that mechanisms for the extraction of interaural difference information are available (cf. Section 3), that localization and lateralization are naturally incorporated into these models, and that the ITD and IID also provide information about binaural detection, binaural pitch, and other phenomena.

The notion that sensitivity to changes in interaural time and intensity differences is based on changes in lateral position was tested directly by Domnitz and Colburn (1977), who measured interaural discriminatin of interaural time and intensity differences (on and off the midline) for tonal stimuli as well as lateral-position judgments for the same stimuli and the same subjects. They showed that their data were consistent with the notion that changes in lateral position provide the cues for interaural time and intensity discrimination. In their quantitative evaluation, the variance of lateral position is specified by the intensity discrimination data and used to predict the time discrimination results. The results supported the general notion of a decision variable based on lateral position for most values of the reference interaural parameters.

As we have noted, interaural differences vary in time during the presentiation of some stimuli, and models based on interaural differences must specify which samples of interaural differences are used and how they are combined. For example, Webster (1951) calculated predictions from an approximation based on the maximal ITD sample (i.e., the ITD value for the tone-noise phase resulting in the maximal ITD within the distribution). Domnitz and Colburn (1976) demonstrated that the observed dependence on target interaural parameters would be obtained by *any* model based on ITD, IID, or a fixed combination of these differences. As noted previously (Colburn and Durlach 1978), Webster assumed a sensitivity to a 100-μsec deviation for the detection performance, a value that is significantly larger than the observed sensitivity to ITD (which leads to threshold values closer to 10 μsec). This difficulty, which is also seen with other formulations of interaural difference models (e.g., Colburn et al. in manurscript), could of course be related to the fact that, in discrimination experiments, the subject is judging shifts in the mean of the distribution of interaural differences whereas in masked detection experiments the subject is judging the width of the distributions of interaural delays.

Zurek and Durlach (1987) pointed out that temporal averaging of the interaural differences leads to a reduction in the internal noise relative to the shift in the mean, but does not change the relative variability in estimates of the width of the distribution. This suggests that the temporal smearing ("binaural sluggishness") seen in binaural experiments, if modeled as a forced averaging of interaural differences in time, may lead to a relative degradation in binaural detection compared with interaural discrimination. These ideas were quantitatively modeled in Gabriel (1983) and in Colburn et al. (in manurcript). Spectfically, if one assumes that there are separate estimates of ITD and IID with independent noise sources, one can show that the relation of SNR_0, the detection threshold in the N_0S_π case (in decibels), to the jnds in IPD and IID, $(\Delta\varphi)_0$ and $(\Delta\alpha)_0$, respectively, is given by

$$\text{SNR}_0 = (1/C_0)10\log\left[\frac{\text{const} * T_{sw}}{[(C_1/(\Delta\varphi)_0)^2 + (C_2/(\Delta\alpha)_0)^2]}\right] \qquad (7)$$

where T_{sw} is the duration of the smoothing window and the constants C_0, C_1, and C_2 are specified by the relationship between the signal-to-noise ratio (SNR), and the probability and the probability distributions of IPD and IID (Zurek 1991). This formulation shows that increases in the binaural sluggishness parameter T_{sw} result in increases in the detection threshold relative to the jnds in ITD and IID and also allow the description of both detection and discrimination with a common internal noise.

Interaural difference models have also been applied to situations in which the effects of remote frequency components on interaural parameter discrimination are investigated. Buell and Hafter (1991) and Woods and Colburn (1992) formulated models that postulate a nonoptimal combination of information in different frequency bands. These postulates are generally consistent with the notion that components perceived as coming from a common source are processed in common, even when some of these components contain no information relevant for the discrimination task. These models are consistent with some data and inconsistent with other data, and this remains an active area of investigation.

4.2.2 Cross-Correlation Models

Models based on the interaural cross-correlation function (ICF) have a long history in models of binaural phenomena (Sayers and Cherry 1957; Licklider 1959). These models were discussed at some length in Colburn and Durlach (1978), and the discussion here is focused on the work since 1974. Much of the current work has been in a context of models based on explicit physiological models and is presented in the Section 4.3 addressed to pink-box models. In the next few paragraphs, we discuss work that is most directly related to the classic work on cross-correlation models of binaural interaction, notably the work of Sayers and Cherry and of Licklider. Both Sayers and Cherry (1957) and Licklider (1959) specified a running cross-correlation operation function as a representation of binaural information on the stimulus. Sayers and Cherry (1957) were particularly concerned about binaural fusion of coherent stimulus components and their perceived laterality, and Licklider (1959) was particularly concerned with pitch phenomena.

Licklider noted explicitly that the ICF of corresponding frequency bands would be consistent with the bandpass filtering of the cochlea and with a Jeffress-like mechanism. Licklider included both interaural cross-correlation and autocorrelation within his triplex theory (Fig. 8.7), which includes many ideas that remain viable hypotheses for auditory processing

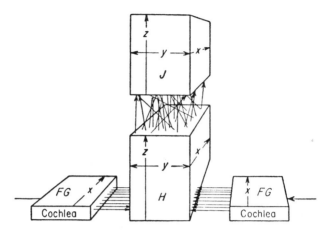

FIGURE 8.7. Triplex model containing interaural cross-correlation suggested by Licklider (1959). "The signals enter the two cochleas. The cochlear frequency analysis (F transformation) maps stimulus frequency into the spatial dimension x, and the ordinal relations along x are preserved in the excitation of the neurons of the auditory nerve (operation G). The time-domain analyzer H preserves the order in x but adds an analysis in the y dimension, based mainly on interaural time differences, and an analysis in the z dimension, based on periodicities in the wave envelope received from FG. In the projection from H to J, the x analysis is largely preserved, but each point in any frontal plane of H is, initially, connected to every point in the corresponding frontal plane of J. The H–J transformation organizes itself, according to rules imposed by the dynamics of the neuronal network, under the influence of acoustic stimulation. The patterns in J thereby acquire the properties that are reflected in pitch perception." (Reproduced from Licklider 1959 with permission.)

(as discussed in Section 4.3). The most interesting aspect of this figure from Licklider is that he has taken a wide view of auditory processing. The block labeled H includes a network of Jeffrfess-like coincidence elements that are arrayed in frequency. In additon, the H block includes autocorrelation connections in the z direction. Notice also that the cochlear transformations to auditory-nerve firing patters is explicit (as blocks labeled FG) and that the block labeled J includes high-level functions.

Blauert and Cobben (1978) specified a computational model based on the running cross-correlation of filtered, rectified, and smoothed versions of the stimulus waveforms. The authors noted that these processed wave-forms are analogous to the rate of firing of peripheral neurons and hypothesized that the average ICF from a stochastic version of this model would not differ substantially from the deterministic ICF generated by the model as presented. The model is applied to input waveforms that were recorded in the ear canals of human subjects in free-field listening situations. When the model is applied without the cochlear filters, the wideband running ICF reflects the source direction in a relatively straightforward way. The ICF with the cochlear filters shows initial peak movements that led the authors to predict anomalies in the precedence effect for low-frequency narrowband stimuli. The predicted anomalies were observed in experiments that were run to test the hypothesis, and the authors concluded that the cross-correlation model is useful for modeling complex localization experiments and that the horizontal angle is predictable by the location of peaks in the ICF.

Lindemann (1986a,b) extended the running cross-correlation model by incorporating a dynamic lateral inhibition mechanism and monaural channels into the ICF. These extensions allow the model to incorporate a time–intensity trading mechanism *within* the ICF function and provide a basis for the precedence effect. When these are coincidences at one value of delay, the model inhibits coincidences at other values of delays. As in the Blauert and Cobben model just described, the model can be interpreted as a representational model that predicts the lateral position of images by reading maxima or centroids from the ICF display. Neither model addresses the variability question or the sensitivity to changes in parameter values; only mean lateral position judgments are predicted. Although predictions are restricted to lateral positon judgments, the amount of data that are predicted by the model is extensive. For stationary signals, the model is applied to lateral position judgments for low-frequency sinusoids that vary in ITD and IID, including time–intensity trading data and observations of multiple images (that are assumed to correspond to multiple peaks in the display), as well as to similar judgments for wideband noise stimuli with varying degrees of interaural coherence (statistically decorrelated noise). For dynamic signals, the model simulates peaks in the running ICF that show properties compatible with the precedence effect.

The detailed mechanisms by which these phenomena are incorporated into the Lindemann (1986a) model are difficult to describe simply. The computational structure of the model is shown in Figure 8.8. The left and right input waveforms (in discrete form) are identified as $l(M, n)$ and $r(-M, n)$, respectively. This figure shows the inhibition and its dependence on both the correlation $k(m, n)$ and the contralateral input. It does not show how the monaural channels are created. The underlying idea of a laterally inhibited ICF is understandable and directly related to the predictions for the dynamic signals. Similarly, the incorporation of monaural channels into the ICF display relates to the representation of IID information and can represent multiple images when an IID is present. On the other hand, the model as specifically represented has a relatively complex structure with many parameters, and it is difficult to ascertain the importance of the assumptions and parameter choices. The model is an elegant demonstration of a processor that is consistent with a relatively peripheral

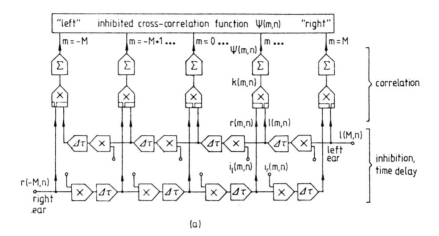

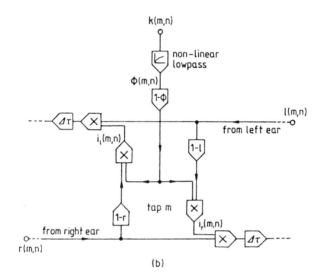

FIGURE 8.8A,B. Inhibited interaural cross-correlation model (Lindemann 1986a).
(A) Input signals are represented as $r(-M, n)$ and $l(M, n)$, which are functions of
the discrete time variable n. The delayed and attenuated input signals are represented at location m as $r(m, n)$ and $l(m, n)$, where the location index m is also a
discrete internal interaural delay variable (the input signals are delayed by $\Delta\tau$
between successive values of m). The attenuation of the input signals as they pass
along their individual delay lines is indicated by the boxes containing the symbol
$\times$. The attenuation imposed on the signals as they pass by location m is derived
(as indicated in B) such that attenuation increases with the level of the contralateral signal at the corresponding location and with the strength of the output of
the processor at that location. The output of the processor at a given location
is roughly the correlation between the left and right signals at that location in the
delay line. This scheme allows the waveforms from each side to cancel each other
as they pass along the delay line and allows a strong output to inhibit outputs at
other locations. (Reproduced from Lindemann 1986 with permission.)

basis for the precedence effect and that incorporates IID and ITD information within a common mechanism.

The future usefulness of this model will depend on the generality of its applicability, which has not been tested. It is notable that the recent physiological data (Yin and Litovsky 1993) recorded from the IC using stimuli designed to correspond to the precedence effect are not consistent with the Lindemann model. Specifically, the Lindemann model predicts that the maximum suppression at position m_0 would be generated by preceding stimuli with lateral positions that are lateral to m_0 and the minimum inhibition would be generated by a preceding stimulus that would stimulate position m_0. This relationship is not observed in the IC data, at least not consistently.

Gaik (1993) extended the model of Lindemann to include a mechanism for making natural ITD–IID combinations more compact (without multiple peaks along the ICF axis) than unnatural combinations, consistent with psychophysical data that he presents. Natural combinations are taken to be those that are measured in each frequency band in a free-field (anechoic) environment with an impulsive sound source. This extended model adds attenuation factors (that are frequency- and delay-specific) along the delay lines in the ICF mechanism. These factors are chosen such that the signals from each ear are of approximately equal magnitude when they coincide at the natural delay for that frequency band. In other words, the attenuation factors are chosen such that natural ITD–IID combinations result in internal signals (from the left and right) that are of equal strength when they arrive at their coincidence point along the internal delay line. This results in an output-inhibited correlation function at each frequency that has a single peak for natural combinations and multiple peaks for very unnatural combinations (which listeners perceive as multiple images).

Lyon (1983) described a computational model for binaural localization and source separation that is built upon a set of frequency-specific cross-correlation functions. The inputs to the cross-correlators in the model are outputs of a cochlear model. For each frequency band, the stimulus is bandpass filtered, half-wave rectified, and amplitude compressed. The cochlear model outputs are cross-correlated with a very short window (1-msec time constant) to generate a correlogram display (over a limited range of interaural delay) from which a local estimate of time delay is made. These local (in frequency and time) delay estimates are used to choose frequency- and time-dependent weights for the left and right cochleagrams. In this way, output cochleagrams are generated for each source and for the echoes. The idea behind the algorithm is to weight the left input cochleagram when the left source is dominant, etc. The algorithm successfully improves the output displays of separate images when multiple sources are present.

The preceding paragraphs summarize a variety of models based on cross-correlation operations. It should be clear that these models are fundamentally part of a common thread of development. In addition, although they are discussed as black-box models, it is clear that they are compatible with many aspects of available physiological data from the auditory nerve and brainstem nuclei and they could be considered as pink-box models. However, their formulation and focus in publication has been oriented almost exclusively toward perceptual phenomena, and they are a natural extension of the original black-box models of Sayers and Cherry (1957) and Licklider (1959). The family of models of Blauert and Cobben (1978), Lindemann (1986a,b), and Gaik (1993) can be regarded as the extended development of one model. This development is continuing with the recent work of Bodden (1993), who applies this model to sound source segregation and intelligibility performance ("the cocktail party effect"), topics also addressed by Lyon (1983).

4.2.3 The Equalization-Cancelation Model

The equalization-cancellation (EC) model (Durlach 1960, 1963, 1972) is a paradigmatic example of a black-box model. It has a very small set of parameters (two) and successfully describes a wide variety of data. Further, applying the model to most situations is relatively straightforward and extensive computations are not required. This model was described and discussed at length in our previous review chapter (Colburn and Durlach 1978), and the reader is directed to that source for coverage of that model. Here we limit discussion to a description of the basic idea of the model and its recent applications.

The basic idea of this model comes from the binaural detection context. If the interaural relationships of the noise and target are different, it is generally possible to eliminate much of the noise energy (by interaurally matching, or equalizing, the noise signals and then subtracting) without eliminating a comparable amount of the signal energy; one thus obtains a gain in signal-to-noise ratio. The detailed description of the model constrains the set of allowable operations for the equalizing operation (such as a limited repertoire of internal delays) and specifies the internal noise that ultimately limits performance. In the usual versions of the model, the internal noise is specified simply as random time jitter and random amplitude jitter applied to each of the two input waveforms. The jitters are assumed to be statistically independent of each other and at each ear and are described as Gaussian random variables with zero means and constant standard deviations. With this simple description, the model describes successfully a wide variety of binaural detection, interaural discrimination, and binaural pitch data. As noted previously (Colburn and Durlach 1978), the EC operation is related to cross-correlation because the energy in the difference of two waveforms is equal to the difference between the sum of the individual energies and twice the unnormalized crosscorrelation. The model was recently applied to binaural detection data from experiments with reproducible noise (Gilkey, Robinson, and Hanna 1985) and to identification of the side of the monaural target in the presence of a masking noise (Schneider and Zurek 1989).

4.2.4 Models of Sound Localization

The process of sound localization can be described as a matching of characteristics of the received signals with the stored knowledge of the characteristics of sounds from given directions. The binaural characteristics are usually considered to be the interaural phase and magnitude spectra, and the monaural characteristics are taken to be the received monaural magnitude spectra. This zeroth-order model of localization is consistent with the general aspects of the data but requires substantial development to provide quantitative predictions. Specifically, it is necessary to specify the limitations on the accuracy with which the interaural differences and the monaural magnitude spectra can be measured as well as an indication of the logic of the pattern-matching algorithm. A more interesting requirement of the model is the incorporation of a priori assumptions about the stimulus spectrum and the consideration of multiple sound sources, each with its own location.

This type of model was explored by Middlebrooks (1992) in a study of human localization of narrowband, high-frequency noises. Middlebrooks found compatibility between human behavior and predictions based on IIDs and spectral analysis. This result is compatible with the earlier results and with interpretations from Blauert (1983). These models naturally include front–back and up–down confusions. Although they do not explicitly include plasticity and memory effects, the experiments are assumed to reflect the normal steady state of the system, and memory effects for these conditions are implicit in the model formulation.

A more abstract analysis of sound localization was reported by Searle et al. (1976). In this study, the various potential cues for localization were analyzed for their contribution to specific stimulus situations. This study was able to account for a large variety of available data obtained in many different circumstances by assuming that the cues that are used in any experiment vary according to the circumstances, as one would expect for a smart processor.

There have been few computational models of sound localization for circumstances in which multiple sources are simultaneously present. Culotta (1988) investigated a model for multiple source localization based on temporal sequences of estimates of ITD and IID for each frequency band. Her approach was similar to the notion underlying the single-source modeling of Gaik (1993) just discussed: natural combinations of ITD and IID were used to resolve ambiguities at each frequency. Culotta also included the obvious notion that consistency of estimated location across frequency would be expected for each source and that multiple sources would generally correspond to different locations. The multiple-source localization problem has also been discussed by Blauert (1983), by Bronkhorst and Plomp (1988, 1989, 1992), and by Bodden (1993).

4.3 Pink-Box Models of Binaural Processing

4.3.1 A Time–Intensity Trading Model

The first binaural computational model that explicitly included a description of the probability of firing of real neurons is the model of Hall (1965), which addressed both interaural time discrimination and time–intensity trading for click stimuli. This study combined an empirical study of the responses of binaural neurons that were judged to be located in the accessory nucleus of the superior olive (the MSO) on the basis of gross potentials recorded from the electrode. Although the location of the cells can be questioned because of the lack of histological verification (and the difficulties of localizing an electrode from the gross potential as reported by Guinan, Norris, and Guinan in 1972), the incorporation of the physilogical data into models for psychophysical behavior resulted in an interesting and insightful example of a pink-box model. Hall measured the probability of firing in response to a binaural click as a function of its ITD and IID and average intensity. As discussed more fully in Colburn and Durlach (1978), Hall used a count-comparison model to provide the basis for psychophysical decisions. A notable aspect of this work is that it included explicit estimates for the variance of the decision variables based on the stochastic nature of the observed firing patterns. The agreement between predictions and observations was very good, with reasonable assumptions about the number of cells.

4.3.2 The Auditory-Nerve-Based Model

The first attempt to formulate a general model of the processing of binaural information with explicit descriptions of the activity of real neurons was described in the series of papers by Colburn and his colleagues (Colburn 1973, 1977; Colburn and Latimer 1978). Their model incorporated an explicit description of the firing patterns of auditory nerve fibers and modeled binaural processing in terms of operations on the auditory nerve patterns. The description of the firing patterns is stochastic (nonhomogeneous Poisson process) and therefore incorporates a loss of information in the peripheral transformation. The analysis was conducted analytically and therefore required approximations whose consequences have not yet been fully evaluated, as we describe more fully later.

Early work on this model (Colburn 1973) demonstrated that there is more than enough information available in the firing patterns of auditory-nerve fibers to allow performance at levels significantly superior to those observed in binaural tasks. That is, even when the random nature of these firing patterns is incorporated into the analysis, there is more than enough information for observed performance. Although the firing patterns of individual neurons show temporal uncertainties of the order of milliseconds and one might speculate that there would not be sufficient information for microsecond time resolution, the large number of active fibers and the multiple periods within a single presentation effectively allow averaging to eliminage most of this variability. Colburn argued that a significantly reduced set of information was adequate, specifically, the information contained in the number of coincidences generated from a very limited set of comparisons. Each fiber was assumed to be compared with only one other fiber with the same characteristic frequency, and it was further assumed that the only available binaural information was the number of coincident inputs (suitably defined) after a delay that was fixed for each fiber pair, but distributed over the array of fibers. A processor that would exploit this information can be represented by a set of processors of the form shown in Figure 8.9 and is closely related to the coincidence detector array suggested by Jeffress (1948).

Binaural processing is represented as a "binaural displayer" followed by a "central processor" in Colburn (1977). Consistent with the conclusion of the earlier analysis, the "binaural displayer" is a neural representation of the information that was postulated to be available to the rest of the brain (the "central processor") for the generation of perceptions and for decision making. Mathematically, the binaural displayer is a set of decision variables L_m where each is defined by

$$L_m(f_m, \tau_m) = \sum_{i,j} d(t_i^{Lm} - t_j^{Rm} - \tau_m) \tag{8}$$

where m is a fiber index for a particular pair of input fibers, $d(\bullet)$ is a function that defines the width of the coincidence window, t_i^{Lm} and t_j^{Rm}

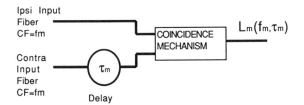

FIGURE 8.9. Model based on auditory nerve firings (Colburn 1973, 1977). Input fibers were modeled explicitly as auditory nerve fibers even though actual inputs to binaural processors are from cochlear nucleus neurons. Note that only input fibers with a common characteristic frequency f_m are explicitly compared and that each fiber pair has associated a fixed interaural delay that can be thought of as reflecting different pathlengths of innervating fibers. The distribution of τ_m values over the population is described by $p(\tau_m)$, which was assumed to be independent of f_m.

are the firing times on the right and left input fibers, respectively, each with characteristic frequency f_m, and τ_m is the characteristic delay, which is fixed for each m but varies over the set of m. For a given stimulus, the displayer outputs depend on the characteristic frequency f_m of the input fibers and on the characteristic delay τ_m. In the formulation of Colburn (1969, 1977), the distribution of internal delays is represented by a density function $p(\tau_m)$ that is independent of f_m and which includes values in a range of about ± 1 msec. It should be apparent that this model is a quantitative formulation of the network suggested by Jeffress (1948). The $p(\tau_m)$ function corresponds to Jeffress's statement that "cells are less dense away from the median plane."

This information is conceptualized as a representation of binaural timing information as a pattern of counts in the (characteristic frequency, characteristic delay) plane, which we refer to here as the (f_m, τ_m) plane. The display can also be thought of as a set of estimates of the band-by-band ICFs. Binaural discrimination experiments are predicted by the minimal differences in the pattern that can be discriminated. Binaural detection experiments are predicted by the ability to distinguish between the masker-alone patterns and the masker-plus-target pattern. Subjective experiments correspond to estimation of the relevant parameter taken from this display. This display is essentially equivalent to the integrated output (counts) from a Jeffress coincidence network as discussed above, and it is also essentially the same as the "central activity pattern" in the central spectrum model developed for binaural pitch phenomena and described in Section 4.3.3. The (f_m, τ_m) plane at its most basic level is a display of statistics computed from random point processes, and the statistical characterization of the display is required to predict performance based on this display. The specific evaluation of this statistical behavior is both a virtue and a difficulty of this effort. The random nature of the display and the limitations that this display imposes on psychophysical performance arise from the random nature of the auditory-nerve activity. The auditory-nerve description used in these calculations constitutes a nonhomogeneous Poisson process for each fiber so that the description is determined by the set of intensity functions of the statistically independent Poisson processes. The intensity functions of the processes (i.e., the time-varying rate functions) were generated by bandpass filtering and exponentially rectifying the stimulus waveforms.

This model assumes that the randomness imposed by the processing that follows the auditory-nerve patterns is negligible in comparison to the randomness of the patterns. The modeling of the randomness on the auditory nerve is thus a central part of the model. The combined constraint of an analytical description of the processing and the complexity of the auditory nerve statistics resulted in an oversimplified model, and a final analysis of the adequacy of the model is not complete. The performance levels predicted for interaural time discrimination are comparable to those observed, which indicates that the representation might be minimal for binaural performance. The predicted dependence of binaural masked threshold on the parameters of the signal and noise, including the interesting dependence on interaural phase shifts and time delays in the masking noise, are treated satisfactorily by this model but the overall level of binaural detection thresholds are predicted to be better than those observed. This difference is probably related to the fact that the model assumes that the variability of the decision variables caused by the variability of the noise waveforms is negligible relative to the internal variability (that arises from the statistics of the auditory nerve activity).

This assumption was made for three reasons: there were ways to remove the noise variance by more complex processing of monaural normalizing statistics, the assumed exponential rectifier in the auditory nerve model tended to exaggerate the external noise variability, and finally the auditory nerve model lacked a short-time envelope normalization that would have reduced external noise variability but that could not be imposed without compromising the ability to compute predictions analytically. That this assumption is inadequate is demonstrated by the reproducible noise studies described in Section 2.3 (e.g., Gilkey, Robinson, and Hanna 1985). The modifications to this model explored by Stern and Shear (unpublished manuscript) and described in Section 4.3.4 are in part motivated by these difficulties. It appears that model studies of binaural detection that incorporate a description of the activity of auditory nerve fibers will require a more sophisticated model of the firing patterns such as the one recently described by Carney (1993). The Carney model includes the nonlinear effects of stimulus intensity on the sharpness of tuning, adaptation effects, and refractory effects, all of which would have significant impacts on psychoacoustic performance.

4.3.3 The Central Spectrum Model

Bilsen, Raatgever, and colleagues (Bilsen 1977; Raatgever and Bilsen 1977, 1986; Frijns, Raatgever, and Bilsen 1986) developed the "central-spectrum model," which is primarily concerned with binaural pitch phenomena (see Section 2.2.6). This model is also based on an (f_m, τ_m)-plane representation of binaural information called the central auditory pattern (CAP). The CAP, which is closely related to the "binaural displayer" just discussed, is described explicitly in the next paragraph. The central spectrum model can be thought of as a set of rules for extracting information from the (f_m, τ_m) display. Although they do not explicitly model physiological data, the primary representation of the binaural information within their model is sufficiently similar to the representation just discussed that is it easiest to treat the central spectrum model in close apposition to the auditory-nerve-based model. The basic idea of the central spectrum model (see the discussion in Bilsen and Goldstein 1974) is that, within the (f_m, τ_m) plane, contours of constant τ_m can be considered as frequency spectra that are generated centrally (they may not be present in the stimuli at either ear). These "central spectra" are associated with many of the binaural pitch pheonomena in the same way that monaural spectra are associated with monaural pitch pheonomena.

In the most complete presentation of the central spectrum theory (Frijns, Raatgever, and Bilsen 1986; Raatgever and Bilsen 1986), comparisons are made to pitch-matching experiments, lateralization matching experiments, and binaural detection experiments. The CAP is generated from the monaural stimuli by bandpass filtering, imposing a network of delays at each frequency, and then computing the power in the sum of the delayed and filtered inputs. The CAP consists of this power output in the (f_m, τ_m) plane, weighted by a pair of functions that emphasize particular regions of the time and frequency plane. These functions, which are chosen empirically, correspond to the $p(\tau_m)$ function described in Section 4.3.2 (which constrains available internal time delays to be within about a millisecond of zero delay) and to the frequency dominance region described in Section 2.2.6. (Some frequency weighting is also present in the auditory-nerve-based model by virtue of the loss of synchrony in the auditory nerve fibers as the frequency increases above 800 Hz.)

The pitch and lateralization matching experiments are well described by the central spectrum theory whenever the intensities at the two ears are balanced. When the intensities are unbalanced, an interesting dichotomy occurs. The lateral position of the dichotic pitch percept is unaffected by the intensity difference in the stimulus, whereas the lateral position of the total stimulus (as normally reported in lateralization experiments) shows the usual time–intensity trading. The interpretation of these data is that the (f_m, τ_m)-plane representation is not significantly affected by the interaural intensity difference, that the position of the perceived binaural pitches are determined by the value of τ_m for which the central spectrum is computed, and that the lateral position of the stimulus is determined by an interaction of the time-based position read off the display and an intensity factor that follows the display. This general question is addressed again in connection with the models of Stern and his colleagues in Section 4.3.4.

The binaural detection data addressed by the central spectrum theory were generated with stimuli that are related to the binaural pitch stimuli. The theory is used to predict the detection thresholds for conditions in which the target signal is diotic (identical at the two ears) and the masking noise is one of the binaural pitch conditions shifted by τ. The thresholds were measured in such a way as to position the relevant region of the display along the $\tau_m = 0$ axis in the (f_m, τ_m) plane and to keep attention focused on this region (Frijns, Raatgever and Bilsen 1986; Raatgever and Bilsen 1986). Predictions are made relative to the noise-out-of-phase ($N_\pi S_0$) condition, the condition for which the minimum threshold is found for a diotic signal. As the predictions are made relative to performance for the $N_\pi S_0$ condition, the internal noise within this model is specified implicitly by assuming that this condition is predicted. With this approach, the CAP for each binaural pitch combintion generates a set of threshold predictions for $f-\tau$ combinations. The agreement between predictions and results is quite good.

4.3.4 The Position-Variable Model

The general approach of the model based on auditory nerve data has been applied by Stern and his colleagues (Stern and Colburn 1978, 1985; Stern, Zeiberg, and Trahiotis 1988; Stern, Zeppenfeld, and Shear 1991; Stern and Shear unpublished manuscript) to model binaural phenomena, particularly lateralization data. Stern's models incorporate the effects of interaural intensity differences into the network of coincidence counters. The mechanism chosen by Stern is an intensity weighting function applied along the internal time delay axis. This can be thought of as a weighting of the cross-correlation function. A lateral position variable is then calculated as the centroid of this weighted cross-correlation function. Stern's early work showed that the lateral position data for tones, including cue-reversal phenomena and time–intensity trading, could be described by this model (Stern and Colburn 1978) and that most of the interaural discrimination data can be explained as optimum performance using the lateral position as the decision variable (Stern and Colburn 1985). Of course, discrimination performance depends on the variance of the position variable, which arises from the randomness of the auditory-nerve firings.

Stern and Shear (unpublished manuscript) extended the model and demonstrated that a modified model is able to predict a wider class of low-frequency lateralization and detection phenomena. They improved the description of the auditory nerve input, and the description of the $p(\tau_m)$ function was also modifed so that the width of the distribution of available delays is frequency dependent. With these modifications, the model is able both to predict frequency dominance in pitch experiments (e.g., Raatgever and Bilsen 1986) without a separate frequency weighting function and to describe the lateralization of tonal stimuli as a function of the interaural time delay (ITD) for the complete range of low frequencies (up to 1200 Hz), without losing the ability to describe the detection of tones in wideband noise.

Some of the recent work on the position-variable model was designed to evaluate more selectively the rules for extracting information from the (f_m, τ_m) plane. Stern, Zeiberg, and Trahiotis (1988) addressed the lateralization of noise bands as a function of bandwidth, sinusoidally amplitude-modulated (SAM) low-frequency tones, and low-frequency transients. Many of these phenomena involve interactions or comparisons across frequency in the (f_m, τ_m) plane, and an explicit weighting scheme is suggested for the combination of information in this plane. According to this scheme, more emphasis is given to regions in the (f_m, τ_m) plane that show straight contours of maximal activity (a straightness weighting) in addition to the weighting of regions that are closer to the $\tau_m = 0$ axis [through the $p(\tau_m)$ distribution]. The model without the straightness weighting is unable to predict the data satisfactorily, while the straightness-weighted comparisons are shown to provide a satisfactory fit to the data.

In a related study, Stern and Trahiotis (1992) suggested a mechanism for the implementation of this straightness weighting, one that is based on a second level of coincidences. They pointed out that this mechanism is a way of incorporating the consistency of timing over frequency into the strength of input of each point in the (f_m, τ_m) plane. Although they have not published the details of the algorithm for this second coincidence mechanism, they showed simulation results that demonstrate a successful description of the data for lateralization of noise bands versus bandwidth at low frequency. Trahiotis and Stern have argued that this second level of coincidence is also consistent with the interesting perceptions generated by a stimulus in which frequency components are amplitude modulated, holding interaural relationships constant. The observed sensitivity to the synchronization in the maxima across frequency is consistent with their proposed across-frequency mechanism.

4.3.5 The Stereausis Model

Shamma, Shen, and Gopalaswamy (1989) suggested a binaural processing scheme based on a network of coincidence detectors, similar to those suggested by Jeffress (1948) and incorporated into the models already discussed, with the significant change that the interaural time delays are implemented by comparing locally mistuned filter outputs instead of time-delayed versions of identically tuned filter outputs (an idea suggested by Schroeder 1977). The fact that highly tuned filters have phase functions that are strong functions of frequency allows a small mistuning to generate a significant phase shift. Shamma, Shen, and Gopalaswamy noted that a mistuning of 200 Hz for a filter (fiber) tuned to 1 kHz would be enough to generate a delay of 700 μsec, which is approximately equal to the largest naturally occurring delays for a distant source in a free field.

The inputs to the stereausis network are generated by a cochlear model that has been applied to several types of phenomena (Shamma 1985; Shamma et al. 1986), but that is outside the purview of the current chapter. The (f_m, τ_{eq}) display in their model is followed by a lateral inhibition network that makes the interesting aspects of the display more visually prominent. The importance of this sharpening network for performance predictions would depend on the nature of the internal noise that would be postulated in the model. A noise source after the display would have fundamentally different consequences than a noise source before the display. Because there is no specification of noise within the current model, the importance of the lateral inhibition network is primarily visual in this version of the model.

Although there are no quantitative predictions from the stereausis model because the internal noise is unspecified and because the complexity of the cochlear model would make the quantitative predictions somewhat arbitrary, there are several interesting aspects to the output displays. First, as one would expect by analogy to models based on the usual coincidence network, the model output patterns are consistent with the lateral position created by time delays and with differences corresponding to the presence of a target in detection stimuli. Second, and more interestingly, this network generates a lateral position cue as a result of interaural level differences and shows time-intensity trading at low frequencies. This capability is apparently related to the fact that the chosen interaction rule (the square of the sum of the monaural inputs) and display mechanism generate a meaningful output in the case of a purely monaural input that is not centered. As we have noted, simple coincidence mechanisms with point processes as inputs are more like correlation functions and do not naturally generate output patterns that shift with interaural level differences.

4.3.6 The Averaged Cross-Correlation Model

Another computational model of lateralization that is similar to those already described is presented and evaluated for lateralization by Shackleton, Meddis, and Hewitt (1992) This model uses a relatively complete computational model of the auditory periphery (Meddis, Hewitt, and Shackleton 1990; Meddis and Hewitt 1991) to characterize inputs to the binaural mechanism and a running cross-correlation function to calculate output patterns in the (f_m, τ_m) plane. Further, and still similar to other models discussed above (notably Stern's), the cross-correlation function at each point in the (f_m, τ_m) plane is weighted: by a function that weights with respect to τ_m and by a function which weights with respect to f_m. The weighted cross-correlation display is then simply integrated across frequency (78 points equally spaced on an equivalent rectangular bandwidth (ERB)-rate scale [Moore and Glasberg 1986] over the range from 50 to 3000 Hz) to generate the "summary corss-correlogram." The lateral position is estimated by the delay that generates the peak value of this cross-correlogram or, for cases in which "there is evidence that subjects are using judgement averaging," the lateral position is estimated by the delay corresponding to the centroid of the summary cross-correlogram. It is interesting that this model predicts the observed (Trahiotis and Stern 1989) image movement across the head as stimulus bandwidth increases for large ITDs. As noted, Stern and colleagues assumed a straightness weighting to describe these data. Shackleton, Meddis, and Hewitt (1992) suggested that the primary difference in the models is the width of the τ_m-weighting functions; specifically, they imply that these lateralization data require a function wider than Stern's $p(\tau_m)$ (originally used by Colburn [1977] to match binaural detection data). On the other hand, Trahiotis and Stern (1994) stated that their more complicated, second-coincidence model *is* required to describe the details of the lateral position data.

4.3.7 The MSO as an Interaural Time Processor

An investigation of the temporal information provided by the firing patterns of MSO cells was presented by Colburn and Isabelle (1992). The output patterns of an array of model MSO cells (Han and Colburn 1993) were postulated to be the only source of information about interaural time delay for low-frequency sinusoidal stimuli. Statistical decision theory was used to determine how much information about interaural time delay was provided by the set of MSO cells. When reasonable assumptions were made about the number of cells excited for sinusoidal stimulation and the distribution of internal interaural delays [i.e. the $p(\tau_m)$ function of Colburn (1977) described above], interaural time discrimination performance comparable to that observed (about 10 μsec) was predicted. In terms of the models already discussed, this modeling was consistent with an (f_m, τ_m)-plane representation that is generated from outputs of MSO cells. This is evidence for the general compatibility of the physiological models in Section 3 and the models described in this section.

4.3.8 The LSO as an Interaural Intensity Processor

A structural model for the processing of interaural intensity difference (IID) information was proposed by Reed and Blum (1990). They postulated a linear array of cells in the LSO such that the pattern of activity along the array is simply related to the IID. The individual cells in the array are specified in the manner described in Section 3.2. The input patterns to the cells in the array are arranged such that the outputs are a monotonic function of position in the array; that is, the firing rate of the cells in the network decreases along the cellular array until the rate reaches zero, after which point the remaining cells are inactive. Information about the IID of the stimulus is contained in the position along the array beyond which the cells are inactive.

The model postulates a systematic distribution of thresholds for the input neurons of each type so that the point at which the rate of the model LSO cells goes to zero is a sigmoidal funciton of IID. This is essentially a mechanism for comparing the rates of activity of input excitatory and inhibitory fibers and depends on the regular array of cell parameters to provide information about the IID. Reed and Blum showed that the numbers of available cells are reasonable and that estimates for the activity of individual cells is consistent with the resolution and distribution of sensitivity of the overall system. Efforts to predict quantitatively psychophysical abilities in intensity discrimination or time–intensity trading have not been undertaken.

4.3.9 Function-Based Modeling

Johnson and his colleagues suggested an approach (Johnson, Dabak, and Tsuchitani 1990; Dabak and Johnson 1992) that they call "function-based modeling" of binaural processing. They applied this approach to sound source localization with superior olivary neurons in mind. Specifically, the firing patterns of the inputs to the olivary nuclei are characterized as Poisson processes, and the dependence of the intensity functions (i.e., the cosntant rate functions of the processes) on the sound source angle θ is described with simple equations. An algorithm for optimally estimating the angle θ is then derived and performance using this algorithm is calculated. In the first paper, expressions for a sufficient statistic for the angle estimation task are calculated for the case that only level information is available, and in the second paper, expressions are derived for the case that only phase information is available.

These researchers concluded that LSO neurons operate consistently with what is required for optimal estimations based on IIDs and that MSO neurons operate consistently with what is required for optimal estimations based on interaural phase differences at low frequencies. These conclusions are consistent with our foregoing discussion and with speculations in the literature. These results are also closely related to the analysis of Colburn (1973), who computed the variance of the optimum estimates of ITDs and IIDs based on Poisson descriptions of primary auditory nerve fibers. It would be of interest to consider the relationship of the source angle θ to the interaural differences in the stimulus and the importance of this relationship for the function-based model analyses. (That this is a significant relationship can be appreciated by referring to Equations 1–3 in Section 2, which relate changes in source angle to changes in ITD.) It would also be of interest to consider the representation and combination of information across neurons with different tuning characteristics.

4.3.10 Localization Model Based on Individual Cell Responses

Essentially all the neural-based models of psychophysical abilities that we have considered have been concerned with a relatively peripheral representation of temporal information and the performance that one would predict for discrimination, lateralization, and detection experiments. Another approach to the modeling of localization is to describe the activity of a cell population that might nbe involved in sound localization and to relate this activity to localization behavior. Brainard, Knudsen, and Esterly (1992) have taken such an approach incorporating measurements from the optic tectum of the barn owl. This species is dependent on its localization ability for its feeding behavior, and its capabilities have been documented. Many aspects of its peripheral physiology have also been studied as we discussed in Section 3.3. The barn owl differs from mammalian species in several ways; one important difference for the present discussion is that the nerve fibers in the owl show synchronization to the fine structure of the stimulus for frequencies up to 9 kHz. In addition, it appears that the owl is sensitive to interaural phase up to this frequency as well. Finally, it should be kept in mind that the barn owl uses ITD and IID for both azimuth and elevation cues.

Brainard, Knudsen, and Esterly (1992) presented a descriptive model for receptive fields of neurons in the tectum (superior colliculus) that they fit to results for the receptive fields measured with narrowband (sinusoidal) stimuli. The measurements and the modeling were confined to the frontal hemifield. The model describes the response of a particular nerve cell on the basis of the interaural differences in phase (IPD) and level (ILD) that are generated by the stimulus as a function of position in space. This descriptive model assumes that the "response" of a neuron is given by the following expression:

$$\text{Response} = 1 - \{[(\text{ILD}_{opt} - \text{ILD})/\alpha]^2 + [(\text{IPD}_{opt} - \text{IPD})/\beta]^2\}^{1/2} \tag{9}$$

where ILD_{opt}, IPD_{opt}, α, and β are parameters that are chosen separately for each frequency and for each cell. Clearly, the parameters ILD_{opt} and IPD_{opt} correspond to the interaural differences that define the "center" of the receptive field and α and β determine the relative importance of deviations in ILD and IPD in influencing the response. Parameters are chosen to give a qualitatively good fit to the predicted and measured receptive fields. This fitting process includes measurements for all directions of stimulation in the frontal plane.

The examples that are provided in Brainerd, Knudsen, and Esterly (1992) show relatively good fits to the receptive fields for the sinusoidal stimuli. Little information is provided about the quality of the fits for the 44 fields (that were measured from 26 recording sites, 10 of which were single units). The examples show relatively clearly that, for a given neuron, different frequencies may require significantly different pairs of values of IPD and ILD for generation of a response. It is also shown that narrowband stimuli presented from some regions of space generate inhibition of spontaneous responses and of excitatory responses which are generated from the same position by other frequencies, as one would predict from the negative value of "response" generated by some sets of interaural differences in Eq. 9. Finally, it is noted that the wideband receptive fields are consistent with the intersection of the narrowband fields, and it is concluded that these neurons combine information across frequency with inhibitory inputs from regions outside the receptive fields. There is no discussion of the realtionship between these results and the analyses of human lateralization judgments related to bandwidth effects, phase ambiguity, and onset time information, however, nor is there a consideration of multiple sound sources or overall level effects.

5. Summary and Comments

This chapter reviews computational models of binaural processing, including those that address only physiological data, those that address only psychoacoustic data, and those that address both types of data. Although the number of models that explicitly incorporate both types of data is still small, as the amount of relevant data increases it is increasingly likely that useful insights will be generated from physiology which are useful for understanding behavior, and vice versa. Further, although this chapter does not address results from physiological or psychophysical experiments on subjects with known hearing impairments, information about binaural processing in these situations is also increasing and should also provide useful insights. The ultimate computational model will eventually incorporate all these phenomena, physiological and psycho-acoustical results from both normal and impaired auditory systems. Incidentally, note that the importance of species differences for binaural processing (Brown 1994) has been practically ignored in most models and in this chapter. Current modeling efforts are either restricted to a single species or assume implicitly that binaural mechanisms are similar for different species.

Although psychoacoustic models are here separated (somewhat artificially) into models that address processing of the acoustic stimuli without attention to physiological data (black-box models) and models that explicitly incorporate physiological evidence into their formulations (pink-box models), models within these categories are closely related. Most models in both categories are based on the overlapping concepts of interaural coincidence networks and interaural correlation functions applied to individual frequency bands. All these models, whether black box or pink box, can be related naturally to the physiological data from EE cells and the models for these data as described in Sections 3.1 and 3.3. These models are all fundamentally similar, and the concept of an internal display of activity on a center frequency versus interaural time delay plane is almost universally used for insight about binarual phenomena.

One of the ways in which models differ is in their incorporation of IIDs. Models based on interaural coincidence or correlation do not naturally include the effects of IIDs so that additional assumptions or mechanisms must be included to incorporate these effects. In some cases (e.g., the models of Lindemann and Gaik in Section 4.2.2 and the EC model in Section 4.2.3), the effects of IIDs are incorporated into a common mechanism. In other cases (all the models in Section 4.3 and the interaural difference models in Section 4.2.1), IID information is extracted separately and then combined with ITD information. There are advantages and arguments for both approaches. The models of sound localization in Sections 4.2.4 and 4.3.10 do not explicitly address mechanisms for the extraction of ITD or IID information; they are primarily concerned with the spectral variation in overall intensity and interarual differences that relate to source position.

In considering the ability of models to describe available psychoacoustic data, we note that most models have restricted their application to a relatively small subset of the available data. Very few models attempt to describe a large variety of data. Most models address one type of data even though many of them could be applied to a significantly broader set of data, usually with additional assumptions. In particular, predictions are often only made for mean responses, such as mean lateralization judgments, and comparisons with the resolution limits on performance have not yet been made. Predictions for resolution limits require modeling the internal variability that has been addressed by few models. Because few models have been applied to a wide variety of data, there is no simple summary of the virtues and deficiencies of the various models. One set of data for which models appear to be inadequate in a fundamental way are the reproducible noise detection data. Another area that has been resistant to adequate modeling is related to auditory scene analysis. This area almost certainly spans topics from peripheral information limitations to central interpretative functions. It is likely to involve learning and plasticity as well as cognitive aspects of perception such as the set of expectations of the subject.

Future directions of research will almost certainly continue to address the issues we have raised in this summary. Basic information processing issues will be addressed with computational models that will integrate physiological and psychoacoustic results, take note of species differences, and incorporate results from impaired auditory systems. Models that seek to explain more complex phenomena such as auditory scene analysis, including cocktail-party effects, the varieties of types of interference, and the simulation of complex acoustic environments will have to incorporate better models of neural plasticity and perceptual learning. Our increased computational capacities make possible much more complex models that can include these multiple factors.

Acknowledgments. Work on this chapter was supported by a grant from the National Institute for Deafness and Communication Disorders (grant No. 5 R01 DC00100). I appreciate the assistance and useful suggestions of my students and colleagues, particularly Laurel Carney, who suffered through an early draft. I thank Yuda Albeck for many helpful comments, including the suggestion that I try to avoid long sentences and awkward constructions that interfere with the readers' abilities to follow the content. Finally, I thank Monica Hawley; without her thoughtful assistance, the chapter would probably still be under revision.

List of Symbols

Symbol or Acronym		Section in Which Introduced
AVCN	anteroventral cochlear nucleus	3
CAP	central auditory pattern	4.3.3
EC	equalization cancellation	4.2.3
EE	excitatory-excitatory (cell type)	3
EI	excitatory-inhibitory (cell type)	3
f_m	internal frequency coordinate	4.3.2
IC	inferior colliculus	3
ICF	interaural correlation function	4.1
IE	inhibitory-excitatory (cell type)	3
IID	interaural intensity difference	2.1.1
ITD	interaural time difference	2.1.1
jnd	just-noticeable difference	2.2.3
LSO	lateral superior olive	3
MAA	minimum audible angle	2.1.1
MNTB	medial nucleus of trapezoid body	3.2.6
MSO	medial superior olive	3.1.1
N_0S_π	classic binaural detection condition	2.2.5
$p(\tau_m)$	distribution of internal delays	4.3.2
τ	interaural time delay	2.1.1
τ_m	internal interaural time delay	4.3.2
θ	source azimuth angle	2.1.1

References

Adams JC, Mugnaini E (1990) Immunocytochemical evidence for inhibitory and disinhibitory circuits in the superior olive. Hear Res 49:281–298.

Anderson DJ, Rose JE, Hind JE, Brugge JF (1971) Temporal position of discharges in single auditory nerve fibers within the cycle of a sinewave stimulus: Frequency and intensity effects. J Acoust Soc Am 49:1131–1139.

Bernstein LR, Trahiotis C (1982) Detection of interaural delay in high-frequency noise. J Acoust Soc Am 71:147–152.

Bernstein LR, Trahiotis C (1985) Lateralization of sinusoidally amplitude-modulated tones: Effects of spectral locus and temporal variation. J Acoust Soc Am 78:514–523.

Bilsen FA (1977) Pitch of noise signals: Evidence for a central spectrum. J Acoust Soc Am 61:150–161.

Bilsen FA, Goldstein JL (1974) Pitch of dichotically delayed noise and its possible spectral basis. J Acoust Soc Am 55:292–296.

Blauert J (1983) Spatial Hearing. Cambridge: MIT Press.

Blauert J, Cobben W (1978) Some consideration of binaural cross correlation analysis. Acustica 39:96–103.

Blauert J, Col J-P (1989) Etude de quelques aspects temporels de l'audition spatiale. Note-laboratoire LMA, No. 118. Marseilles: Centre National de la Recherche Scientifique.

Blum JJ, Reed MC (1991) Further studies of a model for azimuthal encoding: lateral superior olive neuron response curves and developmental processes. J Acoust Soc Am 90:1968–1978.

Bodden M (1993) Modeling human sound-source localization and the cocktail party effect. Acta Acustica 1:43–55.

Bonham BH, Lewis ER (1993) Development of sound source localization by interaural time/phase difference—a model. Soc Neurosci Abstr 19:887.

Boudreau JC, Tsuchitani C (1968) Binaural interaction in the cat superior olive S segment. J Neurophysiol (Bethesda) 31:442–454.

Brainerd MS, Knudsen EI, Esterly SD (1992) Neural derivation of sound source location: resolution of spatial ambiguities in binaural cues. J Acoust Soc Am 91:1015–1027.

Bregman AS (1990) Auditory Scene Analysis. Cambridge: MIT Press.

Bronkhorst AW, Plomp R (1988) The effect of head-induced interaural time and level differences on speech intelligibility in noise. J Acoust Soc Am 83:1508–1516.

Bronkhorst AW, Plomp R (1989) Binaural speech intelligibility in noise for hearing-impaired listeners. J Acoust Soc Am 86:1374–1383.

Bronkhorst AW, Plomp R (1992) Effect of multiple speechlike maskers on binaural speech recognition in normal and impaired hearing. J Acoust Soc Am 92:3132–3139.

Brown CH (1994) Sound localization. In: Fay RR, Popper AN (eds) Comparative Hearing: Mammals. New York: Springer-Verlag.

Buell TN, Hafter ER (1991) Combination of interaural information across frequency bands. J Acoust Soc Am 90:1894–1900.

Buell TN, Trahiotis C, Bernstein LR (1991) Lateralization of low-frequency tones: relative potency of gating and ongoing interaural delays. J Acoust Soc Am 90:3077–3085.

Butler RA (1969) Monaural and binaural localization of noise bursts vertically in the median sagittal plane. J Aud Res 3:230–235.

Caird D, Klinke R (1983) Processing of binaural stimuli by cat superior olivary complex neurons. Exp Brain Res 52:385–399.

Cant NB, Hyson RL (1992) Projections from the lateral nucleus of the trapezoid body to the medial superior olivary nucleus in the gerbil. Hear Res 58:26–34.

Carney LH (1993) A model for the responses of low-frequency auditory nerve fibers in cat. J Acoust Soc Am 93:401–417.

Carney LH, Yin TCT (1989) Responses of low-frequency cells in the inferior colliculus to interaural time differences of clicks: excitatory and inhibitory components. J Neurophysiol (Bethesda) 62:144–161.

Carr CE, Konishi M (1988) Axonal delay lines for time measurement in the owl's brainstem. Proc natl Acad Sci USA 85:8311–8315.

Carr CE, Konishi M (1990) A circuit for detection of interaural time differences in the brainstem of the barn owl. J Neurosci 10:3227–3246.

Cherry EC (1961) Two ears—but one world. In: Rosenblith WA (ed) Sensory Communication. Cambridge: MIT Press, pp. 99–117.

Clifton RK (1987) Breakdown of echo suppression in the precedence effect. J Acoust Soc Am 82:1834–1835.

Clifton RK, Freyman R (1989) Effect of click rate and delay and breakdown of the precedence effect. Percept Psychophys 46:139–145.

Colburn HS (1969) Some Physiological Limitations on Binaural Performance. Ph.D. dissertation, Massachetts Institute of Technology, Cambridge, MA.

Colburn HS (1973) Theory of binaural interaction based on auditory-nerve data. I. General strategy and preliminary results on interaural discrimination. J Acoust Soc Am 54:1458–1470.

Colburn HS (1977) Theory of binaural interaction based on auditory-nerve data. II. Detection of tones in noise. J Acoust Soc Am 61:525–533.

Colburn HS, Durlach NI (1978) Models of binaural interaction. In: Carterette EC, Friedman M (eds) Handbook of Perception, Vol. IV. New York: Academic Press, pp. 467–518.

Colburn HS, Ibrahim H (1993) Modeling of precedence-effect behavior in single neurons and in human listeners. J Acoust Soc Am 93:2293.

Colburn HS, Isabelle SK (1992) Models of binaural processing based on neural patterns in the medial superior olive. In: Cazals Y, et al. (eds) Auditory Physiology and Perception. Oxford: Pergamon Press, pp. 539–545.

Colburn HS, Latimer JS (1978) Theory of binaural interaction based on auditory-nerve data. III. Joint dependence on interaural time and amplitude differences in discrimination and detection. J Acoust Soc Am 64:95–106.

Colburn HS, Moss PJ (1981) Binaural interaction models and mechanisms. In: Syka J, Aitkin L (eds) Neuronal Mechanisms of Hearing. New York: Plenum Press, pp. 283–288.

Colburn HS, Han Y, Culotta CP (1990) Coincidence model of MSO responses. Hear Res 49:335–346.

Cramer EM, Huggins WH (1958) Creation of pitch through binaural interaction. J Acoust Soc Am 30:413–417.

Culotta CP (1988) Auditory Localization of Multiple Sound Sources. M.S. Thesis, Boston University, Boston, MA.

Dabak A, Johnson DH (1992) Function-based modeling of binaural interactions: interaural phase. Hear Res 58:200–212.

Davenport W, Root W (1958) Random Signals and Noise. New York: Wiley.

Davis JB (1985) Remote frequency masking: differential effects in binaural versus monaural detection. Senior project, Dept. of Biomedical Engineering, Boston University, Boston, MA.

Davis MF (1980) Computer Simulation of Static Localization Cues. Ph.D. dissertation, Massachusetts Institute of Technology, Cambridge, MA.

Diranieh YM (1992) Computer-Based Neural Models of Single Lateral Superior Olivary Neurons. M.S. thesis, Boston University, Boston, MA.

Domnitz RH, Colburn HS (1976) Analysis of binaural detection models for dependence on interaural target parameters. J Acoust Soc Am 59:598–601.

Domnitz R, Colburn HS (1977) Lateral position and interaural discrimination. J Acoust Soc Am 61:1586–1598.

Durlach NI (1960) Note on the equalization and cancellation theory of binaural masking-level differences. J Acoust Soc Am 32:1075–1076.

Durlach NI (1963) Equalization and cancellation theory of binaural masking-level differences. J Acoust Soc Am 35:1206–1218.

Durlach NI (1972) Binaural signal detection: Equalization and cancellation theory. In: Tobias JV (ed) Foundations of Modern Auditory Theory, Vol. 2. New York: Academic Press, pp. 369–462.

Durlach NI, Colburn HS (1978) Binaural phenomena. In: Carterette EC, Friedman M (eds) Handbook of Perception, Vol. IV. New York: Academic press, pp. 405–466.

Durlach NI, Rigopoulos A, Pang XD, Woods WS, Kulkarni A, Colburn HS, Wenzel EM (1992) On the externalization of auditory images. Presence 1:251–257.

Dye RH (1990) The combination of interaural information across frequencies: Lateralization on the basis of interaural delay. J Acoust Soc Am 88:2159–2170.

Ericson MA, McKinley RL (1992) Experiments involving auditory localization over headphones using synthesized cues. J Acoust Soc Am 92:2296.

Florentine M (1976) Relation between lateralization and loudness in asymmetrical hearing loss. J Am Audiol Soc 1:243–251.

Foster S, Wenzel EM (1992) The Convolvotron: Real-time demonstration of reverberant virtual acoustic environments. J Acoust Soc Am 92:2376.

Frijns JHM, Raatgever J, Bilsen FA (1986) A central spectrum theory of binaural processing. The binaural pitch revisited. J Acoust Soc Am 80:442–451.

Gabriel KJ (1983) Binaural Interaction in Impaired Listeners. Ph.D. dissertation, Massachusetts Institute of Technology, Cambridge, MA.

Gaik W (1993) Combined evaluation of interaural time and intensity differences: Psychoacoustic results and computer modeling. J Acoust Soc Am 94:98–110.

Gaumond RP, Molnar CE, Kim DO (1982) Stimulus and recovery dependence of cat cochlear nerve fiber spike discharge probability. J Neurophysiol (Bethesda) 48:856–873.

Gilkey RH, Robinson DE, Hanna TE (1985) Effects of masker waveform and signal-to-masker phase relation on diotic and dichotic masking by reproducible noise. J Acoust Soc Am 78:1207–1219.

Gilliom JD, Sorkin RD (1972) Discrimination of interaural time and intensity. J Acoust Soc Am 52:1635–1644.

Goldberg JM, Brown PB (1968) Functional organization of the dog superior olivary complex: An anatomical and electrophysiological study. J Neurophysiol (Bethesda) 31:639–656.

Goldberg JM, Brown PB (1969) Response of binaural neurons of dog superior olivary complex to dichotic tonal stimuli: Some physiological mechanisms of sound localization. J Neurophysiol (Bethesda) 32:613–636.

Grantham DW (1982) Detectability of time-varying interaural correlation in narrow-band noise stimuli. J Acoust Soc Am 72:1178–1184.

Grantham DW (1984) Discrimination of dynamic interaural intensity differences. J Acoust Soc Am 76:71–76.

Grantham DW, Wightman FL (1978) Detectability of varying interaural temporal differences. J Acoust Soc Am 63:511–523.

Grantham DW, Wightman FL (1979) Detectability of a pulsed tone in the presence of a masker with time-varying interaural correlation. J Acoust Soc Am 63:511–523.

Grothe B, Sanes DH (1993) Inhibition influences time difference coding by MSO neurons—an in vitro study. Assoc Res Otolaryngol Abstr 16:108.

Grün S, Aertsen A, Wagner H, Carr C (1990) Sound localization in the barn owl: A quantitative model of binaural interaction in the nucleus laminaris. Soc Neurosci Abstr 16:870.

Guinan JJ Jr, Guinan SS, Norris BE (1972) Single auditory units in the superior olivary complex. I. Responses to sounds and classifications based on physiological properties. Int J Neurosci 4:101–120.

Guinan JJ Jr, Norris BE, Guinan SS (1972) Single auditory units in the superior olivary complex. II. Location of unit categories and tonotopic organization. Int J Neurosci 4:147–166.

Hafter ER (1971) Quantitative evaluation of a lateralization model of masking-level differences. J Acoust Soc Am 50:1116–1122.

Hafter ER, Carrier SC (1970) Masking-level differences obtained with a pulsed tonal masker. J Acoust Soc Am 47:1041–1047.

Hafter ER, Carrier SC (1972) Binaural interaction in low-frequency stimuli: The inability to trade time and intensity completely. J Acoust Soc Am 51:1852–1862.

Hafter ER, Buell TN, Richards VM (1988) Onset-coding in lateralization: Its form, site and function, In: Edelman GM, Gall WE, Cowan WM (eds) Auditory Function: Neurobiological Bases of Hearing. New York: Wiley, pp. 647–678.

Hall JL II (1965) Binaural interaction in the accessory superior-olivary nucleus of the cat. J Acoust Soc Am 37:814–823.

Han Y, Colburn HS (1993) Point-neuron model for binaural interaction in MSO. Hear Res 68:115–130.

Hausler R, Colburn HS, Marr E (1983) Sound localization in subjects with impaired hearing. Spatial-discrimination and interaural-discrimination tests. Acta Otolaryngol Suppl 400:1–62.

Held R (1955) Shifts in binaural localization after prolonged exposures to atypical combinations of stimuli. Am J Psychol 68:526–548.

Henning GB (1983) Lateralization of low-frequency transients. Hear Res 9:153–172.

Houtsma AJM, Goldstein JL (1972) The central origin of the pitch of complex tones: evidence from musical interval recognition. J Acoust Soc Am 51:520–529.

Irvine DRF (1992) Physiology of the auditory brainstem. In: Popper AN, Fay RR (eds) The Mammalian Auditory Pathway: Neurophysiology. New York: Springer-Verlag, pp. 153–231.

Isabelle SK, Colburn HS (1991) Detection of tones in reproducible narrowband noise. J Acoust Soc Am 89:352–359.

Jeffress LA (1948) A place theory of sound localization. J Comp Physiol Psychol 41:35–39.

Jeffress LA (1958) Medial geniculate body—a disavowal. J Acoust Soc Am 30:802–803.

Johnson DH (1980) The relationship between spike rate and synchrony in responses of auditory-nerve fibers to single tones. J Acoust Soc Am 68:1115–1122.

Johnson DH, Williams J (1992) Simulation of single LSO unit responses. Assoc Res Otolaryngol Abstr 15:28.

Johnson DH, Dabak A, Tsuchitani C (1990) Function-based modeling of binaural interactions: Interaural level. Hear Res 49:301–320.

Johnson DH, Zertuche E, Pelton M (1991) Computer simulation of single LSO neurons. Assoc Res Otolaryngol Abstr 14:51.

Johnson DH, Tsuchitani C, Linebarger DA, Johnson MJ (1986) Application of a point process model to responses of cat lateral superior olive units to ipsilateral tones. Hear Res 21:135–159.

Joris PX, Yin TCT (1990) Time sensitivity of cells in the lateral superior olive (LSO) to monaural and binaural amplitude-modulated complexes. Assoc Res Otolaryngol Abstr 13:267–268.

Klein MA, Hartmann WM (1981) Binaural edge pitch. J Acoust Soc Am 70:51–61.

Knudsen EI (1984) Synthesis of a neural map of auditory space in the owl. In: Edelman GM, Cowan WM, Gall WE (eds) Dynamic Aspects of Neocortical Function. New York: Wiley, pp. 375–396.

Koehnke J, Durlach NI (1989) Range effects in the identification of lateral position. J Acoust Soc Am 86:1176–1178.

Konishi M, Takahashi TT, Wagner H, Sullivan WE, Carr CE (1988) Neurophysiological and anatomical substrates of sound localization in the owl. In: Edelman Gm, Gall WE, Cowan WM (eds) Auditory Functon: Neurobiological Bases of Hearing. New York: Wiley, pp. 721–746.

Kuhn GF (1987) Physical acoustics and measurements pertaining to directional hearing, In: Yost WA, Gourevitch G (eds) Directional Hearing New York: Springer-Verlag, pp. 3–25.

Kulkarni A, Woods WS, Colburn HS (1992) Binaural recordings from KEMAR in several acoustical environments. J Acoust Soc Am 92:2376.

Kuwada S, Batra R (1991) Sensitivity to interaural time differences (ITDs) of neurons in the superior olivary complex (SOC) of the unanesthetized rabbit. Soc Neurosci Abstr 17:450.

Kuwada S, Batra R, Stanford TR (1989) Monaural and binaural response properties of neurons in the inferior colliculus of the rabbit: Effects of sodium pentobarbital. J Neurophysiol (Bethesda) 61:269–282.

Licklider JCR (1959) Three auditory theories. In: Koch ES (ed) Psychology: A Study of a Science. Study 1, Vol. 1. New York: McGraw-Hill, pp. 41–144.

Lindemann W (1986a) Extension of a binaural cross-correlation model by contralateral inhibition. I. Simulation of lateralization for stationary signals. J Acoust Soc Am 80:1608–1622.

Lindemann W (1986b) Extension of a binaural cross-correlation model by contralateral inhibition. II. The law of the first wave front. J Acoust Soc Am 80:1623–1630.

Litovsky RY, Yin TCT (1993) Single-unit responses to stimuli that mimic the precedence effect in the inferior colliculus of the cat. Assoc Res Otolaryngol Abstr 16:128.

Lyon RF (1983) A computational model of binaural localization and separation. Proc International Conference on Acoustics, Speech and Signal Processing ICASSP '83 3:1148–1151.

MacGregor RJ (1987) Neural and Brain Modeling. New York: Academic press.

Makous JC, Middlebrooks JC (1990) Two-dimensional sound localization by human listeners. J Acoust Soc Am 87:2188–2200.

Manis PB, Marx SO (1991) Outward currents in isolated ventral cochlear nucleus neurons. J Neurosci 11:2865–2880.

McFadden D, Pasanen EG (1976) Lateralization at high frequencies based on interaural time differences. J Acoust Soc Am 59:634–639.

McFadden D, Jeffress LA, Lakey JR (1972) Differences of interaural phase and level in detection and lateralization: 1000 and 2000 Hz. J Acoust Soc Am 52:1197–1206.

Meddis R, Hewitt MJ (1991) Virtual pitch and phase sensitivity of a computer model of the auditory periphery. I. Pitch identification. J Acoust Soc Am 89:2866–2882.

Meddis R, Hewitt MJ, Shackleton TM (1990) Implementation details of a computational model of the inner hair-cell/auditory-nerve synapse. J Acoust Soc Am 87:1813–1818.

Middlebrooks JC (1992) Narrow-band sound localization related to external ear acoustics. J Acoust Soc Am 92:2607–2624.

Middlebrooks JC, Green DM (1991) Sound localization by human listeners. Annu Rev Psychol 42:135–139.

Mills AW (1958) On the minimum audible angle. J Acoust Soc Am 30:237–246.

Molnar CE, Pfeiffer RR (1968) Interpretation of spontaneous spike discharge patterns of neurons in the cochlear nucleus. Proc IEEE 56:993–1004.

Moore BCJ, Glasberg BR (1986) The role of frequency selectivity in the perception of loudness, pitch and time. In: Moore BCJ (ed) Frequency Selectivity in Hearing. London: Academic Press, pp. 251–308.

Oppenheim AV, Schafer RW (1989) Discrete-Time Signal Processing. Englewood Cliffs: Prentice-Hall.

Raatgever J (1980) On the Binaural Processing of Stimuli with Different Interaural Phase Relations. Doctoral dissertation, Technische Hogeschool, Delft, Netherlands.

Raatgever J, Bilsen FA (1977) Lateralization and dichotic pitch as a result of spectral pattern recognition, In: Evans EF, Wilson JP (eds) Psychophysics and Physiology of Hearing. London: Academic Press, pp. 443–453.

Raatgever J, Bilsen FA (1986) A central spectrum theory of binaural processing: evidence from dichotic pitch. J Acoust Soc Am 80:429–441.

Rakerd B, Hartmann WM (1985) Localization of sound in rooms. II: The effects of a single reflecting surface. J Acoust Soc Am 78:524–533.

Rakerd B, Hartmann WM (1986) Localization of sound in rooms. III: Onset and duration effects. J Acoust Soc Am 80:1695–1706.

Rayleigh, Lord (Strutt JW) (1907) On our perception of sound direction. Philos Mag 13:214–232.

Reed MC, Blum JJ (1990) A model for the computation and encoding of azimuthal information by the lateral superior olive. J Acoust Soc Am 88:1442–1453.

Ruotolo BR, Stern RM Jr, Colburn HS (1979) Discrimination of symmetric time-intensity traded binaural stimuli. J Acoust Soc Am 66:1733–1737.

Rupert A, Moushegian G, Whitcomb MA (1966) Superior-olivary response patterns to monaural and binaural clicks. J Acoust Soc Am 39:1069–1076.

Sayers BM (1964) Acoustic-image lateralization judgements with binaural tones. J Acoust Soc Am 366:923–926.

Sayers BM, Cherry EC (1957) Mechanism of binaural fusion in the hearing of speech. J Acoust Soc Am 29:973–987.

Schneider B, Zurek PM (1989) Lateralization of coherent and incoherent targets added to a diotic background. J Acoust Soc Am 86:1756–1763.

Schroeder MR (1977) New viewpoints in binaural interactions. In: Evans EF, Wilson JP (eds) Psychophysics and Physiology of Hearing. New York: Academic Press, pp. 455–467.

Schwartz IR (1992) The superior olivary complex and lateral lemniscal nuclei. In: Webster DB, Popper AN, Fay RR (eds) The Mammalian Auditory Pathway: Neuroanatomy. New York: Springer-Verlag, pp. 117–167.

Searle CL, Braida LD, Davis MF, Colburn HS (1976) Model for auditory localization. J Acoust Soc Am 60:1164–1175.

Shackleton TM, Bowsher JM, Meddis R (1991) Lateralization of very-short duration tone pulses of low and high frequencies. Q J Exp Psychol 43(A):503–516.

Shackleton TM, Meddis R, Hewitt MJ (1992) Across frequency integration in a model of lateralizaton. J Acoust Soc Am 91:2276–2279.

Shamma SA (1985) Speech processing in the auditory system. I: The representation of speech sounds in the responses of the auditory nerve. J Acoust Soc Am 78:1612–1621.

Shamma SA, Shen N, Gopalaswamy P (1989) Stereausisi: Binaural processing without neural delays. J Acoust Soc Am 86:989–1006.

Shamma SA, Chadwick RS, Wilbur WJ, Morrish KA, Rinzel J (1986) A biophysical model of cochlear pressing: Intensity dependence of pure tone reponses. J Acoust Soc Am 80:133–145.

Shinn-Cunningham BG, Zurek PM, Durlach NI (1993) Adjustment and discrimination measurements of the precedence effect. J Acoust Soc Am 93:2923–2932.

Siebert WM (1970) Frequency discrimination in the auditory system: Place or periodicity mechanisms? Proc IEEE 58:723–730.

Smith PH, Banks MI (1992) Intracellular recordings from neurobiotin-labeled principal cells in brain slices of the guinea pig MSO. Soc Neurosci Abstr 18:382.

Smith PH, Joris PX, Yin TCT (1993) Projections of physiologically characterized spherical bushy cell axons from the cochlear nucleus of the cat: Evidence for delay lines to the medial superior olive. J Comp Neurol 331:245–260.

Snyder D, Miller M (1991) Random Point Processes in Time and Space, 2nd Ed. New York: Springer-Verlag.

Stern RM Jr, Colburn HS (1978) Theory of binaural interaction based on auditory-nerve data. IV. A model for subjective lateral positon. J Acoust Soc Am 64:127–140.

Stern RM Jr, Colburn HS (1985) Lateral position-based models of interaural discrimination. J Acoust Soc Am 77:753–755.

Stern RM Jr, Trahiotis C (1992) The role of consistency of interaural timing over frequency in binaural lateralization. In: Cazals Y, et al. (eds) Auditory Physiology and Perception. Oxford: Pergamon Press, pp. 547–554.

Stern RM Jr, Zeiberg AS, Trahiotis C (1988) Lateralizaiton of complex binaural sitmuli: a weighted image model. J Acoust Soc Am 84:156–165.

Stern RM Jr, Zeppenfeld T, Shear GD (1991) Lateralization of rectangularly-modulated noise: explanatons for counterintuitive reversals. J Acoust Soc Am 90:1908–1917.

Stutman E, Carney LH (1993) A model for temporal sensitivity of cells in the auditory brainstem: the role of a slow, low-threshold potassium conductance. Assoc Res Otolaryngol Abstr 16:121.

Sujaku Y, Kuwada S, Yin YCT (1981) Binaural interaction in the cat inferior colliulus: Comparison of the physiological data with a computer simulated model, In: Syka J (ed) Neuronal Mechanisms of Hearing. New York: Plenum Press, pp. 233–238.

Sulivan WE, Konishi M (1984) Segregation of stimulus phase and intensity coding in the cochlear nucleus of the barn owl. J Neurosci 4:1787–1799.

Sullivan WE, Konishi M (1986) Neural map of interaural phase difference in the owl's brainstem. Proc Natl Acad Sci USA 83:8400–8404.

Trahiotis C, Bernstein LR (1986) Lateralization of bands of noise and sinusoidally amplitude-modulated tones: Effects of spectral locus and bandwidth. J Acoust Soc Am 79:1950–1957.

Trahiotis C, Bernstein LR (1990) Detectability of interaural delays over select spectral regions: Effects of flanking noise. J Acoust Soc Am 87:810–813.

Trahiotis C, Stern RM Jr (1989). Lateralizaiton of bands of noise: Effects of bandwidth and differences of interaural time and phase. J Acoust Soc Am 86:1285–1293.

Trahiotis C, Stern RM Jr (1994) Across-frequency interaction in lateralization of complex binaural stimuli. J Acoust Soc Am 96:3804–3806.

von Békésy G (1930) Zur Theorie des Hörens. Physik Z 31:857–868. [Translation in Wever EG (ed) Experiments in Hearing. New York: McGraw-Hill.]

Webster FA (1951) The influence of interaural phase on masked thresholds I. The role of interaural time deviation. J Acoust Soc Am 23:452–462.

Wenzel EM (1992) Localization in virtual acoustic displays. Presence 1:80–107.

Wenzel EM, Arruda MA, Kistler DJ, Wightman FL (1993) Localizaiton using non-individualized head-related transfer functions. J Acoust Soc Am 94:111–123.

Wightman FL, Kistler DJ (1989) Headphone simulation of freefield listening. II: Stimulus synthesis. J Acoust Soc Am 85:858–878

Wightman FL, Kistler DJ (1993) Sound localization, In: Yost WA, Popper AN, Fay RR (eds) Human Psychophysics. New York: Springer-Verlag.

Woods WS, Colburn HS (1992) Test of a model of auditory object formation using intensity and ITD discrimination. J Acoust Soc Am 91:2894–2902.

Yin TCT, Chan JCK (1990) Interaural time sensitivity in the medial superior olive of the cat. J Neurophysiol (Bethesda) 64:465–488.

Yin TCT, Kuwada S (1983) Binaural interaction in low-frequency neurons in inferior colliculus of the cat. II. Effects of changing rate and direction of interaural phase. J Neurophysiol (Bethesda) 50:1020–1042.

Yin TCT, Litovsky RY (1993) Physiological correlates of the precedence effect: Implications for neural models. J Acoust Soc Am 93:2293.

Yost WA (1972) Tone-on-tone masking for three binaural listening conditions. J Acoust Soc Am 52:1234–1237.

Yost WA (1981) Lateral position of sinusoids presented with interaural intensive and temporal differences J Acoust Soc Am 70:397–409.

Yost WA, Harder PJ, Dye RH (1987) Complex spectral patterns with interaural differences: dichotic pitch and the 'central spectrum'. In: Yost WA, Watson CS (eds) Auditory Processing of Complex Sounds. Hillsdale: Erlbaum, pp. 190–201.

Young SR, Rubel EW (1983) Frequency-specific projections of individual neurons in chick brainstem auditory nuclei. J Neurosci 3:1373–1378.

Zacksenhouse M, Johnson DH, Tsuchitani C (1992) Excitatory/inhibitory interaction in an auditory nucleus revealed by point process modeling. Hear Res 62:105–123.

Zurek PM (1980) The precedence effect and its possible role in the avoidance of interaural ambiguities. J Acoust Soc Am 67:952–964.

Zurek PM (1985) Spectral dominance in sensitivity to interaural delay for broadband stimuli. J Acoust Soc Am 78:S121.

Zurek PM (1987) The precedence effect. In: Yost WA, Gourevitch G (eds) Directional Hearing. New York: Springer-Verlag, pp. 85–105.

Zurek PM (1991) Probability distributions of interaural phase and level differences in binaural detection stimuli. J Acoust Soc Am 90:1927–1932.

Zurek PM (1993) A note on onset effects in binaural hearing. J Acoust Soc Am 93:1200–1201.

Zurek PM, Durlach NI (1987) Masker bandwidth dependence in homophasic and antiphasic tone detection. J Acoust Soc Am 81:459–464.

9
Auditory Computations for Biosonar Target Imaging in Bats

James A. Simmons, Prestor A. Saillant, Michael J. Ferragamo, Tim Haresign, Steven P. Dear, Jonathan Fritz, and Teresa A. McMullen

1. Introduction

1.1 Bats and Echolocation

Bats are nocturnal flying mammals classified in the order Chiroptera. These animals have evolved a biological sonar, called *echolocation*, to orient in darkness—to guide their flight around obstacles and to detect their prey (Griffin 1958; Novick 1977; Neuweiler 1990; see Popper and Fay 1995). Echolocating bats broadcast ultrasonic sonar signals that travel outward into the environment, reflect or scatter off objects, and return to the bat's ears as echoes. First the outgoing sonar signal and then the echoes impinge on the ears to act as stimuli, and the bat's auditory system processes the information carried by these sounds to reconstruct images of targets (Schnitzler and Henson 1980; Simmons and Kick 1984; Suga 1988, 1990; Simmons 1989; Dear, Simmons, and Fritz 1993; Dear et al. 1993).

1.2 The Big Brown Bat

The big brown bat, *Eptesicus fuscus* ("dusky house-flier"), is a common, widely distributed North American bat of the family Vespertilionidae (Kurta and Baker 1990). *Eptesicus* is one of many species of insectivorous bats that produce frequency-modulated (FM) echolocation sounds and use echoes to find and intercept flying insects (Griffin 1958; Pye 1980; Simmons 1989; Neuweiler 1990; see also Popper and Fay 1995). Figure 9.1 shows an insect's-eye view of a big brown bat as it approaches a target during an interception maneuver guided by sonar. A wide range of behavioral experiments have been carried out with *Eptesicus* to evaluate aspects of the performance of its sonar (see Popper and Fay 1995). These studies identify fundamental features of the auditory computations underlying FM echolocation by specifying the final output of these com-

FIGURE 9.1. The big brown bat, *Eptesicus fuscus*, approaching a target (photographed from the target's position by S.P. Dear and P.A. Saillant).

putations, the *images*, in relation to the input, *emissions and echoes* (Simmons 1989, 1992; see Popper and Fay 1995).

1.3 Scope of This Chapter

Our chapter describes auditory computations that create the dimension of distance, or *target range*, in the images perceived by the big brown bat and the physiological mechanisms which support these computations. To provide a conceptual framework, we use a model of echolocation (spectrogram correlation and transformation [SCAT]; see Saillant et al. 1993) that assumes the bat's cochlea (1) segments the range of frequencies in the bat's sonar sounds into parallel bandpass-filtered channels, (2) half-wave rectifies and then smooths the resulting frequency segments of the sounds, and (3) triggers neural discharges from these excitation patterns. The bat's sonar sounds are frequency modulated (FM), and the model uses a modified auditory spectrogram format (Altes 1980, 1984) for initial encoding of their frequency sweeps, with several parameters (e.g., scaling of filter center frequencies, sharpness of tuning, integration time) quantified from physiological data. Taking "neural discharges" triggered from the auditory spectrograms of broadcast and received sounds as input to further computations, the model focuses attention on reconstruction of the locations of echo sources along the axis of target range to form images equivalent to the "A-scope" display of a radar or sonar system (essentially a plot of target range versus target strength; see Skolnik 1962).

2. The Sonar System of the Big Brown Bat

2.1 Echolocation Sounds used by Eptesicus

Bats universally seem to use FM echolocation signals as acoustic probes to determine target range from the travel time of echoes (Griffin 1958; Simmons 1973; Schnitzler and Henson 1980; Simmons and Grinnell 1988). *Eptesicus* broadcasts ultrasonic sounds that contain frequencies in the range from about 20 kHz to more than 100 kHz depending on the bat's situation. Figure 9.2 shows spectrograms of echolocation signals recorded during the bat's approach to a small target in an interception maneuver (Popper and Fay 1995). These sounds are harmonically structured FM signals, with first and second harmonics always present (FM$_1$, FM$_2$; terminology after Suga 1988, 1990), plus a segment of the third harmonic (FM$_3$) and often also a segment of the fourth harmonic (FM$_4$) present. Usually, FM$_1$ sweeps from about 50–60 kHz down to about 20–25 kHz, while FM$_2$ sweeps from about 100 kHz down to 40–50 kHz (Fig. 9.2a–d). Only a short segment of FM$_3$ is produced around 75–90 kHz, and FM$_4$ is also restricted to around 80–90 kHz when is appears (Fig 9.2e–g). In Figure 9.2, the shapes of the FM sweeps are curved, with the sweep shape being approximately hyperbolic, making the sweep itself approximately linear with period. That is, as *frequency* sweeps curvilinearly downward from 55 kHz to 23 kHz in FM$_1$, *period* sweeps upward from about 18 μsec to 43 μsec in a linear fashion (see following).

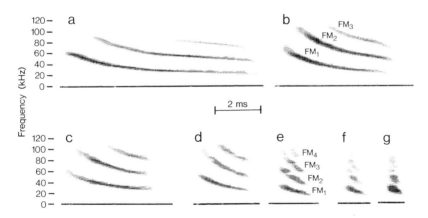

FIGURE 9.2a–g. Spectrograms of multiple-harmonic FM echolocation sounds emitted by *Eptesicus* at different stages of insect pursuit and capture while being photographed and videotaped in laboratory studies of interception (Fig. 9.1). (a) search stage; (b–d) approach or tracking stage; (e–g) terminal stage (see Popper and Fay 1995).

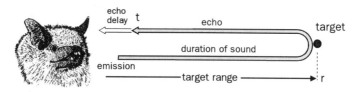

FIGURE 9.3. Duration of bat's sonar sound in relation to target range and echo delay.

2.2 Target Ranging

The bat's sonar sounds travel outward from the bat's mouth to impinge on objects located at different distances and return to the bat's ears as echoes at different times. The bat perceives the distance to objects from this time delay (Simmons 1973). In air, the delay of echoes is 5.8 msec per meter of range, the time required for the sound to travel over the two-way path length of 2 m for a target at a range of 1 m. Figure 9.3 illustrates the relationship between the target's location at a particular range (r) and the duration of the bat's sonar sounds. As a rule of thumb, *Eptesicus* keeps the duration of its sounds slightly shorter than the two-way travel time of echoes (echo delay, t, in Fig. 9.3) by propressively shortening the sounds during interception, so there is little or no overlap of transmissions with echoes from the insect (Griffin 1958; Simmons 1989; Hartley 1992; Popper and Fay 1995). The distance in the air spanned by the sound's duration almost completely fills up the pathlength from the bat's mouth out to the target and back to its ears (Fig. 9.3).

2.3 Operating Range of Echolocation

Eptesicus can detect insect-sized targets as far away as 5 m, which corresponds to an echo delay of about 30 msec (Kick 1982). Small objects located farther away than this maximum operating range return echoes that are too weak to be heard (Pye 1980; Lawrence and Simmons 1982). At long range, *Eptesicus* can detect echoes at levels as low as 0 dB SPL (Kick 1982; Kick and Simmons 1984). The bat's broadcasts are roughly 100–110 dB stronger than the weakest echo that can be detected, so the bat can tolerate considerable attenuation of echoes resulting from the small size or long range of targets before the echoes become inaudible. Consequently, a wide range of different sizes of insects or other objects located at different distances are potentially detectable with echolocation, which means that the bat's target-ranging system must be able to accommodate reception of echoes at a variety of different delays and display objects a variety of different distances. The axis of echo delay (target range) in the "A-scope" images perceived by *Eptesicus* must extend from roughly 0.5 msec (about 10 cm) to roughly 30 msec (about 5m) (Simmons 1989; Dear, Simmons, and Fritz 1993).

3. Targets, Echoes, and Images

3.1 Target Glints and the Structure of Echoes

Objects in air behave as though they consist of a small number of parts that each reflect a more-or-less faithful replica of the incident sonar sound (Simmons and Chen 1989). Consequently, the overall "echo" that actually returns to the bat's ears from a natural target actually consists of several discrete echoes arriving together. Figure 9.4 shows a two-glint target (a dipole) as the simplest example of a complex object. The reflecting surfaces or points in the target are called glints (A and B in Fig. 9.4), and their separation from each other in range (δr) determines the time separation (δt) of the reflected replicas. The flying insects that bats prey upon are small; the different body parts (head, abdomen, wingtips, legs) are separated from each other by only 2–3 cm, so the delay separations between their reflections will be less than 100–150 μsec. The critical factor is the relatively small size of these delay separations in relation to the duration of the incident sonar sound. Because the distance from the bat to the insect is large compared to the separation of the insect's parts, and because the bat uses this overall distance to determine the duration of its sonar sounds (see Fig. 9.3), the multiple reflections sent back to the bat will overlap each other when they arrive (see Fig. 9.4).

3.2 "Auditory Spectrograms" of Emissions and Echoes

3.2.1 Frequency Scale and Integration Time

Figure 9.2 shows spectrograms of echolocation sounds that have a conventional vertical axis that is linear with frequency. The FM sweeps of the bat's sonar sounds appear curved on this scale because the sweep functions are approximately hyperbolic, or linear with period. However, the bat's auditory system does not scale frequency as a linear variable.

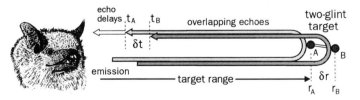

FIGURE 9.4. Duration of bat's sonar sound and overlapping echoes from two parts of the same dipole target at slightly different distances.

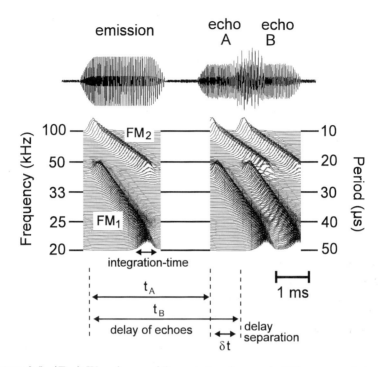

FIGURE 9.5. (*Top*) Waveforms of 2-msec, two-harmonic FM sonar emission and two echoes (A,B) at delays of 3.7 and 4.7 msec. (*Bottom*) Spectrogram correlation and transformation (SCAT) spectrograms representing hyperbolic frequency axis for emission and two echoes at top (81 parallel bandpass filter channels followed by rectification and smoothing with integration time of 300–400 µsec). Although the raw echo waveforms overlap, their spectrograms are separate because the delay separation (1 msec) is more than the integration time (300–400 µsec).

Physiological measurements reveal that frequency scaling in the auditory system of *Eptesicus* is approximately hyperbolic, too (see following). Figure 9.5 shows an example of an "auditory" spectrogram with a hyperbolic frequency axis for a sonar sound of *Eptesicus* (emission) and two echoes (A and B) arriving at different delays ($t_A = 3.7$ msec and $t_B = 4.7$ msec). The duration of the broadcast signal is 2 msec, and it has two harmonics (FM_1, FM_2; see Fig. 9.2). The shape of each sweep in the emission and both echoes is "linearized" in the auditory spectrogram by the hyperbolic vertical frequency scale. These spectrograms serve as the initial signal representations in the spectrogram correlation and transformation (SCAT) model of echolocation (Saillant et al. 1993). They are made by passing the signals through a bank of 81 parallel bandpass filters with constant, 4-kHz bandwidths and center frequencies spaced at fixed period intervals ($1/f$) of 0.5 µsec. The filter outputs are half-wave rectified and then smoothed with a low-pass filter of about 1 kHz to approximate at least some of the physiological events associated with transduction that produce neural discharges from excitation comparable to the horizontal slices in Figure 9.5.

The two echoes in Figure 9.5 are only 1 msec apart, while the broadcast sound is 2 msec long, so the raw waveforms of the echoes overlap each other (top of Fig. 9.5). However, the auditory spectrograms of the echoes do *not* overlap because the time width or integration time of the spectrogram (time width of spectrogram slices in Fig. 9.5) is only a fraction of a millisecond ($\pm300\,\mu$sec to $\pm400\,\mu$sec). The two echoes in Figure 9.5 will always appear as separate spectrograms with discrete FM sweeps as along as their time separation (δt) exceeds the integration time (Beuter 1980; Altes 1984).

3.2.2 Auditory Spectrograms of Overlapping Echoes

Figure 9.6 illustrates hyperbolically scaled auditory (SCAT) spectrograms for a series of overlapping echoes (A and B) at different time separations (δt) from 0 to 1 msec. These echo-delay separations correspond to separations of 0 to about 17 cm along the axis of range. The shorter separations of 0, 60, and 110 μsec are realistic values for the echoes reflected by different parts of a small target such as an insect with dimensions of up to about 2 cm, while larger time separations of 225 μsec to 1 msec correspond better to the range separations associated with different targets, such as an insect located from 4 cm to 17 cm away from background vegetation. Figure 9.6 demonstrates the importance of the integration time associated with the initial stages of auditory processing for determining how information about targets must be represented in the bat's auditory system (Beuter 1980; Altes 1984; Simmons 1989). For time separations of 350 μsec and greater, the two overlapping echoes are represented by separate SCAT spectrograms, with a recognizable sweep for each harmonic (FM_1 and FM_2 in Fig. 9.5). For separations of less than 350 μsec, however, the separate spectrograms merge into only one recognizable sweep for each harmonic (Simmons et al. 1989), and now the amplitude of the spectrogram is modulated at different frequencies by interference between the echoes as they mix together within the integration time of the parallel filter channels. That is, the spectrum of the combined echoes is rippled by the interference. At longer echo separations of 450 μsec and more, the spectrograms in Figure 9.6 appear as smooth, sloping ridges, but at shorter separations of 60, 110, and 225 μsec, the ridges contain peaks and valleys – spectral features whose separation in frequency (vertical axis) provides the only indication that the echoes actually are separated in time (horizontal axis). The integration time of spectrograms thus establishes a boundary between representation of the time separation of the two echoes as a difference in time along

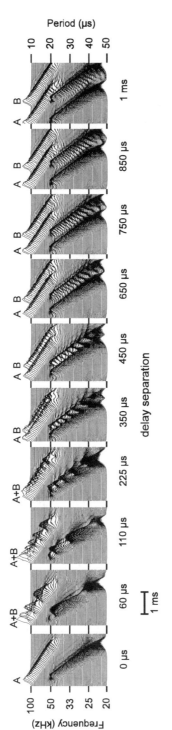

FIGURE 9.6. SCAT (hyperbolic frequency) spectrograms of pairs of overlapping echoes (A,B from Fig. 9.5) at different delay separations in relation to their integration time of 300–400 μsec. They appear as separate spectrograms for separations larger than the integration time and merge into the same spectrogram for separations that are smaller.

the horizontal axis of the spectrograms and representation of the time separation as a difference in frequency between peaks and valleys along the vertical axis (Simmons 1992; see also Popper and Fay 1995).

3.2.3 The Bat's Integration Time for Echo Reception

In clutter-interference experiments, where the bat's task is to detect test echoes at one particular arrival time in the presence of masking echoes at different arrival times, *Eptesicus* fails to detect the test echoes when the time separation is smaller than 350 µsec (Simmons et al. 1989). This result reveals that the bat receives and segregates echoes with an intrinsic integration time of about 350 µsec. Echoes that arrive closer together than this time window become merged into a single sound for purposes of detection. Physiological responses recorded in the bat's auditory system also fail to register overlapping sounds at short time separations. Neurons in the cochlear nucleus and nucleus of the lateral lemnisicus can register the presence of two separate sounds with separate discharges so long as these sounds are more than 300–500 µsec apart (Grinnell 1963; Suga 1964; Covey and Casseday 1991); for shorter time separations the neurons fail to produce discharges in response to the second sound by itself. It thus appears that *Eptesicus* fails to detect test echoes in the presence of cluttering echoes because its auditory system merges the two sounds together and represents both with just one volley of neural discharges.

3.3 Segregation of Echoes Within the Integration Time

Because insects are small objects, the reflected replicas from their glints will arrive closer together than the integration time of 350 µsec and will overlap to interfere with each other (Simmons and Chen 1989; Kober and Schnitzler 1990; Schmidt 1992; Moss and Zagaeski 1994). Do bats perceive the actual range separation of the glints in the insect (Simmons, Moss, and Ferragamo 1990; Simmons 1989, 1992, 1993) or just the spectral effects of interference between the overlapping reflections (Neuweiler 1990; Schmidt 1992)? If the two glints in Figure 9.4 are just two different parts of the same insect, it might seem unnecessary for the bat to perceive that there really are two glints at slightly different ranges to intercept the target. The time delay associated with the spectrogram of the combined reflections in the echo from the insect would be adequate to perceive the insect's overall distance, and the spectrum of the overlapping echo replicas should be adequate to characterize the target's shape and fluttering motion without explicitly perceiving the range separation of its glints (Simmons and Chen 1989; Neuweiler 1990; Schmidt 1992).

However, the same configuration of closely spaced echoes often occurs in situations in which perception of the range to each glint probably is necessary. It is perfectly possible for the two echo replicas in Figure 9.4 to come from different objects rather than from the same object. For example, the nearer glint (A) might be part of an insect while the farther glint (B) might be part of some vegetation that the insect happens to be flying past. In this situation the bat would need to perceive the distance to the insect (r_A) to affect a capture, while it would also need to perceive the distance to the vegetation (r_B) to avoid colliding with it. Just characterizing the overlapping echoes by the interference spectrum would not be enough to accomplish these two tasks; the bat has to perceive *both* ranges to catch the insect while avoiding the obstacle (Popper and Fay 1995). The problem is that information about the ranges of the two glints has been blurred by the integration time of 350 μsec so that both echoes have only one delay that can be determined directly from the time axis of the spectrogram (see Fig. 9.6).

3.4 The Bat's Images of Two-Glint Targets

A crucial aspect of performance in echolocation must be the bat's ability to resolve two closely spaced echoes as arriving at different times. What is the smallest separation in the arrival time of two echoes (δt in Fig. 9.4) that the bat can perceive? Analysis of the behavior of bats in several naturalistic situations (obstacle-avoidance tests, interception of targets in clutter, discrimination of airborne or suspended targets) has revealed that even a flying bat probably can resolve two echoes as having discrete delays for separations as short as 5–20 μsec (Popper and Fay 1995). Figure 9.7 illustrates the results of experiments that demonstrate more directly that *Eptesicus* can perceive the arrival times of both replicas contained in two-glint echoes at small delay separations. The graphs show the performance (% errors) of two bats detecting changes in the arrival time of overlapping pairs of test echoes separated by 0, 10, 20, or 30 μsec (δt) at an overall delay of about 3.2 msec. In this experiment, probe echoes are moved to different arrival times in relation to either of the test echoes (located at 0 μsec and at 10, 20, or 30 μsec on the horizontal axis of graphs in Fig. 9.7); the decline in the bat's performance when the probe echoes are aligned in time with either of the test echoes (peaks in error curves) shows that the bat characterizes these double echoes as having two integral delay values, not just one delay value. (For details of experiments and compound performance plots, see Simmons et al. 1990.) In effect, Figure 9.7 shows examples of the "A-scope" images perceived by *Eptesicus* for two-glint targets with range separations (δr) of 0, 1.7, 3.4, or 5.2 mm.

The most important feature of the images in Figure 9.7 is that the two overlapping echoes are both assigned perceived magnitudes of delay along a scale of delay subdivided into sufficiently fine steps that differences of 10, 20, or 30 μsec can be displayed. The smallest echo-delay separation that *Eptesicus* can perceive with a separate error peak in the

FIGURE 9.7. Compound performance curves (% errors) for two *Eptesicus* in a task that uses a probe echo at different delays (horizontal axis) to locate the delay values perceived for two overlapping test echoes (similar to A and B in Fig. 9.6) at delay separations of 0, 10, 20, or 30 μsec. Bats make errors (peaks) at delays where they perceive test and probe echoes to have same delay (60 trials per data point per bat). Each test echo is perceived at its correct delay, the target has two glints, and the image has two glint components (0 μsec on horizontal axis is equivalent to 3.2-msec overall delay of nearer of the two overlapping echoes).

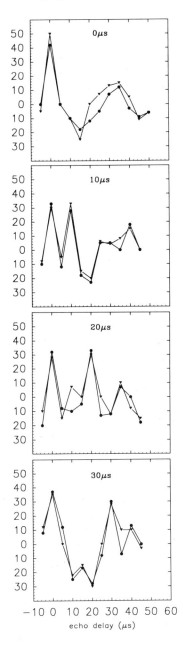

image for each delay (as in Fig. 9.7) is about 2 μsec (Saillant et al. 1993). The scale of delay in the bat's images must therefore be graduated in increments no larger that 2 μsec; otherwise, the two echoes would have been assigned the same delay value. *Eptesicus* is extraordinarily accurate at perceiving small changes in the arrival time and phase of echoes: several experiments have demonstrated that the bat can detect changes smaller than 0.4–0.5 μsec (Simmons 1979; Moss and Simmons 1993; Menne et al. 1989; Moss and Schnitzler 1989), and the smallest detectable change actually measured is about 10–15 nsec (Simmons et al. 1990). It is thus plausible that the spacing of adjacent "units" of delay along the echo-delay axis in the bat's images really could be as small as 1–2 μsec.

4. Computations on Spectrograms to Determine Delay and Recover Resolution

4.1 The Computational Problem in Echolocation

4.1.1 Fine Delay Resolution but Coarse Integration Time

The challenge for understanding the auditory computations that support echolocation lies in the difficulty of achieving fine temporal resolution of 10, 20, or 30 μsec for multiple echoes arriving within the 350-μsec integration time of echo reception (Fig. 9.7). These computations must produce an accurate estimate of the arrival time of the first reflected replica in each echo while also overcoming the limitations imposed by the blurring effects of the integration time to also produce an estimate of the arrival time of the second reflected replica (Simmons 1989; Simmons et al. 1990; Saillant et al. 1993). Because echoes separated by 10, 20, or 30 μsec are merged into only one spectrogram (see Fig. 9.6), an estimate of delay for only one of the echoes can be made directly from the spectrograms. The estimate of delay for the other echo has to be derived instead from the effects of overlap and interference on the merged spectrograms. Crucially, the inputs to the computations that yield the two delay estimates in each image in Figure 9.7 must have different numerical formats, one essentially a measurement of time between the emission and echo spectrograms and the other essentially a measurement of frequencies for spectral peaks and notches. However, both estimates of delay are manifested along the same dimension of the images (Fig. 9.7), so the outputs of the underlying computations ultimately must converge upon the same numerical scale in the bat's images. Moreover, the graduations of the echo-delay axis in these images must be finely divided enough to allow echo-delay resolution down to about 2 μsec (Saillant et al. 1993). Our goal is to learn how the bat's images are created, and identification of the computational locus for this convergence of formats would place us close to the site of image formation itself.

4.1.2 Convolution to Form Spectrograms

During bandpass filtering by the inner ear, the FM waveform is *convolved* with the impulse response of each filter, and the output, which consists of a segment of the waveform of the FM sweep in the neighborhood of the filter's center frequency, is then half-wave rectified and smoothed by a low-pass filter with a cutoff frequency of about 1 kHz to produce discharges of auditory nerve fibers. The smoothing filter is the most significant limiting component in this peripheral auditory signal-conditioning regime; for most practical purposes, the time of occurrence of each frequency in the FM sweep comes to be represented by the impulse response of this smoothing filter, which is about 300–400 μsec wide. Subsequently, in the nervous system these peripheral, receptor-generated impulses are themselves replaced by the neural discharges they trigger; these discharges also are spike shaped pulses several hundred microseconds wide. Each horizontal slice of the auditory spectrogram for the emission or the echo is approximately the width of this impulse response (see integration time in Fig. 9.5). When two echoes arrive close enough to each other that these impulse responses collide and form a single impulse, the volleys of neural discharges they produce also merge into one volley, and the bat can no longer detect one echo in the presence of the other. The clutter-interference experiment measures the time window for the collision of these impulses as viewed by the bat, giving a value of about 350 μsec as the minimum separation required for the bat to detect one echo as a separate sound in the presence of the other.

4.1.3 *Deconvolution* to Form Images

The result of convolution is to replace the series of frequency–time points in each FM sweep with a sloping ridge about 300–400 μsec wide (see Fig. 9.5). To segregate echoes whose sweeps overlap and merge into just one set of ridges in the spectrogram (see Fig. 9.6), the bat has to deblurr these sweeps-as-ridges to recover the temporal resolution that was lost during convolution. This deblurring operation is called deconvolution, and it requires explicit knowledge of the transmitted waveform so that the presence of multiple replicas can be recognized even when they arrive so close together that they merge into one spectrogram (Fig. 9.6). The key to understanding the computations at the heart of echolocation lies in knowing what is meant by this "knowledge" and how it can be used to reverse the blurring effects of convolution.

4.2 The SCAT Model

Figure 9.8 is a diagram of the principal computational stages required to convert the raw time-series waveform of a sonar emission and two overlapping echoes into an A-scope sonar image depicting the delay of both echoes along the same scale of time. This diagram shows the signal processing operations of the SCAT model as a guide to identifying what has to be learned about echo processing in the bat. The model's first stage is its "cochlea" (Saillant et al. 1993), which uses 81 bandpass filters in

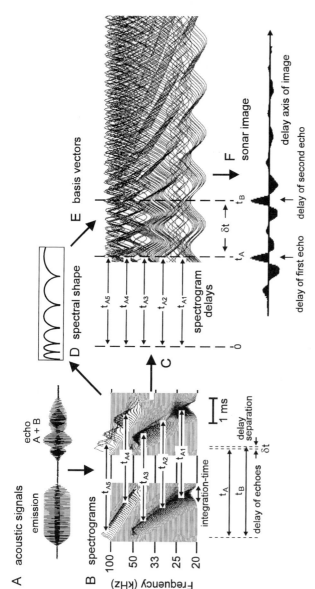

FIGURE 9.8A–F. SCAT computational model of convolution-deconvolution echo processing for target range and fine range separation by *Eptesicus* (Saillant et al. 1993). (A) Waveform of FM sonar emission and two echoes (A+B) separated by 60 µsec. (B) 81-Channel SCAT spectrograms of waveforms in A, with spectrogram delays (t_{A1-A5}) delineating time offset of echo. (C) Alignment and averaging of spectrogram delays to determine overall delay (indicated as t_A). (D) Shape of spectrum for overlapping echoes, with notches and peaks caused by interference (see Section 4.4.1). (E) Cosine-phase basis vectors individually phase aligned to start at time specified by spectrogram delay for each channel. (F) Echo-delay ("A-scope") sonar image formed by summing basis vectors. Image contains delay for echo A originally from spectrogram delays and echo B from summation and cancellation of different basis vectors phase aligned to each channel of the spectrogram. The shapes of the image components resemble emission-echo cross-correlation functions weighted by the hyperbolic frequency regime.

parallel to transform the raw input waveforms into hyperbolically scaled SCAT spectrograms (see Figs. 9.5 and 9.6) for further processing. This component of the model emulates the most critical features of the bat's inner ear and then generates "neural discharges" registering successive frequencies in the FM sweeps of the sounds. The model's remaining stages are two parallel pathways: a spectrogram correlation system for determining the time separation between the spectrogram of the emission and the spectrogram of echoes (Altes 1980, 1984) and a spectrogram transformation system for converting the pattern of peaks and notches in the spectrogram of overlapping echoes (Beuter 1980; Altes 1984) into an estimate of the time separation of the merged replicas. The outputs of these two processing pathways converge to write values of echo delay along a single delay axis in the final images.

4.3 Delay Lines for Spectrogram Correlation to Determine Echo Delay

4.3.1 Storing the Shape of the FM Sweeps in Emissions for Comparison with FM Sweeps in Echoes

Figure 9.8A shows the raw waveforms of the input signals, a sonar transmission with a duration of 2 msec and two overlapping echoes (A and B). The delay of the first echo (t_A) is 3.7 msec and the delay of the second echo (t_B) is only 60 μsec larger. This short delay separation ($\delta t = 60$ μsec) results in the two echoes merging to form just one spectrogram in Figure 9.8B. The amplitude of the echo spectrogram at different frequencies contains peaks and notches reflecting the interference that takes place within the 350-μsec integration time of the spectrograms. As a first step, the SCAT model determines the arrival time of the compound echo (A + B) from the time intervals between the spectrogram of the emission and the spectrogram of the echo at different frequencies (that is, the horizontal time displacement of the echo spectrogram to the right of the emission spectrogram is Fig 9.8B). These spectrogram delays (shown as $t_{A1} - t_{A5}$ in Fig. 9.8B) are extracted using delay lines that register the time of occurrence of each frequency in the emission and then compare it with the time of occurrence of the corresponding frequency in the echo. In effect, the delay lines store the shape of the spectrogram for the emission and slide it to the right (in Fig. 9.8B) until it lines up with the spectrogram for the echo, a process equivalent to correlation of the echo and emission spectrograms (Altes 1980). An "event" travels along each delay line from one delay tap to the next to register the ocurrence of one specific frequency in the broadcast sweep, and the relative position of similar events across all the delay lines preserves the shape of the sweep as the events propagate along the delay lines. This property of the spectrogram is crucial because the sweep shape really represents information about the phases of the different frequencies in the broadcast sound. If the shape of the sweep in the echo is distorted in the course of reflection from the target (the frequencies in the echo undergo different phase shifts), and this change in shape is detected, then whatever target feature caused the change in shape can be incorporated into the range image.

4.3.2 Reading Echo Delay from Delay Taps

At each frequency, the delay of the echo ($t_{A1} - t_{A5}$) is represented by the specific delay tap in the delay line that is active at the same moment that the echo arrives. This "moment" is judged by detecting coincidences between events taking place at the delay taps and events triggered by the incoming echo. The spectrogram delays ($t_{A1} - t_{A5}$), which are represented by the active delay taps in different delay lines, are then averaged across all the delay lines in Figure 9.8C to estimate the delay of the echo as a whole (t_A). This overall delay value is obtained from measurements of the timing of discharges and represents the distance to the object that contains the two glints; it usually is interpreted to be the distance to the nearer of the two glints (Simmons et al. 1990; Simmons 1993; Saillant et al. 1993). In the absence of noise, all 81 channels normally register their delay estimates at the same delay value (or delay-line tap); the addition of noise merely broadens the distribution of active delay taps around the mean value, and, at high levels, noise sometimes displaces the mean, too. Registration of echo delay is very precise by this method when averaged across a number of parallel delay lines, and *Eptesicus* is also very accurate at determining the delay of echoes to within 10–15 nsec from the timing of neural discharges (Simmons et al. 1990). Delay lines and delay coincidence devices are commonly used in radar systems to display the arrival times of echoes. Furthermore, this part of the model is equivalent to the delay–coincidence correlation process used in some models of auditory pitch coding (Licklider 1951; Langner 1992), and a delay coincidence model specifically tailored to echolocation is widely assumed to be the basis for perception of target range in bats (Sullivan 1982; Suga 1988, 1990; Park and Pollak 1993).

If the processing of echo information were to stop at this point, the range image would depict just the overall distance to the target. No distinction would be made about the distances to the two glints. Further information about the target is contained in the shape of the spectrum of the overlapping echoes (Fig. 9.8D), and one widely accepted hypothesis is that the bat classifies targets in terms of the spectral coloration supplied by the peaks and notches at different frequencies (Neuweiler 1990; Schmidt 1992). The bat does not appear to stop at this point, however, because it perceives both delays associated with the two-glint target in the same image (Fig. 9.7).

4.4 Spectrogram Transformation to Determine Delay Separation

4.4.1 Knowledge of Signals for Deconvolution

The capacity to deconvolve two overlapping echoes that have been merged into one spectrogram (Fig. 9.8A,B) depends on being able to translate the pattern of peaks and notches at diferent frequencies in the echo spectrum (Fig. 9.8D) into an estimate of the delay separation required to create these peaks and notches by interference. It is not sufficient to know just these frequency values, however, deconvolution requires knowledge of the values for the periods of the frequencies corresponding to the tops of the peaks and the bottoms of the notches. To be complete, deconvolution also requires knowledge about the detailed shape of the ridges in the echo spectrogram in the vicinity of the peaks and notches. (This is the previously mentioned phase information inherent in the shape of the sweeps.) The frequencies of the spectral peaks (f_p) are related to the reciprocal of the time separation of the overlapping echoes (δt):

$$f_P = n/\delta t \quad \text{(where } n = 1,2,3\ldots)$$

Similarly, the frequencies of the notches (f_n) are related to the echo time separation (δt):

$$f_n = (2n + 1)/2\delta t \quad \text{(where } n = 0,1,2,\ldots)$$

For example, in Figure 9.8A,B, where $\delta t = 60\,\mu\text{sec}$, the spectral peaks (Fig. 9.8D) fall at 17, 33, 50, 66, 83, and 100 kHz (even frequency intervals), while the notches fall at 8.3, 25, 42, 58, 75, and 92 kHz (odd frequency intervals). The frequency spacing of the peaks and the notches is the reciprocal of the time separation ($df = 1/\delta t$) itself. Moreover, the even-frequency or odd-frequency spacing of these peak or notch frequencies specifies whether there is a phase shift accompanying the time separation, as when one glint returns an echo that is 0° or 180° relative to the other echo. (Objects in air are so discontinuous from the air itself in acoustic impedance that the echoes they return are usually either 180° or 0° relative to the incident sound.)

4.4.2 Basis Vectors

The most complete implementation of the frequency-to-period knowledge required for deconvolution is reconstruction of the waveform of echoes at each frequency within the integration time window for convolution. this can be achieved even after spectrograms have been formed (that is, after convolution) by using the spectrogram delays at individual frequencies (t_{A1}-t_{A5} in Fig. 9.8C) as time-marking events to trigger the start of oscillatory signals, or basis vectors, that represent the original echo frequencies themselves. That is, each delay line is used to register the arrival time of one specific frequency in the echo, and the moment a coincidence between the echo and the emission at that frequency is detected, the basis vector begins to oscillate. In the SCAT model, the basis vectors (Fig. 9.8E) are cosine functions with durations sufficient to cover the interval of time across which the glint structure of echoes is to be reconstructed (usually the integration time; but see scaling factor, following). However, the model is robust and works reasonably well for any periodic function used as basis vectors, even square waves (Saillant et al. 1993). The horizontal slices in the spectrograms (Fig. 9.8B) correspond to the frequency channels of the SCAT model, and each channel is tuned to a specific frequency in the emission or echo. Each channel then produces its own basis vector at a frequency that matches (or is proportional to) its original tuned frequency (Fig. 9.8E) The amplitude of the basis vector in each frequency channel is scaled according to the shape of the spectrum for the echo (from Fig. 9.8D to Fig. 9.8E).

Two effects are achieved by using the spectrogram delays in different channels to start these oscillations in "cosine phase" separately for each channel. First, the slopes of the FM sweeps in the harmonics (FM_1,FM_2) are removed from subsequent concern by "dechirping" the signals. This permits the transmitted signals to be changed in duration and sweep shape without having to keep track of this change. (One SCAT receiver processes all FM signals.) Second, information about changes in the shape of the sweep, or the phase of the different frequencies in echoes relative to emissions, is retained in differences in the starting time across the basis vectors. These starting times vary by as much as one full period at each basis vector frequency (compare basis vectors with each other in Fig. 9.8E).

4.5 Formation of SCAT Images

4.5.1 Reconstruction of the Glint Structure in Echoes

Once the basis vectors begin to oscillate, the next stage in the imaging process is simple: The arrival times of the overlapping echoes (A and B) are reconstructed by summing the basis vectors across all 81 parallel channels (Fig. 9.8F) to form an average basis waveform. This average waveform is the image of the target's glint structure along the axis of echo delay or target range. Because of reinforcement and cancellation of peaks and troughs in the basis vectors across channels, the original arrival times of the echoes (t_A and t_B) appear as positive-going peaks in the resulting image even though no correspondingly well-resolved pair of events in the original echoes registers their arrival-time separation. The locations of the tips of the peaks can be estimated with considerable accuracy, depending on the signal-to-noise ratio for the echoes. (Note that the SCAT model's reconstructed image in Fig. 9.8F resembles the shape of the images perceived by the bat in Fig. 9.7.) Because the frequencies, amplitudes, and phases of the basis vectors are controlled by the frequencies, amplitudes, and phases of the coresponding frequencies in the echo relative to the emission, the internal temporal organization of the echo can be reconstituted by itself. All other factors, FM sweep shape, harmonic structure, and propagation delay to the target, are removed to reveal the contribution of the target in isolation. Moreover, because the basis vectors are aligned to start oscillating in cosine phase at the echo-delay values specified by the delay lines, the entire image is displayed in absolute units of echo delay or target range.

4.5.2 Time Stretching of Basis Vectors and Images

One particularly significant feature of the SCAT model is the capacity to extract estimates of spectrogram delay (for echoes A + B together) and delay separation (from echo A to echo B) using processing elements that have different time scales. It is convenient to introduce this feature by equating the frequency of each basis vector with the center frequency in each frequency channel of the spectrogram. In this case, the reconstructed image (Fig. 9.8F) has the same time scale as the original signals; that is, the time between the first and second delay estimates is 60 μsec in the time series signal formed by summation of the basis vectors, just as it was 60 μsec in the original ultrasonic echoes (δt in the echoes [Fig. 9.8A,B] equals δt in the image [Fig. 9.8F]). However, this requires the basis vectors to be oscillations at ultrasonic frequencies, which is physiologically implausible. Oscillatory responses observed in the mammalian auditory system typically have frequencies of 100 Hz to about 1–2 kHz (Langner and Schreiner 1988; Langner 1992).

An alternative is to scale the frequencies of the basis vectors to be lower than the original ultrasonic signals, keeping the frequencies of the basis-vector oscillations proportional to the center frequencies of the bandpass filters rather than equal to them (Saillant et al. 1993). In this case, the ultrasonic frequencies in emissions and echoes extend hyper-bolically from 20 to 100 kHz in Figure 9.8B, while the frequencies of the basis vectors extend hyperbolically from some fraction times 20 to 100 kHz in Figure 9.8E. This fraction is a scaling factor for the frequencies of the basis vectors; it lowers the frequencies in the reconstructed image and lengthens the spacing of the image components. That is, the time interval between the ultrasonic echoes (δt in Fig. 9.8A,B) is 60 μsec while the corresponding time interval between the first and second delay estimates in the time-series signal created by adding the basis vectors together (δt in Fig. 9.8F) could be 600 μsec (for a scaling factor of 1:10) or 6 msec (for a scaling factor of 1:100). In the example in Figure 9.8, the original echo-delay separation is 60 μsec, while the separation of the peaks in the reconstructed image is about 2.3 msec, so the scale factor is about 1:38. These longer time intervals might realistically be represented by the timing of successive neural responses whereas the original 60-μsec interval could not be represented directly by two successive neural responses only 60 μsec apart.

5. Neural Responses in the Bat's Auditory System

The previous sections described the acoustic stimuli received by echolocating bats, the images they perceive, and the computational requirements for creating these images from an initial representation consisting of hyperbolically scaled spectrograms with an integration time of about 350 μsec. The critical feature of these images is their display of the arrival times of both replicas of the sonar signal contained in overlapping echoes at time separations (δt) substantially smaller than this integration time. We now turn to the problem of whether responses in the bat's auditory system manifest properties that are consistent with the requirements of deconvolution by the SCAT process or some near equivalent to it.

5.1 Principal Auditory Centers in the Bat's Brain

5.1.1 The Auditory Pathways

The auditory system of echolocating bats is much like the auditory systems of other mammals, except that is has been adapted to serve as a sonar receiver (Henson 1970; Suga 1988, 1990; Pollak and Casseday 1989; Haplea, Covey, and Casseday 1994). Figure 9.9 is a diagram showing the principal routes taken by neural responses to sounds as they follow the auditory pathways ascending from the cochlea through the bat's central nervous system (redrawn from Schweizer 1981). This diagram shows the principal auditory tracts leading from the auditory nerve of the bat's left ear to the major auditory nuclei depicted in cross-sections at four levels of the brain. The principal brain sites for processing acoustic information depicted in Figure 9.9 are (1) the *cochlear nucleus* (designated CN in figures), which is the first synaptic step in auditory processing beyond the periphery; (2) a group of small nuclei (*trapezoid nuclei; medial and lateral superior olivary nuclei*) located along the ventral surface of the brain stem; (3) the *nucleus of the lateral lemniscus* (NLL), which receives the output from the cochlear nucleus and other brain stem sites through the lateral lemniscus, a large fiber tract connecting the brain stem with the midbrain; (4) the *inferior colliculus* (IC), which is the major midbrain auditory center and a much-enlarged structure in bats; (5) the *medial geniculate*, which lies in the thalamus just below the cerebral cortex; and

(6) the *auditory cortex* (AC), which is usually described as the highest level of auditory processing and the presumed site of the physiological events that actually cause perception to happen.

5.1.2 Bilateral Connections of Auditory Centers

The diagram in Figure 9.9 shows only the main excitatory projections from the left ear, which enter the left cochlear nucleus and then cross to the right side of the brain. This arrangement of crossed connections appears to make each auditory nucleus have its input from the contralateral ear. Information from the ipsilateral ear projects to most sites as well, however. Most structures above the level of the cochlear nucleus receive both contralateral and ipsilateral inputs, and the responses of their neurons exhibit varying degrees of *binaural* interactions. In the inferior colliculus of *Eptesicus*, for example, about 75% of neural responses recorded from single cells involve contralateral excitation and ipsilateral inhibition, often in combination with facilitation at some stimulus levels (Haresign et al. in press).

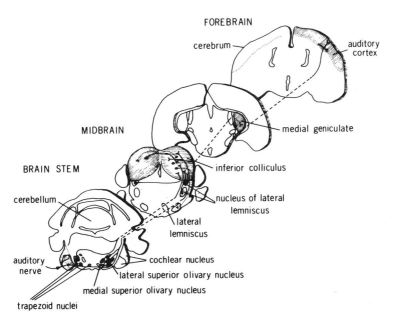

FIGURE 9.9. Principal auditory centers in the bat's brain. These anatomical structures receive their inputs from auditory stimulation approximately in succession (after Schweizer 1981; see Section 5.1.1 and Fig. 9.17).

5.2 Frequency Tuning of Neural Responses

The most pervasive feature of auditory coding in mammals is the tuning of neural responses at all levels of the auditory system to specific frequencies in sounds (for bats, see Henson 1970; Suga 1988, 1990; Pollak and Casseday 1989). Figure 9.10 illustrates tuning curves recorded from four levels in the auditory system of *Eptesicus* (see Fig. 9.9). All the neurons at these sites respond selectively to frequencies used for echolocation (see Fig. 9.2).

5.2.1 Auditory Nerve, Cochlear Nucleus, and Nucleus of the Lateral Lemniscus

Figure 9.10A shows tuning for five representative cells from the anteroventral cochlear nucleus (AVCN), and Figure 9.10B shows five neurons from the posteroventral cochlear nucleus (PVCN). From what is presently known, the tuning curves from the AVCN can be taken as most representative of the frequency selectivity of primary auditory neurons at the input of the bat's auditory nervous system. Figure 9.10C–E shows tuning curves for representative cells from the two enlarged monaural divisions of the nucleus of the lateral lemniscus in *Eptesicus* (intermediate nucleus of the lateral lemniscus, INLL; ventral nucleus of the lateral lemniscus, VNLLm, VNLLc) (Covey and Casseday 1991). Most of these tuning curves are V shaped, with a sharp tip and progressively wider tuned regions above the tip. In contrast, many cells in the VNLLc (Fig. 9.10E) instead have very broad tuning curves. These broadly tuned neurons nevertheless are capable of conveying frequency-specific information about echoes because their discharges register the entry of an FM sweep into the tuning curve and can also register amplitude modulations spread across the FM sweep as a whole (Covey and Casseday 1991).

5.2.2 Sharpness of Tuning at the Periphery

Figure 9.11 shows sharpness of tuning expressed as $Q_{10\,dB}$ for cells in the cochlear nucleus of *Eptesicus* (Haplea, Covey, and Casseday 1994). These $Q_{10\,dB}$ values range from about 3 to 15 at frequencies from 10 to 70 kHz. The sharpness of tuning for first-order auditory neurons has been predicted, from the rate of sweep at different frequencies in the bat's sonar sounds, on the assumption that the neurons are tuned to optimize the accuracy of registering echo delay (Menne 1988). These predicted $Q_{10\,dB}$ values, based on more than 400 recorded echolocation sounds, are shown by dashed lines in Figure 9.11 (mean $\pm$ 1 SD). They are in the same range as the measured $Q_{10\,dB}$ values for those frequencies at which tuning has been measured. Figure 9.11 also has a sloping line showing values of $Q_{10\,dB}$ measured from frequency–response curves of the bandpass filters (SCAT filters; Saillant et al. 1993) used in the SCAT model of echolocation that provides a conceptual framework for this chapter. These filters have a frequency selectivity comparable to tuning curves in the bat, at least for the range of frequencies at which $Q_{10\,dB}$ values have been obtained.

5.2.3 Inferior Colliculus

Figure 9.10F–I shows tuning curves from the inferior colliculus of *Eptesicus* (Jen and Schlegel 1982; Poon et al. 1990; Casseday and Covey 1992; Ferragamo, Haresign, and Simmons in press). Tuning curves appear narrower in the inferior colliculus (Fig. 9.10F–I) because they have different shapes, with steeper, more nearly vertical, high-frequency and low-frequency slopes. A significant proportion of cells in the IC also have "closed" tuning curves, with an upper limit to the response area (Fig. 9.10I). Echoes typically reach the bat's ears at amplitudes that are less than the amplitudes of the transmissions picked up directly at the ears. Many cells with upper limits to their tuning curves would be unresponsive to the loud outgoing sound and thus would be selective for responding to echoes rather than transmissions. In the band of frequencies from 22 to 30 kHz, some of the tuning curves in the IC of *Eptesicus* (Fig. 9.10H) appear narrower than those at 22–30 kHz in more peripheral nuclei. These especially sharply tuned cells, (called filter neurons; Casseday and Covey 1992), provide a narrow segment of the frequency axis from 22 to 30 kHz with an exaggerated sharpness of tuning probably produced by juxtaposition of excitatory and inhibitory inputs at slightly different frequencies (Suga and Schlegel 1973; Casseday and Covey 1992).

5.2.4 Auditory Cortex

Figure 9.10J illustrates tuning curves recorded from neurons in the auditory cortex of *Eptesicus* (Jen and Schlegel 1982). In anesthetized bats, cortical cells readily respond to tone bursts, are tuned to a specific frequency, and have V-shaped tuning curves resembling those found in the cochlear nucleus (Fig. 9.10A,B) or the nucleus of the lateral lemniscus (Fig. 9.10C,D). However, in awake bats, most cortical neurons are relatively unresponsive to tone bursts; they require instead combinations of different frequencies and specific timing of multiple sounds to evoke an appreciable response. For example, neurons in a sizable population from the cortex of *Eptesicus* are tuned to more than one narrow frequency region (Dear et al. 1993). Figure 9.10K illustrates a typical multipeaked tuning curve, with tuned frequencies centered at 15 kHz and 45 kHz, and a fairly level-tolerant tuning curve around each tuned frequency. These cells mostly have a lower tuned frequency in the range of 10–40 kHz and a higher tuned frequency in the range of 30 to 80–90 kHz. Multipeaked cells are relatively unresponsive to frequencies

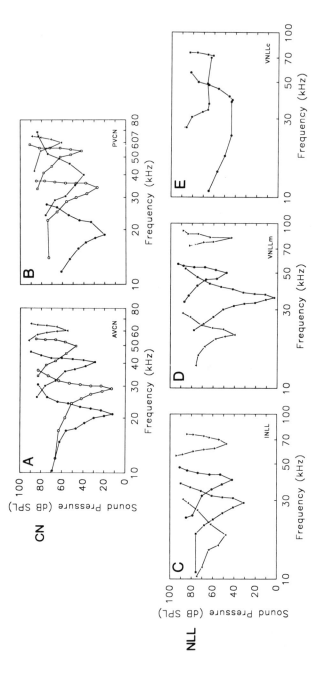

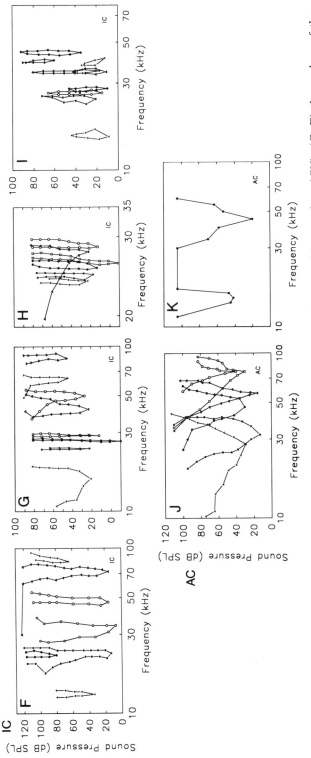

FIGURE 9.10A–K. Representative frequency tuning curves for *Eptesicus* from (A, B) the cochlear nucleus (*CN*); (C–E) the nucleus of the lateral lemniscus (*NLL*); (F–I) the inferior colliculus (*IC*); and (J,K) the auditory cortex (*AC*) (see Fig. 9.9). INLL, intermediate nucleus of LL; VNLLm, VNLLc, ventral nucleus of LL.

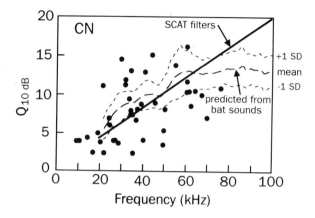

FIGURE 9.11. Sharpness of frequency tuning ($Q_{10\,dB}$) in the cochlear nucleus of *Eptesicus* (data points are from Haplea, Covey, and Casseday 1994). *Dashed lines*, predicted tuning values, mean $\pm$ 1 SD (Menne 1988); *solid sloping line*, tuning of SCAT filters measured from frequency–response curves (Saillant et al. 1993).

in the interval between their tuned frequencies. A few cells with multiple tuned frequencies are found in the IC of *Eptesicus* (Casseday and Covey 1992), but the multipeaked cells in the cortex are more common.

Figure 9.12 shows the distribution of frequency ratios (f_2/f_1) for the high-frequency tuned region (f_2) to the low-frequency tuned region (f_1) from cortical multipeaked neurons in *Eptesicus*. A large proportion of the multipeaked cells have their two tuned frequencies in a frequency ratio around 2:1 or 3:1, with a smaller proportion of cells having intermediate ratios. The equations in Section 4.4.1 suggest that these multipeaked responses may embody frequency-domain information ("knowledge") about the locations of spectral features necessary for deconvolution of overlapping echoes. The frequency spacings of peaks and notches fall at specific frequency ratios that actually appear to be represented physiologically.

5.3 Distribution of Tuned Frequencies

5.3.1 Density of Frequency Tuning

Figure 9.13 illustrates the distribution of tuned frequencies for neurons at the cochlear nucleus (Haplea, Covey, and Casseday 1994), nucleus of the lateral lemniscus (Covey and Casseday 1991), inferior colliculus (Jen and Schlegel 1982; Poon et al. 1990; Casseday and Covey 1992; Ferragamo, Haresign, and Simmons in press), and auditory cortex (Dear et al. 1993). These histograms show the emphasis placed on encoding

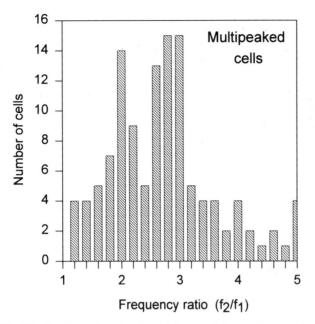

FIGURE 9.12. Distribution of ratios for higher and lower frequencies (f_2/f_1) in auditory cortical neurons with two widely separated tuned frequencies (f_1, f_2). Ratios are clustered around 2:1 and 3:1.

information at different frequencies, with an especially large proportion of cells tuned to 20–40 kHz and a secondary proportion tuned to about 60–70 kHz.

Figure 9.14A replots the density of frequency tuning in the inferior colliculus with histogram bin widths of 1 kHz from the 2-kHz bin width shown in Figure 9.13C. This density distribution has two segments, one with a peak at 25–30 kHz and steady decline to 55–60 kHz and the other with a peak at 60–63 kHz and a decline to 95 kHz. (Neurons tuned to frequencies of 22–30 KHz include the sharply tuned "filter" neurons in Fig. 9.10H). The data shown along the horizontal frequency axis in Figure 9.14A are replotted along a period scale in Figure 9.14B by taking the reciprocal of the frequency value for each bin in Figure 9.14A. Figure 9.14B shows two jagged curves, with a break at 16.7 µsec (the reciprocal of 60 kHz, which is approximately the location of the break between the two parts of the distribution in Fig. 9.14A). The curves in Figure 9.14B thus show a natural segmentation into two parts at about 16.7 µsec. In Figure 9.14B, separate regression lines (a,b) are plotted for the short-period segment (12–16.7 µsec) and the long-period segment (16.7–40 µsec). When these two regression lines are transposed back onto the frequency scale of Figure 9.14A (curves a and b), they outline the shape

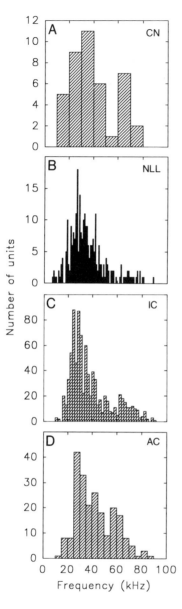

FIGURE 9.13A–D. Density of frequency tuning in *Eptesicus* at different frequencies for the cochlear nucleus (*CN*), the nucleus of the lateral lemniscus (*NLL*), the inferior colliculus (*IC*), and the auditory cortex (*AC*). Tuning is predominantly at frequencies of 20–40 kHz and secondarily at 60–70 kHz.

of the original density distribution along the frequency axis. The shape of the regression curves in Figure 9.14A is hyperbolic because of their reciprocal relation to the straight regression lines in Figure 9.14B. Because the bat's sonar sounds have FM sweeps that are approximately hyperbolic in shape (see Fig. 9.2), these curves trace the dwell time of the bat's sonar sounds at each frequency for FM_1 (a) and FM_2 (b) (Fig. 9.2A–D).

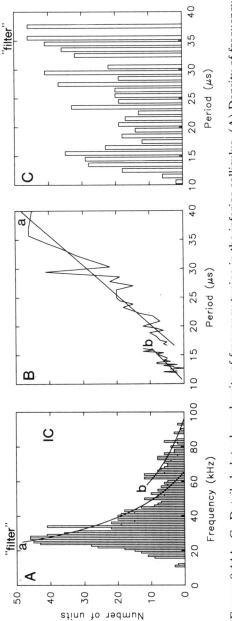
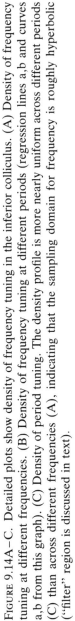

FIGURE 9.14A–C. Detailed plots show density of frequency tuning in the inferior colliculus. (A) Density of frequency tuning at different frequencies. (B) Density of frequency tuning at different periods (regression lines a,b and curves a,b from this graph). (C) Density of period tuning. The density profile is more nearly uniform across different periods (C) than across different frequencies (A), indicating that the sampling domain for frequency is roughly hyperbolic ("filter" region is discussed in text).

5.3.2 Overrepresentation of Low Frequencies

In Figure 9.13 and 9.14A, frequencies of about 20–40 kHz are over-represented (Casseday and Covey 1992) by neurons tuned to these frequencies compared to frequencies of 40–100 kHz. One aspect of this overrepresentation is the presence of filter neurons tuned to 25–30 kHz with especially sharp tuning curves (Fig. 9.10H). (*Eptesicus* probably uses these neurons to enhance detection for echoes of the relatively shallow FM sweeps in the range of 28–25 kHz that it broadcasts when searching for prey in open areas.) Approximately 24% of the neurons tuned to frequencies of 22–30 kHz are identified as filter neurons (Casseday and Covey 1992). However, even if the heights of the histogram bars in Figure 9.14A at 22–30 kHz are reduced by 24% to remove these specialized filter cells from the density distribution, the overrepresentation of frequencies in the 20- to 40-kHz region still exists because these histogram bars remain substantially higher than those at other frequencies.

The regression curves (a,b) in Figure 9.14A suggest that the density of frequency tuning might be numerically related to the period of each tuned frequency rather than directly to the frequency values themselves. To test this possibility, the entire data set is transformed from the distribution of cells at each tuned frequency (1-kHz bins) in Figure 9.14A to the distribution of cells at each tuned period (1-μsec bins) and plotted yet again in Figure 9.14C. Now the histogram density is more nearly uniform across different bins. In Figure 9.14C, the proportion of cells tuned to each period from 12–13 μsec to about 35 μsec varies by only a factor of roughly 2 from one histogram bar to another, whereas the proportion of cells tuned to each frequency in Figure 9.14A varies by a factor of as much as 8 from one bar to another. Furthermore, the nonuniformity remaining in the distribution of Figure 9.14 is chiefly a higher proportion of cells tuned to periods of 33–38 μsec compared to shorter periods. Periods of 33–38 μsec correspond to frequencies of 26–30 kHz, which are frequencies to which the specialized filter neurons are tuned. If the heights of the histogram bars in Figure 9.14C at 33–38 μsec are reduced by 24% to remove the proportion of filter neurons from the data, then the density of period tuning does appear to be approximately uniform across all periods from 12–13 μsec to 35–40 μsec. Thus, there is a genuine overrepresentation created by the presence of specialized filter neurons at low frequencies combined with a gradual skewing of the distribution toward lower frequencies as a consequence of the nearly uniform representation of ultrasonic periods across "nonfilter" neurons.

5.4 Tonotopic Organization of Tuned Frequencies

5.4.1 Nucleus of the Lateral lemniscus

Figure 9.15A shows the mapping of frequencies along the frequency axis for one region of the nucleus of the lateral lemniscus in *Eptesicus*

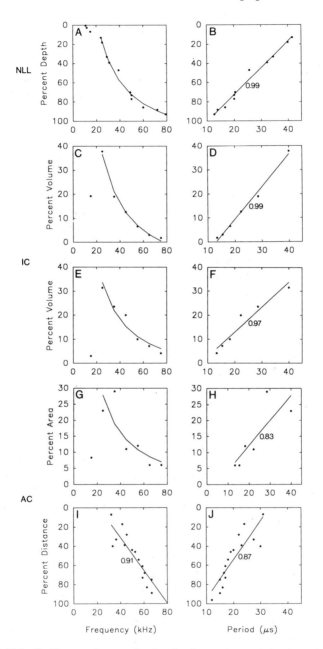

FIGURE 9.15A–J. Tonotopic organization in the nucleus of the lateral lemniscus (*NLL*), the inferior colliculus (*IC*), and the auditory cortex (*AC*). (A,C,E,G,I) Anatomical location plotted against frequency; (B,D,F,H,J) anatomical location plotted against period.

(VNLLc) (Covey and Casseday 1986). This plot yields a curvilinear relationship from about 20 kHz to 80 kHz. The same data are replotted in Figure 9.15B in terms of the period corresponding to each tuned frequency (Simmons et al. 1990). This relationship is a straight line that can then be replotted back on frequency coordinates in Figure 9.15A as a regression curve. The tonotopic axis in the nucleus of the lateral lemniscus (VNLLc) of *Eptesicus* appears approximately hyperbolic with frequency, or linear with period.

5.4.2 Inferior Colliculus

Figure 9.15C shows a tonotopic axis for the inferior colliculus of *Eptesicus* taken from a rough three-dimensional reconstruction of its frequency-tuned layers (Poon et al. 1990). This relationship between frequency and volume is curvilinear from about 20 to 80 kHz. Figure 9.15D shows the same volumetric relationship in terms of period, and once again, the topographic relationship is approximately linear. Figure 9.15E,F shows similar relationships for frequency and period from a more detailed set of measurements than in Figure 9.15C (Casseday and Covey 1992; J.H. Casseday, personal communication). The plot of period and volume in Figure 9.15F is as linear as the plot in Figure 9.15D. Thus, the tonotopic axis of the inferior colliculus in *Eptesicus* appears to be well described as linear with period or hyperbolic with frequency.

5.4.3 Auditory Cortex

Spatial representation of frequency is the most global characteristic of the auditory cortex in *Eptesicus* (Jen, Sun, and Lin 1989; Dear et al. 1993). However, the cortical frequency contours are quite convoluted, and individual bats differ in their tonotopic maps, which makes it difficult to produce a composite map that reliably depicts all frequency regions. Frequency scales estimated from the two available tonotopic maps are shown in Figure 9.15G (Dear et al. 1993) and 9.15I (Jen, Sun, and Lin 1989). In both cases, there is considerable scatter in the data points compared to the tonotopic axes for the nucleus of the lateral lemniscus (Fig. 9.15A) or the inferior colliculus (Fig. 9.15C,E). When these cortical data are replotted in terms of period (Fig. 9.15H,J), a straight regression line is about as accurate for chracterizing the frequency plot as for characterizing the period plot. It not clear whether the cortical tonotopic map is incomplete (more auditory area awaits recording), or whether the auditory cortex simply does not have a tonotopic organization that matches precisely the organization found at lower centers. For example, the presence of multipeaked tuned neurons that are unique to the cortex (Fig. 9.10K and Fig. 9.12) may affect the frequency organization of the cortex in comparison with other sites containing neurons tuned to just one frequency region.

5.5 On-Responses to FM Stimuli

The bat's auditory system marks the time of occurrence of individual frequencies in the FM sweeps of sonar emissions and echoes with on-responses to these frequencies (Suga 1970; Pollak et al. 1977; Pollak and Casseday 1989). For example, in the inferior colliculus of *Eptesicus*, 93% of the recorded cells in one study respond to their tuned frequencies in a 2-msec artificial FM echolocation sound with just one discharge (Ferragamo, Haresign, and Simmons in press). Figure 9.16A–C illustrates latency (or peristimulus time [PST]) histograms of on-responses in three different single neurons to a 2-msec FM sweep that passes through each cell's tuned frequency. All three of these cells respond with an average of a single on-discharge to each occurrence of the FM stimulus; one discharge invariably occurs, and there is no prominent second or third discharge to distort the histogram from its peaked shape. Moreover, cells in the inferior colliculus typically respond to both the emission and the echo unless their tuning curves have an upper threshold that can block responses to the emission (see Fig. 9.10I) or unless they have recovery times that exceed the delay of the echo. The first cell (Fig. 9.16A) responds to a frequency between 30 and 10 kHz with a latency of 12.5 msec and a standard deviation of 70 μsec; the second cell (Fig. 9.16B) responds to a frequency between 40 and 20 kHz with a latency of 16.8 msec and SD of 390 μsec; and the third cell (Fig. 9.16C) responds to a frequency between 30 and 10 kHz with a latency of 21.6 msec and a standard deviation of about 6 msec. (The fourth latency histogram in Fig. 9.16D shows the on responses of several cells recorded together as a small multiunit cluster. The properties of these local multiunit responses are examined next.) Similar observations of the magnitude and variability of latencies have been made in the FM bats *Myotis lucifugus* (Suga 1970) and *Tadarida brasiliensis* (Pollak et al. 1977).

Many neurons in the cochlear nucleus or the nucleus of the lateral lemniscus of *Eptesicus* will respond to a tone burst with a sustained response that persists for the duration of the sound (Covey and Casseday 1991; Haplea, Covey, and Casseday 1994), and neurons in the inferior colliculus often respond selectively to the duration of a sound with a well-defined response at the end if the stimulus falls within the window of the cell's duration tuning (Casseday, Ehrlich, and Covey 1994). However, for short FM sounds comparable to echolocation signals broadcast when the bat is at distances of less than 1–2 m from a target (see Fig. 9.2), the effective stimulus at each frequency is very short. The sound sweeps through its frequencies so rapidly that the dwell time of the sweep in the vicinity of any particular frequency is typically a fraction of a millisecond. Consequently, the responses of most cells will be brief, too.

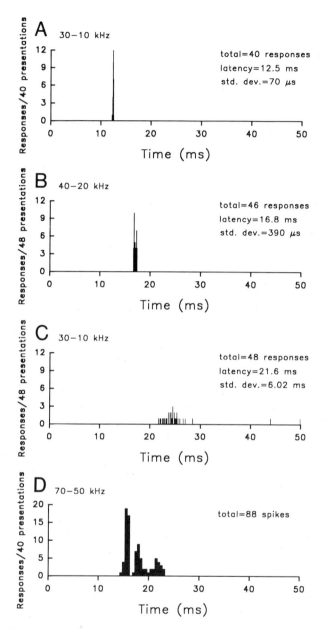

FIGURE 9.16. Responses of neurons in the inferior colliculus of *Eptesicus* to 2-msec FM (frequency-modulated) sounds. (A–C) Latency (or peristimulus time [PST]) histograms of on-responses in three different neurons evoked by 40 presentations of a 2-msec FM sweep that covered each cell's tuned frequency; 93% of these neurons discharge only once per short-duration FM stimulus, but they differ in their characteristic latency and latency variability (standard deviation). (D) Latency histogram for multiunit response dominated by on-discharges in three neurons with latencies at a periodic spacing.

5.6 Latencies of On-Responses at Different Frequencies

Figure 9.9 illustrates the major centers in the bat's auditory system and the order in which they are activated by the onset of a brief sound. Figure 9.17 shows the time of occurrence, or latency, of the on-responses in neurons tuned to different ultrasonic frequencies at the cochlear nucleus (Haplea, Covey, and Casseday 1994), the nucleus of the lateral lemniscus (Covey and Casseday 1991), the inferior colliculus (Haplea, Covey, and Casseday 1994; Ferragamo, Haresign, and Simmons in press), and the auditory cortex (Dear et al. 1993) of *Eptesicus*.

5.6.1 Cochlear Nucleus and Nucleus of the Lateral Lemniscus

The cochlear nucleus is the first site to respond following activation of the auditory nerve, with discharges occurring at latencies of about 1–5 msec (CN in Fig. 9.17). All the cells in the cochlear nucleus of *Eptesicus* are tuned to specific ultrasonic frequencies (see Fig. 9.10A,B), and most cells respond at a latency of about 2–3 msec to tone bursts at their tuned frequencies, with a slightly greater spread of latencies at lower frequencies of 20–40 kHz than at higher frequencies of 50–90 kHz. In the nucleus of the lateral lemniscus, on-responses occur about 2–5 msec after the stimulus (NLL in Fig. 9.17). In addition, on-responses in some cells tuned to lower frequencies of 20–40 kHz have latencies as long as 8–12 msec. The majority of these responses fall 1–2 msec after responses in the cochlear nucleus, which reflects the synaptic delay and propagation time that intervenes between these two centers (see Fig. 9.9).

5.6.2 Inferior Colliculus

In the inferior colliculus, the first on-responses occur at latencies of 3–6 msec (IC in Fig. 9.17). These starting latencies are approximately 1–2 msec longer than at the nucleus of the lateral lemniscus, as would be expected from intervening synaptic and conduction delays (Fig. 9.9). Latencies of 3–6 msec are only the shortest latencies at each frequency, however. The responses of the inferior colliculus are characterized principally by a wide dispersion of latencies from their minimum values of 3–6 msec out to as much as 50 msec (Jen and Schlegel 1982; Pollak and Casseday 1989; Poon et al. 1990; Kuwabara and Suga 1993). In Figure 9.17 (IC), latencies for responses in cells tuned to 20–45 kHz are the most widely spread, being densely distributed from 3–6 msec to 25 msec, and more sparsely distributed to as much as 30–40 msec. At frequencies above 45 kHz, responses in the inferior colliculus have latencies mostly from 3 to about 12 msec, with a sparse distribution to as much as 25–30 msec. These dispersed responses are all on-responses to a brief sound, not sustained responses to a long-duration sound. The occurrence of these delayed responses appears to be regulated by prolonged intervals of

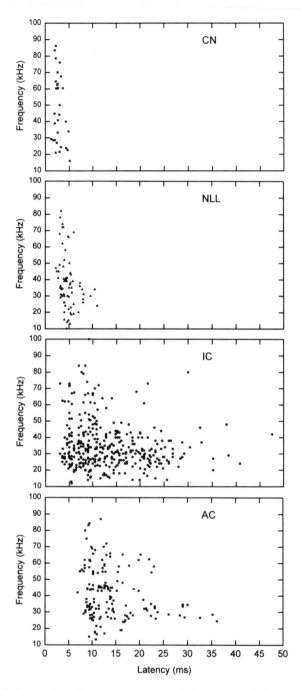

FIGURE 9.17. Latencies of on-responses at different tuned frequencies in the cochlear nucleus (*CN*), the nucleus of the lateral lemniscus (*NLL*), the inferior colliculus (*IC*), and the auditory cortex (*AC*). Responses in lower auditory centers (*CN, NLL*) have a narrow spread of latencies, but at higher centers (*IC, AC*) they are dispersed to as much as 30–35 msec or more.

inhibition initiated by the stimulus followed by abrupt excitation (Suga 1964; Park and Pollak 1993; Casseday, Ehrlich, and Covey 1994). The time at which this discharge occurs may be considerably delayed in relation to the stimulus (Fig. 9.17, IC), but it nevertheless is an on-response to the sound.

Figure 9.17 (IC) shows a strong overrepresentation of dispersed latencies at low frequencies of 20–40 kHz, but these frequencies are themselves overrepresented chiefly because the scale of frequency is nearly uniform with period rather than frequency (Fig. 9.14A–C). The frequency axis for the graphs in Figure 9.17 is linear, which emphasizes this overrepresentation. Figure 9.18 shows the distribution of on-response latencies for the inferior colliculus using a hyperbolically scaled vertical frequency axis, which is more realistic (see earlier). When the low frequencies are spread out to show cells at equal increments of period, the distribution of latencies appears more broad and uniform. In particular, the sparse density of cells with responses at frequencies of 50–80 kHz in Figure 9.17 (IC) is compressed in Figure 9.18 to create a density comparable to that observed at 20–40 kHz.

5.6.3 Auditory Cortex

Neural responses in the auditory cortex mirror the pattern of latency dispersal observed at the inferior colliculus. Cortical on-responses begin at a latency of about 6–8 msec and are distributed rather densely for

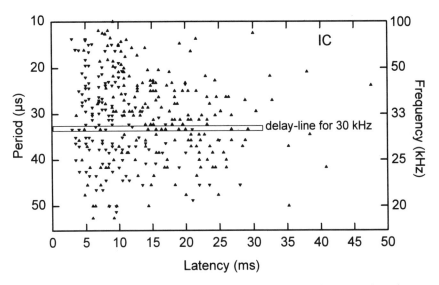

FIGURE 9.18. Latencies of on-responses in the inferior colliculus replotted on a hyperbolic frequency axis to correspond to SCAT spectrograms (Fig. 9.5).

latencies as long as 15–20 msec at most frequencies (AC in Fig. 9.17). In addition, response latencies at lower frequencies of 20–40 kHz extend more sparsely for 30–35 msec or more, while responses at higher frequencies are mostly finished by about 15–20 msec. The initial latencies in the cortex are about 3 msec longer than in the inferior colliculus, but the cortex is separated from the inferior colliculus by another auditory center (see Fig. 9.9). The medial geniculate intervenes in the ascending pathway to add more synaptic and propagation delays, which accounts for the larger latency increment between the inferior colliculus and the auditory cortex in Figure 9.17 than between the earlier stages.

6. Neural Processing of Echo Delay

6.1 Target-Ranging Computations Based on Latencies

Echolocating bats determine the distance to a target by measuring the time that separates the echo from the emission (Simmons 1973). The mechanism for displaying echo delay is a supreme example of a well-defined auditory computation being carried out using neuronal circuits that have been identified, at least in outline (Jen and Schlegel 1982; Sullivan 1982; Suga 1988, 1990; Pollak and Casseday 1989; Casseday and Covey 1992; Kuwabara and Suga 1993; Park and Pollak 1993; Ferragamo, Haresign, and Simmons in press). These circuits operate on the latencies of on-responses at each frequency in the broadcast sound, followed by latencies of on-responses to each frequency in the echo.

6.2 Delay Lines in the Inferior Colliculus

6.2.1 Latencies as Delay Taps in Physiological Delay Lines

The principle that underlies target ranging by bats is to retard neural responses to the emission for a sufficiently long interval of time that the echo comes back and its responses start to occur simultaneously with those for the emission. The broad dispersal of latencies in the inferior colliculus creates these delays. Because the sonar of *Eptesicus* has an operating range of 5 m, responses to the emitted sound have to be delayed for as long as about 33–35 msec to ensure that some will still be occurring when the earliest responses to the echo take place at their shortest latencies of 3–5 msec. The neurons whose responses implement these delays each produce a single discharge to the emission or the echo at a characteristic latency (Fig. 9.16A–C). That is, the delay taps in the physiological implementation of the delay lines are single cells tuned to the same frequency but with different latencies. For example, the responses of neurons tuned to 30 kHz in Figure 9.18 act as a delay line for

30 kHz in the FM sweep. Similar subpopulations of neurons act as delay lines at other frequencies.

The physiological delay system in the bat is not so simple as a set of delay lines creating a long series of responses to the emission with short-term responses to the echo. Most of the cells shown in Figure 9.18 will respond to both the emission and the echo if the sounds are a few milliseconds apart, so each sound triggers a flurry of responses stretched out over 25–30 msec. The extended pattern of responses to the emission depicted in Figure 9.18 thus is followed by a similar extended pattern of responses to the echo. These patterns overlap each other at a time separation equal to the acoustic delay of the echo, and the coincidence comparisons that correlate the shape of the FM sweeps for the emission and the echo actually take place between these patterns of dispersed latencies rather than simply between the spectrograms. Computations for echo delay based on coincidence detection of dispersed latencies appear at first glance to be approximately equivalent to the spectrogram correlation stage of the SCAT model (see Fig. 9.8A–C). However, the patterns of dispersed latencies themselves will presently be seen to have properties as time-series signals that go beyond mere delays to make coincidence detection equivalent in some respects to the spectrogram transformation stage as well.

6.2.2 Accuracy of Time Registration by Delay Taps

The latencies of on-responses in different neurons vary somewhat from one stimulus presentation to another, and this variability is assumed to limit the accuracy with which the arrival time of echoes can be determined (see Pollak et al. 1977; Schnitzler, Menne, and Hackbarth 1985; Haplea, Covey, and Casseday 1994). The widths of the peaks in the latency histograms shown in Figure 9.16A–C are typical of on-responses in the inferior colliculus of *Eptesicus*; some are as narrow as 50–100 μsec (A), and many are as narrow as 300–500 μsec (B), but many also are as much as several milliseconds wide (C). Figure 9.19 shows the distribution of different latency variabilities (standard deviations for widths of histogram peaks) across absolute latencies (Ferragamo, Haresign, and Simmons in press). Responses that have absolute latencies from 3 msec to 25–30 msec also have latency variabilities from about 50–100 μsec to 4 msec, with a wider scattering of variabilities over 4 msec. Latency variabilities from 50–100 μsec to about 2 msec are the most densely represented values across most absolute latencies. Surprisingly, the variability of response timing does not simply increase as latency increases in Figure 9.19, as might be expected if longer latencies leading to responses in the inferior colliculus bring more opportunities for jitter in latency to accumulate. There are numbers of cells with latency variabilities as small as 50–100 μsec at absolute latencies all across the range from 4 msec to 25 msec.

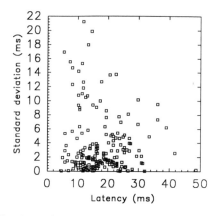

FIGURE 9.19. Distribution of latency variability (standard deviation) for on-responses in the inferior colliculus having different characteristic latencies. There are cells with latencies from 4 to 25–30 msec that have narrow variability (SD, 50–100 µsec or less).

This is a significant result because accurate registration of the timing of sounds depends on retention of narrow latency variability across a wide span of absolute latencies in at least some of the cells.

6.3 Delay-Tuned Responses and Coincidence Detection

6.3.1 Tuning Curves for Echo Delay

Neurons in the inferior colliculus act collectively like a system of multiple-tap delay lines for storing the time of occurrence of each frequency in the broadcast sound or echo (e.g., delay-line at 30 kHz in Fig. 9.18). However, this delayed representation does not by itself create a display of echo delay; it only registers both sounds as events occurring at different times. The display is created by neurons that compare responses to emissions and echoes at the next level of processing, neurons that are specialized for responding only to echoes at certain delays. Figure 9.20A illustrates responses of three neurons in the auditory cortex of *Eptesicus* that are "tuned" to different values of echo delay (Dear, Simmons, and Fritz 1993). One of these cells responds most strongly to echoes at a best delay (BD) of 5 msec, the second cell has a BD of 12 msec, while the third cell has a BD of 20 msec. The first cell thus responds most strongly to a target located 86 cm away, the second cell responds most strongly to a target at 2.1 m, while the third cell responds to a target at 3.4 m. Delay-tuned cells in the auditory cortex of *Eptesicus* provide more-or-less continuous coverage of delays from 2 to 28 msec, which corresponds to target ranges of 34 cm to 4.8 m (Dear et al. 1993).

6.3.2 Detection of Coincidences Between Responses to Emissions and Responses to Echoes

The neurons that compare echoes with emissions respond to the coincidence between a response generated at one latency in the inferior colliculus to the emission and a response generated at another, necessarily shorter, latency to the echo. Figure 9.20B shows a dot-raster plot for a series of responses in a delay-tuned neuron to repeated presentations of an emission and an echo at this cell's BD of 10 msec. This particular neuron responds with a single discharge every time an echo arrives at a delay of 10 msec. Each response is characterized by a fixed latency of 16.2 msec after the emission and a latency of 6.2 msec after the echo. The difference between these two latencies is the 10-msec value of echo delay associated with delay tuning. Coincidence-detecting cells are located in the medial geniculate (see Fig. 9.9), where they receive two inputs from the inferior colliculus at different latencies and then send their response signaling a coincidence between these inputs onward to the auditory

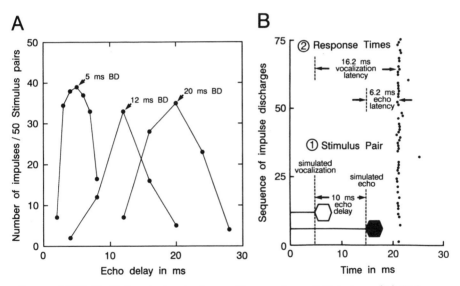

FIGURE 9.20A,B. Delay tuning in the auditory system of *Eptesicus*. (A) Delay-tuning curves for three different cortical neurons with best delays of 5, 12, and 20 msec. Even though the bat's echo-delay acuity is a fraction of a microsecond, the width of delay-tuning curves is in the range of milliseconds. (B) Dot-raster plot of 75 successive delay-tuned responses to an echo at the 10-msec best delay of a neuron, showing the latency from the emission, the latency from the echo, and the difference, which corresponds to best delay. The variability of response latency is only 300 μsec, while this cell's delay-tuning curve is 7.5 msec wide. Even after delay tuning is established, the timing of responses still conveys needed information about delay.

cortex. Cortical cells that receive the coincidence registration as their input are tuned to a BD corresponding to the latency difference for the inputs delivered to the coincidence detectors by the inferior colliculus (Jen and Schlegel 1982; Sullivan 1982; Suga 1988, 1990; Pollak and Casseday 1989; Poon et al. 1990; Casseday and Covey 1992; Kuwabara and Suga 1993; Ferragamo, Haresign, and Simmons in press). Delay-tuned neurons also are found in the midbrain of *Eptesicus*, in a structure located between the inferior colliculus and the superior colliculus (Feng, Simmons, and Kick 1978; Dear and Suga 1995).

6.3.3 Temporal Accuracy of Delay Tuning *Versus* Accuracy of Coincidence Registration

In the examples in Figure 9.20A, the sharpness of delay tuning is proportional to BD. The width of delay tuning (at 50% of maximum response) is ±5 msec at a BD of 10 msec, ±7 msec at a BD of 12 msec, and ±10 msec at a BD of 20 ms. In other cortical neurons, delay-tuning width does not increase but is constant across different BDs, which makes it proportionally sharper at long delays than short delays (Dear, Simmons, and Fritz 1993). However, the degree of selectivity to delay for each cell is still measured in *milliseconds*, while the bat's behavioral echo-delay acuity is a fraction of a *microsecond* and in the limiting case 10–15 *nanoseconds* (see Section 3.5). At the input to delay tuning, the variability in the latencies generated by the inferior colliculus is mostly in the region from hundreds of microseconds to several milliseconds (see Fig. 9.19), which also is substantially greater than the bat's behavioral acuity. Delay-tuning curves thus have widths that correspond more to the latency variability of their inputs than to the bat's perceptual acuity for echo delay. If the bat uses its delay-tuned cells to perceive echo delay, why are there no cells that come within a factor of 5000 of the precision exhibited by the bat as a whole? This discrepancy between physiological and behavioral measures of timing accuracy is very large, which suggests that the tuning curves for echo delay are not the only higher level representation of delay available to the bat.

Figure 9.20B reveals one aspect of delay coding that has greater precision than the width of the delay-tuning curve, that is, the timing of the discharges in the delay-tuned cell. The neuron illustrated in Figure 9.20B has a BD of 10 msec and a delay-tuning width of about ±3.7 msec, which is typical of delay-tuned cells. However, the dot-raster plot of response latencies for repeated presentations of an echo at this cell's BD of 10 msec is very stable from one presentation to the next. Latencies are tightly clustered within a SD of 300 µsec even though the width of this cell's tuning curve for delay is about 3.7 msec. The latency of this cell's response thus betrays about 20 times more delay-coding accuracy than is evident in the tuning curve alone, so the transformation of temporal information about echo delay into "place" by delay tuning evidently is not a complete transformation (Simmons and Dear 1991). Information about delay still is retained in the timing of the response after the selectivity of the response to delay has been set up.

6.3.4 Progression of Delay-Tuned Responses and Buildup of Cortical Range Images

Perhaps the most interesting feature of cortical delay-tuning curves in *Eptesicus* is the evolution of delay tuning across the interval of about 30–35 msec following the broadcast sound. (Recall that this interval is the epoch for return of all usable echoes of the emitted sound from targets as far away as 5 m.) The inferior colliculus of *Eptesicus* supplies a wide range of latencies for its responses to both the emission and the echo (Fig. 9.18), and delay-tuned neurons compare different combinations of emission latencies and echo latencies. Figure 9.21 shows the time after the broadcast (or emission latency) at which cortical delay-tuned neurons respond for different values of BD from 2 to 28 msec (or 34 cm to 4.8 m of target range). The horizontal time axis in Figure 9.21 traces the pro-

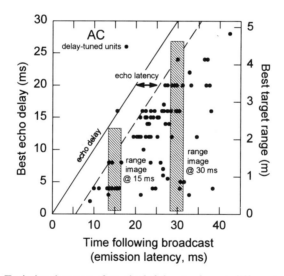

FIGURE 9.21. Emission latency of cortical delay tuning at different values of best delay (BD). The echo arrives at delays indicated by *upward-sloping solid line*, and cortical responses begin to occur as early as 6 ms afterward along *sloping dashed line*. As time progresses, echoes return from targets at longer ranges, to be registered by delay-tuned responses, but delay-tuned responses to targets at shorter ranges that are already registered continue to occur, storing older information about near targets while newer information builds up about far targets. *Vertical shaded bars* delineate best delays in range images occurring 15 and 30 msec after the broadcast.

gression of delay-tuned responses throughout the epoch for echo reception. The first event in the epoch, reception of the broadcast by the bat's ears, releases a sequence of events that enables different delay-tuned neurons to respond if an echo happens to arrive near any particular cell's BD. This sequence is regulated by the passage of on-responses to the emission along the delay lines of the inferior colliculus (see Fig. 9.18).

As time progresses following the broadcast (horizontal axis in Figure 9.21), echoes are received (upward-sloping solid line labeled echo delay) and registered by appropriate delay-tuned neurons after a minimum latency of about 5–6 msec (upward-sloping dashed line offset to right by this minimum echo latency). A curious feature of the system of dispersed latencies for responses to the emission and the echo is that each echo continues to be registered by delay-tuned neurons tuned to its arrival time long after it has been received, up to the length of the whole epoch at 28 msec. This is made possible by the long span of latencies for responses to the emission and to the echo; so long as responses to both sounds occur concurrently and feed into the coincidence-detecting system, there will be delay-tuned responses to register that echo. For example, at a point in time 15 msec after the emission, delay-tuned neurons with BDs from 2 msec to about 8–9 msec are set to respond if an echo arrives (vertical shaded bar labeled range image @ 15 ms). These neurons create an image depicting the ranges of targets out to about 1.5 m. At a time 30 msec after the broadcast, neurons with BDs from 2 to 25 msec are set to respond (vertical shaded bar labeled range image @ 30 ms), so that the range image now extends out to 4 m.

A good way to visualize this evolving range image is to imagine the shaded bar starting at the left of Figure 9.21 and moving to the right as time passes. The bar becomes progressively taller because echoes from further away continue to come in and be registered. In the narrow region around 2–4 msec immediately following the broadcast, the bar is short, but it grows to encompass the full span of 2–28 msec at the end of the epoch of time for reception of all echoes out to the maximum operating range. Beyond this longest practicable delay, responses to the emission cease—they "run off the end" of the delay lines—and delay tuning disappears. The range image itself thus is *dynamic*, moving in the latency-delay space of Figure 9.21 as a kind of vertically spreading, A-scope display that accumulates new targets as their echoes arrive. The persistence of registration of a target in each cell is only a fraction of a millisecond (the duration of a neural discharge), but the persistence of registration across the population of cells at each delay is nearly 30 msec. Then, the image vanishes, to be refreshed by broadcast of the next sonar signal.

7. Temporal Organization of Response Latencies in the Inferior Colliculus

7.1 Reconstruction of FM Spectrograms from Dispersed Latencies

The latency histograms in Figure 9.16A–C illustrate on-responses to FM sounds in three different neurons from the inferior colliculus of *Eptesicus*. More than 90% of the individually recorded neurons respond to a 2-msec FM sound with an average of just one discharge per stimulus at a particular latency (Ferragamo, Haresign, and Simmons in press). Figure 9.18 shows the spread of latencies at each frequency to be quite broad, especially at first-harmonic frequencies of 20–50 kHz, but there does not appear to be any special temporal pattern or organization to these latencies, just a noisy-looking continuous distribution made up of a large number of latencies distributed over the range from about 4–5 msec to 25–30 msec. Nevertheless, if the timing of each on-response reflects the timing of the specific tuned frequency for each neuron (see Fig. 9.10F–I) in the stimulus (Bodenhamer and Pollak 1981; see Pollak and Casseday 1989), then the serial order of responses to different frequencies in an FM sweep should be mirrored in the serial order of responses tuned to different frequencies. In other words, the spectrogram of the FM sweep should reside in the pattern of latencies across neurons tuned to different frequencies; it should, in fact, be present in the data in Figure 9.18, but it is concealed by the overriding effects of latency dispersion itself.

The difference in latency for responses to each neuron's tuned frequency, presented as a tone burst, compared to the latency of the response to an FM sweep containing that tuned frequency, should equal the latency change related to the position of each frequency in the sweep. Figure 9.22 shows this latency subtraction for 41 neurons in the inferior colliculus of *Eptesicus*, with an added correction for the periodic pattern of latencies in the tone-burst responses relative to the more phasic responses to short-duration FM sounds. The stimulus for each neuron is a 2-msec downward FM sweep with a width of 20–25 kHz that contains that cell's tuned frequency. The cells shown in Figure 9.22 have tuned frequencies from 15 to 85 kHz and latency variabilities (SDs) smaller than 2 msec (see Fig. 9.19) to keep them compatible with the sweep duration. The latencies for each neuron in Figure 9.22 are close to the location of the tuned frequency in the FM sweep, with an underlying distribution that is about ±400 µsec wide (50% width). The pattern of response times in Figure 9.22 essentially recovers the time-frequency sweep itself, indicating that the initial auditory spectrogram representation is still present in the system of dispersed responses in the inferior colliculus even though the latency dispersion is larger than the differences in latency at different frequencies along the 2-

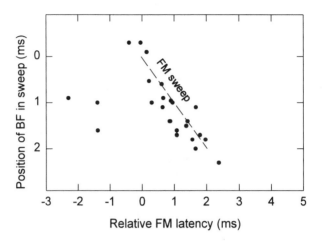

FIGURE 9.22. Relative latency of on-responses in inferior colliculus neurons tuned to different frequencies reconstructs the location of each tuned frequency along FM sweeps. Latencies of on-responses to tone bursts at each cell's tuned frequency are subtracted from FM latencies and then corrected for the periodic spacing of responses to tone bursts relative to FM sweeps. The spectrogram of the sweep is concealed within the dispersed absolute latencies of the responses (Fig. 9.18) (see also Bodenhamer and Pollak 1981).

msec FM sweep. The ±400-μsec width of this reconstructed spectrogram (shown schematically as integration time in Fig. 9.5) is about the same as the behaviorally measured integration time in *Eptesicus*, which is ±350 μsec (Simmons et al. 1989).

7.2 Patterns of Dispersed Latencies as Time-Series Signals

7.2.1 Extracellular Single-Unit Recordings

Each frequency-and-latency data point in Figure 9.18 corresponds to a single cell from the large number that were studied; the graph is made by combining data from numerous different neurons individually recorded during daily recording sessions from recording sites distributed throughout the frequency layers of the inferior colliculus. For all practical purposes, the pooled single-unit data in Figure 9.18 are randomly sampled from the distribution of dispersed latencies actually present in the inferior colliculus. Only one neuron is recorded at a time; the electrode is moved until one cell's discharges are well isolated electrically from the discharges of other cells in the immediate vicinity. After one cell has been recorded, the electrode is moved to pick up discharges of a new cell. The usual practice is to move the electrode through a minimum distance of 100 μm or so before starting to hunt for another single unit to record. This precaution avoids accidental rerecording of the same cell, but it costs the chance of recording responses within the 100-μm zone immediately surrounding the first cell. If there is any temporal organization to the responses of cells grouped within this small zone, it will not be observed in single-unit data.

7.2.2 Multiunit Responses

When the recording electrode is "between cells," the signal it picks up consists of discharges from several neurons mixed together about equally rather than being dominated by discharges from a single, well-isolated cell. This type of record is a multiunit response. Figure 9.16D shows a multiunit response analyzed by the same level-discriminator method as the single-unit responses in Figure 9.16A–C and displayed as a latency (or PST) histogram. In this case there are three neurons whose on-responses appear prominently at the same electrode site. The three main peaks in the histogram correspond to their individual on-response latencies. These peaks are separated by a fixed latency interval of 2–3 msec. Numerous multiunit recordings analyzed with several methods confirm the impression given by the histogram in Figure 9.16D that the local organization of response latencies is not random but structured into a periodic sequence of on-responses. At each recording site, the neurons encountered in multiunit signals have latencies spaced roughly at equal intervals of 1–3 msec and extending over a span of latencies from as little as 4–6 msec to as much as 20–25 msec. The three latency values shown in Figure 9.16D could have been picked up in the course of single-cell recordings but not during the same recording session because the electrode would be moved too far after the first cell's responses are recorded. Then, the latencies would be treated the same as the latencies in Figure 9.16A–C; they would be plotted in Figure 9.18 as completely independent data points with no suggestion that they occur together locally, as the multiunit recordings reveal to be the case.

7.2.3 Local Averaged Multiunit Responses

The most efficient way to process recordings of multiunit responses is to treat them as miniature evoked potentials and average them in synchrony with the stimulus. Figure 9.23A shows a multi-unit response processed with a spike-level discriminator and displayed as a latency histogram. The stimulus is a digitally synthesized, 2-msec FM sound that has the same sweep structure and envelope as the echolocation sounds of *Eptesicus*. Figure 9.23B shows the multiunit response from the same electrode site processed by averaging as an analog signal. The conventional spike-detection response in Figure 9.23A contains several neural discharges at different latencies, most prominently at 6, 8, and 12 msec. There is also a good deal of noise because the level-triggering technique only follows events in the envelope of the recording and cannot cancel these out over repetitive sweeps. The peaks in the histogram in Figure 9.23A must be responses of different neurons because single-unit recordings demonstrate that neurons in the inferior colliculus usually respond with ony one discharge to a short-duration FM stimulus (see Fig. 9.16A–C). The averaged response in Figure 9.23B registers the same peaks at 6, 8, and 12 msec, and it also reveals more structure in the signal, notably at 14, 18, 20–22, and 24 msec. Although the histogram suggests the presence of these other peaks at longer latencies, the averaged response reveals them more sharply because the level of noise intrinsic to the averages is lower. Events that are not correlated with the sound tend to cancel out, whereas they accumulate in the histogram. Another advantage of the averaged signal is the presence of both positive-going and negative-going waves that delineate each other's latencies more effectively than do the exclusively positive-going peaks in the histogram.

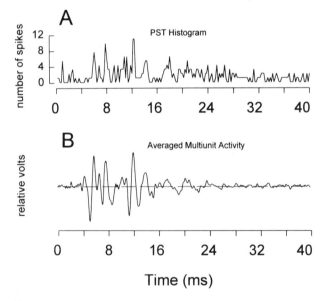

FIGURE 9.23A,B. Comparison of latency histogram (A) and analog-averaged response (B) for multiunit response from the inferior colliculus of *Eptesicus*. The histogram (vertical scale is number of discharges in each 50-μsec time bin for 40 stimulus repetitions) and the averaged response (vertical scale is voltage, with maximum peak height about 100 μV at electrode tip; *n* = 256) both show the same prominent features, but the relative noisiness of the histogram (see Section 7.2.3) conceals finer details and even partially obscures the larger peaks. For responses originating from local clusters of units, the averaged response is a superior index of latency organization, which is composed of on-responses in different cells with periodically staggered latencies.

7.2.4 Latency Dispersion in Multi-Unit Responses

Because the inferior colliculus is organized tonotopically, local multiunit responses are tuned to a specific frequency corresponding to the site of the electrode's tip along the frequency axis of the inferior colliculus (see Fig. 9.15C,E). Figure 9.24 shows a series of averaged multiunit responses recorded from the inferior colliculus of *Eptesicus* at different recording depths in the same electrode track (in 100-μm steps from 200 to 1600 μm; see scale at left in Fig. 9.24). Each recording depth has a tuned frequency, and the frequencies sampled in this particular electrode penetration extend from 24 kHz near the surface of the inferior colliculus (bottom trace in Fig. 9.24) to 76 kHz at the deepest site recorded (top trace). The stimulus for the responses in Figure 9.24 is a 2-msec FM sound with two harmonics to mimic a sonar emission of *Eptesicus*. The averaged responses vary according to the position of the electrode or according to the tuned frequency of each site. They begin at a latency of 4–5 msec and continue as a series of peaks at intervals of about 0.5–3 msec with latencies of at

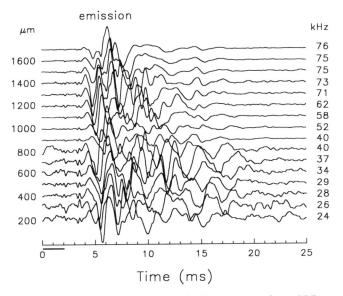

FIGURE 9.24. Series of local averaged multiunit responses ($n = 256$) recorded in the inferior colliculus from different depths (left vertical axis identifying series of responses) tuned to different frequencies (right vertical axis). The stimulus is a 2-msec, multiple-harmonic FM sound simulating a sonar emission of *Eptesicus* (horizontal bar below origin of time axis). Duration of the responses and range of latencies for response peaks correspond to dispersed latencies of single-unit on-responses (Fig. 9.18). The responses exhibit organization across tuned frequencies consisting of periodically spaced ridges that slope downward to the right, with slopes that change gradually as latency increases.

least 12–15 msec. At lower frequencies of 24–40 kHz, the responses are relatively long lasting, with latencies as much as 15–20 msec or more. At progressively higher frequencies the length of the responses, the span of latencies covered by the peaks, becomes progressively shorter.

In general, the pattern of latencies at different frequencies mirrors the dispersion of latencies in single-unit recordings (see Fig. 9.17, IC), which is to be expected if the local multiunit responses consist of clusters of neurons responding together but at different latencies relative to one another. Very few neurons in the inferior colliculus of *Eptesicus* discharge more than one time to the 2-msec FM stimulus, so the pattern of dispersed latencies in Figure 9.18 represents different latencies for each cell, and, correspondingly, each peak in the multiunit response in Figure 9.24 represents the contribution of either one cell or a small group of cells with the same, fixed latency. For the most part, then, different peaks are caused by responses to the FM stimulus in different cells, not different latencies for sequential responses in the same cells.

7.2.5 Periodic Organization of Multiunit Responses

The principal information added by the multiunit responses is the prevalence of a periodic organization to the dispersal of latencies at each site. The latency histograms in Figures 9.16D and 9.23A show multiple peaks in the responses separated by intervals of 0.5 to 3 ms, and the peaks in the averaged response in Figure 9.23B match the latencies of discharges in the corresponding histogram. Crucially, the latencies of the peaks in the averaged multiunit responses change systematically across recording depths with different tuned frequencies. In the upper range of stimulus frequencies (52–76 kHz), the peaks slide to longer latencies as frequency decreases. For example, the peak with a latency of 6 msec in the response at 76 kHz (upper trace in Fig. 9.24) shifts gradually to about 7 msec at lower frequencies of 71–73 kHz, and the following peak with a latency of 7 msec at 76 kHz shifts progressively to nearly 10 msec as frequency decreases from 76 to 58 kHz. Taken together, multiunit responses to higher ultrasonic frequencies in FM sounds consist of three to perhaps six or seven large, nearly parallel ridges roughly 2 msec apart that slope downward to the right in Figure 9.24, with some lower amplitude, narrower peaks spaced 0.5–1 msec apart. The large ridges are not *exactly* parallel, however; the size of the shift in latency across frequencies generally is larger for peaks at longer absolute latencies. Their slope changes because the spacing of the ridges opens up, from 1–2 msec near the start to 2–3 msec near the end. In contrast, at lower stimulus frequencies (24–40 kHz), the responses still contain peaks separated by 1–3 msec, but the peaks do not shift consistently to longer latencies at progressively lower frequencies. These responses contain more peaks over a longer span of time, and most peaks clearly shift gradually from one trace to the next, but their relationship to stimulus frequency is complicated; the latency lengthens as frequency decreases from 40 kHz to 29 kHz and then shortens as frequency decreases further to 24 kHz.

7.3 Echo-Delay Coding by Multiunit Responses

7.3.1 Multiunit Responses as Signal Representations

Taken as a whole, the multiunit responses from the inferior colliculus constitute a two-dimensional system of time-series events having the dimensions of time (latency to each peak; horizontal axis in Fig. 9.24) and frequency (ultrasonic tuned frequency; vertical axis in Fig. 9.24) with respect to the external ultrasonic stimulus. These in fact are the time and frequency dimensions for the spectrogram of the sound (see Fig. 9.5). The chief difference is that the spectrogram consists of a single peak (300–400 μsec wide, the integration time) at a specific time representing the moment at which each of the frequencies in the FM sweep occurs, whereas the system of multiunit responses consists of a series of peaks at fixed latencies, each with a width of (intriguingly) several hundred microseconds. The responses in Figure 9.24 are made of synchronous on-responses to the FM sound (see Fig. 9.23), and these, in turn, contain a spectrogram representation embedded in their dispersed latencies, as shown in Figure 9.22 and implied by earlier studies of response latencies in the inferior colliculus of bats (Bodenhamer and Pollak 1981).

Each one of the local responses has a waveform (a series of peaks) with a specific frequency composition (equivalent to the spacing of their peaks) and phase structure (equivalent to latencies of peaks) that is internal to the response itself. What makes averaged multiunit responses potentially so important for understanding echolocation is that their internal structure changes systematically with the external ultrasonic stimulus. As signals in their own right, they contain frequencies of about 300 Hz to 2 kHz, with a tendency for lower frequencies to occur at the end of the response rather than at the beginning because the spacing of the peaks grows larger at longer latencies (the slope of the ridges located to the right in Fig. 9.24 is more gradual than the slope of the ridges located to the left). Moreover, their phases and durations change systematically with the stimulus. If the stimulus occurs later, the responses shift uniformly to a later time, and if the stimulus consists of lower ultrasonic frequencies, the responses become longer. Because the internal horizontal and vertical dimensions of the responses carry information about the time and frequency dimensions of the external stimulus, they constitute a signal representation perhaps best described as an expanded spectrogram.

7.3.2 Similarity of Multiunit Responses to Emissions and Echoes

The averaged multiunit responses in Figure 9.24 were evoked by repetitive presentations of an FM stimulus comparable to an echolocation sound used by *Eptesicus*. Figure 9.25 shows multiunit responses collected at the same recording sites for a pair of FM sounds that simulate a 2-msec FM sonar broadcast followed 7 msec later by an echo of the same sound. This time interval is equivalent to the echo delay for a target range of 1.2 m. The multiunit responses to each of the two sounds in Figure 9.25 are similar to the responses evoked by the single FM sound in Figure 9.24, but the responses to the two sounds together overlap considerably. In Figure 9.25, both the "emission" and the "echo" evoke multiunit responses consisting of sequences of peaks at fixed latencies, with an interval of 7 msec between the extended series of peaks representing the emission and the corresponding series of peaks representing the echo. Essentially the entire pattern of sloping ridges evoked by the first sound is

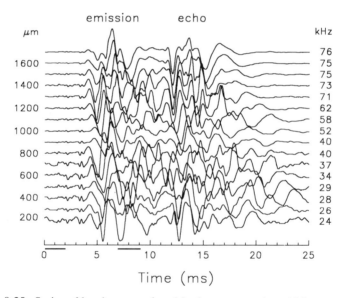

FIGURE 9.25. Series of local averaged multiunit responses ($n = 256$) to a pair of 2-msec, multiple-harmonic FM sounds simulating a sonar emission of *Eptesicus* (horizontal bar below origin of time axis) and an echo at a delay of 7 msec (horizontal bar below 7-msec point on time axis). Both sounds evoke patterns of responses similar to that shown in Fig. 9.24, with overlap of their duplicate series of ridges at emission latencies from about 12 to about 18–20 msec. These responses have their own physiological time-frequency space (horizontal and vertical axes identifying responses) related to time-of-occurrence and frequency of FM sounds; they constitute a type of signal representation that displays echo delay in the overlap of the response patterns.

repeated 7 msec later for the second sound. The responses to both the emission and the echo last for 5 to 25 msec at different recording sites, with longer sequences of peaks at locations tuned to lower ultrasonic frequencies of 24–40 kHz. Because the delay of the echo is only 7 msec, the responses overlap for an appreciable time at most sites, beginning at about 12 msec following the emission (5 msec following the echo) and continuing until about 18–20 msec after the emission (11–13 msec following the echo). At higher ultrasonic frequencies of 73–76 kHz, the responses in Figure 9.25 do not overlap, however, because the response to each sound is shorter than the delay of the echo.

7.3.3 Overlap and "Interference" of Multiunit Responses

The averaged responses to the emission and the echo in Figure 9.25 necessarily will overlap to some degree for all echo delays shorter than 20–25 msec, which corresponds to target ranges up to about 4 m. By thinking of the multiunit responses as signals with a specific waveform (frequency, phase, duration), we can describe the region of overlap between the emission and echo responses as a region of interference between the response waveforms. In Figure 9.25, at sites with tuned frequencies from 24 to 71 kHz the peaks in the two sets of responses are intermingled for latencies from 12 to at least 18–20 msec after the emission, or from 5 msec to 11–13 msec following the echo. The peaks in the overlapping waveforms at each of these sites intersect, sum together, or cancel each other depending on the polarity of the emission response and the echo response at each point in time, creating a pattern of interference that changes according to the delay of the echo after the emission. This pattern of interference carries much detailed information about echo delay that is not evident from considering individual response peaks one at a time, only from viewing the whole responses as periodic signals.

7.3.4 Delay Tuning as "Readout" of Interference Patterns in Multiunit Responses

If the delay of the echo is less than 25–30 msec, then there will still be single-unit responses to the emission taking place in the inferior colliculus (see Fig. 9.18) when the earliest responses to the echo also occur, that is, when the on-responses to the emission and the echo overlap. We thus see that the "region of overlap" between the emission and echo responses has already been described from single-unit data in terms of delay lines at different frequencies (Fig. 9.18) followed by a system of coincidence-detecting neurons that create delay-tuned responses (Fig. 9.20). The coincidence-detecting neurons respond to simultaneously occurring, long-latency responses to the emission and to shorter-latency responses to the echo (Fig. 9.20); in other words, to overlap of one particular component of the response to the emission with a correspondingly particular response to the echo (Fig. 9.21). Viewed in this manner, the delay-tuned neurons can be thought of as monitoring the conditions of overlap between responses to emissions and echoes, registering the simultaneity of small segments of these responses—segments that correspond to individual peaks in the local multiunit responses to emissions and echoes (Fig. 9.25).

Although delay-tuned neurons respond selectively to echoes at delays that equal the difference between their emission and echo latencies (Figs. 9.20 and 9.21), the crucial fact that the local responses are periodic is not yet part of the delay-line model. By incorporating the locally manifested periodicity of these responses into the already-described delay-tuning scheme, the higher level delay-tuned neurons become a mechanism for reading out the interference pattern between responses to the emission and to the echo. Each peak in the multiunit responses corresponds to the latency of on-responses in a small number of neurons (Fig. 9.23). Moreover, the relative latencies of on-responses to emissions and echoes are systematically extracted from the dispersed latencies of the inferior colliculus by delay-tuned neurons in the auditory cortex (see Fig. 9.21). The occurrence or nonoccurrence of delay-tuned responses at specific combinations of ultrasonic tuned frequencies in the emission and the echo (vertical axis of Fig. 9.25) and differences between emission and echo latencies (horizontal axis of Fig. 9.25) directly maps the shape of the overlapping multiunit responses into the activity of the auditory cortex, where it is presumed that patterns of activity lead to the formation of perceived images.

8. Sensitivity of Multiunit Responses to Fine Echo Delay and Phase

8.1 Are Multiunit Responses Functionally Equivalent to Basis Vectors in the SCAT Model?

The averaged multiunit responses are periodic signals originating in the inferior colliculus and containing a series of waves at approximately constant frequencies (allowing for their noisiness and for the progressive widening of the interval between successive peaks at longer latencies, which would appear as a decrease in frequency over the duration of the response; see Fig. 9.25). As such, they are suggestive of the basis vectors in the SCAT model. In the model, the basis vectors are cosine-phase oscillations, but in the bat they have to be "synthesized" from sequences of local, synchronized on-responses at discrete latencies. In all probability no individual neuron in the inferior colliculus actually oscillates at the frequency of the multiunit response (about 300 Hz to 2 kHz); instead, the on-responses of different neurons are interleaved locally to appear as a succession of events at that frequency. The question is whether these responses serve a function comparable to the computations developed in the SCAT process.

In one respect, the discharges making up multiunit responses more or less exactly fulfill one of the functions identified in the SCAT model, that of the delay lines for determining the spectrogram delays (t_{A1-A5} in Fig. 9.8C) from coincidences detected at specific delay taps. This component of the SCAT model correlates the emission and echo spectrograms. The model places additional computational significance on the basis vectors that goes beyond just delays, however, by locking their starting phases to the occurrence of delay-tap coincidences in the various frequency channels. In effect, the phase of the basis vectors is a surrogate for the phase of the ultrasonic frequencies in the FM sweep of the echo relative to the emission. The larger aspect of this "phase" is simply similarity between the slope of the ridges in the spectrogram for the echo relative to the slope of the ridges in the spectrogram for the emission, and this is subsumed into the delay lines and coincidence detectors common to the SCAT model and the bat.

A smaller but still critical part of this "phase" in the SCAT model, however, is the detailed variation in the timing of coincidences in adjacent frequency channels. Small variations in phase across frequency channels are transposed into the basis vectors as variations in the starting time of the cosine oscillations (see Fig. 9.8E), which in turn are critical for accurate reconstruction of the arrival times of overlapping echoes (Fig. 9.8F). While the interference pattern produced by the overlapping multiunit responses in Figure 9.25 seems related to the delay lines and coincidence detectors in the SCAT model, the multiunit responses also have to be sensitive to the phase and fine temporal structure of echoes if they are to qualify as candidates for biological basis vectors like those in the model. The astonishing thing is that these responses in fact *are* sensitive to the phase of echoes, even though the stimuli are at ultrasonic frequencies.

8.2 Phase-Coherent Multiunit Physiological Responses

8.2.1 Responses to Binaural Echo-Delay Differences

Figure 9.26 shows an example from an experiment illustrating the mode for representing echo phase and fine echo delay by multiunit responses in the inferior colliculus of *Eptesicus*. First, the stimuli for these experiments are illustrated in Figure 9.26A,B. To obtain independent control of binaural stimuli, the sounds are produced with earphones inserted into the ear canal of the bat's left and right ears. Figure 9.26A shows the waveform of an electronically delivered, 2-msec FM sonar emission (pulse, P) and echo (E) at a delay of 6 msec (corresponding to a target range of about 1 m) delivered to the bat's ipsilateral ear (same side as inferior

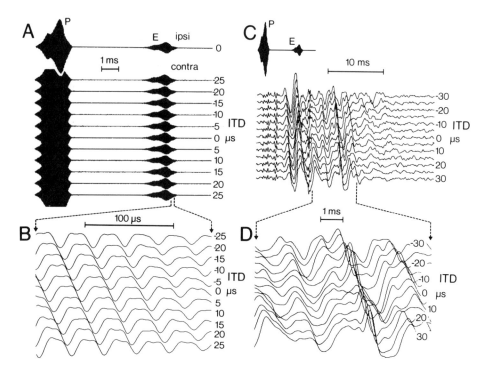

FIGURE 9.26A–D. The latency structure of local averaged multiunit responses (*n* = 256) in the inferior colliculus encodes information about fine temporal structure of echoes; in this example, binaural delay or phase differences are shown. (A) Envelopes of stimuli that mimic 2-msec, FM emission and echo (6-msec overall echo delay, with changes from −25 μsec to +25 μsec in delay of contralateral echo to mimic changes in target azimuth). (B) Expanded view of contralateral echo waveform to show shifts in time or phase in one ear relative to the other. (C) Averaged responses recorded from site tuned to 20–22 kHz for series of binaural echo-delay differences from −30 μsec to +30 μsec (stimulus envelopes on same time scale at top). Latencies of responses shift to right by about 27 times the binaural delay difference in the echo. (D) Expanded view of multiunit responses to show detailed structure and how it resembles a time-expanded reconstruction of echo acoustic waveform in B. (Note different time scales for each part of the figure).

colliculus from which the recording was made). Below this ipsilateral trace is a series of echoes delivered at the contralateral ear at slightly different delays around 6 msec. In successive contralateral traces, the echo varies in delay from 25 μsec *before* 6 msec to 25 μsec *after* 6 msec. Each successive trace shows a change of 5 μsec in this contralateral echo delay. The interaural time difference (ITD) for these binaurally delivered echoes varies from −25 to +25 μsec around the absolute delay of 6 msec. Binaural time differences of this magnitude are equivalent to differences in target azimuth from about 18° ipsilateral to about 18° contralateral. Next, Figure 9.26B shows expanded views of the contralateral echo wave-forms in the region where the first-harmonic FM sweep passes through 25 kHz (note time scales of 1 msec in Fig. 9.26A and 100 μsec in Fig. 9.26B). These "zoom" views of the stimulus show individual cycles of the echoes and graphically confirm that their peaks shift to the right in 5-μsec steps as the binaural delay difference (ITD) changes. The electronically produced echoes delivered to the ipsilateral and contralateral ears indeed do differ in their arrival time by amounts from −25 μsec to 25 μsec.

Figure 9.26C shows local averaged multiunit responses from the inferior colliculus of *Eptesicus* recorded for the stimuli shown in Figure 9.26A. (This particular recording site is tuned to 20–22 kHz.) Each trace in Figure 9.26C is similar to one of the overlapping multiunit responses shown in Figure 9.25 at sites tuned to 24–28 kHz, but now there is a new dimension to consider. (In Fig. 9.26C, the responses are stacked vertically to demonstrate how binaural echo-delay differences (ITDs) are manifested in the responses. Again, note the times scales of 10 msec in Fig. 9.26C and 1 msec in Fig. 9.26D.) The peaks in these responses shift to longer latencies as the binaural echo-delay difference changes from −30 μsec to 30 μsec in steps of 5 μsec. This range of binaural delay differences cor-responds to target azimuths of about 22° ipsilateral to 22° contralateral. The significant feature of this effect is that the time shifts in the responses are many times larger than the binaural echo-delay changes themselves.

Figure 9.26D shows an expanded section of the neural responses from Figure 9.26C to illustrate the magnitude of the response time shift in relation to the original stimuli in Figure 9.26B. In this example, the amount of time expansion in the responses is by a factor of about 27; that is, a 5-μsec change in binaural delay difference leads to a 135-μsec change in response latency. Moreover, it is completely surprising that these responses appear in some respects to reconstruct the stimulus waveform on this stretched time scale (compare Fig. 9.26B with Fig. 9.26D). Notice also that the shift in response latency affects the entire series of peaks evoked by both the emitted pulse (P) and the echo (E) as a function of the binaural echo-delay difference (ITD). As the delay difference changes from −30 μsec to +30 μsec, all the peaks in the responses slide to the right, that is, to longer latencies, even those peaks that we would initially regard as part of the response to the emission because they precede the arrival of the echo. Evidently the responses to the emission and the echo belong to a single system of events that are collectively altered just by changes in the delay of the echo, presumably because responses to one pulse–echo pair are modified by the previous echo.

8.2.2 Responses to Echo Phase Shifts

Not only do the latencies of local multiunit responses in the inferior colliculus represent binaural echo-delay differences on an expanded time scale, but they also represent the phase of the echo waveform itself on an expanded time scale. Figure 9.26D shows segments of the neural responses to illustrate the effect on latencies of what amount to binaural phase changes. Figure 9.27A now illustrates the same response segment at a binaural echo-delay difference (ITD) of zero (target straight ahead) for two added conditions, 0° and 180° echo phase shifts relative to the phase of the simulated emission. Note how the local multiunit response is both shifted to the right and expanded slightly just as a result of the 180° echo phase shift. Figure 9.27B goes even further by showing a series of neural responses for different binaural delay differences combined with echo phase shifts of 0° or 180°. The latencies of the response peaks represent both interaural echo-delay differences and the phase of the echo in both

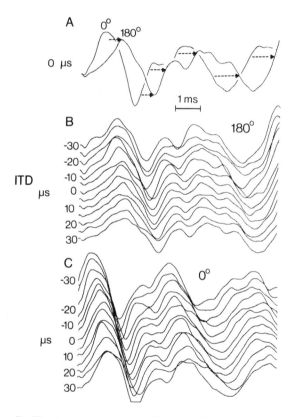

FIGURE 9.27A–C. The latency structure of local multiunit responses ($n = 256$) in the inferior colliculus encodes information about echo phase. (A) Segments of responses (same as D in Fig. 9.26) show latency shifts of peaks for 180° echo phase shift. Binaural echo-delay difference is 0 μsec. (B,C) Series of responses for echoes at 0° or 180° phase at different binaural delay differences from −30 μsec to +30 μsec show joint coding of binaural delay difference and phase on expanded time scales. *ITD*, interaural time difference.

ears together. Furthermore, the size of the latency shift is much larger than the size of the temporal shift in features of the ultrasonic echoes, where the phase shift amounts to a displacement of the cycles in the echo waveform of only 5–25 µsec and the binaural difference is only −30 µsec to +30 µsec.

8.2.3 Multiple Time Scales in Multiunit Responses

The changes in response latency illustrated in Figures 9.26 and 9.27 reveal a new scheme of representation in the inferior colliculus. First, the time intervals between responses to the emission and responses to the echo represent the delay of echoes on a time scale that closely matches real time in the stimuli (see Fig. 9.25). That is, peaks in the multiunit response to the echo lag the corresponding peaks in the multiunit response to the emission by an interval approximately equal to echo delay. Thus, the echo-delay axis of an ordinary A-scope display prevails along the horizontal time axis of the multiunit responses. However, the entire pattern of latencies in the local responses also appears to represent details of the fine phase and delay structure of the echoes on a wholly different time scale. Shifts of a few microseconds in the binaural delay or the phase structure of the waveform of echoes lead to shifts of hundreds of microseconds in the multiunit responses. This discovery leads to a very novel conclusion: the bat may be able to "read" information from these responses at time scales quite different from that actually observed electrically. Time-scale magnifications of this sort may account for aspects of the bat's performance that until now have appeared impossible to reconcile with physiological results, including echo-delay acuity of 10–15 nsec, two-point echo-delay resolution of about 2 µsec, and sensitivity to changes in echo phase. While the latency axis appears to be associated with variabilities of hundreds of microseconds (the width of the response peaks), the true variability in the latencies of the responses is only a few microseconds as far as the fine delay and phase of echoes is concerned. The bat's practice of packing multiple time axes into the same neural signals is, to say the least, unexpected from previous physiological considerations, although it is anticipated by the time-scaling feature of the basis vectors in the SCAT model.

9. Cortical Responses to Echo Waveforms

9.1 Binaural Echo-Delay Differences

It is widely presumed, but not of course demonstrated, that the content of perceived images is generated by neural activity taking place in the cerebral cortex. It therefore is important to know whether the unusual latency shifts and their associated time expansions seen in the inferior colliculus (Figs. 9.26, 9.27) also occur in cortical responses. Figure 9.28 shows averaged multiunit responses recorded from the auditory cortex of *Eptesicus* that confirm the presence of these time-domain events at the highest level of auditory representation. (The format for this figure is similar to that in Figs. 9.26 and 9.27.) The acoustic stimulus consists of a 2-msec FM emission and then an echo arriving 15 msec later. This recording shows a large peak (*) corresponding to the discharge of a single, well-isolated, delay-tuned cell (tuned to a delay of 15 msec) accompanied by smaller peaks that most likely originate from other cells in the vicinity. The complex of peaks in this part of the response shifts in latency by relatively large amounts in response to small changes in the binaural delay difference ($-50\,\mu$sec to $+50\,\mu$sec) delivered around the overall echo delay of 15 msec.

As in Figures 9.26 and 9.27, the magnitude of change in latency appears magnified with respect to the magnitude of the change in the original echo waveform. The expanded view of this latency shift at the bottom of Figure 9.28 shows how the magnification effect "rides on top" of what would otherwise just be considered an ordinary delay-tuned response from a cortical single unit (see Fig. 9.20). Not shown here is the effect of changing the phase of echoes relative to emissions by 0° or 180°; the latency of cortical multiunit responses also changes by a large amount for echo phase shifts, as already shown for the inferior colliculus in Figure 9.27. Both these effects are consistent with a possible role for delay-tuned, coincidence-detecting neurons in reading out information about the overlap of local multiunit responses to the emission and echo (Fig. 9.25), but it is only an indication of what might be happening as a mechanism for echo processing, not a convincing proof of basis vectors in the brain. (At this stage of our knowledge about the mechanisms of perception, even unconvincing evidence is a step forward.)

9.2 Delay Separation of Overlapping Echoes

The chief concern of this chapter is the bat's ability to perceive the arrival times of closely spaced echoes (see Fig. 9.7), and the last question to consider is whether the same magnification of time scales seen in the latencies of multiunit responses is also seen in response to changes in the delay separation of closely spaced echoes. Figure 9.29 illustrates this type of response: The curves show a series of multiunit responses evoked by FM stimuli that mimic a sonar emission (12-msec duration) and a two-glint echo at a delay of 24 msec. This series shows the effect of changing the delay of the second of the two overlapping components of the echo from 0 to 200 μsec in steps of 25 μsec while the first component remains fixed at a delay of 24 msec. The latency of the principal peaks in the response shifts by about 3 msec for a 175-μsec change in separation of the echo components, which is a magnification factor of about 17. This value

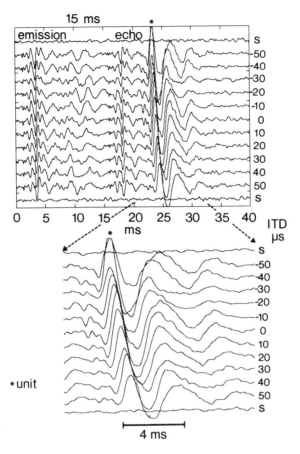

FIGURE 9.28. The latency structure of local multiunit responses ($n = 256$) in the auditory cortex encodes overall echo delay on a coarse scale and binaural delay or phase on an expanded time scale. (*Top*) Responses averaged from recording of a delay-tuned cortical neuron to a 2-msec, FM emission followed by an echo at the cell's best delay (15 msec) at different binaural delay differences from $-50\,\mu$sec to $+50\,\mu$sec. Peak marked * (asterisk) at emission latency of about 23–24 msec is cortical response; earlier peaks are brainstem and midbrain responses to the emission (latency of 2–5 msec) and echo (latency of 16–18 msec) that "leak" into recording from a distance. Traces labeled *S* show spontaneous activity in absence of stimulus. (*Bottom*) Expanded view of cortical response to show size of latency shift induced by small binaural delay differences. This cell's delay-tuning curve determined from its isolated discharges as a single unit represents delay on a scale of milliseconds (see Fig. 9.20); the latency of the local averaged response, which contains contributions from neighboring cells as well, encodes binaural delay differences on an expanded time scale of microseconds. These responses also undergo latency shifts in response to 0° or 180° echo phase shifts (similar to effect in Fig. 9.27).

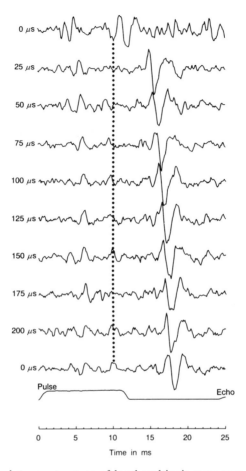

FIGURE 9.29. The latency structure of local multiunit responses ($n = 128$) in the auditory cortex encodes information about fine delay separation of overlapping echoes within the bat's 350-μsec integration time for echo reception. A series of responses over a 25-msec span of emission latencies are shown for echoes simulating two reflected replicas at delay separations of 0–200 μsec at 25-μsec steps (horizontal axis shows time after the onset of the envelope of the 12-msec FM emission and also the very beginning of the echo at a delay of 24 msec). Main peaks in responses shift in latency by about 3 msec for the total 200–μsec span of delay separations, with a larger initial latency shift of about 3.5 msec from delay separation of 0 to 25 μsec. This series of responses demonstrates an expanded time-domain representation of information that initially was represented by the spectrum of the overlapping echoes.

is well within the range of magnification factors seen in the inferior colliculus for echo phase and binaural delay differences. For all practical purposes, the responses in Figure 9.29 may be part of the bat's A-scope display of the second of the two overlapping echoes (at delay t_B in Fig. 9.8). Thus, as predicted at least in a loose way by the SCAT model, the big brown bat seems to encode the delay separation of closely spaced, overlapping echoes by the timing of neural responses. It presently is unclear how many cells contribute to the multiunit responses or what mechanism creates the time-scale magnification, but the physiological reality of these effects is well demonstrated by Figures 9.26–9.29.

10. Summary and Conclusions

10.1 The SCAT Model and the Bat

The SCAT model has four functions: (1) it represents the waveform of FM sonar emissions and echoes (see Fig. 9.8A) as spectrograms with 81 parallel frequency channels (Fig. 9.8B); (2) it determines the time separation beween the emission and the echo by measuring spectrogram delays in each channel using delay lines that register coincidences between delayed representations of the emission and immediate representations of the echo ($t_{A1–A5}$ in Fig. 9.8C); (3) it triggers the occurrence of oscillatory basis vectors in each channel from the coincidences detected in the delay lines (Fig. 9.8D,E; and (4) it creates an image of echo delay by summing the basis vectors across all the channels (Fig. 9.8F). The multiunit responses suggest that a similar process may underlie echolocation: perhaps the bat (1) represents the waveform of FM sonar emissions and echoes as spectrograms with numerous parallel frequency-tuned receptors and sharpens the temporal registration of each frequency in the spectrogram with on-responses in lower auditory centers (CN, NLL in Fig. 9.17); (2) disperses this spectrogram representation by triggering sequences of responses over the course of at least 20–25 msec after the emission or the echo in the inferior colliculus (Figs. 9.18 and 9.22); (3) organizes these dispersed latencies at periodic intervals (Figs. 9.23–9.25); and (4) determines the time separation of emissions and echoes from coincidences of responses at different latencies using the dispersed latencies as the equivalent of delay lines to create delay-tuned neurons (see Fig. 9.21). The chief difference between the SCAT model and this hypothetical description of echolocation is that the model triggers the periodic basis vectors off the coincidences detected at specific delay taps in the delay lines (which themselves were activated by the initial registration of each frequency in the spectrogram of the FM sweep), while the bat triggers the periodic responses of local groups of neurons off the initial registration of each frequency in the spectrogram of the FM sweep and then uses these periodic responses as delay lines for subsequently detecting coincidences between emissions and echoes using delay-tuned neurons. Essentially, the order in which the delay lines and the periodic oscillations are introduced into the computations may be reversed for the bat relative to the model.

10.2 Physiological Representation of Echo Waveforms

Physiological events in the inferior colliculus of *Eptesicus* do not consist just of lots of on-responses; they consist of bursts of neurally simulated "oscillations" when viewed as local averaged potentials recorded at different electrode sites in the inferior colliculus (Figs. 9.24 and 9.25). The phase of these multiple-peaked responses encodes the arrival time and phase of echoes with surprising precision in a format not previously considered (Figs. 9.26 and 9.27). The principal new finding in the physiological data is that the time scale of the bat's phase-coherent physiological responses to FM echoes is not only real time but also an expanded or zoom time scale that stretches significant elements of the echo waveform as time-series signals in the brain. These neural responses are sensitive to both binaural echo-delay differences and also echo phase, and they appear easily able to account for the bat's otherwise puzzling ability to perceive echo phase and submicrosecond echo delay. These properties are repeated in multiunit responses recorded from the auditory cortex, with the added observation that the delay separation of closely spaced echoes is included in this representation, too (Figs. 9.28, 9.29). The presence of both normal time scales and expanded time scales in responses evoked within the auditory cortex greatly strengthens the likelihood that these multiple time scales are involved in creating the perceived images. It thus appears as though the big brown bat may perceive A-scope images of targets at different ranges on a coarse time scale, plus "zoom A-scope" images of target azimuth and shape on expanded, finer time scales. These possibilities have not emerged from recent physiological studies by themselves but instead with the guidance of the SCAT model to point out parameters of neural responses that deserve more attention as candidate coding dimensions. The utility of the SCAT model in offering the basis vectors as a conceptual guidepost has been especially valuable for breaking new experimental ground in understanding the auditory computations for echolocation.

All three principal ways of influencing the waveform of sonar echoes at the bat's ears—changing their phase, changing their arrival times at the left and right ears, and changing the time separation of overlapping components (A and B in Fig. 9.6)—are manifested as large shifts in response latency. Some means for neurally representing these dimensions of echo waveform seems necessary to explain the bat's performance in critical tasks (see Popper and Fay 1995), even though it is frequently thought that bats cannot perceive these features because of physiological limitations on the speed and accuracy of neural discharges (Schnitzler,

Menne, and Hackbarth 1985; Pollak 1988, 1993). It appears as though some property of the colliculocortical system in *Eptesicus* can rerepresent, actually *reconstruct*, time-series acoustic information delivered to the cochlea as time-series information made up of neural responses. It is astonishing that this reconstruction incorporates a magnification of the time scale of the original acoustic waveform. From the point of view of this chapter, the most significant feature of these results is that the neural responses register acoustic time-series information at lower frequencies of the order of 300 Hz to 2 kHz, which are reasonable rates for neurons to operate at (Langner 1992), rather than at ultrasonic frequencies of 20–100 kHz, which presumably are beyond the capacity of neurons to encode directly.

Acknowledgments. This chapter summarizes a program of research on the auditory mechanisms of echo processing in bats supported by U.S. Office of Naval Research (ONR) grant no. N00014-89-J-3055, by National Institutes of Mental Health (NIMH) Research Scientist Development Award no. MH00521, by NIMH Training grant no. MH19118, by National Science Foundation (NSF) grant no. BCS 9216718, by McDonnell-Pew grant no. T89-01245-023, by NIH grant no. DC00511, and by a grant from the System Development Foundation, Palo Alto, CA.

References

Altes RA (1980) Detection, estimation, and classification with spectrograms. J Acoust Soc Am 67:1232–1246.

Altes RA (1984) Texture analysis with spectrograms. IEEE Trans Sonics-Ultrasonics SU-31:407–417.

Beuter KJ (1980) A new concept of echo evaluation in the auditory system of bats. In: Busnel R-G, Fish JF (eds) Animal Sonar Systems. New York: Plenum Press, pp. 747–761.

Bodenhamer RD, Pollak GD (1981) Time and frequency domain processing in the inferior colliculus of echolocating bats. Hear Res 5:317–355.

Casseday JH, Covey E (1992) Frequency tuning properties of neurons in the inferior colliculus of an FM bat. J Comp Neurol 319:34–50.

Casseday JH, Ehrlich D, Covey E (1994) Neural tuning for sound duration: role of inhibitory mechanisms in the inferior colliculus. Science 264:847–850.

Covey E, Casseday JH (1986) Connectional basis for frequency representation in the nuclei of the lateral lemniscus of the bat, *Eptesicus fuscus*. J Neurosci 6:2926–2940.

Covey E, Casseday JH (1991) The monaural nuclei of the lateral lemniscus in an echolocating bat: parallel pathways for analyzing temporal features of sound. J Neurosci 11:3456–3470.

Dear SP, Suga N (1995) Delay-tuned neurons in the midbrain of the big brown bat. J Neurophysiol (Bethesda) 73:1084–1100.

Dear SP, Simmons JA, Fritz J (1993) A possible neuronal basis for representation of acoustic scenes in auditory cortex of the big brown bat. Nature 364:620–623.

Dear SP, Fritz J, Haresign T, Ferragamo M, Simmons JA (1993) Tonotopic and functional organization in the auditory cortex of the big brown bat, *Eptesicus fuscus*. J Neurophysiol (Bethesda) 70:1988–2009.

Feng AS, Simmons JA, Kick SA (1978) Echo detection and target-ranging neurons in the auditory system of the bat, *Eptesicus fuscus*. Science 202: 645–648.

Ferragamo MJ, Haresign T, and Simmons JA (in press) Response properties in the inferior colliculus of the echolocating bat, *Eptesicus fuscus*: I. Frequency and latency dimensions of auditory spectrograms. J Comp Physiol A.

Griffin DR (1958) Listening in the Dark. New Haven: Yale University Press. (Reprinted by Cornell University Press, Ithaca, NY, 1986.)

Grinnell AD (1963) The neurophysiology of audition in bats: temporal parameters. J Physiol 167:67–96.

Haplea S, Covey E, Casseday JH (1994) Frequency tuning and response latencies at three levels in the brainstem of the echolocating bat, *Eptesicus fuscus*. J Comp Physiol A 174:671–683.

Haresign T, Wotton JM, Ferragamo MJ, and Simmons JA (in press) Sound localization by the big brown bat, *Eptesicus fuscus*. In: CF Moss and S Shettleworth (Eds.) *Neuroethological Studies of Cognitive and Perceptual Processes*, Westview Press, Boulder, CO.

Hartley DJ (1992) Stabilization of perceived echo amplitudes in echolocating bats: II. The acoustic behavior of the big brown bat, *Eptesicus fuscus*, while tracking moving prey. J Acoust Soc Am 91:1133–1149.

Henson OW Jr (1970) The ear and audition. In: Wimsatt WA (ed) Biology of Bats, Vol. 2. New York: Academic Press, pp. 181–263.

Jen PHS, Schlegel PA (1982) Auditory physiological properties of neurons in the inferior colliculus of the big brown bat, *Eptesicus fuscus*. J Comp Physiol A 147:351–363.

Jen PHS, Sun X, Lin PJJ (1989) Frequency and space representation in the primary auditory cortex of the frequency modulating bat *Eptesicus fuscus*. J Comp Physiol A 165:1–14.

Kick SA (1982) Target detection by the echolocating bat, *Eptesicus fuscus*. J Comp Physiol 145:431–435.

Kick SA, Simmons JA (1984) Automatic gain control in the bat's sonar receiver and the neuroethology of echolocation. J Neurosci 4:2725–2737.

Kober R, Schnitzler H-U (1990) Information in sonar echoes of fluttering insects available for echolocating bats. J Acoust Soc Am 87:874–881.

Kurta A, Baker RH (1990) *Eptesicus fuscus*. Mamm Species 356:1–10.

Kuwabara N, Suga N (1993) Delay lines and amplitude selectivity are created in subthalamic auditory nuclei: the brachium of the inferior colliculus of the mustached bat. J Neurophysiol (Bethesda) 69:1713–1724.

Langner G (1992) Periodicity coding in the auditory system. Hear Res 60: 115–142.

Langner G, Schreiner CE (1988) Periodicity coding in the inferior colliculus of the cat: I. Neuronal mechanisms. J Neurophysiol (Bethesda) 60:1799–1822.

Lawrence BD, Simmons JA (1982) Measurements of atmospheric attenuation at ultrasonic frequencies and the significance for echolocation by bats. J Acoust Soc Am 71:585–590.

Licklider JCR (1951) A duplex theory of pitch perception. Experientia 7:128–134.

Menne D (1985) Theoretical limits of time resolution in narrow band neurons. In: Michelsen A (ed) Time Resolution in Auditory Systems. New York: Springer-Verlag, pp. 96–107.

Menne D (1988) Is the structure of bat echolocation calls an adaptation to the mammalian hearing system? J Acoust Soc Am 83:2447–2449.

Menne D, Kaipf I, Wagner I, Ostwald J, Schnitzler HU (1989) Range estimation by echolocation in the bat *Eptesicus fuscus*: trading of phase versus time cues. J Acoust Soc Am 85:2642–2650.

Moss CF, Schnitzler H-U (1989) Accuracy of target ranging in echolocating bats: acoustic information processing. J Comp Physiol A 165:383–393.

Moss CF, Simmons JA (1993) Acoustic image representation of a point target in the bat, *Eptesicus fuscus*: evidence for sensitivity to echo phase in bat sonar. J Acoust Soc Am 93: 1553–1562.

Moss CF, Zagaeski M (1994) Acoustic information available to bats using frequency-modulated sounds for the perception of insect prey. J Acoust Soc Am 95:2745–2756.

Neuweiler G (1990) Auditory adaptations for prey capture in echolocating bats. Physiol Rev 70:615–641.

Novick A (1977) Acoustic orientation. In: Wimsatt WA (ed) Biology of Bats, Vol. 3. New York: Academic Press, pp. 73–287.

Park TJ, Pollak GD (1993) GABA shapes a topographic organization of response latency in the mustache bat's inferior colliculus. J Neurosci 13:5172–5187.

Pollak GD (1988) Time is traded for intensity in the bat's auditory system. Hear Res 36:107–124.

Pollak GD (1993) Some comments on the proposed perception of phase and nanosecond time disparities by echolocating bats. J Comp Physiol A 172: 523–531.

Pollak GD, Casseday JH (1989) The Neural Basis of Echolocation in Bats. New York: Springer-Verlag.

Pollak GD, March DS, Bodenhamer R, Souther A (1977) Characteristics of phasic on neurons in inferior colliculus of unanesthetized bats with observations relating to mechanisms for echo ranging. J Neurophysiol (Bethesda) 40: 926–942.

Poon PWF, Sun X, Kamada T, Jen PHS (1990) Frequency and space representation in the inferior colliculus of the FM bat, *Eptesicus fuscus*. Exp Brain Res 79:83–91.

Popper A, Fay RR (1995) Handbook of Auditory Research, Vol. 5, Hearing by Bats. New York: Springer-Verlag.

Pye JD (1980) Echolocation signals and echoes in air. In: Busnel R-G, Fish JF (eds) Animal Sonar Systems. New York: Plenum Press, pp. 309–353.

Saillant PA, Simmons JA, Dear SP, McMullen TA (1993) A computational model of echo processing and acoustic imaging in frequency-modulated echolocating bats: the spectrogram correlation and transformation receiver. J Acoust Soc Am 94:2691–2712.

Schmidt S (1992) Perception of structured phantom targets in the echolocating bat, *Megaderma lyra*. J Acoust Soc Am 91:2203–2223.

Schnitzler H-U, Henson OW Jr (1980) Performance of airborne animal sonar systems: I. Microchiroptera. In: Busnel R-G, Fish JF (eds) Animal Sonar Systems. New York: Plenum Press, pp. 109–181.

Schnitzler H-U, Menne D, Hackbarth H (1985) Range determination by measuring time delay in echolocating bats. In: Michelsen A (ed) Time Resolution in Auditory Systems. New York: Springer-Verlag, pp. 180–204.

Schweizer H (1981) The connections of the inferior colliculus and the organization of the brainstem auditory system in the greater horseshoe bat (*Rhinolophus ferrumequinum*). J Comp Neurol 201:25–49.

Simmons JA (1973) The resolution of target range by echolocating bats. J Acoust Soc Am 54:157–173.

Simmons JA (1979) Perception of echo phase information in bat sonar. Science 207:1336–1338.

Simmons JA (1989) A view of the world through the bat's ear: the formation of acoustic images in echolocation. Cognition 33:155–199.

Simmons JA (1992) Time-frequency transforms and images of targets in the sonar of bats. In: Bialek W (ed) Princeton Lectures on Biophysics. River Edge, NJ: World Scientific, pp. 291–319.

Simmons JA (1993) Evidence for perception of fine echo delay and phase by the FM bat, *Eptesicus fuscus*. J Comp Physiol A 172:533–547.

Simmons JA, Chen L (1989) The acoustic basis for target discrimination by FM echolocating bats. J Acoust Soc Am 86:1333–1350.

Simmons JA, Dear SP (1991) Computational representations of sonar images in bats. Curr Biol 1:174–176.

Simmons JA, Grinnell AD (1988) The performance of echolocation: the acoustic images perceived by echolocating bats. In: Nachtigall P, Moore PWB (eds) Animal Sonar: Processes and Performance. New York: Plenum Press, pp. 353–385.

Simmons JA, Kick SA (1984) Physiological mechanisms for spatial filtering and image enhancement in the sonar of bats. Annu Rev Physiol 46:599–614.

Simmons JA, Moss CF, Ferragamo M (1990) Convergence of temporal and spectral information into acoustic images of complex sonar targets perceived by the echolocating bat, *Eptesicus fuscus*. J Comp Physiol A 166:449–470.

Simmons JA, Ferragamo M, Moss CF, Stevenson SB, Altes RA (1990) Discrimination of jittered sonar echoes by the echolocating bat, *Eptesicus fuscus*: the shape of target images in echolocation. J Comp Physiol A 167:589–616.

Simmons JA, Freedman EG, Stevenson SB, Chen L, Wohlgenant TJ (1989) Clutter interference and the integration time of echoes in the echolocating bat, *Eptesicus fuscus*. J Acoust Soc Am 86:1318–1332.

Skolnik MI (1962) Introduction to Radar Systems. New York: McGraw-Hill.

Suga N (1964) Recovery cycles and responses to frequency modulated tone pulses in auditory neurons of echolocating bats. J Physiol 175:50–80.

Suga N (1970) Echo-ranging neurons in the inferior colliculus of bats. Science 170:449–452.

Suga N (1988) Auditory neuroethology and speech processing: complex-sound processing by combination-sensitive neurons. In: Edelman GM, Gall WE, Cowan WM (eds) Auditory Function. New York: Wiley, pp. 679–720.

Suga N (1990) Cortical computational maps for auditory imaging. Neural Networks 3:3–21.

Suga N, Schlegel P (1973) Coding and processing in the nervous system of FM signal producing bats. J Acoust Soc Am 84:174–190.

Sullivan WE (1982) Neural representation of target distance in auditory cortex of the echolocating bat *Myotis lucifugus*. J Neurophysiol (Bethesda) 48:1011–1032.

10
Further Computations Involving Time

EDWIN R. LEWIS

1. Introduction

The concept of temporal order, as embedded in the notions of *before*, *after*, and *simultaneous*, is useful for considering the activities of the individual organism and the effectiveness with which it interacts with the world in which it lives. The potential prey organism will be more likely to survive, for example, if it detects the presence of the predator *before* the predator is close enough for an effective attack; and, in that case, the motor actions involved in flight from the predator will be effective if they are carried out in a particular temporal order, but ineffective otherwise. Therefore, the organism whose nervous system is able to embody temporal order in its computations will have distinct selective advantages over an organism whose nervous system is unable to do so. One expects that nervous systems have evolved with that ability.

When they are incorporating temporal order into their own computations, scientists sometimes take *time* to be the label of a metric space that consists of the set of real numbers and the distance function (or metric):

$$d(t_1,t_2) = |t_1 - t_2| \qquad (1)$$

where t_1 and t_2 are real numbers. The order property of the real numbers (e.g., one and only one of the following statements is true: $t_1 < t_2$, $t_1 = t_2$, $t_1 > t_2$) makes them a convenient system with which to describe temporal order (i.e., to construct formal definitions of before, after, and simultaneous). Furthermore, one can construct a line to represent an open or closed subset of the set of real numbers and display along it, in graphical form, a sequence of temporal events (Fig. 10.1). Such graphs represent spaces in which time is one of the dimensions. In such spaces, one can use the metric d to construct formal definitions of such temporal notions as duration, rate, period, frequency, and rhythm. On the other hand, when graphing the dynamics of a nonlinear system, scientists often employ a *state space*, in which time is not one of the dimensions. Such graphs explicitly represent the order of events as the dynamics unfold, but

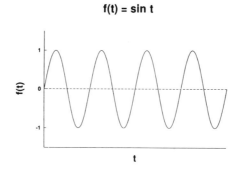

f(t) = sin t

get in line

[gɛt ɪn laɪn]

FIGURE 10.1. Graphs that represent time explicitly, following the Western convention with left to right representing earlier to later. *Top*: graph of a sinusoidal function (e.g., a pure tone). *Middle*: two graphs of an English phrase, one with standard spelling (nine letters) and the other a phonemic transcription (based on Ladefoged 1982) for a common American dialect. Phonemic transcriptions of speech are based on the following hypotheses: (a) a spoken word, as perceived sound, comprises a sequence of discrete, irreducible, perceptual elements (consonant sounds and vowel sounds) and the identity of the word (hence its range of meanings) is represented entirely by the identities of those elements and their order; (b) a spoken word, as physical sound, comprises a set of irreducible elements (*phonemes*), each of which will be identified by the auditory system of the listener as one of the discrete perceptual elements; and (c) all changes in meaning are represented by changes in the organization of phonemes in the physical sound. Any change in an irreducible element of physical sound that conveys a change of meaning is taken to be a transition from one phoneme to another. Thus the two words /thigh/ and /thy/, for example, demonstrate the existence of two phonemes associated with the letter combination /th/ in English;

FIGURE 10.1. *Continued*

and in American English, the two words /allusion/ and /Aleutian/ differ by one phoneme.

Bottom: part of a familiar tune represented with five-line staff notation. The vertical coordinate represents pitch with eight discrete locations per octave, conforming to a diatonic scale. It approximates a logarithmic coordinate. If the music includes any of the other 4 pitches of the 12-pitch octave, then more than 8 pitches must be mapped to eight locations, which is accomplished by using accidentals (e.g., sharp and flat signs). Time is treated as a discrete variable (the unit of which is the *beat*). Music is depicted as a sequence of discrete tones (each represented by a note) and discrete pauses (each represented by a rest). The relative duration of each element is indicated by its shape (e.g., the half note is open and has a stem, the quarter note is solid and has a stem). The element assigned a duration equal to one beat is the one whose name corresponds to the reciprocal of the lower number in the *time signature*, which usually appears at the left end of the first staff (after the clef and the key signature). Thus, a *4* indicates that a duration of one beat is assigned to the quarter note and to the quarter rest. In that case, a whole note (or whole rest) is sustained for four beats, a half note (or half rest) for two, and so forth. The duration of the beat usually is indicated approximately by an expression for tempo (e.g., *largo, andante, presto*), but it sometimes is indicated more precisely with a *metronome marking* above the staff, over the time signature. The number is the reciprocal of the duration (in minutes) of the note printed next to it. The horizontal axis is divided by vertical lines (*bars*) into *measures*, each of which is assigned the same number of beats, given by the upper number in the time signature. The number of beats to the measure and the distribution of accents over those beats together are called the *meter*. In European and American music the meter typically is based on multiples of two, with the first member of each pair of beats emphasized, or on multiples of three, with the first of every three beats emphasized. Thus the upper number in the time signature typically is a multiple of two or a multiple of three. Designating the element assigned the beat, the lower number is an integral power of two. Defined, as it commonly is, to be purely a matter of beats in time, *rhythm* is the more-or-less periodic temporal structure that emerges from the concatenation of measures, the measure-by-measure succession of emphasized and unemphasized beats.

Because the principal purpose of written music is reproduction, one might take it to be a representation of physical sound. On the other hand, when reproduction is accomplished in the traditional way, by a human performer, the translation from written to acoustic form is calibrated, ultimately, by the human auditory system. During rehearsal, for example, the performer adjusts the amplitude and timing of each tone or group of tones according to his or her own perception of loudness and time. Each pitch in the repertoire of a musical instrument is represented by one or more resonances. In some instruments, the tuning of resonances is fixed in advance. In other instruments, tuning is continuously controllable by the performer and is adjusted during rehearsal according to the performer's perception of pitch. Thus, there is considerable merit in the notion that written music represents sound the sensation more directly than it represents sound the physical process. In that regard, written music seems to be analogous to written speech.

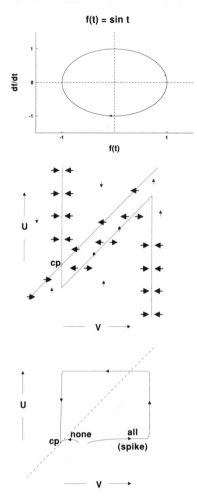

FIGURE 10.2. Graphs that represent time implicitly, employing state-space conventions. *Top*: a state-space representation of the sinusoidal function of Figure 10.1. As time progresses, the function moves clockwise around the elliptical trajectory. Middle: state-plane representation of a modified version of the two-time-constant model of neural spike triggering (Rashevsky 1933; Monnier 1934; Hill 1936). The state variables represent excitation (V) and accommodation (U). The line (*vertical isocline*) shaped like an inverted N is the locus of points along which dV/dt is zero. Above and to the right of that line, dV/dt is negative (i.e., trajectories are leftward); below and to the left of the line, dV/dt is positive (trajectories are rightward). The diagonal line (*horizontal isocline*) is the locus of points along which dU/dt is zero. It separates a region (above and to the left) in which dU/dt is negative (trajectories are downward) from a region (below and to the right) in which dU/dt is positive (trajectories are upward). The intersection of the two lines is a critical point (*cp*) at which the system rests unless it is disturbed. The leftward and rightward motion is fast (indicated by *large arrows*); the upward and downward motion is slow (indicated by *small arrows*). After selecting an arbitrary starting point, one can easily infer the general path of the subsequent trajectory. *Bottom*: two such trajectories, both ending at the critical point. One represents a full spike or action potential (*all*); the other represents a subthreshold response (no spike, or *none*).

represent durations, rates, periods, frequencies, and rhythms only implicity (Fig. 10.2).

A theme running through the final section (Section 4) is the possibility that the temporal order used in each neural computation is retained only implicitly in the result of that computation. This would lead to a progressive increase in the coarseness (i.e., loss of resolution) in the metric d as one progresses from more peripheral computations to more central ones (i.e., from cochlea to cerebral cortex), which evidently does occur (Schreiner and Langner 1988). This trend presumably is reversed as one moves from central computations to peripheral actions (e.g., vocalizations) on the motor side. Thus, the details of motor patterns would be represented only implicitly in the cortex and they would be represented increasingly explicitly as one moves toward the periphery.

When considering how a nervous system might embody temporal order in its computations, one should remember that real numbers and spaces based upon them are human inventions. On the other hand, all hypotheses that humans put forth to account for natural pheomena, such as neural computations, also are human inventions. It is not unreasonable, therefore, to consider existent inventions that already have been used effectively by humans to communicate, compute, or engineer temporal order, and to search among those inventions for ideas that might be relevant to computational processes in nervous systems. In that spirit, in the original draft of this chapter I had included discussions of various human representations of the temporal structure of sound, including written speech and written music, and various computational processes that humans have invented to analyze that structure, including various linear transforms and signal descriptors derived from them. That was to be the historical-factual part of the chapter. Unfortunately, the resulting draft was too long. The historical-factual part had to be reduced considerably, leaving largely the more conjectural discussions.

2. Sound

The word *sound* has two meanings: (1) physical motion (e.g., vibration) of matter, and (2) the sensation caused by such motion when it reaches the ears. The earlier chapters in this book have demonstrated how important it is for people thinking about the ear and hearing to separate the notion of *sound the sensation* (or *perceived sound*) from that of *sound the physical process*. All of the qualities (such as loudness, pitch, and timbre) that the human listener attributes to perceived sound are computed by the nervous system. It is clear from the preceding chapters that the algorithms underlying those computations still are not completely understood. The computations are carried out on the patterns of activity over the 30,000 afferent axons of each cochlear nerve, but they are affected by concurrent activities in other sensory channels, and many of them are shaped by past experience.

We attribute a separate set of properties to sound the physical process. These are observed with calibrated physical devices or are computed from such observations with algorithms that we know. A major part of the study of hearing is exploration of the relationships between the computed qualities of sound the sensation and the properties of sound the physical process. An important part of hearing itself is the ability of the listener to compute—in sound the sensation, reliable inferences about sound the physical process.

2.1 Inferences

Sounds seem to fall into two broad classes, those that are generated by discrete physical events and thus are distinctly transient (such as a footstep, the snapping of a twig, the crushing of leaves, the flapping of canvas in a breeze, a fish jumping) and those that are more or less continuous with slowly varying amplitude (such as those generated by turbulence in a flowing fluid as in a breeze or a stream, those of an engine or of tires rolling over a rough road, those of a teakettle whistle, and those of insects such as bees and mosquitoes). In either case, the listener may derive four kinds of inferences from the perceived version of the sound: (1) identification of the general class to which the source belongs (e.g., flowing air, flowing water, motor car, teakettle, mosquito); (2) inferences regarding the internal state of the source (e.g., the volume velocity of fluid flow, whether the engine is running well, whether the water is boiling); (3) inferences regarding the location and motion of the source relative to the listener; and (4) inferences regarding the state of the environment in which the sound is propagating. In the case of transient sounds, the attentive listener extracts information not only from the sounds themselves, but also from the silences separating them. For example, a single hoofbeat might be misidentified or overlooked, but the rhythmic pattern of hoofbeats of a walking or running animal probably would be identified as such, and the durations of the silences would be important cues for estimating the speed of walking or running.

2.2 Communication

From sounds (such as vocalizations) produced for communicative purposes by animals, including music and speech produced by humans, the attentive listener may derive not only the four inferences mentioned in Section 2.1 but also inferences about the message carried in the communication. Some ethologists consider the message to be a reflection of the internal state of the animal producing the sound, which, strictly speaking, must be true. On the other hand, the message may reflect directly the sending animal's inferences about its surroundings (e.g., presence of a predator). In that case, the flow of information to the receiving animal begins outside the sender; the information is processed by the sender; and the inferences are communicated to the receiving animal. In this way, communicating animals form coalitions. In many instances in speech, the internal state of the sender, per se, is irrelevant to the receiver. Through speech and its extension, written language, humans have formed gigantic coalitions—which have extended across generations and have had huge impacts on the survivorship and general well-being of every member.

It is important to distinguish between information and inference. Inferences, for example, can be constructed by means of computations applied to an incoming stream of information. Whereas numerous neural modelers currently, and in the past, have focussed their attention on information flow in neural signals and on the efficiency of neural codes in carrying information, it seems to me that much more interesting questions are those related to the information-processing computations of the nervous system and the efficiency with which inferences can be drawn from the information stream.

2.3 Perceptual Source of Sound

Lyon and Shamma (Chapter 6, this volume) discussed computations underlying the listener's inference of the existence of an acoustic entity. From the combinations of physical sounds from various sources, all converging into the motions of the two oval windows, the human listener often is able to segregate the perceptual components corresponding to one source from those corresponding to the others, and then to integrate the components for that one source into an auditory entity, an auditory image of the source (e.g., Hartmann 1988). This auditory entity then can be tracked in time (as one might track the perceived sound of a single musical instrument in a performance by an ensemble, or the perceived voice of a single speaker in a crowded room). It becomes a unitary model of perceived sound, and to that model the listener attaches several computed attributes as labels (e.g., perceived rhythm, perceived duration, loudness, pitch, timbre, perceived direction to source).

2.4 Perceived Duration and Rhythm

According to Michelsen, Larsen, and Surlykke (1985), "the vast majority of studies of sound communication in insects support the notion that the behavioral information is carried by the gross rhythmicity of the songs." Studying seismic and auditory communication in a tropical frog species, my colleagues and I came to the same conclusion about the advertisement call of that animal (Moore et al. 1989). It is clear that duration of physical sound elements and rhythmicity in their presentation both are important to the human listener as well, and that they are represented in the perceptual models that the listener computes for the auditory entities he or she is tracking. Furthermore, considerable evidence suggests that distinct rhythmicity in physical sound helps the human listener store and recall incoming temporal sequences in short-term memory (Sturges and Martin 1974). Glenberg and Jona (1991), for example, found that when long and short sound bursts were presented in random sequences, each comprising nine bursts, the listener was able to reproduce the ordinal relationship in each sequence more faithfully if the sequence had been presented with a distinct rhythm. Their results were consistent with those of Povel (1981, 1984) and Povel and Essens (1985), who concluded that the nervous system of the listener computes a perceptual model of an incoming, rhythmic sequence in three steps: (1) computing a beat interval and imposing the beat on the incoming sequence, (2) computing a classification scheme with a simple model (analogous to a label) for each of the various patterns that unfold over the course of an entire beat interval, and (3) computing the identity of the pattern associated with each successive beat and attaching the appropriate label to that pattern. The sequence thus is envisioned as entering short-term memory as an ordinal arrangement of labels. Each label is envisioned not as a complete abstraction, but as a perceptual model that facilitates reproduction of the pattern being labeled. For the simplest patterns, the label might be a single computed attribute, such as pitch or a perceived duration. For more complicated patterns, the label might be a set of such attributes. Computation of a classification scheme involves a balance between lumping (grouping similar patterns into a single classification element, so that each of them is represented by the same label) and splitting (sorting dissimilar patterns into separate classification elements). Hierarchical organization clearly is a central theme of this hypothesis. It is achieved by the combination of steps 2 and 3, which is called *chunking* (Miller 1956).

2.5 Loudness, Pitch, and Timbre

In the earlier chapters, we saw that loudness, pitch, and timbre are computed attributes that the nervous system attaches to a subset of perceived sound components that it has identified as arising from a single source. In other words, loudness, pitch, and timbre are labels (models) that the nervous system assigns to an auditory entity. They also are labels (models) that could be used in classification schemes for chunking. Although, superficially, the loudness assigned to an entity seems to represent the amplitude of the corresponding physical sound, the computation of loudness by the human auditory system involves duration and spectral content as well as amplitude (Brüel and Baden-Kristesen 1985; Green 1985; Moore 1989). For a sinusoidal tone burst with duration less than approximately 200 msec, perceived loudness seems to increase in proportion to the total energy of the signal. For a sustained sinusoid, perceived loudness seems to increase in proportion to the rms amplitude of the signal.

The pitch assigned to an entity seems to represent fundamental periodicity (see Lyon and Shamma, Chapter 6, this volume). The computation of pitch, however, evidently involves spectral content and amplitude in addition to periodicity (Moore 1989). Furthermore, the pitch computed for an entity formed by binaural fusion may be very different from the pitch computed separately for each of its monaural components when presented alone. In fact, two monaural stimuli to which the nervous system assigns indefinite pitch can combine through binaural fusion to form an entity with definite pitch. A dramatic example of this is provided by psychophysical experiments involving dichotic presentation of broadband noise (Cramer and Huggins 1958).

The timbre assigned to an entity seems to reflect its spectral composition. As with pitch, the timbre computed for an entity formed by binaural fusion may be very different from that computed separately for each of its monaural components. When sound is perceived as arising from a particular musical instrument, the timbre label seems to be transformed into a label for that instrument. When a sound is perceived as speech, the timbre label seems to be transformed at least in part to a vowel label. Perceptual representations of a single component of physical sound (e.g., a single sinusoid) occasionally will be integrated into two entities, one perceived as speech and the other perceived as nonspeech (Rand 1974; Mattingly and Liberman 1988).

2.6 Perceptual Phonemes

Physical sound that is perceived as a vowel, for example, can be synthesized by combining pure tones at frequencies with a distribution pattern that matches the distribution pattern of formants for that vowel. When one of those tones is subjected to vibrato and the others are not, the sound of the frequency-modulated tone can be perceived as a separate nonspeech entity and also continue to be used in the computation of the vowel label for the speech entity (Gardner and Darwin 1986). This apparently simultaneous participation of a single element in (perceptual) speech sounds and (perceptual) nonspeech sounds is called *duplex perception*. It has been taken to imply that the human auditory system has two separate computational systems, one that processes speech sounds or putative speech sounds and another that processes nonspeech sounds (Mattingly and Liberman 1988).

Some of the kinds of attributes (e.g., loudness and pitch) computed for speech entities seem to be very similar to the kinds computed for nonspeech entities. Those who take phonemes to be the basic elements of speech would argue that speech-sound entities are subjected to another kind of computation, which is not applied to nonspeech entities, namely phoneme identification. According to this hypothesis, the perceived phoneme is a label that the speech-processing part of the auditory nervous system attaches to entities identified as speech sounds. Classification schemes and chunking then would be based on phoneme sequences (e.g., syllables, words).

McGurk and MacDonald (1976) have demonstrated that the outcome of phoneme computation can be strongly affected by visual images of moving lips. A consonant being articulated silently by the lips can override a different consonant being presented at the same time in physical sound (i.e., the perceived phoneme is the one being articulated silently). This is taken by some investigators as supporting the hypothesis that phoneme computation in audition is strongly linked to the neural computation involved in speech production (i.e., in phoneme generation) (Mattingly and Liberman 1988). I shall return to this notion in Section 4.5.

2.7 Order of Computation

Other than the existence of duplex perception, there seems to be no evidence indicating that the neural computations described in Section 2.3 through 2.6 are separate. Loudness, pitch, and timbre computations for an entity are not necessarily carried out after the segregation and integration computations that yielded the entity. All of those neural computations may be carried out simultaneously and inextricably, with the end result being inference (perception) of an entity with loudness, pitch, and timbre labels already attached to it. Loudness, pitch, timbre, and segregation/integration computations all involve synthesis of binaural inputs and may be integrated with the binaural computations that generate perceptual models of source direction.

The hypothesis that the various computation tasks are integrated is supported by observations of times required to complete various computations when they are observed individually in psychophysical experiments (Green 1985). Computations of loudness, pitch, and spatial localization all seem to unfold over approximately 200 msec. Perceptions of loudness, pitch, and source location emerge well before 200 msec, but the discriminability of perceptions of similar but nonidentical sounds improves in each case up to approximately 200 msec.

Once an entity has been inferred and is being attended, the computations involved in tracking changes of attributes (pitch, loudness, etc.) for that entity evidently are approximately 10 times faster, of the order of 20 msec (de Boer 1985). At this point, as the resulting stream of attributes passes through short-term auditory memory, other inferential computations are carried out, such as interpretations of communicative sounds.

3. Entry to Short-Term Memory

Many of the computations described in earlier chapters of this volume, and summarized briefly in Section 2, thus evidently are incorporated into the process of inferring a perceptual source (Lyon and Shamma, Chapter 6 [this volume], and Section 2.3). Introspection suggests that one may view this computation as a screening process through which the auditory perceptual components inferred to have arisen from a single source or single entity are isolated and passed into short-term auditory memory unencumbered by components inferred to have arisen from other sources.

Thus one may track the perceived voice of one speaker in the presence of other speakers and other sounds, or the perceived voice of a single musical instrument in the complex sound of a small ensemble. Clearly, an essential inferential task performed by the auditory nervous system is isolation of a perceptual source, and the isolation process evidently includes translation of the (perceived) sound from that source into an ordered sequence of pitches, timbres, durations, loudnesses, and (in some cases) phonemes. The perceived sound of the source evidently enters auditory short-term memory as such a sequence (or stream) of attributes.

3.1 Properties and Computations Associated with Short-Term Memory

Some investigators distinguish two kinds of memory: (1) *procedural* memory, the readout of which is reflexive or automatic and the contents of which often are reflected in performance improvement in various motor or cognitive tasks, and (2) *declarative* memory, the readout of which involves cognitive processes and the contents of which are revealed in declarative statements (e.g., verbal statements, drawn sketches, mimicry) (see Kupfermann 1991). Applying this dichotomy, one would say that any sequence of events in a dream, any thought sequence, any sensory input sequence, or any motor activity sequence that subsequently can be described verbally, sketched, or imitated by a subject has been encoded in declarative memory.

In the act of verbal description, however, the phonological rules employed by the subject would have been encoded in procedural memory, as would the coordinated motor patterns employed in sketching or mimicry. Motor activities that have been rehearsed or repeated so many times that they can be performed automatically, even while the cognitive processes of the performer are focused on something else, would be said to be encoded in procedural memory. Such activities include sufficiently rehearsed sequences of speech or musical performance. Sound patterns that through repeated association automatically evoke certain emotional feelings (e.g., those evoked by the sound of martial music) or certain cognitive responses (as in the sudden shifting of attention that follows the hearing of one's own name spoken) also would be said to be encoded in procedural memory. Thus temporal sequences for generating sounds and temporal sequences associated with perceived sounds that are familiar to the listener can be encoded in procedural memory. Of course, many of the same sequences must be encoded in declarative memory as well.

According to the commonly cited organizational models for memory, incoming sensory material usually enters short-term memory. In the absence of rehearsal, it may persist in that state for at most a few minutes. With rehearsal, it gradually is transferred through a sequence (perhaps a continuum) of memory states, successively more persistent, and eventually to essentially permanent, long-term memory. Sensory input that accompanies especially startling events evidently can enter permanent long-term memory directly, without rehearsal. Introspection reveals that chunks of various sorts (e.g., elements of speech) may be drawn from long-term declarative memory, placed into an attended state and consciously manipulated in that state, and that the results of the manipulation may be rehearsed and gradually moved through successively more persistent memory states. Thus, there seems to be a strong connection between the attended state and the short-term memory state. In fact, organizational models of memory often identify short-term memory as the cognitive system element "that stores information for current attention and in which actual information processing is carried out" (Dempster 1981). Recently this structure has been redefined somewhat and labelled working memory (Baddeley 1992). The short-term memory that is viewed in this way would be declarative.

3.2 Short-Term Declarative Auditory Memory Has Limited Span

The span of short-term declarative auditory memory has been the subject of numerous investigations since the late nineteenth century (see Dempster 1981 for a review of the older literature). Estimates of span invariably have been based on readin and readout of temporal sequences of discrete acoustic objects (e.g., words, letters, nonsense words, digits) or on visual readin and vocal readout of spatial or temporal sequences of pronounceable or nameable visual objects (presumed to be recorded from short-term visual memory to short-term auditory memory by a process of subvocalization). Such estimates are stated in terms of numbers of encoded objects (e.g., the length of a sequence of digits that can be reproduced perfectly 50% of the time). In adults, the span for digits ranges approximately from 5 to 9, with a mean of approximately 7. Miller (1956) took this to imply the presence of 7 ± 2 discrete slots in short-term memory. For objects other than digits, however, the span is smaller (e.g., means are 6 for letters and 5 for short words). Furthermore, the span as measured by numbers of discrete nameable objects has been found to be inversely proportional to the time required by the subject to pronounce the names of the objects (Baddeley, Thomson, and Buchanan 1975; Standing et al. 1980; Hulme and Muir 1985), implying that it should be measured in terms of time rather than numbers of objects (Schweickert and Boruff 1986).

For a set of 18 student subjects, Schweickert and Boruff (1986) estimated the mean span (which they called *verbal trace duration*) to be 1.88 sec. This number agreed with estimates they derived from previously published work of others. If one subscribes to either the discrete-slot theory or the trace-duration theory, short-term declarative auditory memory would be viewed as a cognitive device connected to a sink. An encoded stream of auditory inputs, or a recoded (subvocalized) stream of inputs from other sensory modalities, or an encoded (subvocalized) stream of thoughts can flow into the device and subsequently be dissipated in the sink. If the duration of the encoded or recoded stream is very long, then only the most recent part of it is retained in the device. On the other hand, if the duration of the stream is short, then items near the beginning of the stream and items near the end of the stream are more faithfully recalled than those in the middle (Harcum 1967; Murdock 1974; Healy 1974; Greene and Crowder 1988; Frick 1989).

3.3 Some Operations Associated with Short-Term Declarative Memory

3.3.1 Translation to Actions

Although subjects in experiments related to short-term auditory memory span typically are asked to produce a verbal readout of the encoded sequence of objects, they could be asked to produce other sorts of declarative readouts (e.g., to mimic a sequence of hand positions). The availability of declarative readout implies that one operation performed on short-term (declarative) memory codes is translation to declarative actions or gestures. Introspection suggests that these actions or gestures may be visualized or imagined (as in subvocalization) or carried out (as in vocalization). According to the phoneme hypothesis, declarative readout in terms of speech would imply translation of memory code to the vocal gestures associated with phonemes. Readout in terms of imitative sounds (e.g., the sound of the wind, the sound of hoofbeats, the sound of a simple melody) would imply translation to an expanded set of sound-generating gestures.

3.3.2 Rehearsal

With respect to short-term memory, rehearsal (sometimes called refreshment) seems to involve iteratively translating a stored sequence of codes into a sequence of declarative gestures, each of which reestablishes the short-term memory code for the object it represents. The short-term memory of the sequence of objects thus is repeatedly reestablished, even though the original sequence was presented to the sensory system only once. For the case in which the declarative gestures are subvocalizations, this iterative process has been labeled the "phonological loop" (Baddeley, Costanza, and Norris 1991; Baddely 1992; Baddeley and Logie 1992; Belleville, Peretz, and Arguin 1992; Jones and Morris 1992).

3.3.3 Grouping

When a list of discrete items with pronounceable names (e.g., the digits 8–4–6–3–7–8–5) is placed into short-term auditory memory in serial order, the items can be grouped into pronounceable segments (e.g., eight–four–six and three–seven–eight–five). Considerable experimental evidence suggests that grouping of discrete items in short-term auditory memory tends to facilitate ordered recall of those items (Dempster 1981; Frick 1989). For auditory presentations, optimal group sizes seem to be three or four items.

3.3.4 Chunking

Chunking is the recoding of a familiar sequence of two or more items into a single item in short-term memory. As with grouping, experimental evidence suggests that chunking tends to facilitate ordered recall of items (Miller 1956; Chi 1976; Dempster 1981). Thus, for many current speakers of American English the sequence of letters q–e–d p–d–q t–g–i–f, when grouped as shown and presented aurally, would be recalled more easily than would a random sequence of 10 letters (or 21 phonemes) grouped in the same way. Chunking clearly involves a rapid interplay between short-term memory and long-term memory (where the encoded names of the chunked groups are stored). Chunking presumedly also involves evaluation, inasmuch as the chunked labels for members of homonym sets such as Sue (for Susan), sue (the verb), and Sioux (the Indian nation) presumedly are different. Evidently, chunking is strongly affected by grouping, and grouping is strongly affected by the rhythm of the aural presentation. Thus, the timing of the presentation of a sound sequence can strongly affect the ability of the listener to manipulate the sequence in short-term declarative memory (Sturges and Martin 1974; Dempster 1981).

Grouping and chunking must be especially important in the processing of speech and music in short-term auditory memory. Perceived speech sounds evidently are encoded in short-term memory as phonemes, which in turn are grouped and chunked into morphemes, which in turn are grouped and chunked into words or phrases, and so forth. For digital speech processing, this would correspond to the familiar and difficult problem of segmentation. Again, the rhythm of the aural presentation can strongly affect the grouping and chunking process. This is illustrated, for example, by an invocation familiar to many who at one time or another were young campers: "Awa tagoo Siam."

3.3.5 Segment Boundaries

Experiments on short-term auditory memory typically are based on the assumption that each item in a sequence presented to a subject is clearly separate. When it is the perception of connected speech that is flowing into short-term auditory memory, however, the separations between words may be obscure. In that case, grouping and chunking involve estimation of segment boundaries (e.g., see von der Malsberg and Buhmann 1992). Even in connected speech, however, the speaker typically provides physical cues that facilitate the boundary estimation process. When speech (the physical sound) is carefully enunciated, the junctures between phonemes within words (close junctures) may be distinguishable (physically) from those between words (open junctures), as in the difference between *nitrate* and *night rate*. In connected speech that is rapidly produced, distinction between close and open junctures may disappear, but rhythm and intonation typically remain as cues for boundary estimation. The estimation process probably involves on-the-fly comparison (hypothesis testing) against candidate sequences encoded in long-term memory. The rhythm and intonation as well as the general context of the speech undoubtedly affect the selection of candidate sequences. The rhythm in Western music typically provides bold cues for estimation of the boundaries between figures.

3.3.6 Comparison and Hypothesis Testing

An encoded sequence of items flowing through short-term memory can be compared, on the fly, with a familiar sequence encoded in long-term memory. Thus an encoded sequence of sounds that would be much too long to be retained in short-term memory can be compared, item by item, with a familiar sequence. For example, an encoded sequence of musical figures can be compared with the encoded sequence from a familiar musical phrase. One can consider this to be a form of hypothesis testing, the hypothesis being that the incoming sequence is the same (should be assigned the same label or same interpretation) as the sequence encoded in long-term memory. As does chunking, comparison implies a rapid interplay between short-term and long-term memories.

3.3.7 Inference

In Section 2, five classes of inferences were described for perceived sound: (1) inferences regarding the identity of the source or of the general class to which the source belongs, (2) inferences regarding the internal state of the source, (3) inferences regarding the location and motion of the source relative to the listener, (4) inferences regarding the state of the environment in which the sound is propagating from source to listener, and (5) for an auditory communication, inferences regarding the message carried in the communication. The availability of each of these inferences implies that the auditory system has constructed (through neural computation) a perceptual image of the putative source and is tracking the perceptual sounds associated with that image. The processes of segregation and integration (which may be inextricably linked with computations of loudness, pitch, and timbre, as well as with the binaural computations that generate perceptual models of source direction) that produce the perceptual image thus may compose, in effect, both a filter and an encoder. As a filter, segregation/integration determines which portion of the total auditory input is encoded into short-term memory and which is not. Thus, for example, the filtering accomplished by stream segregation (and integration) makes it extremely difficult for human subjects to track certain sequences of sounds (Bregman and Campbell 1971). As an encoder, segregation/integration attaches labels (e.g., pitch, loudness, timbre, direction, phonemic identity) to the perceived sounds entering short-term memory. Thus, some inferences regarding the source already have been computed on entry to short-term memory, with the computations involved in those inferences being completed in 200 msec or less (see Section 2.7). Inferences that are based on temporal sequences of such labels (e.g., sequences of pitches, loudnesses, timbres, phonemes) evidently are computed, at least in part, within short-term auditory memory. Based on temporal sequences but apparently occurring before entry into short-term memory, stream segregation seems to be an exception (Bregman 1990).

4. Conjectures About Realization with Neural Circuitry

If one considers just one aspect of perceived speech, namely inferences regarding its verbal content (i.e., ignoring inferences regarding identity, internal state, location, and motion of the speaker, and inferences regarding the environment), then the inferential computations that take place within short-term auditory memory imply the presence of at least the following elements: (1) a mechanism for segmenting the perceived stream of speech sounds (according to the phoneme theory, a stream of phonemes) into verbal items (e.g., syllables or words); (2) a single, immediately accessible representation (in long-term memory) of each item or chunked sequence of two or more items; and (3) a mechanism for temporary retention of the serial order of an unfamiliar sequence of learned items or chunks (e.g., "did-you-feed poi to-the platypus?").

4.1 Accessible Representations

The representation of a learned item or chunked sequence presumably is embedded in the central nervous system and corresponds to a unique perceptual label. We can take the assignment of a perceptual label (e.g., a word label or a label for a chunked sequence of words) to be the result of inferential computation, involving some sort of neural circuitry. For speech sounds the label will be subvocalizable, allowing short sequences of labels to be recycled through the phonological loop (Baddely 1992; Belleville, Peretz, and Arguin 1992; Jones and Morris 1992). Introspection suggests that during the recycling process, in response to newly perceived context, the labels assigned to a given sequence of speech sounds may be changed. It also seems clear that a given label can be attached to a wide variety of perceived sounds (e.g., the same written word read aloud by a wide variety of speakers can be given the same perceptual label by a listener). Thus one envisions an inferential computation that (1) is embedded in neural circuitry, (2) has been developed (and continues to be developed) through a learning process, (3) is subject to review in the phonological loop, and (4) effects a many-to-one mapping of perceived speech sounds to a single item label or chunk label. It seems likely that segmentation and label assignment are inextricably connected, in which case the inferential computation accomplishes both at once: identifying a putative speech item or a putative chunked sequence of speech items and connecting it to an immediately accessible representation (i.e., a label) in the central nervous system. What form such neural circuitry might take and the algorithms that might be used to adapt it to its task both are subjects of much speculation in the current literature.

4.2 Classes of Realizations

The realizations proposed so far can be divided into the following classes: (1) those that are based on explicit representations of history, so that the inferential process is begun only after an entire candidate, ordered sequence of sounds has been accumulated and stored with its serial order intact, or (2) those that carry the history of a candidate sequence of sounds implicitly in the current state of the processor, allowing the inferential process to begin before the entire candidate sequence is present and thus to be carried out on the fly. The former seem to be natural extensions of our use of graphical models in which time is represented as one of the coordinates (as in Fig. 10.1) and are based on ordered sequences viewed in their entirety (Spinelli 1970; Grosberg 1974; Stanley and Kilmer 1975; Grossberg and Stone 1986; Sejnowski and Rosenberg 1986; Tank and Hopfield 1987). The essential ingredient of such models is independent storage of each member of the ordered sequence, such as one might accomplish with the electronic circuit known as a FIFO (First-In-First-Out circuit).

It may be significant that the center of gravity of speech-processing technology has moved away from this class of processor, toward the second (Rabiner and Juang 1986; Waibel et al. 1988). In that it retains only an implicit representation of the history of an ordered sequence of sounds, the second class of processor might be considered analogous to state-space representations such as those in Figure 10.2. Although interest among speech-processing scientists and neural modelers currently is focussed on nonlinear members of this class, the class itself includes the widely used linear analog and digital filters.

Proposed realizations of the inferential computations in short-term memory can be divided further into two subclasses: (a) those that map a sequence of sounds (i.e., a speech item or sequence of speech items) into a static pattern of activity, and (b) those that map a sequence of sounds into a dynamic pattern of activity. Both of these subclasses have been simulated in recent modeling studies employing parallel distributed processing.

4.3 Parallel Distributed Processing

Consider the structure depicted by the box in Figure 10.3, which encodes a static pattern of activity distributed over two input lines (A,B) into a static pattern of activity distributed over three output lines (a,b,c). Imagine, for example, that this encoder is synthesized as a network of nonlinear elements (depicted by the circles in Fig. 10.3; see Section 4.6.3) with the following properties: (1) the output of each nonlinear element is a monotonically increasing function of the sum of the inputs to that element; (2) the output approaches 1.0 as the weighted sum of inputs approaches infinity; and (3) the output approaches 0 as the weighted sum of inputs approaches negative infinity. This sort of nonlinear relationship between input and output is known as a squashing function. For illustrative purposes, imagine that for each nonlinear (squashing-function) element in Figure 10.3 the transition from 0 to 1.0 at the output occurs abruptly when the weighted sum of inputs is 1.0, and let the weight for each input be the number next to the arrowhead depicting that input. In that case, the codes for the networks in Figure 10.3 would be those listed in Tables 10.1 and 10.2.

$$Y = \{a,b,c\}$$
$$X = \{A,B\} \tag{2}$$
$$Y = f(X)$$

For squashing-function elements as defined here, the middle layer is necessary for achievement of certain relationships, such as that between b

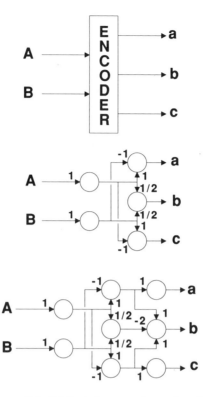

FIGURE 10.3. Simple parallel distributed processing networks that serve as static encoders. Codes for the *middle* and *bottom* circuits are given in Tables 10.1 and 10.2.

and {A,B} in Table 10.2. The squashing-function elements in this layer commonly are called hidden units (Hinton and Sejnowski 1983). With or without a layer of hidden units, the networks of Figure 10.3 yield static outputs for static inputs.

TABLE 10.1. Truth table for the two-layer network of Figure 10.3.

	$B \geqslant 1$	$B < 1$
$A \geqslant 1$	$a = 0$ $b = 1$ $c = 0$	$a = 1$ $b = 0$ $c = 0$
$A < 1$	$a = 0$ $b = 0$ $c = 1$	$a = 0$ $b = 0$ $c = 0$

TABLE 10.2. Truth table for the three-layer network of Figure 10.3.

	B ≥ 1	B < 1
A ≥ 1	a = 0 b = 0 c = 0	a = 1 b = 1 c = 0
A < 1	a = 0 b = 1 c = 1	a = 0 b = 0 c = 0

In Figure 10.3, one has created a nonlinear vector function that maps a two-element vector (A,B) to a three-element vector (a,b,c). By adding one or more elements with memory and introducing feedback, one could convert the networks of Figure 10.3 into dynamic structures that yield sequences of output patterns in response to static input patterns (Jordan 1986). In the network depicted in Figure 10.4, for example, imagine that the memory operates with a clock: during each clock interval, the output of the memory is the value that b held during the previous clock interval (one can easily imagine a realization of this with a very short FIFO and an astable multivibrator; Section 4.6). The input pattern now is given by the value of A alone, and B receives feedback (through memory) of part of the output pattern. The feedback has converted the network into a structure known as a recurrent network. For each value of A, the

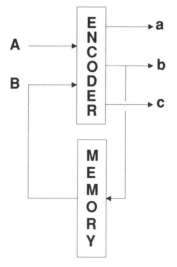

FIGURE 10.4. Configuration of a recurrent parallel distributed processing network that serves as a dynamic encoder. An example code is given in Table 10.3.

TABLE 10.3. Truth table for the three-layer network of Figure 10.3 in the recurrent configuration of Figure 10.4.

	A ≥ 1				A < 1			
a	0	1	0	1	0	0	0	0
b	0	1	0	1	1	0	1	0
c	0	0	0	0	1	0	1	0

network produces a different sequence of values of the output vector {a,b,c} (see also Table 10.3).

In Figures 10.3 and 10.4, the instantaneous relationship between the input pattern and the output pattern generated by the encoder is determined by the connections between the layers of squashing-function elements and by the distribution of weights over those connections. Over the years, investigators have developed training algorithms that automatically, through a systematic process of trial and error known as gradient descent, adjust the weights of generalized two-layer networks (Widrow and Hoff 1960; Rosenblatt 1962; Kohonen 1977) and generalized three-layer networks (Rumelhart, Hinton, and Williams 1985) until specified functional relationships between input and output are achieved. Thus the designer can specify a function, $f(\mathbf{X})$, and then allow the training algorithm to modify a generalized network to achieve a realization of that function in a network. In the last half of the 1980s, networks and training algorithms were extended to include recurrent connections (Jordan 1986; Williams and Zipser 1988, 1989). With these extensions, the designer can specify an action sequence, $\mathbf{Y}(t)$, and a corresponding input pattern, $\mathbf{X}_0$, and then allow the training algorithm to modify a generalized network so that it produces the desired sequence whenever $\mathbf{X}_0$ is presented at the input. Jordan (1986) found that once a recurrent network had been trained to produce a given sequence $\mathbf{Y}(t)$, a stable limit cycle corresponding to $\mathbf{Y}(t)$ had been created in the network dynamics (see Section 4.4).

In an alternative configuration proposed by Elman (1988), the memory input is a duplicate of the states of the hidden units (rather than the output-layer units), which is transferred (one time step later) to a set of units in the first layer (separate from those receiving the input). In Jordan's recurrent network, the input pattern was taken to be fixed, and for that fixed input pattern the present output pattern was a function of the history of previous output patterns. In Elman's recurrent network, the input comprises a sequence of patterns, and the output is a function of the present input pattern and the history of previous input patterns. By applying training algorithms to adjust the weights of the encoder in Elman networks (with various numbers of units in each layer), several investigators have shown that these networks can be trained not only to identify structure in ordered sequences (i.e., to predict the next member of the sequence) (Servan-Schreiber, Cleeremans, and McClelland 1989; Elman 1990, 1991), but also, explicitly, to identify subsequences (i.e., to chunk) (Elman 1990; Norris 1990). From the successes of these investigations and those on similar recurrent networks (Dehaene, Changeux, and Nadal 1987; Stornetta, Hogg, and Huberman 1988; Brown 1989; Wang and Arbib 1990), it is clear that the inferential computations underlying chunking can be carried out on ordered sequences by networks that carry the history of the sequence in their internal states and do not require display of the entire sequence intact.

4.4 Nonlinear Dynamics

The Jordan and Elman networks are examples of nonlinear dynamic systems. Among other things, such systems may exhibit stable limit cycles. A limit cycle is a fixed, ordered sequence of states through which the system proceeds (Fig. 10.5). The limit cycle is stable if all nearby sequences of states converge on it (i.e., if the state of the system is sufficiently close to a stable limit cycle, then the dynamics of the system will carry it toward that limit cycle). The results of application of training algorithms can be considered in terms of shaping the state-space representation (e.g., as creating or modifying limit cycles). This opens another possibility, which has been explored by Port and his colleagues (Port and Anderson 1989; Port 1990; McAuley, Anderson, and Port 1992; see also Baird 1987; Amit 1988). An ordered sequence applied to the input of a nonlinear dynamic system possessing many limit cycles could cause the system to proceed through a corresponding sequence of limit-cycle neighborhoods. Thus, the neighborhood in which the system currently is operating (i.e., the limit cycle that the system currently is approaching) can be considered to be a representation of the history of the input sequence up to the present time. The inferential computation underlying the labeling of a speech item or a chunked sequence of speech items thus would be carried out on the fly. Among other things, Port and his colleagues have found that the sequence of limit-cycle neighborhoods through which a well-trained network proceeds is strongly dependent on the ordered input sequence, but is not strongly dependent on the rate at which the input sequence is presented.

4.5 Linkage to Motor Systems

Inasmuch as a speech item or a chunked sequence of speech items can be represented by a limit-cycle neighborhood and by the ordered sequence of states composing that limit cycle, it seems an obvious step to sculpt the limit cycle so that its ordered sequence of states also represents a declarative readout (vocalization or subvocalization) of the item or

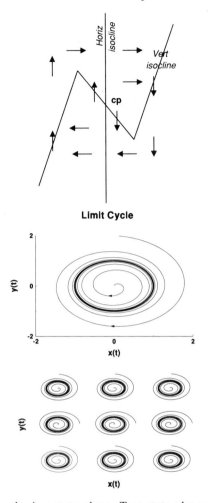

FIGURE 10.5. Limit cycles in a state plane. *Top*: state-plane representation of the van der Pol model of a nonlinear oscillator, with the vertical isocline constructed in a piecewise linear manner. *Middle*: state-plane representation of two trajectories for the van der Pol model. Motion around every trajectory is clockwise and converges on a limiting, elliptical path, the *limit cycle*. *Bottom*: sample trajectories in a state plane with multiple limit cycles, some clockwise and some anticlockwise. Each trajectory corresponds to a sequence of states (in this simple two-dimensional state space, a sequence of values of *x* and *y*). This sequence of states, in turn, could represent the neural command sequence for a learned motor pattern (e.g., a learned pattern of vocalization or subvocalization). The vicinity of each limit cycle is the *region of attraction*, within which all trajectories approach that one limit cycle. Neighboring regions of attraction are separated by boundaries from which trajectories diverge. It could be possible for an external stimulus to bump the state of the system from one region of attraction to another (Port 1990). One can imagine arrangements of neighboring regions of attraction in higher dimensional spaces.

chunked sequence. In other words, the training algorithm might include structuring of the limit cycle (in the manner of the Jordan network) so that it produces a sequence of declarative gestures (such as subvocalizations) corresponding to the sequence of perceived sounds being labeled (see Mannes and Dorffner 1989). In this manner a large part of the phonological loop could be one well-trained neural circuit, a circuit that infers the identity of input vocalizations on the sensory side and translates the inferences into declarative gestures on the motor side.

The possibility that the same central neural circuit might be involved in temporal pattern detection on the acoustic sensory side and temporal pattern articulation or generation on the acoustic motor side is not a new idea. It has been proposed by students of acoustic communication in insects, amphibians, and birds (Hoy, Hahn, and Paul 1977; Hoy 1978; Doherty and Gerhardt 1984). Evidence for this hypothesis includes, for some species, essentially identical changes in acoustic pattern production on the motor side and acoustic pattern sensitivity on the sensory side in response either to temperature variation and or to genotype variation through hybridization. For other species, however, experiments with temperature variation or genotype variation have implied that pattern generation and pattern detection are provided by separate neural circuits.

4.6 Neural Circuits Based on Spike-Trigger Models

4.6.1 Spike-Trigger Models

For many decades, the all-or-none action potential, or spike, was considered to be a key feature of the nervous system, and numerous studies were devoted to developing a theory of spike triggering. Spike triggering traditionally has been described in terms of two variables, which one might label "excitation" and "threshold." When the level of excitation equals the threshold, a spike occurs. The early models of spike triggering were attempts to describe the triggering of a single spike in response to an electric current stimulus applied to a nerve cell that previously had been unstimulated for a long time (and thus presumably was in its normal resting state). In the simpler models of this sort, the threshold was taken to be constant. In a more complicated version (Rashevsky 1933; Monnier 1934; Hill 1936), threshold was taken to vary slowly in response to changes in excitation levels. Subsequent models were attempts to describe repetitive spike triggering, and often included a temporary increase in threshold ("relative refractoriness") following each spike. A feature common to many descriptive models of spike triggering (and common to many real neuronal spike triggers as well) was an ability to produce periodic spikes in response to a constant-amplitude stimulus. If the excitation were taken to be reset to zero after each spike, then the "integrate-and-fire model" without relative refractoriness would yield the following relationship between spike frequency, f_{spike}, and stimulus intensity:

$$f_{spike} = 0 \quad I_0 < 0$$

$$f_{spike} = \frac{I_0}{I_0 T_{spike} + \theta} \quad I_0 \geq 0 \tag{3}$$

where I_0 is the stimulus amplitude (taken to be constant); T_{spike} is the duration of each spike; and θ is the threshold (also taken to be constant). Under the same assumption, the "one-time-constant model" without relative refractoriness would yield:

$$f_{spike} = 0 \quad I_0 < \theta$$

$$f_{spike} = \frac{1}{T_{spike} + T\log_e\left[\dfrac{I_0}{I_0 - \theta}\right]} \quad I_0 \geq \theta \tag{4}$$

where T is the time constant of the model. For both models the spike rate ranges from 0 to $1/T_{spike}$ as I_0 goes from negative infinity to positive infinity. This sort of squashing-function nonlinearity is typical of descriptive models of spike triggers (and of spike triggers themselves). With appropriate means (e.g., a synaptic model) to translate the spike rate of one spike-trigger model to the stimulus amplitude of another, one could use spike-trigger models as the squashing-function elements in Section 4.3 and Figures 10.3 and 10.4.

4.6.2 Neural Oscillators and Clocks

Driven by a constant-amplitude stimulus, a spike trigger can become a neural oscillator or clock. The same function also can be provided by a large number of neural circuit models known as central pattern generators (Delcomyn 1980; Calabrese and De Schutter 1992; Church and Broadbent 1992), including a pair of spike triggers (e.g., in separate neurons) coupled through mutual inhibition (as depicted in Figure 10.6). Consider a pair of spike-trigger circuits, both driven by the same extrinsic input, I_{in}. Let the two circuits have slightly different thresholds and be coupled to each other in a mutually inhibiting fashion [so that the total input, $I_0(i)$, to each circuit is $I_{in} - g\{f_{spike}(j)\}$, where $f_{spike}(j)$ is the spike rate of the other circuit and $g\{.\}$ is a functional]. Let

$$g\{f_{spike}(j)\} = K_1 f_{spike}(j) - K_2 \int_0^t f_{spike}(j, t - \tau)e^{-\alpha_s\tau}d\tau \tag{5}$$

where K_1, K_2, and α_s are positive constants. This functional mimics synaptic adaptation. When I_{in} is below threshold, neither circuit fires. If I_{in} suddenly were stepped to a superthreshold value, then the spike-

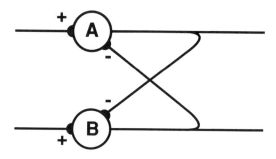

FIGURE 10.6. A diagram depicting two mutually inhibiting neurons. Reiss (1962) introduced modern neural modelers to the possibility of such a circuit serving as a multivibrator. The circuit also can serve as an asynchronous flip-flop.

trigger circuit with the lower threshold would be the first to begin to produce spikes, and if the constant K_1 were sufficiently large, it would completely suppress spike production by the other [because $g\{f_{spike}(j)\} \approx K_1 f_{spike}(j)$ would immediately become very large for the other circuit, holding the total input to that circuit well below threshold]. Gradually, $g\{f_{spike}(j)\}$ would decline, owing to its second term. If K_2 were sufficiently large or α_s were sufficiently small, the total input to the suppressed spike-trigger circuit eventually would exceed threshold, allowing that circuit to begin firing. Doing so, it rapidly would suppress the activity of its suppressor, further releasing itself from inhibition by that other circuit. This regenerative (positive-feedback) process would lead quickly to a switching of dominance to the previously suppressed circuit. Eventually, however, with large K_2 or small α_s, that circuit too would yield its dominance and again become suppressed. Thus the spike-trigger circuits would alternate with one another in production of spike bursts as long as I_{in} remained at a superthreshold level. The rate of alternation of dominance between the two spike-trigger circuits tends to increase with increasing values of I_{in}, but the detailed relationship between the alternation pattern (a metered rhythm) and I_{in} depends strongly on the values of the various parameters of the trigger circuits (Reiss 1962; Pearson 1976; Wang and Rinzel 1992).

With I_{in} above threshold, the basic sequence of events in this circuit is the same as that in many pulse and timing circuits from electrical engineering (e.g., flip-flops, one-shots, and multivibrators), namely, alternation between two *switching states** (circuit A on/circuit B off and

---

*They are called switching states because the fact that A is on and B is off only partially describes the actual state of the multivibrator (a complete description in the present case would be given by the values of the two functionals $g\{f_A(t)\}$ and $g\{f_B(t)\}$).

circuit B on/circuit A off), with positive feedback during the transitions from one switching state to the next. In the presence of constant, super-threshold input, this pair of trigger circuits behaves in the same way as a multivibrator. It could serve as a clock to move stored representations through a FIFO memory (as described in Sections 4.3 and 4.6.4), or as a pattern generator to drive reciprocally innervated muscle groups, or as part of a larger pattern generator (imagine, for example, a hierarchical array of multivibrators) or for any of many other neural functions. Such circuits could, for example, produce the motor command patterns for generation of acoustic communication patterns in insects, amphibians, and birds. In that case, as proposed by several authors (see Section 4.5), they also could be involved in detection of those same patterns. In much the same way, such a circuit also might be used to generate a pattern of subvocalized beats that could be adjusted (e.g., by adjusting I_{in}) to fit an incoming sequence of sounds. In this way it might provide the first step in the chunking scheme proposed by Povel and Essens (1985) (Section 2.4).

4.6.3 Flip-Flop Circuit

Next, consider a pair of trigger circuits, A and B, each driven by its own input, $I_{in}(A)$ and $I_{in}(B)$, respectively, and by its own internal excitatory current, I_E. Let the two circuits have the same thresholds and be coupled to each other in a mutually inhibiting fashion with little or no adaptation (so that the total input to circuit i is $I_E + I_{in}(i) - K_1 f_{spike}(j)$, where $f_{spike}(j)$ is the spike rate of the other circuit). If I_E and K_1 were sufficiently large and circuit A initially were firing, then its internal excitatory current would allow it to continue to do so and circuit B would remain completely suppressed until $I_{in}(B)$ were sufficiently large and sufficiently greater than $I_{in}(A)$ to cause the switching state to change. The switching state would change back again when $I_{in}(A)$ became sufficiently large and sufficiently greater than $I_{in}(B)$. Thus the switching state of this pair of trigger circuits would represent the recent history of the absolute and relative amplitudes of the two input waveforms. The transition between switching states again involves positive feedback. This combination of trigger circuits behaves very much the same way as an asynchronous electronic flip-flop (Jespers 1982).

Imagine N distinguishable neural flip-flops of this sort, each with its two switching states (A on/B off and vice versa). If the switching state of each flip flop is independent of the switching states of all the others, then the system of N flip-flops could store, temporarily, an N-element representation of an item of speech or some other auditory perception. Presently, as far as I know, there is no evidence for this sort of temporary memory in nervous systems. On the other hand, the fact that it could be achieved so easily with available neural elements makes it thoroughly plausible. Its plausibility, in turn, makes the celebrated neural circuit concepts of

McCulloch and Pitts (1943) plausible biological hypotheses (see also Pitts and McCulloch 1947).

4.6.4 FIFOs

If neural flip-flops were connected in an ordered cascade, so that the current switching state of flip-flop i is the previous state of flip-flop i−1, then the representation could be an N-element history of a stream of perceived sounds. For example, an ordered sequence of long and short tones could be represented by the ordered sequence of states, such as A *on/B off* for long, A *off/B on* for short. To operate as advertised, the flip-flops in this circuit must be connected in such a way that when a new stimulus event or some other timing event (such as a transition of switching states in a neural clock) occurs, the switching state of each member of the cascade is transferred to the next member. The old switching state of the Nth flip-flop will be lost and the switching state of the first member of the cascade will be new. This is a one-bit FIFO (First-In First-Out); obviously, it can be used to represent order explicitly. Connected to the input end of the FIFO must be an encoding circuit that translates some measure of the stimulus waveform into the appropriate code in the first flip-flop of the FIFO (Fig. 10.7).

A simple way to accomplish a FIFO configuration with flip-flops constructed from spike-trigger circuits is to sum the output of circuit A of flip-flop i with the output from a neural clock and apply the sum as the input to circuit A of flip-flop i + 1. Repeat the process for trigger B and for all N-1 connected flip-flop pairs. The thresholds all should be adjusted so that the output from a flip-flop that is on and the output from the neural clock both are insufficient by themselves to cause a switching-state transition, but the two occurring together will cause a transition. This arrangement transforms the asynchronous flip-flops into synchronous flip-flops. There are numerous ways to avoid switching-state ambiguity in this process. For example, the timing waveform could be applied through a series of time delays so that it reaches the input to the Nth flip-flop first and progresses downward through the ordinal sequence, reaching the input to the first flip-flop last.

With an appropriate encoder at its input end, a one-bit FIFO could store an ordered sequence of feature values coded to binary resolution (e.g., long, short, long, long, short, . . . , or loud, soft, loud, loud, soft, . . .). Similarly, with an appropriate encoder at its input, an array of K one-bit FIFOs operating in parallel, all driven by the same neural clock, could store an ordered sequence of feature values coded to finer resolution. Let the switching state of the ith flip-flop of the jth FIFO in the array be X_{ij}. The most recent value of the encoded feature would be represented by the set of switching states $(X_{01}, X_{02}, X_{03}, . . . , X_{0K})$ of first flip-flops, and i cycles of the neural clock will have passed since the

FIGURE 10.7. A system capable of explicit representation of time. A coded input signal is passed through a FIFO (first-in-first-out memory) (in this case a chain of flip-flops operated synchronously with a timing signal). The *circles* with *x*s represent logical *AND* gates (output is 1 when both inputs are 1; output is 0 otherwise). All elements in the figure (encoder, gates, flip-flops) can be synthesized with neural elements.

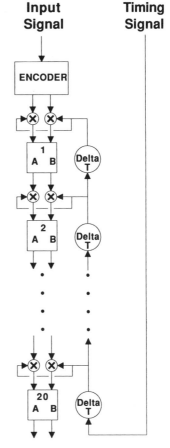

value represented by $(X_{i1}, X_{i2}, X_{i3}, \ldots, X_{iK})$ occurred. Alternatively, the array of FIFOs could be divided into subarrays, each storing the ordered sequence of coded values of one element of a vector. Such a vector might represent a set of features (e.g., loudness, duration, pitch, timbre) coded to various resolutions. In that case, the FIFO would contain a recent history of those features, with serial order preserved. It is not difficult to imagine a structure such as this serving as the memory element in the recurrent neural network of Figure 10.4.

4.7 *Explicit and Implicit Representations of History*

According to the usual model of such a device, a single-input, single-output linear analog filter with N state variables carries N independent representations of the immediately past history of the waveform applied to its input. The time windows used to compute those representations overlap, beginning at the present and extending variously into the past. Thus, the history of the waveform is represented implicitly in the state variables of the analog filter. On the other hand, one could sample such a waveform periodically, employing a clock for timing each sample and an encoder to encode the short-term mean amplitude during each clock cycle. Then one could store, in a FIFO, a coded representation of the recent history of the waveform. In that case, the history of the waveform would be represented explicitly. In some instances, such as in cochlear representations of acoustic waveforms, it is clear that nervous systems employ implicit representations of the history of sensory input. It is not clear, however, that they employ explicit representations of sensory history. The fact that FIFOs are easily realized with neural elements does not imply that they exist in real nervous systems. If, on the other hand, one could find a compelling selective advantage for explicit neural representation of recent sensory history, then the FIFO is available as a candidate realization of that representation in neural circuitry.

It is easy to reorganize the elements of the hypothetical FIFO of Section 4.6.4 to yield a neural network model that could represent history implicitly. For example, let the K times N flip-flops of a FIFO array be rearranged so that the output of each flip-flop is connected to the input of every other flip-flop, but with various weights for the various connections. Thus, at each cycle of the timing waveform, the next switching state of each flip-flop would be determined by its own particular weighted sum of the present switching states of all $K \times N$ flip-flops. If various subsets of the flip-flops also received coded versions of the current values of the elements (e.g., loudness, pitch, timbre) of a time-varying vector as input, then the next switching state of each flip-flop would be determined by the current distribution of switching states over all $K \times N$ flip-flops and the current value of the time function; and the current distribution of switching states over all the flip-flops therefore would represent the entire history of the time-varying vector. In that case, history would be represented implicitly. In other words, given $K \times N$ flip-flops, one could connect them in a way that would yield explicit representation of history (as in a FIFO), or one could connect them in a way that would yield implicit representation of history.

5. Summary

The ability of the listener to separate the sounds from various physical sources must be a quintessential property of hearing. Computations of pitch, timbre, loudness, and source location all may be inextricably combined into the source-separation computation. More general inferential computations that may be applied to the sound from a single source include identification of the source and its state, and, for vocalizing animal sources, interpretation of messages contained in acoustical communications. Some of the latter evidently are carried out in short-term declarative auditory memory. Among the operations and computations associated with that memory are segmentation and grouping, chunking, comparison and hypothesis testing, temporary retention of the serial order of unfamiliar sequences of segmented, grouped and chunked items, translation of such sequences to actions (e.g., subvocalizations or actual vocalizations), rehearsal (e.g., through the phonological loop), and transfer to more persistent memory.

An issue that has arisen in cognitive and neural modeling studies is which, if any, computations in short-term declarative auditory memory are based on explicit representations of sequences (i.e., with short sequences being preserved as they would be with a tape recorder). Studies with models based on parallel distributed processing and nonlinear dynamics suggest that many of those computations could be carried out efficiently with implicit representations of sequences. In one group of modeling studies, the history of a sequence is represented by the current limit-cycle neighborhood in which a nonlinear system is operating. The possibility that the same limit cycle could provide the sequence of neural signals required to produce a declarative reproduction (e.g., subvocalization) of the sequence suggests that identification of sequences by the auditory system may be linked to production of the same sequences by the motor system. Experimental evidence supports this suggestion in some species and rejects it in others.

Using realistic models of spike triggering in neurons, one can construct hypothetical circuits (e.g., neural oscillators and clocks, neural flip-flops, and neural FIFOs) that will provide the hardware bases for the various schemes based on parallel-distributed processing and nonlinear dynamics, and also the hardware bases for computations based on either explicit or implicit representations of sequences.

Acknowledgments. During the preparation of this chapter, the research of Edwin R. Lewis was supported by the Office of Naval Research under grant N0001491J1333 and by the National Institute for Deafness and Communicative Disorders under grant DC 00112.

References

Amit DJ (1988) Neural networks counting chimes. Proc Natl Acad Sci USA 85:2141–2145.

Baddeley A (1992) Working memory. Science 255:556–559.

Baddeley A, Logie R (1992) Auditory imagery and working memory. In: Reisberg D (ed) Auditory Imagery. Hillsdale: Erlbaum, pp. 179–197.

Baddeley A, Costanza P, Norris D (1991) Phonological memory and serial order: a sandwich for TODAM. In: Hockley WE, Lewandowsky S (eds) Relating

Theory and Data: Essays on Human Memory in Honor of Bennet B. Murdock. Hillsdale: Erlbaum, pp. 175–194.

Baddeley AD, Thomson N, Buchanan M (1975) Word length and the structure of short-term memory. J Verb Learn Verb Behav 14:575–589.

Baird B (1987) Bifurcation analysis of a network model of rabbit olfactory bulb with periodic attractors stored by a sequence learning algorithm. In: Proceedings of the First IEEE International Conference on Neural Networks, Vol. II. New York: Institute of Electrical and Electronic Engineers, pp. 147–152.

Belleville S, Peretz I, Arguin M (1992) Contribution of articulatory rehearsal to short-term memory: evidence from a case of selective disruption. Brain Lang 43:713–746.

Bregman AS (1990) Auditory Scene Analysis. Cambridge: MIT Press.

Bregman AS, Campbell J (1971) Primary auditory stream segregation and perception of order in rapid sequences of tones. J Exp Psychol 89:244–249.

Brown GDA (1989) A connectionist model of phonological short-term memory. In: Proceedings of the 11th Conference of the Cognitive Science Society. Hillsdale: Erlbaum, pp. 572–579.

Brüel PV, Baden-Kristensen K (1985) Time constants of various parts of the human auditory system and some of their consequences. In: Michelsen A (ed) Time Resolution in Auditory Systems. Berlin: Springer, pp. 205–214.

Calabrese RL, De Schutter E (1992) Motor-pattern-generating networks in invertebrates: modeling our way toward understanding. Trends Neurosci 15:439–445.

Chi MTH (1976) Short-term memory limitations in children: capacity or processing deficits? Memory Cognit 4:559–572.

Church RM, Broadbent HA (1992) Biological and psychological description of an internal clock. In: Gormezano I, Wasserman EA (eds) Learning and Memory. Hillsdale: Erlbaum, pp. 105–127.

Cramer EM, Huggins WH (1958) Creation of pitch through binaural interaction. J Acoust Soc Am 30:413–417.

de Boer E (1985) Auditory time constants: a paradox? In: Michelsen A (ed) Time Resolution in Auditory Systems. Berlin: Springer, pp. 141–158.

Dehaene S, Changeux JP, Nadal JP (1987) Neural networks that learn temporal sequences by selection. Proc Natl Acad Sci USA 84:2727–2731.

Delcomyn F (1980) Neural basis of rhythmic behavior in animals. Science 210:492–498.

Dempster FN (1981) Memory span: sources of individual and developmental differences. Psychol Bull 89:63–100.

Doherty JA, Gerhardt HC (1984) Acoustic communication in hybrid treefrogs: sound production by males and selective phonotaxis by females. J Comp Physiol A 154:319–330.

Elman JL (1990) Finding structure in time. Cognit Sci 14:179–211.

Elman JL (1991) Distributed representations, simple recurrent networks, and grammatical structure. Mach Learn 7:195–225.

Frick RW (1989) Explanations of grouping in immediate ordered recall. Memory Cognit 17:551–562.

Gardner RB, Darwin CJ (1986) Grouping of vowel harmonics by frequency modulation: absence of effects on phonemic categorization. Percept Psychophys 40:183–187.

Glenberg AM, Jona M (1991) Temporal coding in rhythm tasks revealed by modality effects. Memory Cognit 19:514–522.

Green DM (1985) Temporal factors in psychoacoustics. In: Michelsen A (ed) Time Resolution in Auditory Systems. Berlin: Springer, pp. 122–140.

Greene RL, Crowder RG (1988) Memory for serial position: effects of spacing, vocalization, and stimulus suffixes. J Exp Psychol Learn Memory Cognit 14:740–748.

Grosberg S (1974) Classical and instrumental learning by neurological networks. In: Progress in Theoretical Biology, Vol. 3. New York: Academic Press, pp. 51–141.

Grossberg S, Stone G (1986) Neural dynamics of attention switching and temporal-order information in short-term memory. Memory Cognit 14:451–468.

Harcum ER (1967) Parallel functions of serial learning and tachistoscopic pattern perception. Psychol Rev 74:51–62.

Hartmann WM (1988) Pitch perception and the segregation and integration of auditory entities. In: Edelman GM, Gall WE, Cowan WM (eds) Auditory Function: Neurobiological Bases of Hearing. New York: Wiley, pp. 623–645.

Hearly AF (1974) Separating order from information in short-term memory. J Verb Learn Verb Behav 13:644–655.

Hill AV (1936) Excitation and accommodation in nerve. Proc R Soc Lond B 119:305–355.

Hinton GE, Sejnowski TJ (1983) Optimal perceptual inference. Proc IEEE Comp Sci Conf Computer Vision and Pattern Recognition, Silver Spring, MD, USA: IEEE Computer Soc. Press, pp. 448–453.

Hoy RR (1978) Acoustic communication in crickets: a model system for the study of feature detection. Fed Proc 37:2316–2323.

Hoy RR, Hahn J, Paul RC (1977) Hybrid cricket auditory behavior: evidence for genetic coupling in animal communication. Science 195:82–84.

Hulme C, Muir C (1985) Developmental changes in speech rate and memory span: a causal relationship? Br J Dev Psychol 3:175–181.

Jespers PGA (1982) Pulsed circuits, logic circuits, and waveform generators. In: Fink DG, Christiansen D (eds) Electronics Engineers' Handbook. New York: McGraw-Hill, pp. 16-1–16-55.

Jones D, Morris N (1992) Irrelevant speech and serial recall: implications for theories of attention and working memory. Scand J Psychol 33:212–229.

Jordan MI (1986) Serial order: a parallel distributed processing approach. La Jolla: University of California Institute for Cognitive Sciences, Report 8604.

Kohonen T (1977) Associative memory: a system theoretical approach. New York: Springer-Verlag.

Kupfermann I (1991) Learning and memory. In: Kandel ER, Schwartz JH, Jessell TM (eds) Principles of Neural Science. New York: Elsevier, pp. 997–1008.

Ladefoged P (1982) A Course in Phonetics. New York: Harcourt Brace Jovanovich.

Mannes C, Dorffner G (1989) Self-organizing detectors for spatiotemporal patterns. Osterreichisches Forschungszentrum Seibersdorf, Vienna, Austria.

Mattingly IG, Liberman AM (1988) Specialized perceiving systems for speech and other biologically significant sounds. In: Edelman GM, Gall WE, Cowan WM (eds) Auditory Function: Neurobiological Bases of Hearing. New York: Wiley, pp. 775–793.

McAuley JD, Anderson SE, Port RF (1992) Sensory discrimination in a short-term dynamic memory. In: Proceedings of the 14th Annual Conference of the Cognitive Science Society. Hillsdale: Erlbaum, pp. 136–140.

McCulloch WS, Pitts W (1943) A logical calculus of ideas immanent in nervous activity. Bull Math Biophys 5:115–133.

McGurk H, MacDonald J (1976) Hearing lips and seeing voices. Nature 264: 746–748.

Michelsen A, Larsen ON, Surlykke A (1985) Auditory processing of temporal cues in insect songs: frequency domain or time domain? In: Michelsen A (ed) Time Resolution in Auditory Systems. Berlin: Springer, pp. 3–27.

Miller GA (1956) The magical number seven, plus or minus two: some limits on our capacity for processing information. Psychol Rev 63:81–87.

Monnier A (1934) L'Excitation Électrique des tissus. Paris: Hermann.

Moore BCJ (1989) An Introduction to the Psychology of Hearing. London: Academic Press.

Moore SW, Lewis ER, Narins PM, Lopez PT (1989) The call-timing algorithm of the white-lipped frog, *Leptodactylus albilabris*. J Comp Physiol A 164:309–319.

Murdock BB Jr (1974) Human memory: theory and data. Potomac, Maryland: Erlbaum.

Norris D (1990) A dynamic-net model of human speech recognition. In: Altmann GTM (ed) Cognitive Models of Speech Processing. Cambridge & MIT Press, pp. 87–104.

Pearson K (1976) The control of walking. Sci Am 235(6):72–86.

Pires A, Hoy RR (1992) Temperature coupling in cricket acoustic communication. J Comp Physiol A 171:79–92.

Pitts W, McCulloch WS (1947) How we know universals—the perception of auditory and visual forms. Bull Math Biophys 9:127–147.

Port RF (1990) Representation and recognition of temporal patterns. Connection Sci 2:151–176.

Port R, Anderson S (1989) Recognition of melody fragments in continuously performed music. In: Proceedings of the 11th Conference of the Cognitive Science Society. Hillsdale: Erlbaum, pp. 820–827.

Povel DJ (1981) Internal representation of simple temporal patterns. J Exp Psychol Hum Percept Perform 7:3–18.

Povel DJ (1984) A theoretical framework for rhythm perception. Psychol Res 45:315–337.

Povel DJ, Essens P (1985) Perception of temporal patterns. Music Percept 2:411–440.

Rabiner LR, Juang B (1986) An introduction to hidden Markov models (Speech recognition and processing). IEEE ASSP Mag 3(1):4–16.

Rand TC (1974) Dichotic release from masking for speech. J Acoust Soc Am 55:678–680.

Rashevsky N (1933) Outline of a physico-mathematical theory of excitation and inhibition. Protoplasma 20:42–56.

Reiss RF (1962) A theory and simulation of rhythmic behavior due to reciprocal inhibition in small nerve nets. Proc AFIPS Spring Joint Comput Conf 21:171–194.

Rosenblatt F (1962) Principles of Neurodynamics. Washington, DC: Spartan.

Rumelhart DE, Hinton GE, Williams RJ (1985) Learning internal representations by error propagation. La Jolla: University of California Institute for Cognitive Sciences, Report 8506.

Schreiner CE, Langner G (1988) Coding of temporal patterns in the central auditory nervous system. In: Edelman GM, Gall WE, Cowan WM (eds) Auditory Function: Neurobiological Bases of Hearing. New York: Wiley, pp. 337–361.

Schweickert R, Boruff B (1986) Short-term memory capacity: magic number or magic spell? J Exp Psychol Learn Memory Cognit 12:419–425.

Sejnowski TJ, Rosenberg CR (1986) NETtalk: a parallel network that learns to read aloud. Baltimore: Johns Hopkins University, Dept. of EECS, Report 86/01.

Servan-Schreiber D, Cleeremans A, McClelland JL (1989) Learning sequential structure in simple recurrent networks. In: Touretzky DS (ed) Advances in Neural Information Processing Systems. San Mateo: Kaufman, pp. 643–652.

Spinelli DN (1970) OCCAM: a computer model for a content addressable memory in the central nervous system. In: Pribram KH, Broadbent E (eds) Biology of Memory. New York: Academic Press, pp. 293–306.

Standing L, Bond B, Smith P, Isely C (1980) Is the immediate memory span determined by subvocalization rate? Br J Psychol 71:525–539.

Stanley JC, Kilmer WL (1975) A wave model of temporal sequence learning. Int J Man Mach Stud 7:395–412.

Stornetta WS, Hogg T, Huberman BA (1988) A dynamical approach to temporal pattern processing. In: Anderson DZ (ed) Neural Information Processing Systems. New York: American Institute of Physics, pp. 750–759.

Sturges PT, Martin JG (1974) Rhythmic structure in auditory temporal pattern perception and immediate memory. J Exp Psychol 102:377–383.

Tank DW, Hopfield JJ (1987) Neural computation by concentrating information in time. Proc Natl Acad Sci USA 84:1896–1900.

von der Malsberg C, Buhmann J (1992) Sensory segmentation with coupled neural oscillators. Biol Cybern 67:233–242.

Waibel A, Hanazawa T, Hinton G, Shikano K, Lang K (1988) Phoneme recognition: neural networks vs. hidden Markov models. Proc Intl Conf Acoust Speech Sign Proc 1:107–110.

Wang D, Arbib MA (1990) Complex temporal sequence learning based on short-term memory. Proc IEEE 78:1536–1543.

Wang XJ, Rinzel J (1992) Alternating and synchronous rhythms in reciprocally inhibitory model neurons. Neural Comp 4:84–97.

Widrow G, Hoff ME (1960) Adaptive switching circuits. IRE WESCON Conv Rec 4:96–104.

Williams RJ, Zipser D (1988) A learning algorithm for continually running fully recurrent neural networks. La Jolla: University of California Institute for Cognitive Sciences, Report 8805.

Williams RJ, Zipser D (1989) A learning algorithm for continually running fully recurrent neural networks. Neural Comp 1:270–280.

Index